CONVERSION FAC

Length   (Main unit = meter = m)

1 centimeter (cm) $= 10^{-2}$ m	1 Angstrom (Å) $= 10^{-10}$ m $= 10^{-8}$ cm
1 kilometer (km) $= 10^{3}$ m	1 Fermi (f) $= 10^{-15}$ m $= 10^{-13}$ cm
1 inch (in) $= 0.0254$ m	1 light-year $= 9.4600 \times 10^{15}$ m
1 foot (ft) $= 0.3048$ m	1 parsec $= 3.084 \times 10^{18}$ m
1 mile (mi) $= 1609$ m	1 mil $= 10^{-3}$ in $= 2.54 \times 10^{-5}$ m

Mass   (Main unit = kilogram = kg)

1 gram (g) $= 10^{-3}$ kg

1 slug $= 14.59$ kg

1 atomic mass unit (amu) $= 1.660 \times 10^{-27}$ kg

1 pound $= 0.4536$ kg

1 short ton $= 907.2$ kg

1 ounce $= 0.02835$ kg

> These are actually units of weight; however, 1 pound = 0.4536 kg means that 0.4536 kg weighs 1 pound under standard conditions of gravity ($g = 9.8067$ m/sec^2).

1 long ton $= 1016$ kg

1 eV/c^2 $= 1.782 \times 10^{-36}$ kg

1 MeV/c^2 $= 1.782 \times 10^{-30}$ kg

($m = E/c^2$)

Time   (Main unit = second = sec)

1 minute (m) $= 60$ sec

1 hour (h) $= 3600$ sec

1 day (d) $= 86{,}400$ sec $= 8.6400 \times 10^{4}$ sec

1 year (y) $= 3.156 \times 10^{7}$ sec

Energy   (Main unit = Joule = J = kg-m^2/sec^2)

1 erg $= 1$ g-cm^2/sec$^2 = 10^{-7}$ Joule

1 British thermal unit (Btu) $= 1055$ Joules

1 watt-second $= 1$ Joule

1 kilowatt-hour (kWh) $= 3.6 \times 10^{6}$ Joules

1 calorie $= 4.186$ Joules

1 electron volt (eV) $= 1.602 \times 10^{-19}$ Joule

1 MeV $= 10^{6}$ eV $= 1.602 \times 10^{-13}$ Joule

1 GeV (giga electron volt) $= 1$ BeV (billion electron volt)

                                $= 10^{9}$ eV $= 1.602 \times 10^{-10}$ Joule

1 kg-c^2 $= 8.987 \times 10^{16}$ Joules ($E = mc^2$)

1 Megaton $\approx 4 \times 10^{15}$ Joules (nuclear explosions)

SAUNDERS COMPLETE PACKAGE FOR TEACHING GENERAL PHYSICS

Melissinos and Lobkowicz: *Physics for Scientists and Engineers* —Volume 1

Lobkowicz and Melissinos: *Physics for Scientists and Engineers* —Volume 2

Slides to Accompany *Physics for Scientists and Engineers* —Volumes 1 and 2

Greenberg: *Discoveries in Physics for Scientists and Engineers, A Laboratory Approach* –Second Edition

Serway: *Concepts, Problems and Solutions in General Physics* —Volumes 1 and 2

Davidson and Marion: *Mathematical Preparation for General Physics with Calculus*

FREDERICK LOBKOWICZ Professor of Physics,
University of Rochester

ADRIAN C. MELISSINOS Professor of Physics,
University of Rochester

ILLUSTRATED BY ALEXIS KELNER

SAUNDERS GOLDEN SERIES

PHYSICS
FOR
SCIENTISTS
AND
ENGINEERS

VOLUME II

1975

W.B. SAUNDERS COMPANY / Philadelphia / London / Toronto

W. B. Saunders Company: West Washington Square
                          Philadelphia, PA 19105

                          12 Dyott Street
                          London, WC1A 1DB

                          W. B. Saunders Company Canada Ltd.
                          833 Oxford Street
                          Toronto, Ontario M8Z 5T9, Canada

Front cover reproduction is "Icarus Descended" by Richard Lytle, 1958. Courtesy of The Museum of Modern Art.

Library of Congress Cataloging in Publication Data
Melissinos, Adrian Constantin, 1929–
Physics for scientists and engineers.
(Saunders golden series)
Authors' names in reverse order in v. 2.   Includes index.
1. Physics.   I. Lobkowicz, Frederick, joint author.   II. Title.
QC21.2.M45 1975   530   74–4578
ISBN 0–7216–6267–6 (v. 1) ISBN 0–7216–5793–1 (v. 2)

Physics for Scientists and Engineers—Volume II   ISBN   0–7216–5793–1

Last digit is the print number: 9   8   7   6   5   4   3   2   1

To our wives,
Desy and Joyce

PREFACE

The rapid technological progress of the last decades has created an increasing need for the understanding of basic science and its applications to everyday life. Physics, as the most basic of all natural sciences, plays a key role in most, if not all, fields of science and engineering. It is no overstatement that no student can successfully approach a field of science without a thorough understanding of physics. The engineer, chemist, or biologist deals with matter in its various forms; all these forms of matter are subject to the same basic physical laws.

In this text we attempt to give the student a thorough understanding of physics. The book is "elementary" in the sense that we stress intuition more than mathematical formalism; but we have not cut corners or swept difficulties under the rug. Physics is a rapidly developing science and there are many outstanding questions. In addition, when one tries to apply the basic laws to real problems, one is frequently forced to oversimplify in order to be able to treat the problem at all. We have not evaded such difficulties, but tried instead to show the student how and where to make approximations to a real problem in order to obtain an answer. We believe that the ability to solve real problems is the test of whether a student has mastered the subject.

An introductory course based on the two volumes of "Physics for Scientists and Engineers" can be covered in three or four terms, depending on the emphasis of the instructor and the desired depth of presentation. To make this possible, the material has been divided into separate self-contained parts which can be combined to structure a curriculum in several ways. Conversely, individual parts of the text can be used for specialized one-term instruction.

Physics uses mathematics as a basic tool, just as literature uses grammar. A problem has to be formulated mathematically before it can be solved; indeed, the mathematical formulation is a major part of the solution itself. In this text we assume a knowledge of algebra and trigonometry. The concepts of calculus are developed as the text progresses; it is expected that the student is taking a concurrent course in calculus. If this is not the case, we recommend a thorough study of a self-study book such as Davidson and Marion (Mathematical Preparation for General Physics with Calculus, W. B. Saunders Co., 1973). The reader should also study carefully Supplements I and III in Volume I and Appendix C in Volume II; these cover most of the necessary mathematical tools used throughout the text.

In all activities of human life nothing is bestowed upon us by divine grace. The student should not expect to learn physics without working hard at it. A simple reading of the text cannot create overnight understanding; a text is meant to be studied, and in particular the reader should work out problems . . . many problems. For this reason we have included a large number of worked-out examples, and while on occasion these examples are used to present new applications of the basic principles, their primary purpose is to guide the student in the proper line of approach to a problem and the "method of science." We recommend strongly that the student first attempt to solve the example by himself without reference to the text; only after some "sweating" over the example should he look at the solution. We also have great respect for numerical values and results. After all, the whole essence of physics is to predict a number, so that a formula which cannot be interpreted numerically is worthless. The reader should also be very careful about the correct order of magnitude of a value; this will give him intuition for solving more complex problems, where approximations are necessary.

Certain parts of the text are set in small print. These cover advanced or more detailed topics which are not required for the continuity of the presentation. They can be omitted on a first reading or depending on the time the instructor wishes to spend on the subject. The two volumes also contain some 800 problems which we believe have physical relevance, in that they refer to realistic conditions in the laboratory or in the world. Again, if the reader cannot find the exact solution to the problem, he should not give up, but try to find an approximate solution. It is more important to know that some distance is roughly 30 feet than that it is 31.4159 feet; usually the tacit simplifications made in setting up the problem make such an "exact" answer meaningless anyway. Problems preceded by an asterisk (*) require more complex mathematical treatment or use material set in small print.

At the end of each chapter we have provided a summary of important relations. Our intention is that these relations provide a summary of the most salient and important points presented in the chapter. They are *not* to be memorized; rather they should focus the student's reading until he thoroughly understands them. Furthermore, they will serve as reference material which can be easily retrieved when it is needed by the student in subsequent chapters or in his future use of physics. Some equations in the text have also been enclosed in colored boxes; again this should help the student focus on the most important features of the text.

We have used the MKSC system of units throughout, since it is now widely accepted at all levels of teaching and in many technological applications. Nevertheless, the British system of units is still with us and we have made use of it in several of the examples, particularly in Volume I. We hope this will familiarize the student with conversion from one system of units to another. The CGS system is not used except in very special occasions.

In this second volume, we first discuss waves in their various forms; we stress those properties which are common to all waves, whatever their detailed nature. We show the student how to predict and analyze phenomena which are due to the wave nature of water waves, sound, or light. The treatment is elementary, but trigonometric functions are heavily used. The reader should refer to Appendix C for relations with which he may be less familiar.

Next we introduce the reader to the theory of electromagnetic phenomena, starting with a discussion of charges at rest. We then progress to treating charges in steady motion, and finally we analyze arbitrary charge and current distributions. Throughout we consider both the basic laws and their manifestations in the properties of materials. The student is rapidly led to understand the behavior of conductors and insulators; we constantly stress the connection between the macroscopic behavior of laboratory equipment and the microscopic phenomena occurring at the atomic level.

The discussion of electric and magnetic properties of materials is more thorough than in most introductory texts. We feel that our approach is particularly useful to the student who wants to apply his knowledge in physics to other fields of science. An understanding of the basic laws is necessary but not sufficient; the electrical engineer has also to understand how these laws apply to transistors, while the chemist or biologist has to be able to apply these same basic laws to the behavior of complex molecular systems.

The third part of this volume discusses how quantum mechanics enables us to understand more clearly the nature of matter. We introduce quantum mechanics in an intuitive way, omitting the more cumbersome mathematical details; we then apply the acquired knowledge to study the properties of electrons in atoms and crystals. Finally, we also discuss the interaction of light with matter.

The problems in this second volume reflect our constantly increasing emphasis on the behavior of real (not idealized) matter. In many instances the student is asked to relate his knowledge to actual physical or technological problems. Such problems usually do not have an "exact" answer; instead the student is asked to make estimates—as he would do when planning a new production process or discussing the economic merits of various approaches. We have therefore also abandoned the usual practice of giving the student all necessary numerical data in the problem itself. We have instead provided in Appendix B extensive tables of material properties; it is hoped that these tables will be useful to the student outside the actual physics course as well. After reading the problem through carefully, the student should decide which additional data he needs and then find them in the tables provided in Appendix B or in the text itself. We also sometimes provide in the statement of the problem a numerical value for a quantity which is relevant to the problem, but whose exact value does not enter the final solution. We have found that this method effectively discourages the student from "finding the solution" by searching around in the text until he finds an equation which has most of the symbols given in the problems somewhere in it; instead he now has to decide for himself what data he needs for the solution, much as he will do later on in his career. Of course, such problems are more difficult to grade, since there may be no single numerical answer which the grader can check against a list of "correct answers." We have therefore provided a separate booklet with model solutions, which is available to instructors in courses using this textbook.

In conclusion, we will be pleased if this text is helpful and inspiring to the students who will use it. It is written for them; we will be grateful if they would communicate to us any errors, misprints, or weaknesses in presentation that they may find.

F. Lobkowicz
A. C. Melissinos

ACKNOWLEDGMENTS

We are grateful to the many people who were of essential help in the preparation of this text. The students at the University of Rochester provided the motivation and used preliminary versions of the text. Colleagues and students pointed out inadequacies, and in particular we thank for their constructive comments Professors P. Axel, University of Illinois at Urbana; E. E. Anderson, Clarkson College of Technology; R. M. Cotts, Cornell University; R. C. Davidson, University of Maryland; A. G. DeRocco, University of Maryland; R. I. Eisenstein, University of Illinois at Urbana; L. H. Greenberg, University of Regina; K. F. Kinsey, SUNY at Geneseo; J. B. Marion, University of Maryland; D. J. Schlueter, Purdue University; R. A. Serway, Clarkson College of Technology; and G. A. Williams, University of Utah. Professor E. M. Hafner provided us with some special problems for Volume I. The seemingly interminable job of typing and retyping the manuscript was ably accomplished by Ms. E. Bauer, B. Griffin, and E. Hughes, whom we thank for their patience and dedication. We also acknowledge the enthusiasm and cooperation of our good friends at W. B. Saunders Company and in particular of Mr. J. J. Freedman. Finally, we owe gratitude to our families, who provided the encouragement and understanding that made this work possible.

CONTENTS

CHAPTER 1

WAVES IN ONE DIMENSION

1. RANDOM AND COLLECTIVE MOTION

When we discuss the behavior of one or at most a few small bodies (mass points), all we need to measure is the position of each body at any given instant. If the body (or bodies) is of finite size, we have to know also its orientation in space in order to describe it completely.

This approach of describing the motion of every individual body fails when the number of particles becomes very large, e.g., in a gas, a liquid, or a deformable solid. Each molecule will move around under the influence of all neighboring molecules as well as under the influence of any external forces acting on it. Since it is obviously impossible to try to calculate — or even to observe — the motion of each individual molecule, we have to be satisfied with a discussion only of the average behavior of the medium (e.g., of the gas).

In thermodynamics and in statistical mechanics one discusses those phenomena of complex systems which are due to essentially random motion of the individual molecules. Temperature, to name an example, is a measure of the mean kinetic energy of the molecules in a medium. Pressure is a measure of the number of molecules hitting the wall of the vessel and of the average change of momentum each molecule undergoes.

In a gas at equilibrium, while each molecule moves quite rapidly, any finite (macroscopic) part of the volume V (Fig. 1–1a) is at rest. The average velocity is zero; as many molecules move to the left as to the right, as many up as down, and so forth. However, here we are interested in collective motion of the medium. To illustrate what we mean, let us again consider a closed volume of gas, but let us assume that initially there are more molecules per unit volume at the left than at the right (Fig. 1–1b). We will assume that the temperature (i.e., the mean kinetic energy) is the same throughout the volume. Since there are more molecules on the left, more molecules will move from left to right than from right to left; there is thus a net flow of gas from left to right. The same would happen if the density were the same throughout but the temperature on the left were higher; i.e., the molecules on the left would on the average move around faster. In a small enclosed volume such as that shown in Fig. 1–1b, nothing very interesting would happen. Pretty soon the density would be equal everywhere and any collective motion would cease. But in the open atmosphere an initially localized disturbance can propagate over large distances. A "high pressure system," an area of the atmosphere where the pressure is higher than

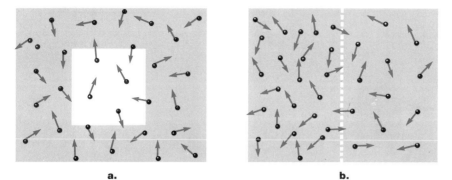

a.   **b.**

Figure 1–1   a. If the density and temperature of a gas are everywhere the same, then on the average as many molecules enter as leave any particular volume (white block in center). **b.** If the gas density is higher at left than at right, the number of molecules leaving the left half is larger than the number entering from the right. Thus, the density difference will decrease with time.

average, will travel across the continent. The collective flow of gas ("wind") at its boundaries carries new, possibly drier and colder air into a town, and we have fair weather. The whole science of meteorology deals with such collective motions of large air masses and their influence on the moisture content of the atmosphere. The motion of such regions of low or high pressure is extremely complicated; it is influenced by other nearby regions, the shape of continents, the season of the year, and so forth. Anybody who has ever followed the seemingly random path of a hurricane can appreciate the difficulties this rather new science poses to its practitioners.

2. WAVES

We are here mainly interested in another type of disturbance, which propagates with a rather well-defined velocity through the medium. Let us first use another example, the disturbance of the surface of a liquid. In equilibrium, because of gravity, the surface will of course be horizontal. If the liquid level is higher on one side of a vessel, water will flow until the level is everywhere the same. But on a lake, where the size of the medium (water surface) is sufficiently large, an initial disturbance (e.g., passage of a boat) will propagate over large distances. It will be noticed at a faraway shore long after the boat has passed.

In physics, we call a wave any disturbance which propagates with a well-defined velocity. We have seen as examples the waves on the surface of a liquid; but waves exist in any material which yields if subjected to outside forces. All materials are to some extent deformable—they will compress under pressure and yield under tension. In a gas (and indeed also in a solid or liquid) there exist compression waves which we call sound waves. An initial disturbance such as a shout produces a localized change in pressure and density in the air right in front of the mouth. The disturbance propagates and can be detected later by any device which is sensitive to rapid pressure changes (e.g., the ear or a microphone). But as we will later see in more detail, there is no need for a medium, such as a gas or a water surface, for the phenomenon of a wave to occur. Light is a wave; the initial disturbance is a localized region of electromagnetic field. In quantum mechanics we will learn that even a stream of particles, e.g., electrons, can be described as a wave. In this section we want to study the general types of phenomena that are due to the wave nature of a system and do not depend on the particular type of wave studied.

3. DESCRIPTION OF A WAVE

We want to discuss how to describe a wave mathematically. We first need to understand the concept of a field. In physics, a field is any physical quantity that we have to specify *at any point in space at any instant of time*. The position of a single particle is not a field; at any instant the particle is at some point and there is nothing we can say about any other different point. An example of a field is temperature; in a gas (or solid or liquid) the temperature can be different in different points in space. Thus, in order to describe a system it is not sufficient to specify a single number T for the temperature; we have to know at any time t the complete function.

$$T(\mathbf{r},t) = T(x,y,z,t)$$

at any point of the medium. Other well-known examples of a field are pressure or density of a gas. But a field does not have to be a scalar; it can be a vector field (Fig. 1–2). In order to specify the flow in a liquid or gas, we have to give the velocity of the gas at any point in space at any instant in time

$$\mathbf{v}(\mathbf{r},t) = \mathbf{v}(x,y,z,t)$$

Frequently, one is not interested in the particular nature — scalar or vector — of a field. We can write for a field

$$\phi(\mathbf{r},t) = \phi(x,y,z,t)$$

without knowing what exact quantity the field ϕ represents — or even whether it is a scalar or a vector. However, the function $\phi(x,y,z,t)$ cannot be completely arbitrary. The temperature will in general be a smooth (continuous and differentiable) function of the coordinates in space as well as of time. The velocity v in a liquid will also not jump arbitrarily from one point to the next. Thus, we will tacitly always assume that the field we discuss is sufficiently smooth.

We now have to decide when we should call a changing field a wave. As we said before, a wave is a disturbance propagating with a well-defined velocity. To make the discussion simpler, let us for the moment assume that we have only one spatial dimension. At the time $t_0 = 0$ the field ϕ can be then described by some function

$$\phi(x,t = 0) = f(x) \tag{1-1}$$

We will say that a wave is propagating in the positive x-direction if at later times

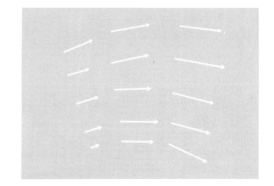

Figure 1–2 A vector field is specified if the magnitude and direction are given at any point in space.

the field has the form

$$\phi(x,t) = f(x - v_0 t) = f(u) \tag{1-2a}$$

where $f(u)$ is the same function as in Eq. (1–1) except that instead of $u = x$ in Eq. (1–1), we have

$$u = x - v_0 t \tag{1-2b}$$

We can see that this indeed corresponds to a disturbance propagating in the positive x-direction if we plot $\phi(x,t)$ at two different times: at $t = 0$ and at a time where $v_0 t = 2$ meters, as is done in Fig. 1–3. If, at $t = 0$, $f(x)$ had the value f_0 at a point x_0 (Fig. 1–3a), then at the later time it will take on the same value f_0 at a point

$$x = x_0 + v_0 t$$

because there the argument of the function $f(u)$

$$u = x - v_0 t = x_0 + v_0 t - v_0 t = x_0$$

is the same. The whole figure of $f(x)$ has been displaced by 2 meters to the right; we say that the wave has traveled 2 meters to the right in the time t. If we want to describe a wave traveling to the left (in the negative x-direction), we have to assume that

$$\phi(x,t) = f(x + v_0 t) = f(w) \tag{1-3}$$

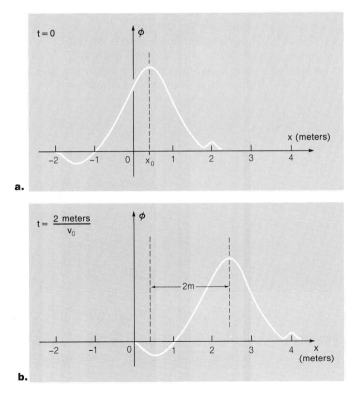

a.

b.

Figure 1–3 A wave is any disturbance in a field which propagates with a well-defined velocity. After the time t the disturbance ϕ has shifted by $\Delta x = v_0 t$.

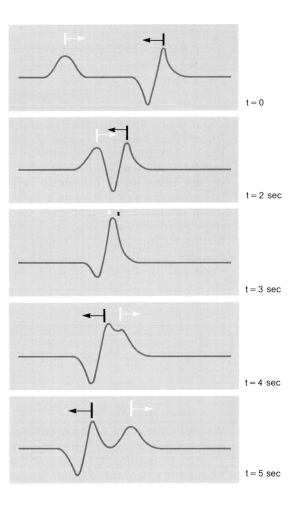

t = 0

t = 2 sec

t = 3 sec

t = 4 sec

Figure 1–4 Superposition of two waves traveling in opposite directions. The black and white tick marks indicate where the maximum of each wave would be if the other was absent.

t = 5 sec

We call v_0 the *wave velocity*. It is the velocity with which the disturbance (i.e., the wave) propagates in the medium. Eqs. (1–2) and (1–3) do not really enable us to do much by themselves. We need another assumption, which fortunately is quite well satisfied in nature; it is called the superposition principle.

Superposition Principle. Two different waves propagate independently of each other through a medium; the resulting disturbance at any point in space at any instant is the superposition (sum) of the disturbances due to each wave.

The superposition principle enables us to treat complicated wave patterns, possibly coming from many directions at once. In Fig. 1–4 we show graphically what happens if two waves going in opposite directions meet each other. You can imagine this situation to be the profiles of two water waves meeting each other head-on. The total disturbance at any time is

$$\phi(x,t) = f_1(x - v_0 t) + f_2(x + v_0 t) \tag{1–4}$$

even when the two waves are meeting, as shown in Fig. 1–4c.

The superposition principle is true even if the two waves are not propagating in the same direction. You can see the phenomenon quite clearly if you watch the bow waves of two motorboats on an otherwise calm lake. Indeed, most phenomena that we will discuss in this section on waves are direct consequences of the superposition principle.

4. HARMONIC WAVES

Of all possible functions $f(x \pm v_0 t)$, there is one whose form particularly suggests the wave in the common everyday sense. Consider

$$f(x - v_0 t) = A \cos [k(x - v_0 t) - \delta] \tag{1-5}$$

where k and δ are arbitrary parameters. We call a wave which is represented by Eq. (1–5) a *harmonic wave*. A graphical representation as given in Fig. 1–5 shows why this function, with its alternating wave crests and troughs (maxima and minima), is a rather close (although by no means perfect) approximation of a surface wave on an otherwise calm water surface. Quite frequently when we talk about a wave, we mean a harmonic wave.

A harmonic wave can also be defined as a wave of precise and well-defined wavelength. We call the wavelength of a harmonic wave the distance between two crests (or troughs). From Eq. (1–5) we see that this distance corresponds to a change of the argument of the cosine by 2π. Thus, if two points x_1 and x_2 are separated by one wavelength λ, we have

$$kx_2 - kx_1 = k\lambda = 2\pi$$

from which we immediately conclude that

$$\lambda = \frac{2\pi}{k} \quad \text{or} \quad k = \frac{2\pi}{\lambda} \tag{1-6}$$

We call k the *wave number* of the harmonic wave; it is the number of crests (or troughs) per 2π meters. Its dimension is (length)$^{-1}$.

Other important parameters of a harmonic wave are the period T, the frequency ν, and the angular frequency $\omega = 2\pi\nu$. The period T is defined as the time difference between the arrival of two crests (or troughs) at the same point x_0. If we plot the harmonic wave [Eq. (1–5)] at a fixed point $x = x_0$, we obtain a plot such as that shown in Fig. 1–5b. We have

$$f(x_0 - v_0 t) = A \cos [k(x_0 - v_0 t) - \delta] = A \cos (-kv_0 t + kx_0 - \delta)$$

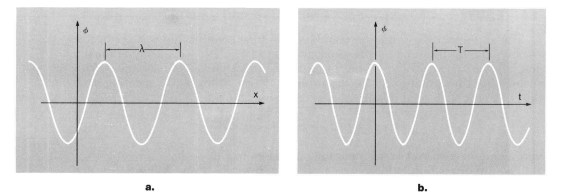

a. **b.**

Figure 1–5 **a.** A harmonic wave at a fixed instant in time. The separation in space between wave crests is the wavelength λ. **b.** The same wave plotted against time at a fixed point in space. The period T is the time difference between the arrival of two wave crests.

and because the cosine is an even function of its argument [i.e., $\cos(-\alpha) = \cos\alpha$], we can write for a fixed point x_0

$$f(x_0 - v_0 t) = A\cos[kv_0 t - (kx_0 - \delta)]$$

The period T is the time difference between two succeeding maxima:

$$kv_0 T = 2\pi$$

or

$$T = \frac{2\pi}{kv_0} \tag{1-7}$$

Using Eq. (1-6), we can write this as

$$T = \frac{\lambda}{v_0} \tag{1-8}$$

The frequency ν is the inverse of the period T:

$$\nu = \frac{1}{T} = \frac{v_0}{\lambda} \tag{1-9a}$$

which we can also write as

$$\nu \cdot \lambda = v_0 \tag{1-9b}$$

The angular frequency ω is defined as

$$\omega = 2\pi\nu = \frac{2\pi v_0}{\lambda} = kv_0$$

so that we can also write

$$\frac{\omega}{k} = v_0 \tag{1-10}$$

The wavelength is measured in meters, or any other unit of length (inches, microns, Ångstroms, and so forth). The frequency has the dimension of inverse time, and its unit is $\sec^{-1}$, called hertz (Hz). This unit replaces the older notation cps (cycles per second):

$$1\ \text{Hz} = 1\ \text{cps} = 1\ \sec^{-1}$$

The angular frequency has also the dimension of inverse time; to distinguish it from the frequency, it is measured in radians per second. Thus, if the frequency is 1 Hz, the angular frequency is

$$\omega = 2\pi\nu = 6.28\ \text{radians/sec}$$

Using our definition of frequency, period, and the other parameters, we can write down the expression for a harmonic wave [Eq. (1–5)] in many completely equivalent ways:

$$f(x,t) = A \cos (kx - \omega t - \delta) = A \cos (\omega t - kx + \delta)$$

$$= A \cos \left(\frac{2\pi}{\lambda} x - 2\pi\nu t - \delta \right)$$

$$= A \cos \left[2\pi \left(\frac{x}{\lambda} - \frac{t}{T} \right) - \delta \right] \qquad (1\text{–}11a)$$

$$= A \cos \left[k(x - v_0 t) - \delta \right]$$

We stress here that all of the expressions in Eq. (1–11a) describe a wave going in the positive x-direction. A wave going in the negative x direction would be described by

$$g(x,t) = A \cos \left[k(x + v_0 t) - \delta \right]$$

$$= A \cos \left(\frac{2\pi}{\lambda} x + 2\pi\nu t - \delta \right) \qquad (1\text{–}11b)$$

$$= A \cos (kx + \omega t - \delta)$$

.

.

.

We call the *amplitude* of a harmonic wave the maximum excursion A. (The cosine has a maximum value of $+1$.) Its unit cannot be defined without knowing the particular field we are discussing. Thus, for water waves, A will be the maximum height of a wave and will have the dimension of length (meters or centimeters); for acoustic waves in air we can take as the field the density, in which case A will be the maximum deviation of air density from that of the quiet air

$$A = \rho_{max} - \rho_0$$

and, of course, A will then have the dimension of density (kg/m³). But we could just as well have taken the air pressure as our basic field, in which case A would have the dimension of force/area or kg/m-sec².

The argument of the cosine in Eq. (1–11a) is called the *phase* or *phase angle*. It is not sufficient to know the wavelength and the period in order to uniquely determine the phase for all points in space at any time. We also have to know the "initial phase" $(-\delta)$, the value for the phase at $t = 0$ and $x = 0$. Two waves are said to be "in phase" if the arguments of the cosines (the phase angles) are the same everywhere; the fields themselves can be quite different, because the two waves can have different amplitudes.

Using the definition of the phase, we can rephrase our earlier definition of the wavelength λ and period T. We thus define the wavelength as the distance λ over which, for a fixed time, the phase changes by 2π. We can also define the period as the time difference T during which the phase advances by 2π at a certain point.

Example 1 The velocity of a light wave in vacuum is $c = 3.00 \times 10^8$ m/sec. What are the frequency, angular frequency, period, and wave number of a light wave of wavelength $\lambda = 6000$ Å $= 6.0 \times 10^{-7}$ m (yellow light)?

This is an easy exercise. We start by calculating the wave number

$$k = \frac{2\pi}{\lambda} = \frac{6.28}{6 \times 10^{-7} \text{ m}} = 1.05 \times 10^7 \text{ m}^{-1}$$

Thus, there are about 10 million waves per 2π meters. The frequency is

$$\nu = \frac{c}{\lambda} = \frac{3.0 \times 10^8 \text{ m/sec}}{6.0 \times 10^{-7} \text{ m}} = 5 \times 10^{14} \text{ Hz}$$

The period is

$$T = \frac{1}{\nu} = 2 \times 10^{-15} \text{ sec}$$

and the angular frequency is

$$\omega = 2\pi\nu = 6.28 \times 5 \times 10^{-14} \text{ Hz} = 3.14 \times 10^{15} \text{ radians/sec}$$

■ Actually, there exists a mathematical theorem due to J. B. Fourier (1768–1830) which says that any function $f(x)$ can be considered to be a superposition of sines and cosines. Specifically, the Fourier theorem says that any function $f(x)$ can be expressed as an integral

$$f(x) = \int_0^\infty A(k) \cos [kx - \delta(k)] dk \qquad (1\text{–}12)$$

where $A(k)$ and $\delta(k)$ are well-determined functions of k; their particular form will of course depend on the specific function $f(x)$ which is to be expressed this way. That Eq. (1–12) really expresses $f(x)$ as a superposition of harmonic functions $\cos (kx - \delta)$ can easily be seen if we remember from mathematics the definition of a definite integral as the limit of a sum (Fig. 1–6):

$$\int_a^b g(x)dx = \lim_{\Delta x_i \to 0} \sum g(x_i)\Delta x_i \qquad (1\text{–}13)$$

■

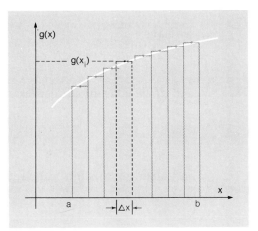

Figure 1–6 A definite integral $\int_a^b g(x)dx$ can be approximated by $[\Sigma \, g(x_i)] \cdot \Delta x$ if we subdivide the interval (a,b) into a number of equal intervals of width Δx.

5. THE WAVE EQUATION

There exists a universal equation obeyed by all wave phenomena. We will derive it for one spatial dimension, where we know that the most general wave possible is

$$\phi(x,t) = f(x - v_0t) + g(x + v_0t) = f(u) + g(w) \tag{1-14}$$

where f and g are *completely arbitrary* functions of the single variables $u = x - v_0t$ and $w = x + v_0t$ and v_0 is again the wave velocity. Our universal wave equation cannot be an algebraic equation; it has to involve derivatives. Let us take the derivatives of $\phi(x,t)$ with respect to x, keeping t constant, and with respect to t keeping x constant. We have to use the chain rule[*]

$$\frac{\partial f}{\partial x} = \frac{df}{du}\frac{\partial u}{\partial x}$$

We then obtain, since

$$\frac{\partial u}{\partial x} = \frac{\partial w}{\partial x} = 1 \tag{1-15a}$$

$$\frac{\partial u}{\partial t} = -\frac{\partial w}{\partial t} = -v_0 \tag{1-15b}$$

the following expressions

$$\frac{\partial \phi}{\partial x} = \frac{df}{du} + \frac{dg}{dw} \tag{1-16a}$$

$$\frac{\partial \phi}{\partial t} = \left(-\frac{df}{du} + \frac{dg}{dw}\right)v_0 \tag{1-16b}$$

If it were not for the minus sign in Eq. (1–16b), our problem would be solved— the two expressions would be proportional to each other. The minus sign is gotten from the minus sign in Eq. (1–15b). However, if we take the derivative with respect to t twice, it disappears. Since

$$\frac{\partial^2 u}{\partial t^2} = \frac{\partial}{\partial t}\left(\frac{\partial u}{\partial t}\right) = \frac{\partial}{\partial t}(-v_0) = 0$$

we have

$$\frac{\partial^2 f}{\partial t^2} = \frac{\partial}{\partial t}\left(\frac{\partial f}{\partial t}\right) = -v_0\frac{\partial}{\partial t}\left(\frac{df}{du}\right) \tag{1-17}$$

and df/du is again a function of u only. Thus

$$\frac{\partial}{\partial t}\left(\frac{df}{du}\right) = -v_0\frac{d^2f}{du^2}$$

and inserting this into Eq. (1–17) we obtain

$$\frac{\partial^2 f}{\partial t^2} = +v_0{}^2\frac{d^2f}{du^2}$$

Performing the same sequence of operations with $g(w)$ and combining with the

---

[*] See R. C. Davidson and J. B. Marion, *Mathematical Preparation for General Physics with Calculus* (W. B. Saunders Co., 1973), p. 100.

preceding result, we have

$$\frac{\partial^2 \phi}{\partial t^2} = v_0{}^2 \left(\frac{d^2 f}{du^2} + \frac{d^2 g}{dw^2} \right)$$

Similarly, twice taking the derivative of ϕ with respect to x, we get

$$\frac{\partial^2 \phi}{\partial x^2} = \frac{d^2 f}{du^2} + \frac{d^2 g}{dw^2}$$

We now see that the two derivatives are proportional to each other:

$$\boxed{\frac{1}{v_0{}^2} \frac{\partial^2 \phi}{\partial t^2} = \frac{\partial^2 \phi}{\partial x^2}}$$

(1–18)

Any wave expressed in the form of Eq. (1–14) is a solution of Eq. (1–18); one can show also that any solution to Eq. (1–18) can be written in the form of Eq. (1–14). Thus, any system obeying the wave equation (Eq. 1–18) will show wave properties.

6. SPECIFIC WAVE SYSTEMS

The velocity with which a wave propagates will depend on the particular system; it can range from about 1 m/sec (water waves) to the velocity of light, $c = 3 \times 10^8$ m/sec. While for many systems it is extremely complicated to derive the wave equation, we can frequently determine the velocity from dimensional considerations, if we can guess—or determine by experiment—the physical quantities which determine the wave velocity.

a. Acoustic Waves in Gases.   Sound waves in a gas are due to local small pressure and density variations. Thus, we would expect the velocity to depend on the pressure p and density ρ of the gas. Other possible parameters would be the ambient temperature T and the molecular composition of the gas. However, we would not expect the sound velocity to depend on any detailed properties of the molecules forming the gas. Since sound is due only to their motion in space, we can guess that only the overall mass of the molecule will determine the sound velocity; the molecular structure should have no effect, or only a minor influence, on the speed of sound. Instead of the mass of the individual molecule, we can take as our parameter the mass of a mole of the gas, the molecular weight M.

The four parameters $p, \rho, T,$ and M on which the sound could depend are not independent of each other. For an ideal gas, we have

$$p \cdot V = R \cdot T \tag{1–19a}$$

where V is the volume of one mole of the gas and $R = 8.3 \times 10^3$ J/(kg-mole °K) is the universal gas constant. Since density is defined as mass/volume, we also have

$$\rho = \frac{M}{V}$$

Combining the two equations we obtain

$$\frac{p}{\rho} = \frac{RT}{M} \tag{1–19b}$$

We are now ready to predict the sound velocity and its dependence on the four parameters p, ρ, M, and T. We have to construct a velocity[*]

$$[v] = \text{m/sec}$$

out of the following quantities, if we include R:

$$[p] = \frac{\text{force}}{\text{area}} = \frac{\text{kg m/sec}^2}{\text{m}^2} = \frac{\text{kg}}{\text{m sec}^2}$$

$$[\rho] = \frac{\text{mass}}{\text{volume}} = \frac{\text{kg}}{\text{m}^3}$$

$$[T] = {}^\circ\text{K}$$

$$[M] = \frac{\text{kg}}{\text{mole}}$$

$$[R] = \frac{\text{Joule}}{\text{mole }{}^\circ\text{K}} = \frac{\text{kg m}^2}{\text{mole} \cdot {}^\circ\text{K} \cdot \text{sec}^2}$$

There exist only two combinations of the five variables which have the dimension of a velocity; in addition, they give the same result because of Eq. (1–19b):

$$v = \text{constant} \sqrt{\frac{p}{\rho}} = \text{constant} \sqrt{\frac{RT}{M}}$$

It is usual to put the constant under the radical, and we thus have

$$v = \sqrt{\frac{\gamma p}{\rho}} = \sqrt{\frac{\gamma RT}{M}} \tag{1–19c}$$

All we know about γ is that it must be a pure number; we also expect it to be not too much different from unity. Actually, γ is not the same for all gases. It takes on the value $\gamma \approx 5/3 = 1.667$ for gases whose molecules consist of single atoms (e.g., the rare gases [He, Ne, A, K, Xe]); for diatomic molecules (e.g., N_2, O_2) one obtains $\gamma \approx 1.40$, and for more complicated molecules (e.g., methane [CH_4] or acetylene [C_2H_4]), γ lies in the range $1.15 < \gamma < 1.35$. Nevertheless, even if we take $\gamma = 1$ for all gases, we have by our crude dimensional consideration determined the speed of sound in all gases to within 30%; that is by no means a small achievement!

b. Transverse Waves on a String.   A stretched string (piano wire, rope, or similar object) allows the passage of transverse waves; we can easily show this by stretching a rope and then wiggling one end briefly (Fig. 1–7). We will then see a wave propagating to the other end of the rope. Experimentally, we can determine that the wave will move faster if we increase the tension (the force pulling at each end) in the rope. Thus, we predict that the velocity depends on the tension in the string.

If the string is thin enough to be completely flexible, we do not expect the wave velocity to depend on the shape of the string cross section, e.g., whether

---

[*] One says that v has the *dimension* of meters per second. Quite generally, we call dimension of a quantity the unit in which it is measured. In equations, one denotes the dimension of a quantity a by $[a]$.

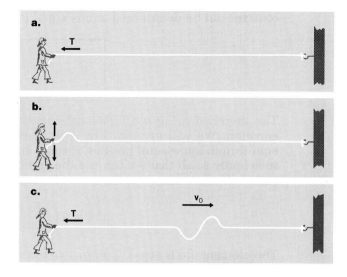

Figure 1-7 Wave on a string under tension.

the cross section is a square, a rectangle, or a circle. However, the velocity might well depend on the cross section area A of the rope; we would also expect it to depend on the density ρ of the rope material.

We can define also a linear density μ of the rope, its mass per unit length. The mass of an element of the rope of length dx is as shown in Fig. 1–8:

$$dm = \rho \cdot A \cdot dx = \mu dx \qquad (1-20a)$$

from which we conclude that

$$\mu = \rho \cdot A \qquad (1-20b)$$

The dimension of the various parameters on which the velocity is likely to depend are

$$[T] = \text{force} = \text{kg m/sec}^2$$
$$[\mu] = \text{linear density} = \text{kg/m}$$
$$[\rho] = \text{density} = \text{kg/m}^3$$
$$[A] = \text{area} = \text{m}^2$$

We see that the only combination which has the dimension of velocity is

$$v_0 = \text{constant } \sqrt{\frac{T}{\mu}} = \text{constant } \sqrt{\frac{T}{\rho A}}$$

where the constant is some pure number. As shown below [Eq. (1–25)], the

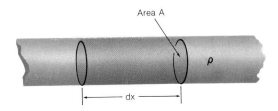

Figure 1-8 Definition of linear density dm/dx.

constant can be determined and is equal to unity. Thus,

$$v_0 = \sqrt{\frac{T}{\mu}} = \sqrt{\frac{T}{\rho A}}$$

(1–21)

The stretched string is a sufficiently simple system that we can derive its wave equation. We will assume (Fig. 1–9) that the deviation of the string from its equilibrium (quiescent) position is small and also that the angle α is everywhere sufficiently small that we can use the approximations $\cos \alpha \approx 1$ and $\alpha \approx \sin \alpha \approx \tan \alpha$.[*] We then have, for the mass of a rope element sketched in Fig. 1–9,

$$dm = \mu \left(\frac{dx}{\cos \alpha} \right) \approx \mu dx$$

The element dm is acted upon by two forces, the tension **T** at each end. But because the element is not exactly straight, the two forces do not exactly cancel each other. If we denote the slightly different angle at the position $(x + dx)$ by

$$\beta = \alpha + d\alpha$$

(note that in Fig. 1–9 we have $d\alpha < 0$), we will have a net force whose horizontal component is

$$F_h = T \cos (\alpha + d\alpha) - T \cos (\alpha)$$

and because $\cos (\alpha + d\alpha) \approx \cos \alpha \approx 1$, this means that $F_h = 0$. However, the vertical component of the net force does not vanish:

$$F_v = T \sin (\alpha + d\alpha) - T \sin \alpha$$

which, because $\alpha \approx \sin \alpha$, we can write as

$$F_v = T \cdot d\alpha$$

---

[*] See Davidson and Marion, p. 59.

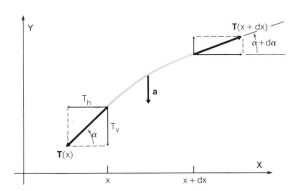

Figure 1–9 Acceleration of small element of string. The tension **T** has slightly different direction on both ends, resulting in a net acceleration **a** downward. Note that in reality the angle α is very small.

and since $\alpha \approx \tan \alpha$, we can just as well write

$$F_v = Td(\tan \alpha) \qquad (1-22)$$

The mass element dm will be accelerated by this force. We denote its instantaneous position by $y(x,t)$ — each element has a different position x, so we can label it by this variable. We then need a function of two variables, x and t, to describe the complete motion of the string. The acceleration of the element is

$$a = \frac{d^2y}{dt^2} = \frac{\partial^2 y(x,t)}{\partial t^2}$$

and Newton's second law yields

$$dm \cdot a = F_v$$

or

$$\mu dx \frac{\partial^2 y}{\partial t^2} = Td(\tan \alpha)$$

This we can rewrite in the form

$$\frac{\partial^2 y}{\partial t^2} = \frac{T}{\mu} \frac{d(\tan \alpha)}{dx} \qquad (1-23)$$

Next we note that $\tan \alpha$ by definition is the slope of $y(x,t)$ at a constant time t:

$$\tan \alpha = \left(\frac{\partial y}{\partial x}\right)_{t=\text{const}}$$

and the derivative $d(\tan \alpha)/dx$ is also to be taken at a fixed instant in time:

$$\frac{d(\tan \alpha)}{dx} = \frac{\partial}{\partial x}\left(\frac{\partial y}{\partial x}\right) = \frac{\partial^2 y}{\partial x^2}$$

Inserting this into Eq. (1–23), we obtain

$$\frac{\partial^2 y}{\partial t^2} = \frac{T}{\mu} \frac{\partial^2 y}{\partial x^2} \qquad (1-24)$$

which indeed is a wave equation identical to Eq. (1–18), if we set

$$v_0{}^2 = \frac{T}{\mu}$$

or

$$v_0 = \sqrt{\frac{T}{\mu}} \qquad (1-25)$$

Thus, we have been able to derive Eq. (1–21) from first principles. The reader should note that we had to assume that the deviation from equilibrium is not too large. This is an assumption which one has to make for any wave motion. If the amplitude of a harmonic wave — or the maximum excursion of any wave — becomes sufficiently large, the wave equation is no longer able to describe the phenomenon. What fails first is the superposition principle; two waves will no longer pass "through" each other without interacting. Such "nonlinear" effects are quite common for many waves. Surface waves in water form "whitecaps" (in deep water) or "breakers" (in shallow water), when their amplitude becomes too large. A sufficiently strong sound wave—the shock wave produced by explosions or sonic booms of airplanes—also behaves differently than ordinary sound. Even light waves passing through vacuum should show such effects. It is predicted that if two light beams of sufficient intensity cross each other in vacuum, then they will not simply continue but will interact and light will emanate in all directions (Fig. 1–10). However, at currently achievable light intensities this effect is unmeasurably small.[°]

  c. Elastic (Acoustic) Waves in a Solid.   If we tap one end of a long rod with a hammer, the rod will be elastically deformed around this end. The disturbance will then propagate through the rod at finite speed. Sound is transmitted through solids in the same way; the pressure variations of the air on one side of a concrete wall produce local deformations in the wall, which then propagate through the concrete.

  Since there is a local deformation of the rod, we would expect the speed of sound to depend on Young's modulus Y, which is a measure of how much the material yields under stress.[†] The speed of sound in a rod could also depend on the density of the material, ρ and on the cross section area—or possibly the diameter—of the rod. Thus, as before, we first write down the dimension for each quantity. Remembering that Young's modulus is the proportionality constant between the stress (force per unit area) and strain (change of length per unit length)

$$Y\frac{\Delta\ell}{\ell} = \text{stress} = \text{force/area}$$

we find for the dimension of Young's modulus

$$[Y] = \frac{\text{kg m/sec}^2}{\text{m}^2} = \text{kg/m sec}^2$$

The density and area have well-known dimensions

$$[\rho] = \text{kg/m}^3$$
$$[A] = \text{m}^2$$

[°] The same effect, however, can be measured if two light beams of sufficient intensity cross each other inside a material.

[†] Compare Vol. I, Chapter 18.

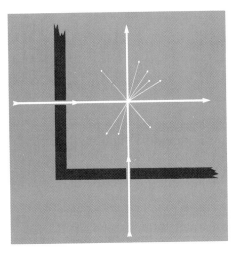

Figure 1–10 Limits of superposition principle; two very intense light beams meeting each other in vacuum should produce scattered light from the interaction region. This experiment has not yet been performed, because currently available light intensities are too small.

We can then easily deduce that the velocity of sound has to be equal to

$$v_0 = \text{constant } \sqrt{\frac{Y}{\rho}}$$

where the constant is a pure number and is the same for all materials. A more detailed study from first principles would show that this constant equals unity; and we therefore have for the speed of sound in a long thin rod

$$v_0 = \sqrt{\frac{Y}{\rho}} \tag{1-26}$$

As a byproduct, we have obtained the result that the speed of sound in a rod cannot depend on the cross section area of the rod—and thus also not on its detailed shape.

d. Surface Waves on Water. This is the phenomenon which originally gave waves in other media (fields) their name; however, it is really a very complicated process. There are actually three different types of waves; a real wave can be due to a combination of these three "wave types."

(1) *Deep sea waves* are the large waves one encounters on the ocean surface, or in the middle of a deep lake or pond. Their wavelength is typically 0.3 to 20 meters. The disturbance exists only near the surface; at great depths the water is completely at rest. These are gravitational waves, in the sense that water flows from the crests to the troughs under the influence of the gravitational attraction of the earth. We therefore naturally expect their velocity to depend on g, the gravitational acceleration of any object on the earth's surface. In addition, the wave velocity could be dependent on the liquid density. There seems to be no other property of the liquid which could influence the wave velocity. Writing down the dimensions

$$[g] = \text{m/sec}^2$$

$$[\rho] = \text{kg/m}^3$$

we then discover that there exists no combination of these two quantities which has the dimension of a velocity.

The answer to our puzzle is that the velocity of deep sea waves depends on their wavelength λ. We can indeed build up a velocity out of g and λ:

$$v = \text{constant } \sqrt{g\lambda}$$

where again we know the constant to be a pure number and independent of the liquid, g, or λ. The exact theory yields the result

$$v_0 = \sqrt{\frac{g\lambda}{2\pi}} \tag{1-27}$$

We encounter here for the first time the phenomenon of *dispersion*. By this we mean that there does not exist a well-defined velocity for a wave of arbitrary shape. Only harmonic waves, because they have a unique wavelength λ, will also have a well defined velocity. We note that because of the basic relation between frequency ν and wavelength λ

$$\nu \cdot \lambda = v_0$$

we can cast Eq. (1–27) into a different form. Substituting $\lambda = v_0/\nu$ into Eq. (1–27) and squaring, we obtain

$$v_0^2 = \frac{gv_0}{2\pi\nu}$$

or

$$v_0 = \frac{g}{2\pi\nu} = \frac{g}{\omega} \tag{1–28}$$

as an alternate expression for the deep sea wave velocity. You can observe the wavelength dependence of water wave velocity on a lake after a storm, when there are waves of many different wavelengths present; the longer waves will move faster than the shorter ones.

(2) *Shallow water waves.* Water waves change their nature if the water depth h is shorter than the wavelength λ. While in deep sea waves the disturbance penetrates about one wavelength into the liquid, in shallow water there is a natural boundary at the bottom. Thus, if the wavelength λ is larger than the water depth h, we would expect the speed of the wave to become independent of λ, but now to become dependent on the depth h. With the help of dimensional analysis, we find easily that the wave velocity then has to be

$$v_0 = \text{constant } \sqrt{gh}$$

and experimentally one finds (Fig. 1–11) that

$$\boxed{v_0 = \sqrt{gh}} \tag{1–29}$$

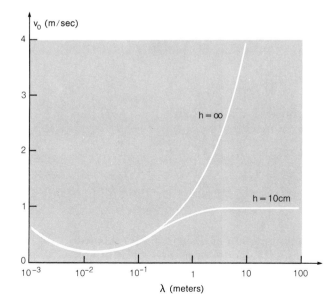

Figure 1–11 The wave velocity of water waves for a range of wavelength λ in shallow water (water depth $h = 10$ cm) and in very deep water.

Of course, it is not quite sufficient that $\lambda \gg h$; it is also necessary that the wave amplitude be much smaller than the depth h:

$$A \ll h$$

If the second condition is not satisfied, the wave becomes a "breaker," as mentioned earlier.

(3) *Surface tension waves.* There exists another type of surface wave on a liquid; one can see them when a sudden wind gust sweeps in on an otherwise calm surface. They are characterized by a very short wavelength (one centimeter or less), and they are due to the surface tension γ of the water. Their speed is

$$v_0 = \sqrt{\frac{2\pi\gamma}{\rho\lambda}} \qquad (1\text{--}30)$$

which again depends on the wavelength.

Actually, from a complete theory of liquid surfaces one can derive a single equation which expresses the velocity of harmonic waves for all wavelengths and water depths[*]:

$$v_0{}^2 = \left(\frac{g\lambda}{2\pi} + \frac{2\pi\gamma}{\rho\lambda}\right) \cdot \tanh\left(\frac{2\pi h}{\lambda}\right) \qquad (1\text{--}31)$$

Our Eqs. (1–27), (1–29), and (1–30) are special limiting cases of the more general Eq. (1–31); we obtain

$$\text{Eq. (1--27) if } h \gg \lambda \quad \text{and} \quad \lambda \gg \sqrt{\frac{\gamma}{\rho g}}$$

$$\text{Eq. (1--29) if } h \ll \lambda \quad \text{and} \quad \lambda \gg \sqrt{\frac{\gamma}{\rho g}}$$

$$\text{Eq. (1--30) if } h \gg \lambda \quad \text{and} \quad \lambda \ll \sqrt{\frac{\gamma}{\rho g}}$$

e. Electromagnetic Waves. Light, radio waves, radar waves, and microwaves (which are used in some ovens) are waves of the electromagnetic field. We have to postpone any detailed discussion of the nature of the electromagnetic fields; but we do not require any particular knowledge if we discuss only effects which are common to all waves. Electromagnetic waves are interesting because their speed is much larger than that of any other type of wave known. In vacuum the velocity of light is a universal constant

$$c = 3.00 \times 10^8 \text{ m/sec} \qquad (1\text{--}32)$$

while in transparent media the light velocity is somewhat smaller

$$v_0 = \frac{c}{n} \qquad (1\text{--}33)$$

---

[*] The hyperbolic tangent, $\tanh(x)$, is defined as $\tanh(x) = (e^x - e^{-x})/(e^x + e^{-x})$.

where n, called the refractive index of the medium, typically has a value of 1.0 to 2.5.

Another special property of electromagnetic waves is the very large range of wavelengths which can be practically used. For sound in air, the usable wavelength range is tens of meters (drum) to about 2 cm (upper range of audibility). Even ultrasound, which can have wavelengths as short as 0.1 mm in air, only extends the range slightly.

Let us compare this with the wavelength range for electromagnetic waves. Radio waves have wavelengths of a few meters to a few kilometers; radar waves are 1 to 10 cm long; infrared ovens use electromagnetic radiation in the range of 1 to 10 microns (10^{-6} m) to heat meat. Visible light has a wavelength range of 3000 Å (violet) to 7000 Å (red), or 0.3 to 0.7 micron. The ultraviolet rays which give us sunburn have a wavelength of 1000 to 3000 Å. X-rays, which we use for medical treatment or diagnostics, have wavelengths of 0.1 to 0.5 Å. The gamma rays of cobalt, which are used for cancer treatment, have a wavelength 0.0012 Å. Even shorter wavelength radiation can be produced by special machines (electron synchrotrons) and is used to investigate the nature of the constituents of atomic nuclei.

Example 2 Let us now compare velocities and frequencies of the various types of waves we have discussed. We will assume that the wave length is $\lambda = 1$ meter, and we will have to assume reasonable properties for the other parameters we encounter.

(i) *Acoustic waves in gases.* We will take air as an example. The "normal pressure" of one atmosphere is

$$p_0 = 1.01 \times 10^5 \text{ newtons/m}^2$$

and the density of air at 20° C (72° F) and normal pressure is

$$\rho_0 = 1.23 \text{ kg/m}^3$$

Since air consists of nitrogen and oxygen—both are diatomic molecules—we take $\gamma = 1.4$. Thus, we have

$$v_0 \text{ (sound in air)} = \sqrt{\frac{\gamma p_0}{\rho_0}} = \sqrt{\frac{1.4 \times 1.01 \times 10^5 \text{ newtons/m}^2}{1.23 \text{ kg/m}^3}} = 339 \text{ m/sec}$$

and as a consequence, a wave of wavelength $\lambda = 1$ m will have the frequency

$$\nu = \frac{v_0}{\lambda} = 3.39 \times 10^2 \text{ sec}^{-1} = 339 \text{ Hz}$$

For comparison we note that the "standard A" tone (which you can sometimes hear the concertmaster sound on his violin before a concert) has a frequency of 440 Hz.

(ii) *Transverse wave on a string.* Let us choose as "string" a piano wire made out of steel with a diameter of $d = 1$ mm (3/64″) and let us choose as tension $T = 1000$ newtons (we ask a 220 lb football player to hang from the wire). Since the density of steel is $\rho = 7.9 \times 10^3$ kg/m³, we have a linear density of

$$\mu = \rho \cdot \pi \cdot \left(\frac{d}{2}\right)^2 = (7.9 \times 10^3 \text{ kg/m}^3) \times 3.14 \times \frac{10^{-6} \text{ m}^2}{4} = 0.62 \times 10^{-2} \text{ kg/m}$$

and the speed of a transverse vibrational wave is

$$v_0 = \sqrt{\frac{T}{\mu}} = \sqrt{\frac{10^3 \text{ newtons}}{0.62 \times 10^{-2} \text{ kg/m}}} = 402 \text{ m/sec}$$

This is of the same order of magnitude as the speed of sound in air. The frequency of vibration at a wavelength of 1 m then is

$$\nu = \frac{v_0}{\lambda} = 402 \text{ Hz}$$

(iii) *Elastic compression in a thin rod.* We take again steel as an example. Its Young's modulus is $Y = 2.0 \times 10^{11}$ newtons/m, and we thus have

$$v_0 = \sqrt{\frac{Y}{\rho}} = \sqrt{\frac{2 \times 10^{11} \text{ newtons/m}}{7.9 \times 10^3 \text{ kg/m}^3}} = 5.03 \times 10^3 \text{ m/sec} \approx 5 \text{ km/sec}$$

The speed of sound in a steel bar is about 15 times the speed of sound in air. The frequency corresponding to a wavelength of 1 m is also 15 times higher:

$$\nu = \frac{v_0}{\lambda} = 5.0 \times 10^3 \text{ Hz} = 5 \text{ kHz}$$

(iv) *Water waves.* We first have to decide which of the three types of water waves we are talking about. The surface tension of water is $\gamma = 7.3 \times 10^{-2}$ newton/m. Using the complete equation, Eq. (1–31), which is valid for all wavelengths, we find for $\lambda = 1$ m

$$\frac{g\lambda}{2\pi} = \frac{9.8 \text{ m/sec}^2 \times 1 \text{ m}}{6.28} = 1.56 \text{ m}^2/\text{sec}^2$$

and since the density of water is $\rho = 10^3$ kg/m³, we have

$$\frac{2\pi\gamma}{\rho\lambda} = \frac{6.28 \times 7.3 \times 10^{-2}}{10^3 \times 1} = 4.6 \times 10^{-4} \text{ m}^2/\text{sec}^2$$

We see that at our wavelength the second term is negligibly small; the wave is a pure gravity wave. In deep water, $h \gg 1$ meter, its speed will be given by Eq. (1–27)

$$v_0 = \sqrt{\frac{g\lambda}{2\pi}} = 1.25 \text{ m/sec}$$

Water waves propagate very slowly; that is why we can follow them with the naked eye. The frequency is then

$$\nu = \frac{v_0}{\lambda} = 1.25 \text{ Hz}$$

A boat would thus rock up and down in a little less than a second if the prevalent waves on a lake have a wavelength of 1 meter. In very shallow water, say in water of a depth of 5 cm = 0.05 m (= 2 inches), the speed will be

$$v_0 = \sqrt{gh} = 0.70 \text{ m/sec}$$

or about one-half the speed in the middle of the lake.

(v) *Electromagnetic waves.* As stated earlier, the wave velocity of light in vacuum (and to a good approximation in air) is

$$c = 3.0 \times 10^8 \text{ m/sec}$$

Electromagnetic waves move so fast that in everyday life we claim to see "instantaneously" what happens even far away. The frequency corresponding to a wavelength of 1 m is

$$\nu = \frac{c}{\lambda} = 3 \times 10^8 \text{ Hz} = 300 \text{ MHz}$$

Because electromagnetic waves are so fast, they also oscillate very rapidly.

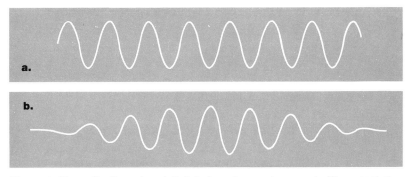

Figure 1–12   a. Section of an infinitely long harmonic wave. **b.** Wave packet.

7.   WAVE PACKETS

When talking about harmonic waves, we disregard the fact that no real wave is ever infinitely long. A water wave will not reach across the ocean. A violin player who plays a note for $\Delta t = 10$ seconds will produce a wave train which has the length

$$v \cdot \Delta t = 339 \text{ m/sec} \times 10 \text{ sec} \approx 3,400 \text{ meters}$$

We call a *wave packet* a wave which is nearly harmonic, but does not last forever as a true harmonic wave would (Fig. 1–12). This will not affect our discussion as long as we are only concerned with phenomena occurring over distances which are small compared to the length of the wave packet, and are studying them only for a time which is short compared to the total time the wave packet takes to pass through a point. We will later discuss the much more complex problem of the motion of the wave packet as a whole. Indeed, much of our study of quantum mechanics will consist of a discussion of the de Broglie wave packet, which makes up a single electron.

8.   ENERGY DENSITY AND ENERGY FLOW

A traveling wave in any physical field (pressure, displacement on a string, and so forth is always associated with a flow of energy through the medium. We can see this by noting that the disturbance associated with the wave is always a deviation from the quiescent equilibrium situation; there is a pressure difference between nearby points in a gas, or a string vibrates, or the water level is not flat. Locally (over the distance of a wavelength or less), we can extract energy from the medium. In a gas we can use the instantaneous pressure difference between two points to deform or to move an object—or even to shatter a window, if the pressure difference is large enough. There is also obviously energy in a vibrating string—the string itself moves, and thus has kinetic energy. A water wave will move an object floating on the water. We could go on forever just mentioning such examples of how to extract energy from a wave. But how do we know that the energy is actually being transported by the wave and that the wave is not somehow "liberating" energy which was locally already there? Let us consider (Fig. 1–13) a wave packet which is coming in from the left, but has not yet reached the point P at the time $t = t_1$. Since the medium at P is quiescent, the second law of thermodynamics tells us that we have no way of extracting energy.

Later on (Fig. 1–11b), when the wave has arrived at P, we can extract energy from the locally varying disturbance of the medium. Even later, at the time $t = t_3$, when the wave has already passed, the medium at P is again quiescent and in the same state as before—but we have already extracted some energy earlier at the time t_2. The medium is in the same state it was in at t_1, but we have some energy left over. This energy had to come from somewhere—and the only possible source is the wave which has passed by.

We now have to introduce two closely related concepts: the energy density ρ_E and the intensity I of a wave. When defining energy density, we have to remember that although we are now discussing only one-dimensional waves (waves going in one direction), the surface of a lake is two-dimensional, and sound propagates in a three-dimensional volume of gas. We will define the energy density, for each type of wave, as the energy content per unit n-dimensional volume. Thus, for a vibrating string the energy density is the energy per unit length; for water waves it is the energy per unit surface area; and for acoustic waves and for electromagnetic waves it is the energy per unit three-dimensional volume.

We define the intensity of a wave as the energy which flows per unit time through a unit "area" perpendicular to the wave direction. Again we have to be careful about the concept of area, which will depend on the type of wave in question. For the transverse waves on a string, where the energy flows only along the string, there is actually no area at all: the intensity is the energy which passes through a point on the string per unit time (Fig. 1–14a). For water waves the intensity is the energy passing through a line of unit length per unit time (Fig. 1–14b). The line is assumed perpendicular to the wave direction. For waves which occur in a three-dimensional medium, the intensity is the energy passing per unit time through a unit area oriented perpendicularly to the wave direction.

It is, in general, quite difficult to calculate the energy density in a wave of arbitrary shape. However, for a harmonic wave (i.e., for a wave of well-defined frequency and wavelength), the *average energy density* $<\rho_E>$ is always pro-

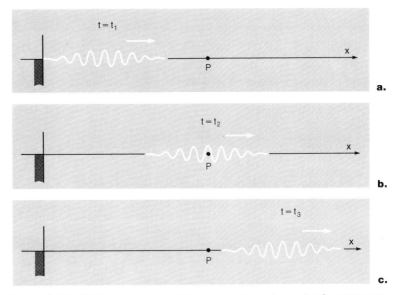

Figure 1–13 While a wave packet is passing through a point P, energy can be extracted; thus, the wave itself has to carry energy.

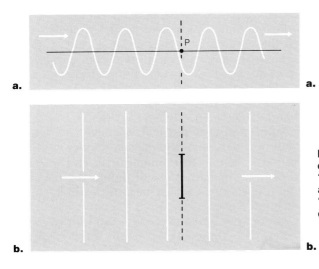

a.

a.

b.

Figure 1–14 The intensity of a wave is the energy flowing per unit time through a unit "area." **a.** For waves on a string, the area is a single point P. **b.** For a water wave, the "area" is a segment of unit length perpendicular to the wave direction.

b.

portional to the square of the amplitude:

$$<\rho_E> = b(\lambda) A^2 \qquad (1–34)$$

where the coefficient b is different for each type of wave and can depend on the wavelength λ and the frequency ν of the harmonic wave. By average energy density, we mean the density averaged over a region of the size of one wavelength; alternately, we can also consider the energy density in a very small region, but average its value over one period of motion. In both cases we obtain Eq. (1–34).

We can, however, derive a general relation between energy density ρ_E and intensity I:

$$I = v_0 \rho_E \qquad (1–35)$$

To prove this relation, let us consider Fig. 1–15, where we have taken the example of water waves. Since the energy flows with the wave velocity v_0, all energy contained in a rectangle of base s and length $v_0 \Delta t$ will pass through the line s in the time Δt. Thus, the intensity (i.e., the energy flow per unit time and unit length) is

$$I = \frac{\rho_E v_0 \Delta t \; s}{\Delta t \; s} = \rho_E v_0$$

In the same way, we can also predict the energy flowing past a line of length s which is oriented in an arbitrary way relative to the direction a wave travels (Fig. 1–15b). The energy flowing through the line per unit time is again all the energy contained in the sketched parallelogram of sides s and $v_0 \Delta t$. The area of the parallelogram is Area $= v_0 \times \Delta t \times s \cos \alpha$, and thus the power—the energy passing through s per unit time—is

$$P = \frac{\rho_E \; \text{Area}}{\Delta t} = \rho_E v_0 s \cos \alpha \qquad (1–36a)$$

Clearly the arriving power will be maximal if the line s is perpendicular to the wave direction.

If we have waves in a three-dimensional medium that are impinging on a screen of area S oriented arbitrarily relative to a wave, the arriving power will also be given by

$$P = \rho_E v_0 S \cos \alpha \qquad (1\text{--}36b)$$

where α is now the angle between the normal to the area A and the direction of the incoming wave. The proof would be identical to our proof of Eq. (1–36a), if we consider the line s in Fig. 1–15b as the projection of the area S on the paper. Eq. (1–36b) explains why a light screen (or a wall) will seem brightest if the light is incident normally; the same screen will seem much darker if it is hit by the same light at an angle.

We will now derive Eq. (1–34) for the particular case of transverse waves on a string. The deviation of a string from its equilibrium (quiescent) position can be described by Eq. (1–11a) for a harmonic wave:

$$y(x,t) = A \cos (kx - \omega t - \delta) \qquad (1\text{--}37)$$

where the ratio of ω and k is the speed of the wave as given in Eqs. (1–10) and (1–21):

$$\frac{\omega}{k} = \sqrt{\frac{T}{\mu}} = v_0 \qquad (1\text{--}38)$$

We now want to calculate the energy per unit time flowing past the point P at $x = x_0$. It is equal to the work per unit time done by the segment of the string

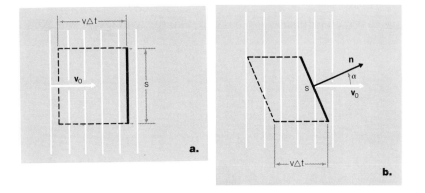

Figure 1–15 All energy within a parallelogram of height $v\Delta t$ and base line s flows through s in the time Δt. The area is $A = s(v\Delta t) \cos \alpha$.

immediately to the left of P on the segment immediately to the right of P. The work is done by the tension T, but it is only its perpendicular component

$$T_\perp = -T \cdot \sin \alpha$$

which does any work (Fig. 1–16). This is so because the string moves only in the perpendicular direction. The motion of the point P in the time interval dt is

$$dy = \frac{\partial y}{\partial t} \cdot dt$$

Thus, the intensity (i.e., the work done per unit time) is

$$I = T_\perp \frac{dy}{dt} = -T \frac{\partial y}{\partial t} \cdot \sin \alpha$$

We again use the small angle approximation $\alpha \approx \sin \alpha \approx \tan \alpha = \frac{\partial y}{\partial x}$. Then we have

$$I = -T \frac{\partial y}{\partial t} \frac{\partial y}{\partial x} \tag{1–39}$$

Taking the derivatives of Eq. (1–37), we obtain

$$\frac{\partial y}{\partial x} = -Ak \sin (kx - \omega t - \delta)$$

$$\frac{\partial y}{\partial t} = +A\omega \sin (kx - \omega t - \delta)$$

and thus

$$I = TA^2\omega k \sin^2 (kx - \omega t - \delta)$$

We see that while the energy flow is always positive, it oscillates between zero and its peak value

$$I_{\max} = TA^2\omega k$$

To obtain the average intensity, we have to take the average of $\sin^2 (kx - \omega t - \delta)$

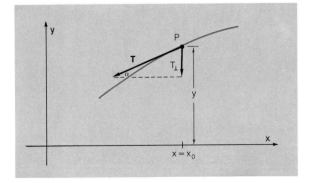

Figure 1–16 Calculating the energy transmitted from the part of the string to the left of P to the part to the right of P. The wave velocity is in the positive x-direction.

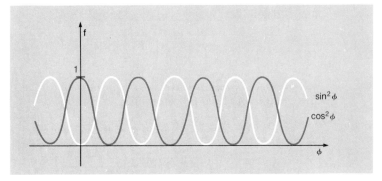

Figure 1–17   The functions $\cos^2 \phi$ and $\sin^2 \phi$.

over one period $T = 1/(2\pi\omega)$. Noting that for any angle ϕ

$$\sin^2 \phi + \cos^2 \phi = 1$$

and that $\sin^2 \phi$ and $\cos^2 \phi$ are really the same function displaced by $\pi/2$ (Fig. 1–17), we average over all values of ϕ between zero and 2π to obtain

$$<\sin^2 \phi> = <\cos^2 \phi>$$

and since the sum of the two equal averages is equal to one, we have

$$<\sin^2 \phi> = <\cos^2 \phi> = \frac{1}{2}$$

or

$$<\sin^2 (kx - \omega t - \delta)> = \frac{1}{2} \tag{1–40}$$

Substituting Eq. (1–40) into the intensity equation, we find that the average intensity is

$$<I> = Tk\omega \frac{A^2}{2}$$

which we can rewrite using Eq. (1–38) as

$$<I> = v_0 Tk^2 \frac{A^2}{2} = \sqrt{\frac{T^3}{\mu}} \left(\frac{2\pi}{\lambda}\right)^2 \frac{A^2}{2}$$

With the help of Eq. (1–35) we can then also obtain an expression for the average energy density, the average energy per unit length

$$<\rho_E> = \frac{<I>}{v_0} = \frac{Tk^2A^2}{2} = T\left(\frac{2\pi}{\lambda}\right)^2 \frac{A^2}{2} \tag{1–41}$$

which indeed is of the form asserted to be correct in Eq. (1–34): The coefficient of A does depend on the wavelength.

We can use a rather general argument to show why the expression for the energy density given in Eq. (1–34) is reasonable. We can obviously write for the average energy density of a harmonic wave of fixed wavelength but arbitrary amplitude:

$$<\rho_E> = f(A)$$

where $f(A)$ is some unknown function of the amplitude A. We can now approximate $f(A)$ by a polynomial

$$f(A) = a_0 + a_1 A + a_2 A^2 + a_3 A^3 + a_4 A^4 + \ldots \qquad (1\text{--}42a)$$

Since for $A = 0$ there is no wave and thus also no energy density, we must have $f(0) = 0$ and therefore also $a_0 = 0$. We can also show that all odd powers of A (i.e., the coefficients $a_1, a_3, \ldots$) all vanish. To see this we compare two waves:

$$y_1 = A \cos (kx - \omega t)$$
$$y_2 = (-A) \cos (kx - \omega t)$$

Since $\cos (\alpha - \pi) = -\cos \alpha$, we can write

$$y_2 = A \cos (kx - \omega t - \pi)$$

We can now set

$$\omega t + \pi = \omega \left(t + \frac{\pi}{\omega} \right) = \omega(t + t_0)$$

and we see that y_1 and y_2 describe the same wave if we start counting our time for y_2 earlier by t_0. However, the wave is periodic, and we are calculating the energy density averaged over one full period; in consequence, $f(A)$ should give exactly the same result for y_1 and y_2. But while for y_1 we have $f(A)$, for y_2 we should have $f(-A)$ describing the density. Thus $f(A) = f(-A)$. If any odd power of A did not vanish, that term would be $a_n(A)^n$ in the expression for $f(A)$, while it would be $-a_n(A)^n$ in the expression for $f(-A)$. However, this contradicts the assertion that $f(A) = f(-A)$, so all odd powers of A must have zero coefficients.

From this we conclude that we can write Eq. (1–42a) in the following form:

$$<\rho_E> = a_2 A^2 + a_4 A^4 + a_6 A^6 + \ldots \qquad (1\text{--}42b)$$

The assertion made in Eq. (1–34) merely states that $f(A)$ consists only of the first term in Eq. (1–42b). This seems reasonable, at least if the amplitude is small; then any higher terms can be neglected.

Summary of Important Relations

Wave in one dimension $\phi(x,t) = f(x - v_0 t) + g(x + v_0 t)$

where f and g are arbitrary functions of one variable

Harmonic wave $\phi(x,t) = A \cos (kx - \omega t)$

wavelength $\lambda = \frac{2\pi}{k}$, k is wave number frequency $\qquad \nu = \dfrac{\omega}{2\pi}, \quad \omega$ is angular frequency

period $\qquad T = \dfrac{1}{\nu} = \dfrac{2\pi}{\omega}$

wave velocity $\quad v_0 = \lambda \nu = \dfrac{\omega}{k}$

Wave equation $\qquad \dfrac{1}{v_0^2}\dfrac{\partial^2 \phi}{\partial t^2} = \dfrac{\partial^2 \phi}{\partial x^2}$

Energy density $\qquad \rho_E = bA^2$

$b(\lambda,\nu)$ different for each type of wave, A is amplitude

Intensity $\qquad\qquad\qquad\qquad\qquad I = v_0 \rho_E$

Questions

1. Give some examples of random and collective processes occurring in (a) the U.S. economy; (b) the ecological system of a forest or a lake; (c) the biological system of a living organism.

2. Discuss the limits of the superposition principle for (a) deep sea waves; (b) transverse waves on a string; (c) sound waves. Estimate at what order of magnitude of the amplitude the superposition principle will start breaking down.

3. Two waves are traveling in opposite directions. At the time $t = 0$ the wave pattern is as shown below. Sketch the pattern at the times $t = 1, 2, 4,$ and 6 seconds if the wave velocity is $v = 1$ cm/sec. Explain in particular what happens at $t = 2$ seconds.

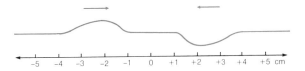

4. A popular way to estimate the distance of a thunderstorm is to count slowly from the instant one sees the flash until one hears the thunder. The rule of thumb is 1000 feet per count. Justify the rule.

5. Discuss why water waves, which seem quite harmonic in the ocean, become "breakers" as they approach a sandy beach. The wave height of the breakers increases — why?

Problems

1. Which of the following functions describe a wave in one dimension? Which describe a wave traveling in one direction?

    (a) $\cos(kx - \omega t)$

    (b) $\cos(k^2 x^2 - \omega^2 t^2)$

    (c) $(k^2 x^2 - \omega^2 t^2)$

    (d) $\cos kx \cdot \sin \omega t$

    (e) $f(\alpha x^2 + 2\beta xt + \gamma t^2)$ [f some function of one variable]

    (f) $e^{-\lambda t} \cos(kx - \omega t)$

    (g) $e^{-(x + \lambda t)^2} + \cos(kx - \omega t)$

2. A wave traveling on a rope has the form

$$y(x,t) = 6 \sin(0.03x - 5.0t)$$

    where y and x are in centimeters and t is in seconds. Calculate the amplitude, wavelength, frequency, period, and velocity of the wave.

3. Two water waves of equal amplitude and wavelength are traveling in the same direction; the phase difference between the two is $\delta = \pi/2$. Calculate the form of the resulting single wave.

4. What is the frequency of electromagnetic radiation which has the wavelength of a typical atomic diameter of 2 Å?

5. The wavelength of a sound wave in a solid cannot be shorter than a few atomic distances. What is the frequency of a sound wave whose wavelength is 10 Å in copper? (Young's modulus $Y = 1.1 \times 10^{11}$ N/m²; for other necessary data, consult Appendix B.)

6. A radio station is transmitting at a frequency of 11.8 MHz (megacycles). What are the wavelength, the angular frequency, and the wave number of the radio wave?

7. A sound wave (velocity 340 m/sec) has a wavelength $\lambda = 1.5$ m. What is the phase difference at a fixed point at two instants 10^{-3} sec apart?

8. The frequency of an F.M. radio station emitting at a mean wavelength of 2.5 m varies by $\pm \Delta \nu = \pm 20$ kHz. By how much does the wavelength itself vary?

9. If a major eruption occurs at the surface of the sun, how long will it take until one sees it on the earth? (Distance from sun to earth $= 9.3 \times 10^7$ miles.)

10. Ground control in Houston is to send a time mark to an astronaut circling the moon, so that he knows exactly when to start his "retro-fire" for landing. How much ahead of the actual time indicated does the man in Houston say "mark"? (Distance from earth to moon $= 3.8 \times 10^5$ km.)

11. A submerged submarine and an airplane are both monitoring an atomic underwater explosion. Both are 500 miles away from the explosion. Determine the time difference between the following three events: (a) airplane pilot sees flash; (b) pilot hears thunder; (c) submarine captain detects explosion on sonar. (The speed of sound in water is 1500 m/sec.)

12. It is reasonable to expect that the ear can no longer hear sound waves if the wavelength becomes smaller than the size of the ear opening. This limit is not reached by humans, but it is reached by dogs. What is the highest frequency a dog is likely to hear?

13. The surface tension of water is $\gamma = 7.3 \times 10^{-2}$ newton/m. At what wavelength is the wave velocity of deep sea waves twice as large as the velocity of a wave with wavelength $\lambda_0 = 10$ cm?

14. The police chief of Honolulu is told that a major underwater earthquake has occurred 2000 miles away, 2 hours ago. He estimates it will take him at least 12 hours to evacuate the low-lying parts of the city. Should he sound the order to evacuate before the tidal wave hits? Justify your answer.

15. At what frequency will sound waves have a wavelength of the typical room size of 20 feet? (These can be bothersome in a room if the walls are reflecting sound.)

16. In a small boat one is most likely to become seasick if the boat moves up and down once in 5 to 15 seconds. What is the range of wavelengths in deep water which corresponds to this period?

17. Steel yields (deforms permanently) if the applied stress is larger than 3×10^9 newtons/m². What is the highest wave velocity one can achieve for transverse waves with a steel wire?

18.° Show explicitly that one obtains Eq. (1–29) from Eq. (1–31) under the approximate limit.

19. Find the wavelength for which the velocity of deep sea water waves is a minimum. Does your result agree with Fig. 1–11?

20. A very long flexible wire of length L and mass M is hanging freely from a support on the top end. Calculate the velocity of transverse waves a distance x from the top of the wire.

21. Assume that the energy density of deep sea waves is (a) independent of wavelength, (b) proportional to the amplitude (wave height) squared. Use a dimensional argument to estimate the energy density. Estimate the energy stored in a square 100 km on a side on the ocean surface, if the waves have a mean height of 4 m.

22. Give the units in which we measure wave intensity for: (a) transverse waves on a string; (b) sound waves in air; (c) water waves; (d) elastic compression waves in a thin rod; (e) electromagnetic waves.

23. A light bulb emits a total power of 100 W in the form of visible light isotropically (the same in all directions). What is the light intensity 3 meters away from the bulb?

24. Using dimensional arguments, estimate the sound wave intensity in terms of the density variation in air. From this, estimate the density variation in a sound wave of intensity 10 W/m². (1 watt = 1 W = 1 Joule/sec.)

25. A transverse harmonic wave of intensity $I = 10$ W and wavelength $\lambda = 0.5$ m is traveling along a steel wire of linear density $\mu = 8$ grams/m under a tension $T = 1000$ newtons. Calculate the maximum excursion of a point on the wire.

CHAPTER 2

INTERFERENCE AND STANDING WAVES

1. INTERFERENCE

In our discussion of waves in Chapter 1, we encountered two facts about waves which are of fundamental importance: (1) *that waves can be superimposed* (added) *to create new wave patterns*, and (2) *that the average intensity of a harmonic wave is proportional to the square of its amplitude.*

As we will soon see, the superposition of two harmonic waves of the same (or nearly the same) wavelength leads again to waves which are harmonic (or nearly harmonic). The above two principles will enable us to discuss and predict an astonishingly wide range of phenomena common to all waves. In this chapter we will limit ourselves to discussing waves in one spatial dimension; the more general case will be treated later.

The most important consequence of the two basic statements is *interference of waves*. From a naïve point of view we would expect that if we have two waves present, the total energy content of their superposition would be the sum of the energy contents of each wave. This is not true; indeed, two waves can add in such a way that the resulting wave has less total energy than either of the individual waves. The energy density of the superposition of two harmonic waves depends on *their relative phase*. We say that the two waves interfere with each other.

We can study this in more detail in a very simple example. Let us consider two waves of equal amplitude, wavelength, and frequency, both traveling in the same direction. The two waves differ only in initial phase. Thus, we can describe the first wave mathematically as

$$u_1(x,t) = A_0 \cos (kx - \omega t) \qquad (2\text{--}1a)$$

and the second wave as

$$u_2(x,t) = A_0 \cos (kx - \omega t - \delta) \qquad (2\text{--}1b)$$

By superposition of the two waves we obtain a total disturbance

$$u(x,t) = u_1(x,t) + u_2(x,t) = A_0 \left[\cos (kx - \omega t) + \cos (kx - \omega t - \delta)\right] \qquad (2\text{--}2)$$

We can reduce the expression in Eq. (2–2) with the help of the trigonometric

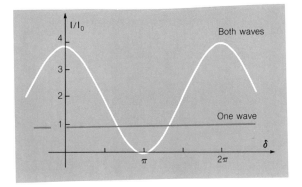

Figure 2–1 The intensity of two superimposed waves of equal amplitude depends on their relative phase δ. It can be as large as four times the single wave intensity, but can also vanish altogether.

identity (see Appendix C, p. 555).

$$\cos \alpha + \cos \beta = 2 \cos \left(\frac{\alpha + \beta}{2}\right) \cos \left(\frac{\alpha - \beta}{2}\right) \qquad (2\text{–}3)$$

Setting

$$\alpha = kx - \omega t$$
$$\beta = kx - \omega t - \delta$$

we obtain for the total disturbance

$$u(x,t) = 2A_0 \cos \left(kx - \omega t - \frac{\delta}{2}\right) \cdot \cos \left(+\frac{\delta}{2}\right) \qquad (2\text{–}4)$$

Thus, we have shown that the superposition of two harmonic waves which differ only in initial phase yields another harmonic wave. The amplitude of the complete wave is, according to Eq. (2–4),

$$A' = 2A_0 \cos \left(\frac{\delta}{2}\right) \qquad (2\text{–}5a)$$

and because the intensity of a wave is proportional to the square of the amplitude, we have for the intensity of the sum of the two waves

$$I' = \text{const} \cdot A'^2 = 4 \cos^2 \left(\frac{\delta}{2}\right) \cdot I_0 \qquad (2\text{–}5b)$$

where we have not specified the proportionality constant between the intensity and the square of the amplitude. However, we know that

$$I_0 = \text{const} \cdot A_0^2$$

is the intensity of each individual wave. The function $\cos^2 (\delta/2)$ varies between $+1$ and 0. Thus, the intensity of the superposed wave can vary between four times the individual wave intensity and zero, as shown in Fig. 2–1. Naïvely we would have expected that "energies add and thus intensities add"—we would have said that $I' = 2I_0$. But this would be incorrect reasoning.

2. PHASORS

We can use a graphical representation which helps us to intuitively understand—and calculate—interference phenomena. A harmonic wave

$$u(x,t) = A \cos (kx - \omega t - \delta) \qquad (2\text{–}6a)$$

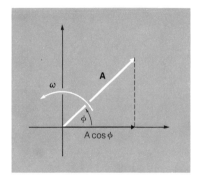

Figure 2-2 Definition of a phasor; its projection on the horizontal axis is equal to the harmonic wave disturbance.

can, for a fixed $x = x_0$, always be considered a function of t alone. Since $\cos(-\alpha) = \cos\alpha$, we can write

$$u(x,t) = A\cos(\omega t + \Phi_0) \tag{2-6b}$$

where we define

$$\Phi_0 = \delta - kx_0$$

Now let us consider a vector (in two dimensions) of length A rotating with angular frequency ω. Its projection on the horizontal axis (Fig. 2-2) is equal to

$$A\cos\phi = A\cos(\omega t + \Phi_0)$$

where Φ_0 is the angle between the vector and the horizontal axis at $t = 0$. We can consider such a rotating vector as representing the harmonic wave in Eq. (2-6). We call the rotating vector a *phasor*, because its angle with the horizontal axis is at any instant equal to the phase of the harmonic wave.

If we have a second harmonic wave, we can represent it by a second phasor; the superposition of the two waves will then be represented by the vectorial sum of the two phasors, as shown in Fig. 2-3. Thus, the two harmonic waves in Eqs. (2-1a) and (2-1b) can be represented by two phasors of equal length; the second is rotated counterclockwise by an angle δ with respect to the first. The superposition of the two harmonic waves is then represented by the sum phasor **C**.

Using the phasor representation, we easily deduce that the phenomenon of interference occurs even if the two amplitudes are not equal. In Fig. 2-4 we show two phasors of unequal length and relative phase δ. The sum **C** of the two phasors **A** and **B** can be determined graphically. However, from plane geometry we can also easily calculate its length. We have, by the law of cosines,

$$C^2 = A^2 + B^2 - 2AB\cos\alpha$$

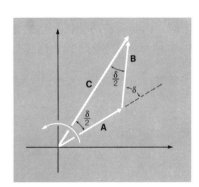

Figure 2-3 Adding two phasors of equal length or amplitude leads directly to Eq. (2-5a).

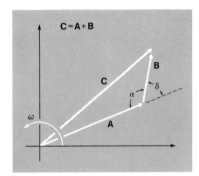

Figure 2-4 Adding two phasors of unequal length (amplitude).

and since

$$\cos \alpha = \cos (180° - \delta) = -\cos \delta$$

we can also write

$$C^2 = A^2 + B^2 + 2AB \cos \delta \qquad (2\text{-}7)$$

Since the intensity of a wave is proportional to the square of the amplitude, we have arrived at a general expression for the intensity of a wave which is the superposition of two harmonic waves of equal frequencies:

$$I_C = I_A + I_B + 2\sqrt{I_A I_B} \cdot \cos \delta \qquad (2\text{-}8)$$

We had to assume that the two phasors—or waves—have the same angular frequency, because otherwise the relative phase angle would vary with time. One should note that Fig. 2-4 shows graphically that the intensity of the sum wave depends only on the *relative* phase of the two phasors or waves; by this we mean that only the difference in phases δ matters. If we have two waves

$$u_1(x,t) = A \cos (kx - \omega t - \delta_1)$$
$$u_2(x,t) = B \cos (kx - \omega t - \delta_2)$$

then the intensity of $u(x,t) = u_1(x,t) + u_2(x,t)$ will depend only on the relative phase

$$\delta_{\text{rel}} = \delta_1 - \delta_2 \qquad (2\text{-}9)$$

Therefore, we can without loss of generality set $\delta_1 = 0$. From Eq. (2-8) we also see that the intensity of the sum wave is the same for a phase difference $-\delta$ as it is for a phase difference $+\delta$.

Example 1   Two waves of equal wavelengths and frequencies propagate in the same direction. Their intensities are I_1 and $I_2 = 2I_1$. What are the maximum and the minimum intensities which can be achieved by their superposition?

From Eq. (2-8) we see that if both intensities are given, the sum intensity depends only on the relative phase δ. We have

$$I_{\text{tot}} = I_1 + 2I_1 + 2\sqrt{I_1 \cdot 2I_1} \cdot \cos \delta$$

or

$$I_{\text{tot}} = I_1(3 + 2\sqrt{2} \cdot \cos \delta)$$

The maximum intensity will occur if $\cos \delta = 1$: then

$$I_{tot} = I_1(3 + 2\sqrt{2}) = 5.83 \, I_1$$

The minimum intensity will occur if $\cos \delta = -1$, in which case we have

$$I_{tot} = I_1(3 - 2\sqrt{2}) = 0.17 \, I_1$$

Example 2 N waves of equal intensity, wavelength, and frequency are to be given such relative phases that the superposition of all N waves has zero intensity. From Fig. 2–5 it is easily seen that if we set the phase for the first wave equal to

$$\delta_0 = \frac{2\pi}{N}$$

and that for the second wave equal to $2\delta_0$, and so on, we obtain a solution. Indeed, the sum of all N phasors has a resultant of zero. However, the question is: Is this the only possible solution? The answer is no. We can just as well subdivide N into $N = N_1 + N_2$ and have the first N_1 phasors add up to zero; then we can also find phases so that the N_2 residual waves superpose so that the resulting intensity also vanishes.

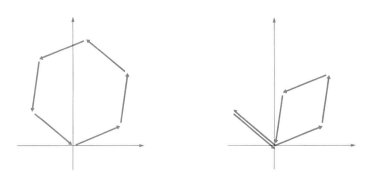

Figure 2–5 Example 2: Two different ways to add six phasors of equal amplitude so as to obtain a zero sum.

3. THE BEAT PHENOMENON

When we strike two tuning forks—or play on two similar instruments—and the frequency of sound from one is nearly—*but not exactly*—identical to the other, we will hear a periodic increase and decrease of the sound intensity. We say that the two instruments are "beating against each other." The same effect can also be made visible with water waves, if one produces two waves going in the same direction with nearly the same wavelength λ (and thus nearly the same frequency $\nu = v_0/\lambda$). The beat phenomenon can in principle occur with any type of wave; it is a consequence only of the wave nature of the process.

To see that, let us assume that two harmonic waves of equal amplitude are traveling in the same direction. We have a disturbance due to the first wave

$$y_1(x,t) = A \cos (k_1 x - \omega_1 t) \tag{2–10a}$$

and the second wave

$$y_2(x,t) = A \cos (k_2 x - \omega_2 t) \tag{2–10b}$$

We will assume that the wave velocity is independent of wavelength:

$$\frac{\omega_1}{k_1} = \frac{\omega_2}{k_2} = v_0 \qquad (2\text{-}11)$$

If both waves are present, the overall disturbance will be

$$y(x,t) = y_1(x,t) + y_2(x,t) = A \left[\cos(k_1 x - \omega_1 t) + \cos(k_2 x - \omega_2 t)\right] \qquad (2\text{-}12)$$

Let us now assume that the two frequencies ω_1 and ω_2 are nearly the same:

$$|\omega_1 - \omega_2| = |\Delta\omega| \ll \omega_1 \quad \text{or} \quad \omega_2 \qquad (2\text{-}13a)$$

Because of Eq. (2–11), this implies that

$$\Delta k = k_1 - k_2 = \frac{\omega_1 - \omega_2}{v_0} \qquad (2\text{-}13b)$$

is also a small quantity:

$$|\Delta k| \ll k_1 \quad \text{or} \quad k_2$$

We can now use again the trigonometric identity

$$\cos\alpha + \cos\beta = 2\cos\left(\frac{\alpha + \beta}{2}\right)\cos\left(\frac{\alpha - \beta}{2}\right) \qquad (2\text{-}14)$$

Inserting for α and β the two phases in Eq. (2–10), we obtain for the disturbance

$$y(x,t) = 2A \cos\left(\frac{k_1 + k_2}{2}x - \frac{\omega_1 + \omega_2}{2}t\right) \cdot \cos\left(\frac{k_1 - k_2}{2}x - \frac{\omega_1 - \omega_2}{2}t\right)$$

Since we have assumed that the two angular frequencies ω_i and the two wave numbers k_i are nearly equal, so that

$$\omega_1 \approx \omega_2 \approx \frac{\omega_1 + \omega_2}{2} = \omega; \quad k_1 \approx k_2 \approx \frac{k_1 + k_2}{2} = k$$

we can write to a good approximation for the disturbance

$$\boxed{y(x,t) = 2A \cos(kx - \omega t)\cos\left(\frac{\Delta k}{2}x - \frac{\Delta\omega}{2}t\right)} \qquad (2\text{-}15)$$

The superposition of two waves of nearly the same frequency and exactly the same amplitude leads to a *modulated* wave, as shown in Fig. 2–6. The modulated wave has an amplitude which varies slowly with time. We can write

$$y(x,t) = A' \cos(kx - \omega t) \qquad (2\text{-}16a)$$

with

$$A' = 2A \cos\left(\frac{\Delta k}{2}x - \frac{\Delta\omega}{2}t\right) \qquad (2\text{-}16b)$$

Let us consider how such a wave would seem to a stationary observer who

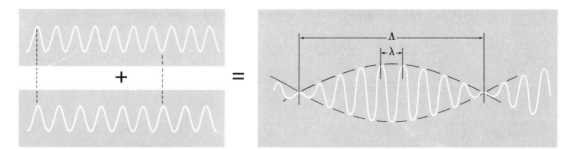

Figure 2-6 Adding two waves of equal amplitudes and nearly equal frequencies results in a modulated wave.

samples the disturbance as a function of time at a point $x = x_0$. He will then see the wave periodically increasing and again decreasing with time. The intensity will be

$$I = \text{const} \cdot 4A^2 \cos^2\left(\frac{\Delta\omega}{2}t + \Phi_0\right) \qquad (2\text{--}16c)$$

with

$$\Phi_0 = -\frac{\Delta k}{2}x_0$$

We can obtain the same result using the phasor technique (Fig. 2–7). We have two phasors, which do not rotate with exactly the same angular frequency. The second phasor rotates with a slightly different angular frequency

$$\omega_2 = \omega_1 + \Delta\omega$$

Thus the angle Φ between the two phasors slowly changes with time:

$$\Phi = \Delta\omega t + \phi_0 \qquad (2\text{--}17)$$

The length of the resulting vector is given in exact analogy to Eq. (2–7) as

$$A'^2 = A^2 + A^2 + 2A \cdot A \cdot \cos\Phi \qquad (2\text{--}18)$$

and using Eq. (2–17) for Φ we have

$$A' = A\sqrt{2 + 2\cos(\Delta\omega t + \phi_0)} \qquad (2\text{--}19)$$

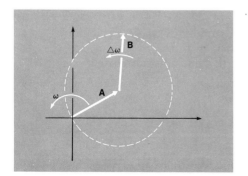

Figure 2-7 Adding two phasors representing waves of unequal frequencies. While the whole figure rotates around the origin with frequency ω, the vector **B** is rotating with frequency $\Delta\omega$ around the tip of **A.**

This seems different from Eq. (2–16b), but is not. Because of the identity

$$\cos \alpha = 2 \cos^2 \frac{\alpha}{2} - 1$$

we have

$$A' = A\sqrt{2 + 2[2 \cos^2 \left(\frac{\Delta \omega t}{2} + \frac{\phi_0}{2}\right) - 1]}$$

or

$$A' = A\sqrt{4 \cos^2 \left(\frac{\Delta \omega t}{2} + \frac{\phi_0}{2}\right)}$$

which is indeed identical to Eq. (2–16b) if we identify the initial phases $\frac{\phi_0}{2} = \Phi_0 = -\frac{\Delta k x_0}{2}$.

Example 3   Calculate the average energy density (energy density averaged over one period $T = \frac{2\pi}{\omega}$) for a superposition of two waves of slightly different wavelengths but equal intensities traveling in the same direction.

Since we are not told the type of wave, all we can say is that the average energy density is proportional to the square of the total amplitude

$$\rho_E \propto A^2$$

We have a wave of the form

$$y(x,t) = A \cos (k_1 x - \omega_1 t) + A \cos (k_2 x - \omega_2 t)$$

and according to Eq. (2–6) we can rewrite this in the form

$$y(x,t) = 2A \cos \left(\frac{\Delta k}{2} x - \frac{\Delta \omega}{2} t\right) \cdot \cos (kx - \omega t)$$

with

$$k = \frac{k_1 + k_2}{2}; \quad \Delta k = k_1 - k_2; \quad \omega = \frac{\omega_1 + \omega_2}{2}; \quad \Delta \omega = \omega_1 - \omega_2$$

We can consider this a wave of angular frequency ω and wavelength $\lambda = 2\pi/k$ which has an amplitude slowly varying with time and position:

$$y(x,t) = A'(x,t) \cdot \cos (kx - \omega t) \tag{2–20a}$$

where

$$A'(x,t) = 2A \cos \left(\frac{\Delta k}{2} x - \frac{\Delta \omega}{2} t\right) \tag{2–20b}$$

Thus, we now have an energy density which slowly changes with time and position (Fig. 2–8):

$$\rho_E = \rho_0 \cdot 4A^2 \cos^2 \left(\frac{\Delta k}{2} x - \frac{\Delta \omega}{2} t\right) \tag{2–21}$$

It is important to note (and we will use this in the next example) that the function $\cos^2 (\alpha)$ has the period π instead of 2π. Indeed,

$$\cos (\alpha + \pi) = -\cos \alpha$$

and thus

$$\cos^2 (\alpha + \pi) = \cos^2 (\alpha)$$

One should also note that we tacitly used the very important assumption that the

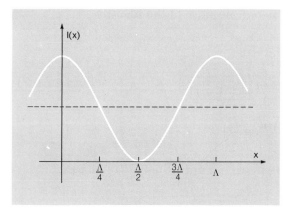

Figure 2-8 The intensity variation of two waves beating against each other. Note the similarity to Fig. 2–1; we have a relative phase which slowly varies with time and position.

separation Λ between two beat maxima is much larger than the wavelength λ. Only then can we consider the amplitude $A'(x,t)$ in Eq. (2–20a) to be a constant while we are averaging. The peak energy density is four times the energy density of a single wave; but there also exists a point P where the energy density at a fixed time t_0 vanishes. If we average over a full beat period, the energy density is equal to the sum of the energy densities of the individual two waves. Since[*]

$$<\cos^2 \alpha> = \frac{1}{\pi} \int_0^\pi \cos^2 \alpha \, d\alpha = \frac{1}{\pi} \left(\frac{\cos \alpha \sin \alpha + \alpha}{2} \right)_0^\pi = \frac{1}{2}$$

we have

$$<\cos^2 \left(\frac{\Delta k}{2} x - \frac{\Delta \omega}{2} t \right)> = \frac{1}{2}$$

and thus for the average density

$$<\rho_E> = \rho_0 \cdot 2A \tag{2–22}$$

Example 4   When a piano tuner strikes the same note (middle A, frequency $\nu = 440$ Hz) on two pianos simultaneously, he hears a beat with a period of 2/3 of a second. One piano is known to be perfectly tuned; what is the sound frequency of the wave emanating from the second piano?

Since the tuner is presumably not moving his head, he is sampling the superposition of the two waves at a fixed position x. We are told that both will have angular frequencies very near to

$$\omega = 2\pi\nu = 6.28 \times 440 = 2.76 \times 10^3 \text{ radians/sec}$$

On the other hand, the beat period T is 2/3 second. From Example 3 we know that the period of $\cos^2 \alpha$ in Eq. (2–16c) is $\Delta \alpha = \pi$, and so we have from Eq. (2–21)

$$\frac{\Delta \omega}{2} T = \Delta \alpha = \pi$$

which in terms of the difference in frequency can be written

$$\frac{2\pi(\nu_1 - \nu_2)}{2} T = \pi$$

_____

[*] Compare also p. 27, where the same quantity $<\cos^2 \alpha>$ is derived in a different way.

Thus the frequency difference of the second piano is

$$\Delta \nu = \nu_1 - \nu_2 = \frac{1}{T} = 1.5 \text{ sec}^{-1} = 1.5 \text{ Hz}$$

However, we can tell nothing about the sign of $\Delta \nu$. Since for any angle α we have $\cos \alpha = \cos (-\alpha)$, we obtain exactly the same beat if the second piano is tuned slightly higher or slightly lower. All we can say is that the second piano is tuned either to a frequency of 441.5 Hz or to a frequency of 438.5 Hz.

4. REFLECTION OF WAVES

A wave propagating in a medium will continue to do so until the medium comes to an end. What happens then will depend on the *"boundary conditions."* The boundary can be such that it will absorb the energy of the wave. A sandy beach will do this to water waves; acoustical tiles will do the same to sound. There is also the possibility that part of the wave will continue into the second medium on the other side of the boundary and part will be reflected; we can hear through the walls of a room a shot fired outside, while a person outside will hear the reflected wave as an echo. These are complex situations; here we will limit ourselves to the simpler case in which all the energy is reflected back into the original medium.

One possible boundary condition which will lead to a complete reflection of all energy occurs if we force the disturbance to vanish at the boundary for all times. We can see this happen, for example, on a stretched rope which is tied down at one end. If we give a quick sidewise jerk on the other end, we will generate a wave which propagates away from us toward the tied end. Experimentally we will observe that the wave travels all the way down to the tied end and then comes back with the same shape, but reversed in sign, as sketched in Fig. 2–9. In order to explain this experiment we note that the rope is fixed at $x = L$; thus, if we denote the disturbance on the rope by $y(x,t)$, we have a boundary condition

$$y(L,t) = 0 \tag{2-23}$$

for all times t. We also know that for $x < L$, the function $y(x,t)$ has to have the form

$$y(x,t) = f(x - v_0 t) + g(x + v_0 t) \tag{2-24}$$

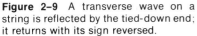

Figure 2–9 A transverse wave on a string is reflected by the tied-down end; it returns with its sign reversed.

Of course, there is no rope at $x > L$ and therefore $y(x,t)$ is not defined there. But we can imagine the rope continuing and search for a solution of the same type as Eq. (2–24) which satisfies the boundary condition, Eq. (2–23). For $x < L$ our solution will then be identical with the real one. We thus demand

$$f(L - v_0 t) + g(L + v_0 t) = 0 \qquad (2\text{–}25)$$

for all t. If we write $L + v_0 t = \xi$, we can rewrite Eq. (2–25) in the form

$$f(2L - \xi) + g(\xi) = 0$$

or

$$g(\xi) = -f(2L - \xi)$$

Thus, if $f(x - v_0 t)$ is given, $g(x + v_0 t)$ is also uniquely specified:

$$g(x + v_0 t) = -f(2L - x - v_0 t)$$

Therefore, the total disturbance has to have the form

$$y(x,t) = f(x - v_0 t) - f(2L - x - v_0 t) \qquad (2\text{–}26)$$

We note that for any time t the disturbance is antisymmetric with respect to the point $x = L$; i.e., for any value a,

$$y(L - a, t) = -y(L + a, t) \qquad (2\text{–}27\text{a})$$

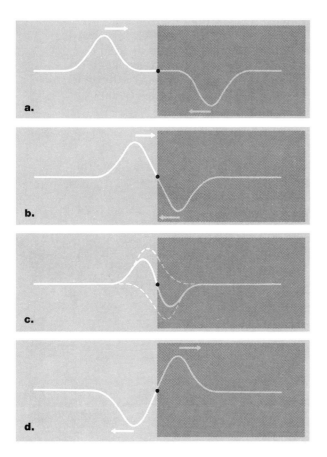

Figure 2–10 Detail sketch of a wave being reflected as in Fig. 2–9; the reflected wave is the continuation of the imagined wave, while the incoming wave "disappears" beyond the end. The time increases from **a** to **d**.

Indeed, if we substitute $(L - a)$ for x in Eq. (2–26) we obtain

$$y(L-a, t) = f(L - a - v_0 t) - f(L + a - v_0 t) \qquad (2\text{--}27\text{b})$$

while if we instead substitute $(L + a)$ for x, we obtain the negative of the expression Eq. (2–27b).

In Fig. 2–10 we draw the whole rope, including its imagined continuation. Whatever the disturbance to the left of $x = L$, it has to have a negative "mirror image" to the right. Thus, as the wave on the left approaches the tied end at $x = L$, the mirror image also approaches the same point from the other side. The two waves meet at $x = L$ in such a way that the disturbance there always vanishes, thus satisfying our boundary condition, Eq. (2–23). Each of the two waves then continues on its way. The wave on the real rope disappears into the imagined continuation, while its mirror image appears and continues as the real wave.

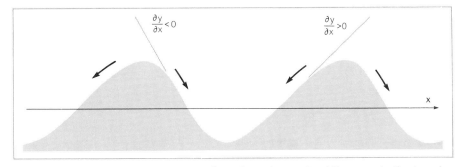

Figure 2–11a In a surface wave, water flows from the crests of the wave to the troughs.

We encounter a different boundary condition if a water wave encounters a vertical wall. In Fig. 2–11a we show a water wave. Near the surface the water flows from the points under the wave crests to the points under the troughs because of the gravitational force on the water molecules. Thus, if we describe the water surface by a function $y(x,t)$, water flows toward the left in Fig. 2–11a wherever $\partial y/\partial x$ is positive and toward the right wherever $\partial y/\partial x$ is negative. However, exactly at the wall there can be no flow either in or out of the wall. Thus, at the wall we have the boundary condition

$$\frac{\partial y(x=L, t)}{\partial x} = 0 \qquad (2\text{--}28)$$

which is valid for all times. The water level at the wall has to be always horizontal.

Again we resort to the trick of continuing the water surface beyond the wall and searching for wave solutions obeying the boundary condition, Eq. (2–28). If we write

$$y(x,t) = f(x - v_0 t) + g(x + v_0 t)$$

then the boundary condition can be written as

$$\left.\frac{\partial f(x - v_0 t)}{\partial x}\right|_{x=L} + \left.\frac{\partial g(x + v_0 t)}{\partial x}\right|_{x=L} = 0 \qquad (2\text{--}29)$$

It is easy to convince ourselves that the boundary condition is satisfied exactly if

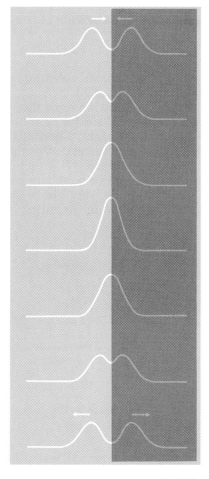

Figure 2–11b Reflection of a water wave from a vertical wall. Note that at the wall the maximum height of the water level is twice the water wave height away from the wall.

$$g(x + v_0t) = f(2L - x - v_0t) \tag{2-30}$$

The rules of differentiation then tell us that

$$\frac{\partial f}{\partial x} = \frac{\partial f(x-v_0t)}{\partial x} = f'(x - v_0t)$$

[we denote by f' the derivative of f with respect to its single argument $(x - v_0t)$] and that

$$\frac{\partial g(x+v_0t)}{\partial x} = \frac{\partial f(2L-x-v_0t)}{\partial x} = -f'(2L - x - v_0t)$$

But for $x = L$ we have

$$x - v_0t = 2L - x - v_0t = L - v_0t$$

and thus

$$\frac{\partial f(L-v_0t)}{\partial x} = -\frac{\partial g(L+v_0t)}{\partial x}$$

which is exactly our boundary condition Eq. (2–29).

What we have shown is that any wave which can be written as

$$y(x,t) = f(x - v_0t) + f(2L - x - v_0t) \tag{2-31}$$

will obey the correct boundary condition. There is again a symmetry around $x = L$, namely

$$y(L+a, t) = +y(L-a, t)$$

for any a and t, as one can easily convince oneself by substituting into Eq. (2–31).

In Fig. 2–11b we show the time development of the wave as we had done in Fig. 2–10. The reflected wave can be thought of as coming out of the imagined side of the water surface, while the incoming wave disappears behind the wall. Of course, this does not really happen; the incoming wave is reflected and nothing happens behind the wall at all. But mathematically the expression in Eq. (2–31) also describes the correct wave for $x < L$.

It is interesting to note that both types of boundary conditions can be rather easily realized for acoustic waves in a pipe full of air. Consider a compressional wave traveling down a pipe, as shown in Fig. 2–12. On the right side the pipe is open to the ambient air, which has a constant pressure. A compression implies a change in pressure – indeed, it is exactly the pressure difference between a point in the pipe and the outside pressure, which is the field exhibiting the wave phenomenon. Because the pipe is open at $x = L$, the boundary condition there is

$$p(x = L,t) = p_0 = \text{ambient pressure}$$

or, since our field is the pressure difference,

$$\Delta p \Big|_{x=L} = (p - p_0) \Big|_{x=L} = 0 \qquad (2\text{–}32a)$$

Thus, the boundary condition at $x = L$ is the same as the one we encountered for a string tied down at one end. On the other hand, at $x = 0$ in Fig. 2–12, the pipe is closed. No air can go across the closure. However, whenever there is a pressure difference between neighboring points, air will stream from the point with higher pressure to the point with lower pressure. Thus, at $x = 0$ we have a boundary condition

$$\frac{\partial \Delta p}{\partial x} \Big|_{x=0} = 0 \qquad (2\text{–}32b)$$

because there must be air flow at any point where $\dfrac{\partial \Delta p}{\partial x} \neq 0$. We see that Eq. (2–32b) is the same type of boundary condition as Eq. (2–28); even the reason is the same – in neither case can there be flow of matter across a rigid boundary.

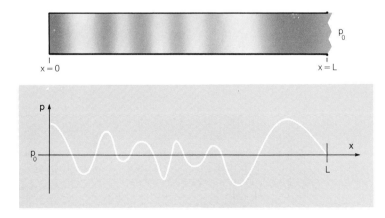

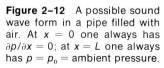

Figure 2–12 A possible sound wave form in a pipe filled with air. At $x = 0$ one always has $\partial p/\partial x = 0$; at $x = L$ one always has $p = p_0 =$ ambient pressure.

■ The reader should note that the mathematical formulation of the boundary condition on the field will in general depend on what quantity we call the field. If instead of the pressure difference Δp we had used as our field the local velocity $v(x,t)$, the boundary conditions would read differently. At $x = L$ we would have

$$\frac{\partial v}{\partial x} = 0 \qquad (2\text{–}33a)$$

while at $x = 0$ the boundary condition would be

$$v(x,t) = 0 \qquad (2\text{–}33b)$$

The physical condition is naturally the same, whatever quantity we call the field. That the two Eqs. (2–32a,b) and (2–33a,b) are equivalent can be seen if we note that

$$\frac{\partial y}{\partial t} \text{ is proportional to } \frac{\partial v}{\partial x} \qquad (2\text{–}34a)$$

and

$$\frac{\partial v}{\partial t} \text{ is proportional to } \frac{\partial p}{\partial x} \qquad (2\text{–}34b)$$

We can even calculate the proportionality factors. The compression of the gas is adiabatic —there is no time to exchange energy with the environment. In thermodynamics,[°] one learns that for adiabatic compression the pressure p and density ρ are related by the relation:

$$p = \text{const} \cdot \rho^\gamma \qquad (2\text{–}35)$$

where $\gamma = c_p/c_v$ is the ratio of specific heats at constant pressure and constant volume. Now consider Fig. 2–13: If the velocity at $x + dx$ is larger than at x, there will be a net outflow of gas from the volume of the pipe between. The change in mass in the time dt will be

$$dm = (\rho v\ dt)_x A - (\rho v\ dt)_{x+dx} A = -\frac{\partial(\rho v)}{\partial x} dx\ A\ dt$$

$$\underbrace{}_{\text{inflow}} \qquad \underbrace{\phantom{(\rho v\ dt)_{x+dx} A}}_{\text{outflow}}$$

if we call A the cross section area of the pipe. But mass equals density times volume, so the change in mass is

$$\frac{dm}{dt} = \frac{\partial \rho}{\partial t} \underbrace{A\ dx}_{\text{volume}}$$

and combining the two equations we have

$$\frac{\partial \rho}{\partial t} = -\frac{\partial(\rho v)}{\partial x} \approx -\rho_0 \frac{\partial v}{\partial x}$$

—————————————

° Compare Vol. I, p. 470.

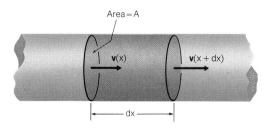

Area = A

v(x) **v**(x + dx)

dx

Figure 2–13 The loss of gas from a volume element of length dx is $dm = -\rho_0\ A\ v(x + dx) + \rho_0\ A\ v(x)$.

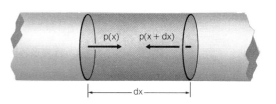

Figure 2–14 The net force on a volume element of gas is due to the difference between the pressures at the two ends.

We have set the density equal to the mean density ρ_0 because we are assuming that all density variations are small. From Eq. (2–35), using the chain rule, we also have

$$\frac{\partial p}{\partial t} = \text{const} \cdot \gamma \rho^{\gamma-1} \frac{\partial \rho}{\partial t}$$

so, dividing both sides by $p = \text{const} \cdot \rho^{\gamma}$, we obtain

$$\frac{1}{p}\frac{\partial p}{\partial t} = -\frac{\gamma}{\rho}\frac{\partial \rho}{\partial t}$$

Thus

$$\frac{\partial p}{\partial t} \approx -\gamma p_0 \frac{\partial v}{\partial x} \qquad (2\text{–}34c)$$

and we have calculated the proportionality constant in Eq. (2–34a). To calculate the proportionality constant in Eq. (2–34b), note that Newton's second law says that the mass in the volume $A\,dx$ is accelerated by the forces acting on it (Fig. 2–14). The net force is due to the difference in pressure between the ends of the element:

$$F = p(x,t)\,A - p\,(x + dx,t)\,A$$

or

$$F = -\frac{\partial p}{\partial x}\,dx\,A$$

But again the mass is the volume times the density of the gas

$$dm = \rho_0\,dx\,A$$

and Newton's second law states that

$$\text{mass} \times \text{acceleration} = \quad \text{force}$$

$$\rho_0\,dx\,A \quad \times \quad \frac{\partial v}{\partial t} \quad = -\frac{\partial p}{\partial x}\,dx\,A$$

Canceling the common volume factor $A\,dx$, we obtain

$$\frac{\partial v}{\partial t} = -\frac{1}{\rho_0}\frac{\partial p}{\partial x} \qquad (2\text{–}34d)$$

■

5. STANDING WAVES

In the last section, where we discussed reflections of a wave, we imagined (at least when drawing the figures) that all waves are localized. This is, of course, not necessary. The incident wave can just as well be a harmonic wave; the reflected wave then will also be harmonic. However, there is now a difference. A harmonic wave continues without end; thus, at no time can we say that we see only the incident or only the reflected wave. Both are always present – and will interfere with each other.

Let us return to the example of the stretched rope or piano wire. If we make one end oscillate up and down, we will see the whole rope vibrate. However, there will be points which never move; at other points, the oscillations will be maximal. While the pattern on the rope is reminiscent of a traveling wave, the pattern does not move: we have produced a standing wave. We can understand this using our knowledge of reflection of waves. From Eq. (2–26) we know that any wave on the string has to have the form

$$y(x,t) = f(x - v_0 t) - f(2L - x - v_0 t)$$

If we choose in particular

$$f(x - v_0 t) = A \cos [k(x - v_0 t)] = A \cos (kx - \omega t)$$

(recall that $v_0 = \omega/k$) we have as the complete solution

$$y(x,t) = A [\cos (kx - \omega t) - \cos (2kL - kx - \omega t)] \qquad (2\text{–}36)$$

We can now use an equation similar to Eq. (2–3)

$$\cos \alpha - \cos \beta = 2 \sin \left(\frac{\alpha + \beta}{2}\right) \sin \left(\frac{\alpha - \beta}{2}\right) \qquad (2\text{–}37a)$$

to simplify Eq. (2–36). Setting

$$\alpha = kx - \omega t$$
$$\beta = 2kL - kx - \omega t$$

we obtain

$$\frac{\alpha + \beta}{2} = kL - \omega t$$

$$\frac{\alpha - \beta}{2} = kx - kL = k(x - L)$$

and thus we can write for the disturbance on the wire

$$y(x,t) = 2A \sin (kL - \omega t) \cdot \sin [k(x - L)] \qquad (2\text{–}37b)$$

The argument of the first sine is independent of the position x, while the argument of the second one is independent of time. We thus have

$$y(x,t) = 2A \, \phi_1(t) \cdot \phi_2(x)$$

with

$$\phi_1(t) = \sin (kL - \omega t)$$
$$\phi_2(x) = \sin [k(x - L)]$$

Such a disturbance is called a standing wave; at each point x_0 the string will oscillate with an amplitude which is a function of position:

$$y(x_0,t) = B(x_0) \sin (\omega t - \delta) \qquad (2\text{–}38)$$

In going from Eq. (2–37b) to Eq. (2–38) we have used the relation $\sin (-\alpha) = -\sin \alpha$, and we have also defined the phase angle

$$\delta = kL \qquad (2\text{–}39a)$$

and the local amplitude

$$B(x_0) = -2A \sin [k(x_0 - L)] = 2A \sin [k(L - x_0)] \qquad (2\text{--}39b)$$

We see that at all points where

$$\sin [k(L - x)] = 0$$

the string will never move at all. Since $\sin (\pm\pi) = \sin (\pm 2\pi) = \sin (\pm n\pi) = 0$, these points are given by the equation

$$k(L - x) = n\pi \qquad n = 0, \pm 1, \pm 2, \ldots \qquad (2\text{--}40a)$$

The points which are always at rest will have coordinates obeying

$$L - x = n\frac{\pi}{k}$$

Because of the relation $k = 2\pi/\lambda$ we can recast Eq. (2–40a) in the form

$$L - x = n\frac{\lambda}{2} \qquad n = 0, \pm 1, \ldots \qquad (2\text{--}40b)$$

The stationary points, called *nodes*, are separated by one half of the wavelength of the harmonic wave. The point at $x = L$ is always a node; since the wire is tied down there, the wire cannot move — and this defines a node.

Let us now choose a slightly different setup — we tie the rope (or piano wire) down at both ends, i.e., at $x = 0$ as well as at $x = L$ (Fig. 2-15). Now both $x = 0$ and $x = L$ have to be node points. Substituting $x = 0$ into Eq. (2–40b), we obtain

$$(L - x)_{x=0} = n\frac{\lambda}{2}$$

or

$$L = n\frac{\lambda}{2} \qquad (2\text{--}41)$$

Thus, a wire which is tied down on both ends can oscillate stably only if its total length is an integer multiple of one half the wavelength. In other words, a wire of given length can oscillate only for a discrete set of wavelengths. This, of course, implies that the wire can oscillate only at a discrete set of frequencies. Recalling the wave velocity on a stretched string

$$v_0 = \frac{\omega}{k} = \lambda \nu = \sqrt{\frac{T}{\mu}} \qquad (2\text{--}42)$$

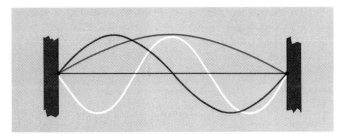

Figure 2-15 Standing waves on a vibrating string. Both ends of the string are always nodes.

where T is the wire tension and μ is the wire mass per unit length, we can write

$$\nu = \frac{v_0}{\lambda} = \sqrt{\frac{T}{\mu}}\,\frac{1}{2L}\,n \qquad\qquad n = 1, 2, 3, \ldots \qquad\qquad (2\text{--}43)$$

Eq. (2–43) gives us all the frequencies at which a piano wire or rope under tension can oscillate stably, if it is tied down at both ends. It also forms the basis for designing all musical instruments using strings. In a piano, a hammer hits a wire whose tension has been adjusted so that it vibrates at the desired frequency. In a violin, the player varies the length of the string by pressing his finger down on the string in the correct location.

Example 5   A steel wire of $d = 1$ mm diameter and $L = 0.5$ meter length is under the tension $T = 1000$ newtons. At what frequencies will it vibrate?

We have to know the mass per unit length of the wire in order to be able to use Eq. (2–43). We go to Table I in Appendix B and find the density of steel to be $\rho = 7.9 \times 10^3$ kg/m^3. The mass per unit length then is

$$\mu = \text{area} \times \text{density} = \frac{\pi d^2}{4}\rho$$

and therefore the frequency at which the wire can vibrate is, according to Eq. (2–43),

$$\nu = \sqrt{\frac{4T}{\rho \pi d^2}}\,\frac{1}{2L}\,n = \sqrt{\frac{T}{\pi \rho}}\,\frac{1}{dL}\,n$$

where $n = 1, 2, 3, 4, \ldots$ The lowest frequency can be obtained by setting $n = 1$. Thus

$$\nu_1 = \sqrt{\frac{1000 \text{ N}}{3.14 \times (7.9 \times 10^3 \text{ kg/m}^3)}} \cdot \frac{1}{(0.001 \text{ m}) \times (0.5 \text{ m})} = 401 \text{ Hz}$$

Of course, the wire can also vibrate at higher frequencies:

$$\nu_2 = 2\nu_1 = 802 \text{ Hz}$$
$$\nu_3 = 3\nu_1 = 1203 \text{ Hz}$$
$$\vdots$$

A string will never vibrate at only one frequency; the differences in how a piano, a guitar, and a violin sound even if they emit the same tone are due to differences in the relative amplitudes of the higher frequencies. We call the lowest frequency—for which $n = 1$ in Eq. (2–43)—the *fundamental frequency;* all the other, higher frequencies are called *overtones.*

Standing waves form the basis not only of string instruments, but also of wind instruments (flute, trumpet) and of organs; however, in this case the standing wave is formed by pressure variations in the air column in the instrument. Let us discuss this for the examples of the closed and the half-open organ pipe, as shown in Fig. 2–16a and b. A standing wave has to obey the boundary condition

$$\Delta p = 0 \text{ at the open end}$$

and

$$\frac{\partial \Delta p}{\partial x} = 0 \text{ at the closed end}$$

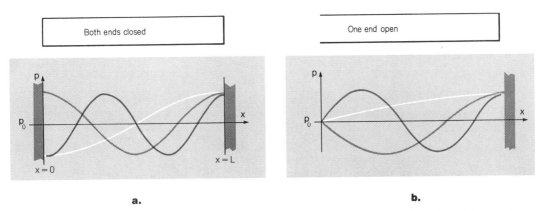

Figure 2–16 Standing waves in an organ pipe which is (a) closed on both ends or (b) open at one and closed at the other end. The different boundary conditions lead to different frequencies for pipes of equal length.

Here $\Delta p(x,t)$ is the pressure difference between the local pressure and the ambient air pressure.

Let us discuss first the pipe which is closed on both ends (Fig. 2–16a). Because the pipe is closed at the right end, it will obey the boundary condition [Eq. (2–28)] there, and thus the pressure will have the form given in Eq. (2–31)

$$\Delta p(x,t) = f(x - v_0 t) + f(2L - x - v_0 t) \tag{2–44}$$

Choosing a harmonic wave as the particular form of $f(\xi)$ we obtain

$$\Delta p(x,t) = A \left[\cos (kx - \omega t) + \cos (2kL - kx - \omega t) \right] \tag{2–45}$$

which because of the trigonometric identity of Eq. (2–3) can be written

$$\Delta p(x,t) = 2A \cos (kL - \omega t) \cdot \cos (kx - kL) \tag{2–46}$$

However, we have up to now satisfied only the boundary condition at $x = L$. The form of the pressure field given in Eq. (2–46) is further restricted by the requirement that at $x = 0$ the pipe is also closed, so

$$\left. \frac{\partial \Delta p}{\partial x} \right|_{x=0} = 0$$

Since only the last factor in Eq. (2–46) depends on x, we have to require that

$$\left. \frac{d}{dx} \cos (kx - kL) \right|_{x=0} = 0$$

or

$$-k \cdot \sin (0 - kL) = k \sin kL = 0$$

Of course, we cannot assume $k = 0$, because we then would have no wave at all. Thus we require

$$\sin kL = 0 \tag{2–47}$$

and, as before, we see that standing waves can exist in the closed pipe only if

$$kL = \pi, \, 2\pi, \, 3\pi, \, \ldots \, n\pi, \, \ldots \tag{2–48a}$$

or, expressed in terms of the wavelength $\lambda = 2\pi/k$,

$$L = n\,\frac{\lambda}{2} \qquad n = 1,\,2,\,3,\,\ldots \tag{2-48b}$$

If (Fig. 2–16b) the pipe is left open, instead of being closed at $x = 0$, then we have to require that

$$\Delta p(x=0,\,t) = 0$$

or, because of the particular form of Eq. (2–46),

$$\cos\,(kL) = 0$$

The cosine is zero whenever its argument equals $\pi/2,\,3\pi/2,\,5\pi/2,\,\ldots$; i.e., the odd multiple of $\pi/2$. Thus, we have

$$kL = (2n + 1)\frac{\pi}{2} \qquad n = 0,\,1,\,2,\,3,\,\ldots \tag{2-49a}$$

or

$$L = (2n + 1)\frac{\lambda}{4} \tag{2-49b}$$

The natural frequencies of the vibrations in a pipe can of course be obtained once the sound velocity is known. Taking the value which we calculated in Example 2 of Chapter 1,

$$v_0(\text{air}) \approx 340 \text{ m/sec}$$

we obtain for the pipe with both ends closed, from Eq. (2–48b),

$$\boxed{\nu = \frac{v_0}{\lambda} = \frac{v_0}{2L}\,n} \qquad n = 1,\,2,\,3,\,\ldots$$

while for the pipe with one end open we have, from Eq. (2–49b),

$$\boxed{\nu = \frac{v_0}{\lambda} = \frac{v_0}{4L}\,(2n + 1)} \qquad n = 0,\,1,\,2,\,\ldots$$

We note that the fundamental frequency $(n = 0)$ of an open-ended pipe is exactly one half of the fundamental frequency $(n = 1)$ of the closed pipe. Also, for the *open-ended pipe* the harmonic frequencies will be an *odd multiple* of the fundamental, while for the pipe which is *closed* at both ends *all multiples* of the fundamental are possible. Thus, two pipes—one completely closed and the other open at one end—that have the same fundamental frequency will nevertheless sound quite different to the human ear, which is very sensitive to the admixture of the harmonic frequencies with the fundamental.

The reader may ask why we have spent so much time discussing standing waves. Strings and pipes, while important to the musician, are not of fundamental importance to modern physics. However, while we used examples which are easily demonstrated in the laboratory, standing waves occur for many other types of waves. Standing electromagnetic waves are produced between two mirrors in a laser. Even the atom can be considered as a standing wave. As we will see in more detail later, an electron which is moving with velocity v — and has therefore the momentum $p = mv$ — can according to quantum mechanics be represented by a wave of wavelength

$$\lambda = \frac{h}{mv} \qquad (2\text{--}50)$$

where $h = 6.624 \times 10^{-34}$ Joule/sec is a fundamental constant, called Planck's constant. According to quantum mechanics, the electron in an atom will be in a stationary state (a state persisting for long times) if and only if the wave representing the electron forms a standing wave. The discrete set of frequencies $\nu_1, \nu_2, \ldots \nu_i$ at which a standing wave can exist is then interpreted to mean that the electron can exist in a stationary state only at well-determined energies

$$E_i = h\nu_i$$

While we have discussed standing waves in detail only in one spatial dimension, standing waves in two and three dimensions will occur whenever the field is two- or three-dimensional. Of course, for a two-dimensional field (e.g., water waves) the boundary condition has to be given on a closed line surrounding the water surface (e.g., the rim of a water cup). One can observe two-dimensional standing waves if one orders a cup of coffee while flying in an airplane and lets the half-filled cup stand on the table. The many small vibrations which are always present in an airplane will excite standing waves in the cup whenever the standing wave frequency coincides with one of the many frequencies with which the plane vibrates. By taking small sips — or adding small amounts of water — one can produce an astonishing variety of standing waves.

For three-dimensional standing waves, the boundary condition has to be given on a closed surface. Such waves occur as electron waves in atoms; but they can also occur as sound waves in rooms. This has been discovered by many architects who designed beautiful halls with naked (nonabsorbing) walls and later had to cover them with sound-absorbing material because the resonating sound waves were driving everybody "up the walls."

Summary of Important Relations

Intensity of sum of two waves (intensities I_1 and I_2, relative phase δ)

$$I = I_1 + I_2 + 2\sqrt{I_1 I_2} \cos \delta$$

Beat phenomenon: $y(x,t) = 2A \cos (kx - \omega t) \cos \left(\frac{\Delta k}{2} x - \frac{\Delta \omega}{2} t \right)$

where $\omega = \dfrac{\omega_1 + \omega_2}{2}$, $k = \dfrac{k_1 + k_2}{2}$

and $\Delta \omega = \omega_1 - \omega_2$, $\Delta k = k_1 - k_2$

Standing waves on string $\nu = \sqrt{\dfrac{T}{\mu}} \dfrac{1}{2L} n$ $n = 1,2,3, \ldots$

Standing wave in air pipe closed at both ends

$$\nu = \frac{v_0}{2L} n$$

with one end open

$$\nu = \frac{v_0}{4L}(2n + 1)$$

Questions

1. If a choir sings a note, you will not hear beats. Explain why; the singers are generally not that precise!

2. You are to perform experiments with water waves in a water tank and you do not want reflections from the walls. How should the tank walls be structured? Explain why.

3. Piano wires which produce the bass tones are usually coiled springs. Explain why.

4. Can Eq. (2–48b) and (2–49b) for the resonant frequencies of pipes be exact? Note that you can hear the sound from the outside.

5. A friend of yours claims that there is dispersion in sound waves and that the speed of sound is 5% higher at frequencies of 3000 to 5000 Hz than at 200 to 300 Hz. What common experiences would contradict this assumption?

Problems

1. You are to measure the speed of sound in water to 0.1% by measuring the time delay from emission of a signal by a loudspeaker to its reception by a microphone a distance d away. The loudspeaker and microphone positions are each measurable to ± 0.5 cm and the time delay to 0.1 msec. How long must the distance d be (at least)? The speed of sound in water is roughly 1500 m/sec.

2. Two harmonic waves of equal frequency have a relative phase of 60° and intensities I_0 and $I_0/2$. What is the total intensity?

3. Four harmonic waves of equal frequency have a relative phase of 0, $\pi/2$, π, and $3\pi/2$ and relative amplitudes of A_0, $A_0/2$, $A_0/4$, and $A_0/3$. Calculate the overall amplitude and phase of the resulting wave.

4. Two harmonic water waves of wavelength $\lambda_1 = 1$ m and $\lambda_2 = 1.05$ m and of equal amplitudes travel in the same direction. Calculate the overall wave pattern and find (a) the separation in time between two maxima passing the same point; (b) the separation in space between two maxima at an instant in time. Does dispersion of water waves affect your result?

5. N harmonic sound waves traveling in the same direction have the same frequency ν and wavelength λ. They all have the same intensity I_0; their relative phase differences are δ, 2δ, 3δ, ... $N\delta$. Calculate the total intensity I_{tot} of the resulting wave. For what values of δ is $I_{tot} = 0$? Plot $I_{tot}(\delta)$ for $N = 5$. (Hint: Draw the phasors first.)

6. The human ear can hear—and finds objectionable—beats which have a period as long as two seconds. To what accuracy (in per cent) does the piano tuner have to adjust the wire tension in two pianos which are to be used in a concerto for two pianos (a) at 400 Hz, (b) at 1500 Hz?

7. Two sound waves of nearly equal angular frequencies ω_1 and ω_2 travel in the same direction. Calculate the ratio between the maximum and minimum intensity registered at a stationary microphone if (a) the intensities of the two waves are in the ratio 2:1, (b) their amplitudes are in the ratio 2:1.

8. Two harmonic waves of exactly the same frequency and same velocity travel in opposite directions. The intensity of one wave is 10% larger than that of the other. Calculate and sketch the resulting wave pattern.

9. In science fiction one sometimes finds mention of a "sound neutralizer" which, by emitting interfering waves, makes it impossible for anybody in the room to hear anything when someone speaks. Show that a localized neutralizer is impossible unless it is in the speaker's mouth. For this purpose consider a point sound source (the neutralizer) a distance d away from the speaker, who is emitting sound at many wavelengths. Show that it is impossible to cancel all sound signals at both P_1 and P_2.

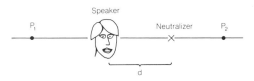

10. The solar cell panels of a satellite are being hit by sunlight at an angle of $\theta = 60°$ from the vertical. The solar cell output, which is proportional to the incident light power, is 3 kW for normal incidence ($\theta = 0$). How much power can one draw at the given value of θ?

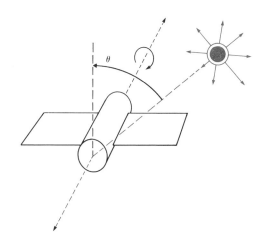

11. The satellite from the previous example is out of control and rotating around an axis perpendicular to the line to the sun (axis is dotted in figure). What average power can the solar cells deliver, if only one side of each solar cell panel is active?

12. Two identical loudspeakers are 50 inches apart and are both connected to the same audio oscillator operating at a frequency of 500 Hz. Find all points along the line connecting the loudspeakers where the sound intensity is a minimum.

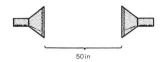

50 in

13. A loudspeaker is emitting toward a reflecting wall at a frequency $\nu = 700$ Hz. Starting from the wall, one finds $n = 3$ maxima of sound intensity. (a) Will there be a maximum at the wall? (b) The maximum nearest to the loudspeaker is 20 cm away from it. How far is the loudspeaker from the wall?

*14. Derive Eq. (1–19c) for the velocity of sound waves in a gas from Eqs. (2–34c) and (2–34d).

15. A metal rod is 0.8 m long and is clamped at one end. If a resin cloth is rubbed along its length, it emits sound at 5000 Hz. What is the speed of sound in the rod?

16. A guitar string is 50 cm long and has a total mass of 2.5 grams. When the string is plucked, it sounds an A-note (440 Hz). Calculate the tension in the string.

17. A wire is strung between two points a distance 1 m apart; its fundamental frequency is $\nu_0 = 300$ Hz. A wedge is put 40 cm away from one end of the wire, touching loosely as in the figure. At what frequencies will the wire vibrate?

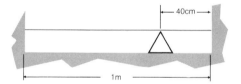

40cm

1m

18. At what frequencies would the air in a narrow concrete hallway, 20 meters long and closed at both ends, resonate? How many such frequencies lie between 500 and 1000 Hz? These are most likely to be excited by somebody speaking.

19. A straight sewer pipe is plugged an unknown distance from one end. The water has been pumped out. Blowing into one end, you hear a resonance of 80 Hz. How far (at least) is the plug?

20. A hanging string of 2 m length and a total mass of 3 g is to resonate at a frequency of 150 Hz. How heavy a weight should be hung from the string? Note that the heavy weight will prevent the bottom end from moving.

21. An open-ended pipe has a fundamental frequency of 300 Hz. At what frequencies will it resonate if the air is replaced by helium?

22. A Kundt's tube consists of a glass tube with a plunger P at one end. At the other end is a rod which is clamped down in the middle and ends with a disk A. Cork dust is sprinkled on the bottom of the tube. Longitudinal vibrations of known frequency ν are induced in the clamped rod, and the plunger P is moved in and out until the dust forms well-defined ridges (along the nodes) in the tube. Calculate the sound velocity in terms of the frequency ν and the mean separation d of the ridges.

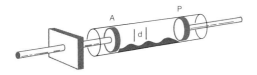

*23. A water channel 20 feet deep has concrete walls 25 feet apart (see figure). Calculate the frequency of (a) the fundamental mode, (b) the tenth harmonic mode of standing waves on the water surface. Can you assume the depth to be infinitely large for either of the two modes?

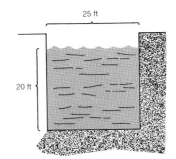

CHAPTER 3

DIFFRACTION

1. PLANE, SPHERICAL, AND CYLINDRICAL WAVES

We will now turn our attention to waves in three-dimensional space. Most of what we will discuss can also be directly applied to two-dimensional waves, e.g., surface waves. We will limit ourselves to a discussion of harmonic waves. Waves of arbitrary shape, as mentioned in Chapter 1, can always be considered as a superposition of harmonic waves. We will apply our theory to sound waves, electromagnetic waves, and matter waves (particles which, according to quantum mechanics, can also be considered waves).

A *plane wave* is a wave which is propagating everywhere in the same direction. For a harmonic wave this means that we have regularly spaced parallel planes where the field is a maximum; halfway between these planes, the field is a minimum. The whole field distribution propagates with velocity **v** in a direction which is perpendicular to the planes of constant phase (Fig. 3–1a).

The wavelength λ is defined in analogy to one-dimensional waves as the separation between two neighboring planes where the field is maximal. It is also usual to define a wave vector **k** whose direction is the direction of propagation of the wave and whose magnitude is

$$|\mathbf{k}| = k = \frac{2\pi}{\lambda} \tag{3-1}$$

The wave vector **k** is a rather obvious generalization of the wave number k from the previous chapter. It is simply the wave number multiplied by a unit vector which points in the direction of propagation of the wave. Because the direction of propagation is perpendicular to the planes of constant phase, the wave vector is also perpendicular to these planes. Any vector **q** which lies completely within one of these planes is also orthogonal to **k**, so the scalar product $(\mathbf{k} \cdot \mathbf{q})$ always vanishes. Thus, as can be seen from Fig. 3–1b, for any **r** whose end point lies in a plane where the field is maximal, one has

$$(\mathbf{k} \cdot \mathbf{r}) = |\mathbf{k}| \, |\mathbf{r}| \, \cos \alpha = \text{constant} \tag{3-2a}$$

because

$$|\mathbf{r}| \cos \alpha = d \tag{3-2b}$$

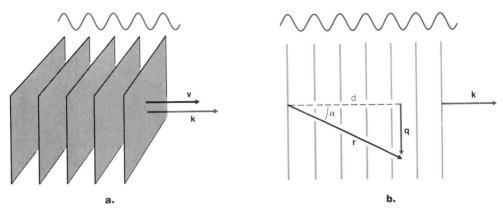

Figure 3-1   (a) In a plane wave, the phase is the same everywhere on a plane perpendicular to the propagation velocity. (b) In a plane wave at an instant in time, the phase depends only on $(\mathbf{k} \cdot \mathbf{r}) = kd$.

is the distance from the origin to the plane. Indeed, in *any plane* for which $(\mathbf{k} \cdot \mathbf{r}) =$ constant, the field has everywhere the same value. Thus, for a harmonic wave, where the position dependence is contained in the argument of a cosine function, the field at any instant can be written as

$$u(\mathbf{r},t) = A \, \cos \, (\mathbf{k} \cdot \mathbf{r} - \omega t) \qquad (3\text{–}3)$$

The velocity of propagation has the magnitude

$$v = \frac{\omega}{|\mathbf{k}|} = \frac{\omega}{k}$$

and points in the same direction as the wave vector **k**. In the special case where the wave vector points in the *x*-direction, i.e., if

$$\mathbf{k} = k\,\hat{\imath} + 0\,\hat{\jmath} + 0\,\hat{k}$$

we have

$$\mathbf{k} \cdot \mathbf{r} = kx \qquad (3\text{–}4)$$

and Eq. (3–3) reduces itself to the familiar expansion for a one-dimensional wave. Indeed, any single plane wave can be considered a one-dimensional wave.

A *spherical wave* is a wave which originates in a single point in space; the surfaces of constant phase are concentric spheres (Fig. 3–2). If the wave is propagating away from the center, we call it an outgoing wave; for an ingoing wave the energy in the wave travels everywhere toward the center.

How do we mathematically describe such a wave? Let us discuss here only outgoing waves; then the first guess would be to write for the outgoing wave

$$u(x,t) \overset{?}{=} A \, \cos \, (kr - \omega t) \qquad (3\text{–}5a)$$

where

$$r = \sqrt{x^2 + y^2 + z^2} \qquad (3\text{–}5b)$$

is the distance from the center. Indeed Eq. (3–5) describes a disturbance which has the same value anywhere on a sphere around the center. However, Eq. (3–5)

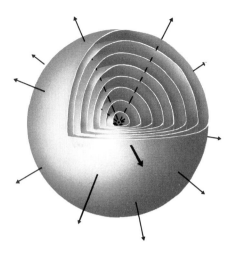

Figure 3−2 In an outgoing spherical wave, the wave fronts are concentric spheres moving away from the center.

cannot be correct. As was shown in Chapter 1, the intensity of the outgoing wave is proportional to the amplitude squared:

$$I = CA^2$$

where the exact value of the constant C will depend on the type of wave under discussion. Let us remember that the intensity is the energy per unit area and unit time flowing through a surface which is perpendicular to the direction of propagation. In the present case such surfaces are sections of a sphere whose center is the origin of the wave. We can then define the energy flux for a surface of arbitrary shape and size as the energy flowing through the surface per unit time. If we choose as the surface the whole sphere of radius R, then the total energy flux through this sphere will be

$$\Phi = \text{intensity} \times \text{area} = I \times 4\pi R^2 = 4\pi CA^2 R^2$$

in other words, proportional to the square of the radius of the sphere. This is clearly impossible, because the total flux through the sphere has to be equal to the energy leaving the single wave source at the center per unit time:

$$\Phi = \text{energy output of source/unit time}$$

But then Φ cannot depend on the radius of the arbitrary sphere that we choose for the calculation. As we see, we have to write for the disturbance

$$u(x,y,z,t) = \frac{A}{r} \cos (kr − \omega t) \qquad (3−6)$$

where again $r = \sqrt{x^2 + y^2 + z^2}$. The wave intensity is then equal to

$$I = \frac{C}{2} \frac{A^2}{r^2} \qquad (3−7)$$

which is proportional to the inverse square of the radius. This means that if we put a piece of paper one meter away from a light bulb, it is illuminated four times more strongly than if it is two meters away; i.e., we need four light bulbs to read the page at two meters, if one bulb was sufficient when we held the page one meter away. If we assume Eq. (3–6) to be correct, the total energy flux through a sphere of radius R is

$$\Phi = 4\pi R^2 \cdot I = 4\pi R^2 \cdot \frac{C}{2}\frac{A^2}{R^2} = 2\pi C A^2$$

and is independent of the radius of the sphere, as it has to be.

Similar, but not quite identical, to spherical waves are *cylindrical waves* (Fig. 3–3). They are in a sense two-dimensional waves, just as plane waves are in a sense one-dimensional waves. We can see them as surface waves if we drop a stone into the water; but they will also be formed by light if it passes through a very narrow long slit. The mathematical expression of a cylindrical wave is

$$u(x,y,z,t) = \frac{A}{\sqrt{\rho}} \cos (k\rho - \omega t) \qquad\qquad (3\text{--}8\text{a})$$

where

$$\rho = \sqrt{x^2 + y^2} \qquad\qquad (3\text{--}8\text{b})$$

is the distance from the "centerline." As we see, for three-dimensional cylindrical waves the source of the wave has to lie along that line. The surfaces of constant phase are coaxial cylinders. The amplitude falls off in proportion to the square root of the distance from the source; this is again required by the law of conservation of energy. If we calculate the energy flux through a cylinder of radius ρ and unit height, we obtain

$$\Phi = \text{area} \times \text{intensity} = 2\pi\rho \times I(\rho)$$

and since the intensity is again proportional to the square of the amplitude,

$$I(\rho) = \text{constant } \frac{A^2}{(\sqrt{\rho})^2} = \text{constant } \frac{A^2}{\rho} \qquad\qquad (3\text{--}9)$$

the total flux is independent of the radius of the cylinder, as required by energy conservation.

Far away from the center in one direction, both the cylindrical wave and the spherical wave can be approximated by a plane wave (Fig. 3–4); the wave vector

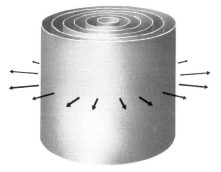

Figure 3–3 In a cylindrical wave, the wave fronts are coaxial cylinders.

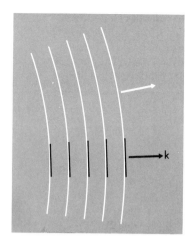

Figure 3–4 Far away from the center, both spherical and cylindrical waves can locally be approximated by a plane wave.

then (for outgoing waves) points away from the (point or line) source. Of course, this is only a *local* approximation: The "planes of constant phase" are in reality curved and if we follow them far enough, the wave vector will change its direction. But also, if we follow the wave outward along the direction of **k,** we will note some distance away that the amplitude has decreased, as it always does for cylindrical or spherical waves. However, over a region which is small compared to the distance to the origin, the plane wave approximation will be a good one. If you drop a stone into water, you will see circular waves traveling away; but an observer on the far shore will see the same wave arrive as a plane wave.

Waves in two or three dimensions can also interfere with each other. Since we have more possibilities than we had in one dimension, an astonishing wealth of wave patterns can occur. But we can always analyze the situation in a way similar to the one-dimensional situation. We calculate the disturbance due to the superposition of two waves by adding the functions of space and time representing each wave. All we have to do then is to analyze the resulting disturbance; we can best show how to do this on a few specific examples.

Example 1 Two plane water waves have the same wavelength and amplitude, but move in directions which are perpendicular to each other. Describe the resulting wave pattern.

Since the waves are traveling in mutually perpendicular directions, we might as well choose our coordinate system so that one goes in the direction of the positive x-axis, while the other travels in the direction of the positive y-axis. The resulting wave then will have the form

$$u = A \cos (kx - \omega t) + A \cos (ky - \omega t) \tag{3–10}$$

Let us first discuss why we did not have to include an arbitrary initial phase; why did we not have to write

$$A \cos (kx - \omega t - \delta)?$$

This is so because we can choose the origin of our coordinate system wherever we want. Given the time $t = 0$, we choose the origin at one of the points where both waves peak; i.e., the phase angles are a multiple of 2π.

By the trigonometric identity Eq. (2–3) we then transform Eq. (3–10) into

$$u = 2A \cos \left[\frac{k}{2}(x + y) - \omega t\right] \cdot \cos \left[\frac{k(x - y)}{2}\right] \tag{3–11}$$

If we define a new wave vector $\mathbf{K}$ by

$$\mathbf{K} = \frac{k}{2}\hat{\mathbf{i}} + \frac{k}{2}\hat{\mathbf{j}}$$

we can write the phase of the first cosine factor as

$$\mathbf{K} \cdot \mathbf{r} - \omega t$$

Note that the wave number is

$$K = |\mathbf{K}| = \sqrt{\left(\frac{k}{2}\right)^2 + \left(\frac{k}{2}\right)^2} = \frac{k}{\sqrt{2}}$$

Thus, the wavelength of the resulting wave is

$$\Lambda = \frac{2\pi}{K} = \sqrt{2}\,\frac{2\pi}{k} = \sqrt{2}\lambda$$

which is $\sqrt{2}$ times larger than that of the original waves. The wave is also not a simple plane wave; it vanishes wherever

$$\cos\left[\frac{k(x - y)}{2}\right] = 0 \tag{3-12}$$

This occurs along strips where the argument in Eq. (3–12) is equal to $(2n + 1)\pi/2$ for $n = 0, 1, 2, 3, \ldots$. That is, the amplitude vanishes wherever

$$(x - y) = \frac{(2n + 1)}{2} \cdot \frac{2\pi}{k} = (2n + 1)\frac{\lambda}{2} \tag{3-13}$$

The strips of vanishing amplitude are separated by $\lambda/\sqrt{2}$ and are parallel to the direction of $\mathbf{K}$. We have sketched the resulting wave pattern in Fig. 3–5.

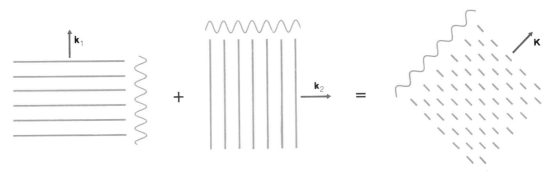

Figure 3–5 Example 1: The superposition of two plane waves traveling in perpendicular directions.

Example 2   On a water surface, two point sources of circular (cylindrical) waves are separated by one wavelength λ as shown in Fig. 3–6. The two sources are exactly in phase and have the same amplitude. Determine the direction in which, at great distances from the sources, there is no flow of energy.

The form of the question leads us to expect that the interference between the two waves will lead to a vanishing intensity in a certain direction. At any point P, the surface will have the height

$$h(x,y,t) = A \frac{\cos(k\rho_1 - \omega t)}{\sqrt{\rho_1}} + A \frac{\cos(k\rho_2 - \omega t)}{\sqrt{\rho_2}} \tag{3–14a}$$

where ρ_1 and ρ_2 are the distances from P to the two point sources (Fig. 3–6). However, we are not asked for a solution close to the source, but only far away. There we can neglect the difference between the two amplitudes. We can set

$$\frac{A}{\sqrt{\rho_1}} \approx \frac{A}{\sqrt{\rho_2}} \approx \frac{A}{\sqrt{\rho}} \tag{3–14b}$$

where $\rho = (\rho_1 + \rho_2)/2$. Then we can use Eq. (2–3) again and obtain

$$\begin{aligned} h(x,y,t) &= \frac{A}{\sqrt{\rho}} \left[\cos(k\rho_1 - \omega t) + \cos(k\rho_2 - \omega t) \right] \\ &= \frac{2A}{\sqrt{\rho}} \cos \left[\frac{k(\rho_1 + \rho_2)}{2} - \omega t \right] \cdot \cos \left[\frac{k(\rho_1 - \rho_2)}{2} \right] \end{aligned} \tag{3–15}$$

Thus, the water surface will be quiet wherever

$$\cos \left[k \frac{(\rho_1 - \rho_2)}{2} \right] = 0 \tag{3–16a}$$

or wherever

$$\frac{k(\rho_1 - \rho_2)}{2} = \pm \frac{\pi}{2}, \frac{3\pi}{2}, \frac{5\pi}{2}, \ldots \tag{3–16b}$$

Since $k = 2\pi/\lambda$, we can write, instead of Eq. (3–16b),

$$(\rho_1 - \rho_2) = \pm \frac{\lambda}{2}, \pm \frac{3}{2}\lambda, \pm \frac{5\lambda}{2}, \ldots \tag{3–16c}$$

However, from Fig. 3–7 we can conclude that for any triangle with sides a, b, and c the inequality

$$|a - b| \leq c \leq a + b$$

holds. Applying this to our problem, we see in Fig. 3–6 that the radii ρ_1 and ρ_2 and the separation λ between the two sources form a triangle. Thus, we must have

$$|\rho_1 - \rho_2| \leq \lambda \tag{3–16d}$$

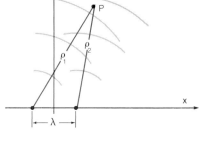

Figure 3–6   Example 2.

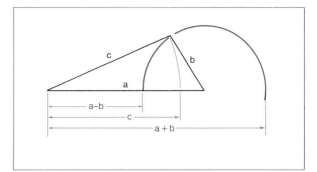

Figure 3–7 For any triangle with sides a, b and c, one has $|a-b| \leqslant c \leqslant a+b$.

because of the laws of plane geometry. Referring to Eq. (3–16c), we see that of the infinitely many solutions, only one is physically possible. Thus, we must have for the region of zero wave amplitude

$$|\rho_1 - \rho_2| = \frac{\lambda}{2} \tag{3-17}$$

This is the equation of a hyperbola. We can show this in the following way. In general, setting

$$\rho_1 - \rho_2 = K \tag{3-18}$$

we can write for the two distances (see Fig. 3–6)

$$\rho_1 = \sqrt{\left(x + \frac{\lambda}{2}\right)^2 + y^2}$$

$$\rho_2 = \sqrt{\left(x - \frac{\lambda}{2}\right)^2 + y^2}$$

and thus we can write

$$\sqrt{\left(x + \frac{\lambda}{2}\right)^2 + y^2} - \sqrt{\left(x - \frac{\lambda}{2}\right)^2 + y^2} = K$$

Transferring the second square root to the right and squaring, after some simplification we obtain

$$\lambda x = K^2 - \lambda x + 2K \sqrt{\left(x - \frac{\lambda}{2}\right)^2 + y^2}$$

or

$$2K \sqrt{\left(x - \frac{\lambda}{2}\right)^2 + y^2} = 2\lambda x - K^2$$

Again we square both sides and obtain, after a little algebra,

$$x^2(4K^2 - 4\lambda^2) + 4K^2 y^2 = K^4 - K^2\lambda^2$$

Because of Eq. (3–17) we know that $K = \lambda/2$. Thus,

$$-3\lambda^2 x^2 + \lambda^2 y^2 = -\frac{3}{16}\lambda^4$$

and we can reduce this to

$$\frac{16x^2}{\lambda^2} - \frac{16y^2}{3\lambda^2} = 1 \tag{3-19}$$

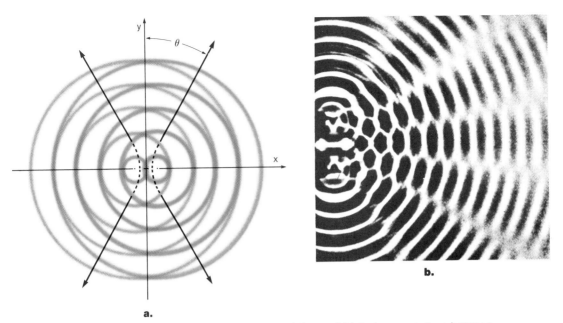

Figure 3–8 (a) The hyperbolas along which there is a minimum. (b) Interference pattern between two sources of cylindrical waves.

which we recognize as one of the standard equations for a hyperbola. If both x and y are much greater than d, the hyperbola can be approximated by its asymptote, whose equation is obtained by dropping the (relatively small) constant term in Eq. (3–19):

$$\frac{16x^2}{\lambda^2} - \frac{16y^2}{3\lambda^2} = 0$$

or

$$y = \pm\sqrt{3x} \qquad (3\text{–}20)$$

Thus, there will be no energy flow in the four directions shown in Fig. 3–8a. For the angle θ we have

$$\tan\theta = 1/\sqrt{3} \qquad \theta = 30° \qquad (3\text{–}21)$$

In Fig. 3–8b we show a picture of such an interference pattern, albeit with a distance between the two sources much larger than λ. One can clearly see the asymptotic directions of the hyperbolas, but the quiet zones become less well-defined near the sources. It is easy to explain this. We have used an approximation, Eq. (3–14b), which is no longer valid near the two sources.

Let us also discuss briefly why we could use the approximation

$$\rho_1 \approx \rho_2 \qquad (3\text{–}22)$$

for the amplitude as we did when writing Eq. (3–14b), while we did not—and could not!—use the same approximation for the phase. If we had set $\rho_1 \approx \rho_2$ in Eq. (3–16a), we would have obtained always

$$\cos\left[\frac{k(\rho_1 - \rho_2)}{2}\right] = 1$$

and we would have concluded wrongly that there is no direction in which there is no energy flow. The answer to our dilemma is quite simple if we choose a numerical example. Let us assume $\rho_1 = 100\lambda$ and $\rho_2 = 99.5\lambda$. Then

and

$$\frac{1}{\sqrt{\rho_1}} = \frac{1}{10\sqrt{\lambda}}$$

$$\frac{1}{\sqrt{\rho_2}} = \frac{1}{9.975\sqrt{\lambda}}$$

and the difference is indeed small:

$$\frac{1}{\sqrt{\rho_1}} - \frac{1}{\sqrt{\rho_2}} = \frac{1}{\lambda}\left[\frac{1}{10} - \frac{1}{9.975}\right] \approx -\frac{0.025}{100\lambda}$$

so that

$$\frac{1}{\sqrt{\rho_1}} - \frac{1}{\sqrt{\rho_2}} \approx 0.25\% \frac{1}{\sqrt{\rho_1}}$$

On the other hand,

$$\frac{k(\rho_1 - \rho_2)}{2} = \frac{2\pi}{\lambda} \cdot \frac{0.5\lambda}{2} = \frac{\pi}{2}$$

so that $\cos\left[\frac{k(\rho_1 - \rho_2)}{2}\right] = 0$, which is very different from $\cos(0) = +1$. In order to be able to set $\rho_1 - \rho_2 \approx 0$ in the phase, we would need

or

$$k(\rho_1 - \rho_2) << 1$$

$$|\rho_1 - \rho_2| << \frac{\lambda}{2\pi} \tag{3-23}$$

which is a much stronger requirement than

$$|\rho_1 - \rho_2| << \rho = \frac{\rho_1 + \rho_2}{2} \tag{3-24}$$

Eq. (3–24) is, however, sufficient to make the approximation

$$\frac{1}{\sqrt{\rho_1}} \approx \frac{1}{\sqrt{\rho_2}} \approx \frac{1}{\sqrt{\rho}}$$

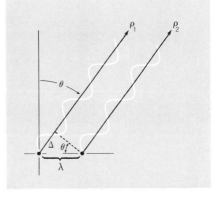

Figure 3–9 A simple determination of maxima and minima in a two-wave interference pattern.

Actually, we could have obtained the final result [Eq. (3–20)] in a much simpler way using the following reasoning: Consider Fig. 3–9, where we show ρ_1 and ρ_2 leading to a very far point P. Since P is so far away, the lines ρ_1 and ρ_2 are practically parallel. The difference in path length is equal to

$$\Delta = \rho_1 - \rho_2$$

If $\Delta = \lambda$, then the two waves from the sources A and B are in phase—the two waves will maximally reinforce each other; on the other hand, if

$$\Delta = \frac{\lambda}{2}$$

the two waves are exactly out of phase. Their phase difference is equal to π; where one has a maximum, the other has a minimum. Thus, the two waves cancel exactly and no energy is radiated in this direction. We thus have a minimum if $\Delta = \lambda/2$. From Fig. 3–9 we see that

$$\Delta = \lambda \sin \theta$$

and thus the minimum condition is

$$\lambda \sin \theta = \frac{\lambda}{2} \tag{3–25}$$

or $\sin \theta = 1/2$, giving $\theta = 30°$, identical to Eq. (3–20). The advantage of our simple reasoning is that we can very easily obtain the maxima and minima of wave intensity very far away, if we have only two sources or a regular array of sources. However, our simple method does not enable us to calculate details of the far-away angular distribution of the radiated intensity. For this purpose we have to resort to adding and manipulating trigonometric functions, as we did in Examples 1 and 2.

2. HUYGENS' PRINCIPLE

Plane, spherical, and cylindrical waves are important because any wave can be decomposed into a superposition of some combination of the three types. Thus, for example, a spherical wave can be decomposed into a superposition of infinitely many plane waves (Fig. 3–10), or also into a superposition of infinitely many cylindrical waves. However, there exists one particular decomposition, namely the one into spherical waves or cylindrical waves, which will be of highest importance to us. It is known as Huygens' principle* and can be stated as follows:

---

1. In a harmonic wave, any of the points P in space can be considered as the source point of a spherical wave; the initial phase of this source is always identical to the phase of the harmonic wave at this point.
2. We can imagine a surface S cutting the region of space into two parts, in such a way that there is an "upstream" and a "downstream" side of the surface: the energy flows through the surface S always from the upstream side to the downstream side. Then the downstream wave can be replaced by the superposition of all the outgoing spherical waves originating on the surface S.

---

*Christian Huygens (1629–1695), Dutch physicist, founder of the modern theory of light. Also discovered the rings of Saturn and the Orion nebula, using a telescope built by himself.

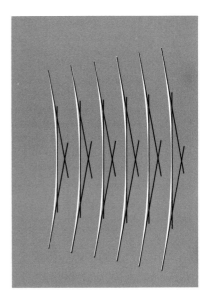

Figure 3-10 A spherical wave can be considered to be a super-position of plane waves.

We can show more clearly what is implied by Huygens' principle if we look at Fig. 3–11a. If we choose an arbitrary point P, we can consider it to be the source of an outgoing spherical wave of the same frequency and wavelength as the plane wave. If the plane wave has the form at the point P

$$A \cos (kx_0 - \omega t - \delta_0)$$

then the outgoing wave will have exactly the same phase. It can be written as

$$\frac{B \cos (kr - \omega t + [kx_0 - \delta_0])}{r} = \frac{B \cos (kr - \omega t + \delta)}{r}$$

where r is the distance from the source P to an arbitrary point P'. The initial phase of the outgoing wave is identical to the phase of the plane wave at the point P; i.e.,

$$\delta = kx_0 - \delta_0 \qquad (3\text{-}26)$$

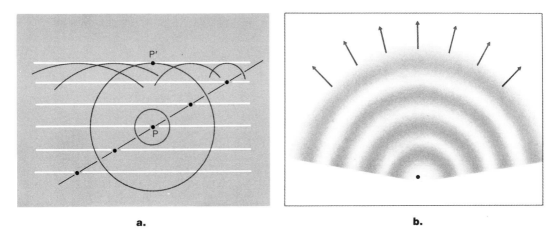

a.   **b.**

Figure 3-11 (a) Each point P can be considered, according to Huygens' principle, to be the starting point of a spherical wave. (b) A single spherical wave.

Let us now consider the superposition of an infinity of such spherical waves and show graphically that Huygens' principle is really true. In Fig. 3–11b we show a single spherical wave. You should imagine the wave crests to be symbolized by the dark line. In Fig. 3–12a we show the superposition of many spherical waves originating on a plane which is, at the moment of viewing, on a crest of the wave. Where the figure is darkest, all the wave crests add and the resulting wave will be maximal. Similarly, the lightest regions denote a minimum in the wave. As we can easily see, the result is the continuation of the plane wave. In Fig. 3–12b we show a similar situation; however, now the spherical waves originate only over a finite area AA' (shown as a line). While nearby and in front of AA' we again have a nearly plane wave, to each side the plane wave becomes

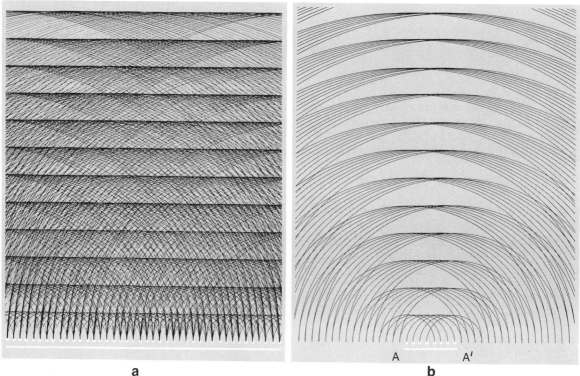

a   b

c

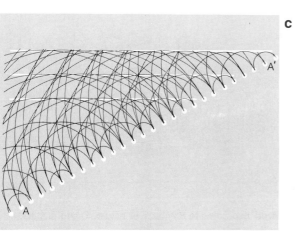

Figure 3–12 (a) Graphical demonstration of Huygens' principle: the circular waves with their centers on the line AA' superimpose to form a plane wave. (b) If the line AA' has only a finite width, the resulting superimposed wave shows a diffraction broadening. (c) Same as Fig. 3–12a, except that the starting line AA' is not parallel to the wave fronts.

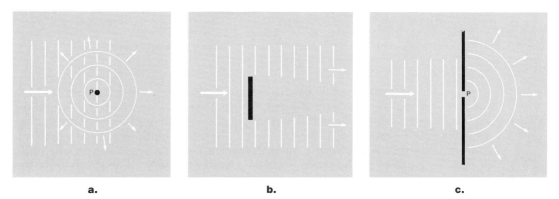

Figure 3-13 (a) A very small obstacle will generate a circular wave starting from the obstacle's position. (b) A larger obstacle will create a shadow zone; the edges are not sharp because the wave nature leads to diffraction. (c) A plane wave incident on a very narrow slit; downstream, a circular wave is seen.

fuzzy and rapidly disappears. If we start the spherical waves along a whole plane which is inclined relative to the wave direction, as shown in Fig. 3–12c, we again obtain the continuation of the plane wave downstream. Huygens' principle is just as true if we replace "spherical wave" everywhere by "cylindrical wave." Fig. 3–12a through c actually depicts cylindrical waves in space — or circular waves on the surface of water.

Huygens' principle will enable us in the next few sections to correctly predict a surprisingly large range of wave phenomena. Let us here just discuss qualitatively the "shadow" which appears whenever we put an obstacle into the path of a plane wave. If the obstacle is small compared to the wavelength, it will cast no shadow at all; instead a circular wave will seem to emanate from the obstacle in all directions (Fig. 3–13a). If, on the other hand, the object is much larger than the wavelength, we will see a reflected wave and a shadow behind the object (Fig. 3–13b). However, the shadow will not be completely sharp, but will have fuzzy edges.

The same basic phenomenon can be seen if water waves arrive at an opening in a wall. If the opening is small, one will see a circular wave on the other side (Fig. 3–13c). If, on the other hand, the opening is much larger than the wavelength, a broad beam will penetrate beyond the wall and propagate in the direction of the original incident wave. Again, its limits will not be sharply defined.

The two situations in which we have either a small obstacle or a small opening are really quite similar. When the wave hits the small obstacle, some of the wave is absorbed. Thus, the wave pattern downstream, instead of being built up by circular waves emanating everywhere on the line AA' (Fig. 3–12b), is built up from circular waves everywhere except at the point of the obstacle. The resulting wave now is a plane wave minus a circular wave originating at P:

$$u_1(x,y,t) = A \cos{(\mathbf{k} \cdot \mathbf{r} - \omega t)} - \frac{B \cos{(kr - \omega t)}}{r} \qquad (3\text{–}27\text{a})$$

On the other hand, if there is only a small opening (Fig. 3–13c), the wave pattern downstream is only the circular wave emanating from the point P; all other waves have been reflected by the wall:

$$u_2 = \frac{B \cos{(kr - \omega t)}}{r} \qquad (3\text{–}27\text{b})$$

Note that since $\cos(\alpha + \pi) = -\cos(\alpha)$, we can recast Eq. (3–27a) in the form

$$u_1(x,y,t) = A \cos(\mathbf{k} \cdot \mathbf{r} - \omega t) + \frac{B \cos(kr - \omega t + \pi)}{r} \qquad \text{(3–27c)}$$

Now, when we compare Eq. (3–27b) for a small opening and Eq. (3–27c) for a small obstacle, we see that Huygens' principle predicts that the situations should be very similar. In both cases, it is as if a circular wave were emanating from the small obstacle or hole. Of course, the two situations are not identical: If there is only a small obstacle, most of the wave continues undisturbed, while in Fig. 3–13c most of the wave does not penetrate the wall and thus only the circular wave is seen downstream.

Since we are discussing water waves, Huygens' principle was stated in terms of circular waves. But it is clear that the same basic phenomenon will occur if sound passes through a small hole or meets a small obstacle. In principle, the same could be said of visible light; however, because the wavelength is so small (less than $1\ \mu = 10^{-3}$ mm), it is nearly impossible to make holes which are much smaller, so we cannot produce such a phenomenon easily in the laboratory. Of course, we can do it without much difficulty using radio waves or even microwaves, whose wavelength is of the order of centimeters to kilometers. As we see, "small" depends on what wavelength we are using.

The situation is more complex if the shadowing object is larger than the wavelength of the wave. We have now removed a whole range of spherical (or circular) waves from the incident plane wave. Similarly, if we have a large opening in a screen or wall, spherical waves will emanate from every point of the opening. Most of this chapter will be devoted to the study of such phenomena.

3. DIFFRACTION FROM A SLIT

Let us consider a light beam, as shown in Fig. 3–14, hitting a black screen with a slit. On a white wall screen beyond, we will see the illuminated image of the slit. However, if we look closely at the edges we will see that they are not completely sharp, but are somewhat fuzzy. Even if we try to improve on the light source, e.g., by making it smaller or placing it further away, we will never achieve a perfectly sharp image. This phenomenon is called *diffraction*. The particular case of *Fraunhofer diffraction** occurs when the wall screen is very far from the slit. By "far away" we mean that the distance L between the slit and the viewing screen is much larger than the slit width d:

$$L \gg d$$

Diffraction can be explained – and the detailed intensity distribution predicted – with the help of Huygens' principle. The incoming plane wave from the left can be considered a superposition of cylindrical waves. The black screen prevents most of these cylindrical waves from propagating beyond the screen. Only those cylindrical waves whose source lies inside the slit will be able to pass through the screen and will influence the illumination of the white wall at S.

Because $L \gg d$, we can assume that all light arriving at a point P on the wall arrives at the same angle θ from the normal to S. We can also neglect the difference in distance and thus the difference in amplitude between cylindrical waves originating at individual points of the slit and the point P. As we did in Eq.

---

* Joseph Fraunhofer (1787–1826), German physicist, was first to build diffraction gratings.

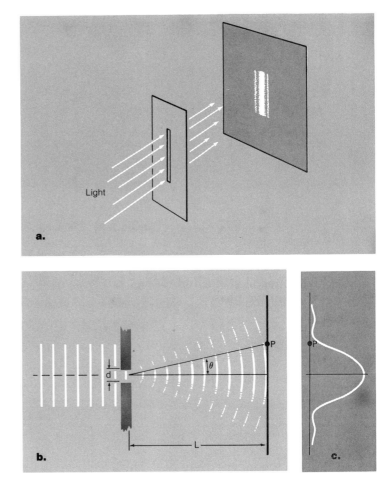

Figure 3-14 (a) A narrow slit will produce a lighted spot on a screen; the edge will be fuzzy because of diffraction. (b) Intensity distribution on the screen.

(3–14b), we write:

$$\frac{A}{\sqrt{\rho}} \approx \frac{A}{\sqrt{\rho_0}}$$

In Fig. 3–15 we show an enlarged view of the slit. We assume the slit to be very long in comparison with its width d. We want to obtain the superposition (sum) of all the cylindrical waves emanating from within the slit in the direction θ. If we call ρ_0 the distance from the lower edge of the slit to the very faraway point P, then the distance $\rho(x)$ between an arbitrary point x and P is given by

$$\rho(x) - \rho_0 = \Delta = x \cdot \sin \theta \qquad (3\text{--}28)$$

Thus, at the faraway wall the amplitude will be

$$A(\theta) = \frac{A_0}{\sqrt{\rho_0}} \int_0^d \cos\left(k\rho(x) - \omega t\right) dx \qquad (3\text{--}29a)$$

which because of Eq. (3–28) can be written

$$A(\theta) = \frac{A_0}{\sqrt{\rho_0}} \int_0^d \cos\left[k(\rho_0 + x \sin \theta) - \omega t\right] dx \qquad (3\text{--}29b)$$

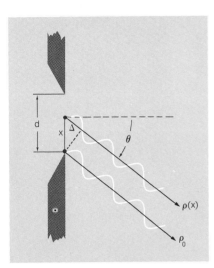

Figure 3–15 Diffraction from a slit: derivation of Eq. (3–29).

The integral seems forbidding, but is actually quite easy to perform. We can write for the phase

$$k(\rho_0 + x \sin \theta) - \omega t = k\rho_0 - \omega t + k \sin \theta \ x$$
$$= \quad \alpha \quad + \quad \beta x \tag{3–30a}$$

where we have defined the quantities

$$\alpha = k\rho_0 - \omega t \tag{3–30b}$$

$$\beta = k \sin \theta \tag{3–30c}$$

which are independent of x. Thus, we have to perform the integral

$$\mathscr{K} = \int_{x=0}^{d} \cos (\alpha + \beta x) dx = \frac{1}{\beta} (\sin [\alpha + \beta x]) \Big|_{x=0}^{x=d}$$

which we can write out as

$$\mathscr{K} = \frac{\sin (\alpha + \beta d) - \sin \alpha}{\beta}$$

Using the first of the "inverse addition theorems" in Appendix C, we obtain for the integral

$$\mathscr{K} = \frac{2}{\beta} \cos \left(\alpha + \frac{\beta d}{2}\right) \sin \left(\frac{\beta d}{2}\right)$$

Substituting our result into Eq. (3–29b) and using the definition of α and β from Eq. (3–30), we obtain

$$A(\theta) = \frac{A_0 d}{\sqrt{\rho_0}} \cos \left[k\left(\rho_0 + \frac{d \sin \theta}{2}\right) - \omega t\right] \frac{\sin \left(\frac{kd \sin \theta}{2}\right)}{\frac{kd \sin \theta}{2}} \tag{3–31}$$

As can be seen from Eq. (3–31), the expression for $A(\theta)$ has three factors with well-defined meanings. The first factor, $A_0 d/\sqrt{\rho_0}$, is the amplitude one would obtain by ignoring the phase differences between waves emanating from different points of the slit. One can verify this easily by ignoring the x-dependence of $\rho(x)$ in Eq. (3–29a). Then the cosine term under the integral is independent of x, and we would obtain for $A(\theta)$

$$A(\theta) = \frac{A_0}{\sqrt{\rho_0}} \cos(k\rho_0 - \omega t) \int_0^d dx = \frac{A_0 d}{\sqrt{\rho_0}} \cos(k\rho_0 - \omega t)$$

The second factor is the familiar wave expression; note that the effective distance is

$$\rho = \rho_0 + \frac{d}{2} \sin\theta$$

i.e., the distance from the center of the slit to the faraway point P. The last factor predicts the intensity distribution which we will see on the wall

$$I = I_0 \cdot \frac{\sin^2\left(\dfrac{kd \sin\theta}{2}\right)}{\left(\dfrac{kd \sin\theta}{2}\right)^2} \qquad (3\text{–}32a)$$

The intensity I_0 that we would have obtained by ignoring interference effects is modified by a function of the angle θ. Let us discuss this function briefly. Setting

$$\gamma = \frac{kd \sin\theta}{2} \qquad (3\text{–}32b)$$

we have to discuss the function

$$f(\gamma) = \frac{\sin^2\gamma}{\gamma^2}$$

Although the sine vanishes at $\gamma = 0$, note that $f(\gamma)$ will not vanish. For small $\gamma \approx 0$ we have $\sin\gamma \approx \gamma$ and thus $f(0) = 1$. However, $f(\gamma)$ will be zero for all $\gamma \neq 0$ for which $\sin\gamma = 0$. Thus, we have $f(\gamma) = 0$ for $\gamma = \pm\pi, \pm 2\pi, \pm 3\pi, \ldots$. Since $0 \leqslant \sin^2\gamma \leqslant 1$, we also have everywhere

$$f(\gamma) \leqslant \frac{1}{\gamma^2}$$

The function $f(\gamma)$ is sketched in Fig. 3–16. We see that the intensity distribution exhibits not only a large central maximum, but also an infinity of other proportionately weaker secondary maxima.

Of course, for any real slit the range of the variable γ is finite. The angle θ cannot be larger than 90°. Since $\sin\theta \leqslant 1$, we have, if we substitute $k = 2\pi/\lambda$:

$$\frac{-\pi d}{\lambda} < \gamma < \frac{\pi d}{\lambda} \qquad (3\text{–}33)$$

Although diffraction from a single slit is most easily demonstrated with visible

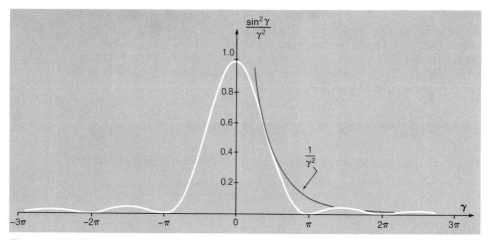

Figure 3–16 Diffraction from a slit: the intensity is a function of the variable $\gamma = kd\sin\theta/2$.

light, it occurs whenever waves pass through a slit. Thus, the same phenomenon can be observed for sound waves, radio waves, water waves, and all other types.

Finally, let us discuss the situation if the slit is very narrow, i.e., if $d < \lambda$. Then there is no minimum at all; the range of $\gamma = (kd\sin\theta)/2$ is

$$-\pi < \gamma < +\pi$$

and therefore $f(\gamma) = \sin^2\gamma/\gamma^2$ never vanishes. From such a narrow slit light is emitted in all directions. In the limit of $d/\gamma \to 0$ (or $d << \lambda$), the wave beyond the slit is a cylindrical wave. In the same way, if we have a pinhole of radius $r << \lambda$, the wave beyond is a spherical wave. This we could have predicted from Huygens' principle without any calculation: Each point of the incoming plane wave in Fig. 3–14 can be considered as the origin of a spherical wave. We are letting through only the wave emanating from nearly one point (over a distance $\pm r < \lambda$); thus we see only the spherical wave emanating from this point.

Example 3 A wall is $L = 1$ m away from a slit of width $d = 0.2$ mm (Fig. 3–17). What will be the separation of the minima (dark stripes), if light of wavelength $\lambda = 6000$ Å is impinging on the slit?

We note first that the wavelength is much smaller than the slit width. Using

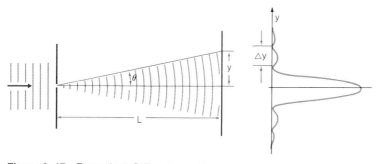

Figure 3–17 Example 4: Diffraction pattern on a faraway screen.

Eq. (3–33) as a measure, we find

$$\frac{\pi d}{\lambda} = \frac{6.28 \times 10^{-4}}{6 \times 10^{-7}} = 1.05 \times 10^{3}$$

Thus the range of γ is very large. Minima will occur for $\gamma = \pm\pi, \pm 2\pi, \ldots$, and from Eq. (3–32b) we obtain as the condition for a minimum

$$\frac{\pi d}{\lambda} \sin \theta = n\pi$$

or

$$\sin \theta = n\frac{\lambda}{d} \qquad (3\text{–}34)$$

Since the angle θ is much less than 90° ($\theta \ll \pi/2$), we can use the approximation $\sin \theta \approx \theta \approx \tan \theta$. This helps us, because from Fig. 3–17 we see that

$$\tan \theta = \frac{y}{L}$$

Thus, within our approximation we have for the location of the minima

$$\frac{y}{L} = n\frac{\lambda}{d} \qquad (3\text{–}35)$$

where n is an integer $n = \pm 1, \pm 2, \ldots$. The minima will be separated by the distance

$$\Delta y = \frac{\lambda L}{d} = \frac{6 \times 10^{-7} \text{ m} \times 1\text{m}}{2 \times 10^{-4} \text{ m}} = 3 \times 10^{-3} \text{ m}$$

or by about 1/8 inch. However, because $n = 0$ corresponds to the center of the large primary maximum of intensity, the two minima on either side of it will be separated by twice that distance.

Example 4   Water waves of wavelength $\lambda = 0.5$ meter hit a gap in a breaker wall. Estimate the size of the gap if there are five maxima and four minima in the diffraction pattern (Fig. 3–18).

We are told to estimate, which gives us the clue that we probably cannot calculate the width exactly. Indeed we cannot, because from Fig. 3–16 we can deduce that in order to have five maxima and four minima between $-\gamma_0$ and $+\gamma_0$

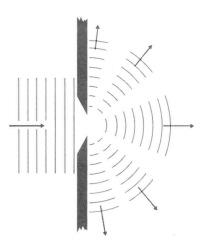

Figure 3–18 Example 5: Diffraction pattern due to a break in harbor wall.

we must have

$$2.5\pi < \gamma_0 < 3\pi$$

Thus $\theta = 90°$, which is the maximum angle of diffraction possible, corresponds to this range of γ_0. We have, in consequence,

$$2.5\pi < \frac{\pi d}{\lambda} < 3\pi$$

and thus

$$2.5\lambda < d < 3\lambda$$

or

$$1.25 \text{ m} < d < 1.5 \text{ m}$$

In other words, the gap in the wall is about 1.3 meters or about 4 feet wide.

In principle, one can calculate the diffraction pattern for an arbitrarily shaped hole in a screen in the same way as we have done it for a long narrow slit. All one has to do is to calculate the superposition of all spherical waves originating at any point in the opening. However, in general the integration corresponding to the transition from Eq. (3–29a) to Eq. (3–31) becomes very complicated. Let us mention here only that for a circular hole of radius R the Fraunhofer diffraction pattern consists of concentric rings, which are the exact analogy of the stripes we have obtained for the slit. In Fig. 3–19 we show the intensity as a function of the angle variable

$$\gamma = kR \sin \theta$$

where again $k = 2\pi/\lambda$ and R is the radius of the circular aperture. The first minimum appears for the value

$$\gamma_1 = 1.220\pi$$

Thus if $\lambda << R$, the opening angle for the first minimum is

$$\boxed{\theta_0 = \frac{1.22\pi}{kR}}$$

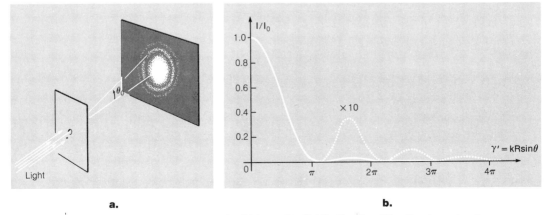

Figure 3–19   (a) Diffraction by a circular hole. (b) Intensity distribution for diffraction by a circular hole.

which we can also write as

$$\theta_0 = 0.61 \frac{\lambda}{R} \tag{3-36}$$

Eq. (3–36), as well as Eq. (3–34), are really special cases of the more general law that if we have an irregular hole of typical size d, then the central maximum will have the width

$$\Delta\theta \approx C\frac{\lambda}{d} \tag{3-37}$$

where C is a number which depends on the exact shape of the opening, but is always of the order of unity.

Let us now assume that we have a screen with a large number of identical apertures which are distributed randomly over the screen. Each aperture will lead to a diffraction pattern. However, the individual apertures will not interfere with each other because of the random distribution of the holes. This will have the effect that the relative phases between waves emanating from different apertures will have no fixed relationship, but will be distributed randomly. If we add two waves of equal amplitude but nonzero relative phase δ:

$$u = A \cos(kx - \omega t) + A \cos(kx - \omega t - \delta)$$

then the overall wave will have the form

$$u = 2A \cos\frac{\delta}{2} \cdot \cos\left(kx - \omega t - \frac{\delta}{2}\right)$$

and the resulting intensity is

$$I = C \cdot 4A^2 \cos^2\frac{\delta}{2} = 4I_0 \cos^2\frac{\delta}{2}$$

where the proportionality constant C depends on the type of wave. If δ now varies randomly between zero and 2π, then the intensity I will also vary with δ. Since the average value of the square of the cosine is

$$<\cos^2\frac{\delta}{2}> = \frac{1}{2}$$

the sum intensity of the two waves on the average will be

$$<I> = 4I_0 \cdot \frac{1}{2} = 2I_0$$

i.e., the sum of the two intensities. If we now have a large number of such waves, each with a randomly distributed phase δ_i,

$$u_i = A \cos(kx - \omega t - \delta_i)$$

the superposition will lead to a total intensity which is the sum of the individual intensities:

$$I(\Sigma u_i) \cong \Sigma I_i$$

We say that waves which have no well-defined relative phase can be added *incoherently.* By this we mean that instead of first having to add the amplitudes, we can directly add the intensities.

The screen with a large number of randomly distributed holes is one example of such an incoherent addition of waves. But the same phenomenon can explain why we never see an interference pattern between two light bulbs, and why we cannot see the interference between waves emanating from different spots of the wire in a single light bulb. The light is being emitted from all points at the wire surface; but there is no defined phase relationship between the spherical waves emanating from each point.

> *Example 5*   An electron beam consists of electrons moving with a velocity $v = 10^6$ m/sec. What would be the size of a circular pinhole which would produce a diffraction pattern with the first minimum at $\theta = 1°$? What is the kinetic energy of these electrons in electron volts (eV)?

This example will convince us that we cannot see the wave nature of electrons — or particles in general — using macroscopic equipment. This is because the de Broglie wavelength from Eq. (2–50)

$$\lambda = \frac{h}{mv}$$

is typically much too small. For our example

$$\lambda = \frac{6.6 \times 10^{-34} \text{ J-sec}}{0.91 \times 10^{-30} \text{ kg} \times 10^6 \text{ m/sec}} = 7.3 \times 10^{-10} \text{ m}$$

or 7.3 Ångstroms. The first minimum for a diffraction pattern from a circular hole, according to Eq. (3–36), lies at

$$\theta = 1.22 \frac{\lambda}{d}$$

where $d = 2R$ is the diameter of the hole. If $\theta = 1° = 0.017$ radians, we need

$$d = \frac{1.22\lambda}{\theta} = \frac{1.22 \times 7.3 \times 10^{-10} \text{ m}}{0.017} = 5.2 \times 10^{-8} \text{ m}$$

or $d = 520$ Å $= 0.002$ mils (thousandths of an inch). Considering that atoms in a typical solid are spaced 2 to 4 Å apart, we are talking about a circular hole which is ~ 200 atomic spacings in diameter. While holes that small can be produced by using special etching processes, it is extremely difficult to produce sufficient intensity in electron beams to measure the diffraction pattern from such a small hole.

We could try to increase the wavelength λ by decreasing the velocity v. However, the kinetic energy of the electrons

$$E = \frac{mv^2}{2} = \frac{0.91 \times 10^{-30}\text{kg} \times (10^6 \text{ m/sec})^2}{2} = 4.55 \times 10^{-19} \text{ Joule} = 2.84 \text{ eV}$$

is quite low already. Electron beams in a TV tube or an oscilloscope have typically kinetic energies of 10,000 to 20,000 eV. It is very difficult to produce electron beams of only a few electron volts kinetic energy. Also, the wavelength λ is inversely proportional to the velocity v of the electrons ($\lambda \sim 1/v$), but the kinetic energy is proportional to the square of the velocity ($E \sim v^2$). Thus, we would have to decrease the kinetic energy by a factor of 100 if we wanted to increase the wavelength tenfold. Considering that at room temperature the kinetic energy of an electron which has only random thermal motion is of the order° $kT = 0.025$ eV, it is

---

° Here k is the Boltzmann constant $k = 1.38 \times 10^{-23}$ J/°K and T is the absolute temperature. Compare Vol. I, Chapter 15.

quite reasonable to say that it would be extremely difficult to produce electron beams going in a well-defined direction, if their kinetic energy was significantly less than a few electron volts.

This example shows that it is impossible to build macroscopic diffraction slits for waves of wavelength $\lambda \lesssim 100$ Å. However, such waves exist in nature; matter waves of electrons and X-rays commonly have wavelengths as short as 0.01 to 0.1 Å. For such waves, the individual atoms in a lattice can be used to obtain the interference patterns. We will discuss this later in this chapter, when we treat diffraction gratings.

4. BABINET'S PRINCIPLE

We have so far discussed only diffraction by a screen with apertures (holes). At least as interesting a problem, and at first glance a much more difficult one, is the fuzziness of shadows caused by diffraction. If we have a small object (Fig. 3–20) and are interested in the diffraction pattern far away, we have to calculate the diffraction amplitude $A(\theta)$. The "obvious" way would be to imagine a screen which has a single very large aperture everywhere except where the object is, as shown in Fig. 3–21a. We would then have to add the spherical waves originating everywhere in the aperture of the screen—the whole plane AA' except the small portion where we have the object.

However, we do not need to solve this rather formidable problem. Instead, we can use *Babinet's principle,*[*] which states:

Consider two complementary screens, A and B, which are such that wherever screen A has an aperture (is open), screen B has no aperture (covers) and vice versa. Then the diffraction pattern $I(\theta)$ for the two screens will be identical except at exactly $\theta = 0°$.

Figures 3–21a and 3–21b show two such complementary screens. One can imagine them quite generally as two flat screens which, when superimposed, completely block the incident wave; but they also have no overlapping portions. We can then derive Babinet's principle with the help of Huygens' principle. The first screen will produce a certain diffraction amplitude as a function of the dif-

---

[*] Jacques Babinet (1799–1872), French physicist, worked mainly on problems of optics.

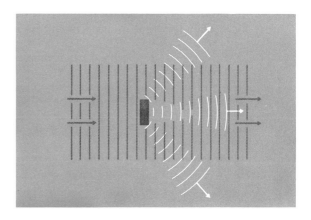

Figure 3–20 Diffraction pattern of small obstacle.

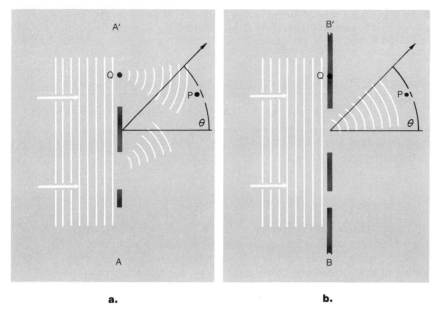

a.   **b.**

Figure 3–21 Babinet's principle.

fraction angle θ; we call this amplitude $A_A(\theta)$. It is due to the superposition of all spherical waves coming from points where screen A is open. Quite generally, Huygens' principle tells us that the disturbance downstream at a point P

$$u_A(\mathbf{r},t) = u_A(P,t)$$

can be considered the superposition of spherical waves originating at all points where the screen is open.

Now consider the complementary screen B (Fig. 3–21b). It will produce downstream a disturbance $u_B(P,t)$ which is due to spherical waves originating from all points where screen B has an aperture. Finally, let us consider

$$u_{\text{tot}}(P,t) = u_A(P,t) + u_B(P,t) \qquad (3\text{–}38)$$

It can be considered to be produced by the superposition of all spherical waves originating wherever either screen A or screen B is open. But since at any point Q either screen A or screen B is open (but never both!), the disturbance u_{tot} in Eq. (3–38) will be the same as if there were no screen at all. Thus, u_{tot} has to be simply the continuation of the incident plane wave:

$$u_{\text{tot}} = \text{plane wave} = A \cos{(\mathbf{k} \cdot \mathbf{r} - \omega t)} \qquad (3\text{–}39)$$

The plane wave propagates only in the direction $\theta = 0$. Thus, the diffraction amplitude $A_{\text{tot}}(\theta)$ which describes the disturbance far away has the property that

$$A_{\text{tot}}(\theta) = 0 \quad \text{wherever} \quad \theta \neq 0 \qquad (3\text{–}40)$$

If we now denote the diffraction amplitudes produced by screen A and screen B by $A_A(\theta)$ and $A_B(\theta)$, we have from Eq. (3–40)

$$A_B(\theta) = A_{\text{tot}}(\theta) - A_A(\theta) = -A_A(\theta) \quad \text{wherever} \quad \theta \neq 0 \qquad (3\text{–}41a)$$

The intensities $I_A(\theta)$ and $I_B(\theta)$ are proportional to the squares of the amplitudes. Thus, we have from Eq. (3–41a)

$$I_A(\theta) = I_B(\theta) \quad \text{for} \quad \theta \neq 0 \tag{3–41b}$$

In Fig. 3–22a and b we show the intensity distributions $I_A(\theta)$ and $I_B(\theta)$ which a faraway observer sees when viewing a circular flat disk and a circular hole, respectively. The two intensity distributions are identical except at $\theta = 0$. An observer viewing a hole will not see the incident beam. If he is looking at the disk, most of the incident wave will bypass the disk and thus the observer will see a sharp spike in intensity at $\theta = 0$.

The diffraction patterns produced by objects or pinholes limit our ability to see the shape of small objects. Certainly we cannot see any details of an object if its size is smaller than the wavelength of the light we are using. All we will see is a diffuse diffraction pattern which obscures the details. On the other hand, diffraction can enable us to measure the sizes of very small objects, as can be seen from the next example.

> *Example 6*   A narrow laser beam (red light, wavelength 6500 Å) passes through a dusty glass plate. The beam spot seen on a screen 1 m away has a halo with a diameter of 1 inch (2.5 cm). What is the average size of the dust particles?
>
> Of course, a good fraction of the beam passes through the glass as a plane wave, independently of whether the plate is dusty or not. The halo around the main beam spot is the result of diffraction by the individual dust particles. We do not know their detailed shape; each dust particle will be shaped differently. But the diffraction "pattern" will have its first minimum roughly at an angle
>
> $$\theta \approx \frac{\lambda}{d} \tag{3–42}$$
>
> where d is the typical diameter of the object (Fig. 3–23). We have seen this for the case of openings (apertures); but Babinet's principle tells us that the same is true for small objects. Indeed, all that Eq. (3–42) says is that the first minimum occurs

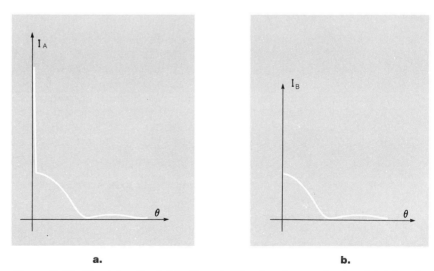

Figure 3–22 The intensity distribution of diffraction from a circular disk (a) and a hole of same radius (b) are the same except exactly in the direction of the incident waves ($\theta = 0$).

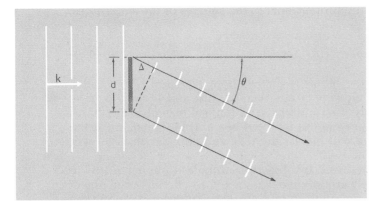

Figure 3–23 Diffraction by a small object of arbitrary shape will always have a central maximum of roughly the width $\theta \approx \lambda/d$; the path difference Δ is then about the same as the wavelength.

at an angle θ for which the path difference between the edges of the object is of the order of one wavelength of the incident light. Then

$$\lambda = d \sin \theta \approx d\theta$$

This is equivalent to Eq. (3–42) for small angles θ.

Returning to our specific example, we see that since the diameter of the halo is 2.5 cm, its radius R is equal to 1.25 cm. Since the screen is $L = 1$ m away, the first minimum occurs at an angle

$$\tan \theta \approx \theta = \frac{R}{L} = \frac{1.25 \times 10^{-2}\ \text{m}}{1\ \text{m}} = 1.25 \times 10^{-2}$$

The typical size of the dust particles is then, according to Eq. (3–42),

$$d \approx \frac{\lambda}{\theta} = \frac{6.5 \times 10^{-7}\ \text{m}}{1.25 \times 10^{-2}} = 5.2 \times 10^{-5}\ \text{m}$$

or about 5/100 mm. This is quite reasonable for fine dust particles.

5. DIFFRACTION BY GRATINGS

Up to now we have considered only single apertures or irregular arrangements of isolated apertures or absorbing objects. However, a *regular* arrangement of apertures (or absorbing objects) leads to new phenomena, because now the waves emanating from the apertures can interfere with each other.

Let us consider first two very narrow long slits of width a placed a small distance d apart from each other (Fig. 3–24a). If the two slits are very narrow ($a \ll \lambda$), each will be the source of a cylindrical wave. Thus, we have exactly the same situation as we discussed earlier in Example 2 on p. 63. Two sources of cylindrical waves are separated by a distance d. There will be a maximum if the difference

$$\Delta = d \sin \theta = \lambda, 2\lambda, 3\lambda, \ldots n\lambda \qquad \text{(3–43a)}$$

while a minimum will occur if

$$\Delta = d \sin \theta = \frac{\lambda}{2}, \frac{3\lambda}{2}, \frac{5\lambda}{2}, \ldots \frac{(2n+1)}{2}\lambda \qquad \text{(3–43b)}$$

It is important to note that while we have assumed the two slits to be narrow, their separation d can be many times larger than the wavelength λ. Thus, in particular for light waves, where λ is very small, we can in principle measure the wavelength λ. All we have to do is first measure (e.g., under a microscope) the separation d between the two slits, and then observe the diffraction pattern. If $d \gg \lambda$, then $\sin \theta$ will be very small and we can write $\sin \theta \approx \theta$. Thus, maxima will occur at

$$\theta = \frac{\lambda}{d}, \frac{2\lambda}{d}, \cdots \frac{n\lambda}{d} \qquad (3\text{-}44)$$

If we put a viewing screen (white wall or similar object) a distance L beyond the two slits, then we will see parallel light lines (maxima) separated by dark lines (minima). From Fig. 3–24b we see that

$$\frac{x}{L} = \tan \theta$$

and since θ is very small, we have $\tan \theta \approx \theta \approx \sin \theta$; thus, the maxima will be separated by a distance Δx, where

$$\frac{\Delta x}{L} = \frac{\lambda}{d}$$

or

$$\Delta x = \frac{L}{d} \lambda \qquad (3\text{-}45)$$

We can even relax the requirement that the width of the slits themselves be very narrow. This is practically not possible anyway, if we want to study the interference pattern of visible light whose wavelength is λ = 4000 to 7000 Å. Thus, let us consider a more realistic arrangement: Two slits of width a have their centers separated by a distance d. We will choose $a = 0.2$ mm and $d = 1.5$ mm. A distance $L = 5$ m away from the slits we put a viewing screen. What will be the diffraction pattern we expect to observe on the screen if light of wavelength λ = 7000 Å is incident on it?

Since the distance L is much larger than the separation of the two slits, we can consider the two rays ρ_1 and ρ_2 to be parallel (Fig. 3–24a); thus, we label both angles as θ. Each slit will emit a diffraction pattern given by Eq. (3–31);

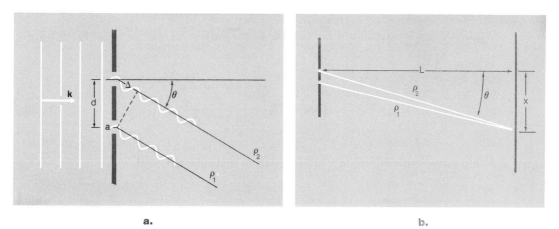

Figure 3–24 (a) Interference between two slits: detail view. (b) Interference between two slits: large scale view.

since ρ_1 and ρ_2 are measured from the center of the slits, we have from slit 1

$$A_1(\theta) = \frac{A_0 a}{\sqrt{\rho_1}} \frac{\sin\left[(ka \sin \theta)/2\right]}{\frac{ka \sin \theta}{2}} \cdot \cos(k\rho_1 - \omega t) \qquad (3\text{--}44\text{a})$$

and from slit 2

$$A_2(\theta) = \frac{A_0 a}{\sqrt{\rho_2}} \frac{\sin\left(\frac{ka \sin \theta}{2}\right)}{\frac{ka \sin \theta}{2}} \cdot \cos(k\rho_2 - \omega t) \qquad (3\text{--}44\text{b})$$

The total amplitude can be obtained by adding the two expressions. But we can set $A_0/\sqrt{\rho_1} \approx A_0/\sqrt{\rho_2}$ because the screen is very far away. Furthermore, there is a common factor in both expressions which describes the fact that both slits are of finite and equal width a. The two expressions are thus different only in the last factor; this means that both slits emit identical waves, but the two waves arrive at the screen with different phases because they have traveled the not quite equal distances ρ_1 and ρ_2. Since we have

$$\cos(k\rho_1 - \omega t) + \cos(k\rho_2 - \omega t) = 2\cos(k\rho - \omega t) \cdot \cos\left(k\frac{\rho_1 - \rho_2}{2}\right)$$

with

$$\rho = \frac{\rho_1 + \rho_2}{2}$$

we can write the total wave as

$$A(\theta) = A_1(\theta) + A_2(\theta) = \underbrace{\frac{2A_0 a}{\sqrt{\rho}}}_{\substack{\text{overall} \\ \text{normal-} \\ \text{ization}}} \cdot \underbrace{\frac{\sin\left(\frac{ka \sin \theta}{2}\right)}{\left(\frac{ka \sin \theta}{2}\right)}}_{\substack{\text{single-slit} \\ \text{pattern}}} \cdot \underbrace{\cos\left(k\frac{\rho_1 - \rho_2}{2}\right)}_{\substack{\text{two-slit} \\ \text{interference}}} \underbrace{\cos(k\rho - \omega t)}_{\text{wave}} \qquad (3\text{--}45\text{a})$$

Thus, the resulting intensity distribution can be written

$$I(\theta) = I_1(\theta) \cdot I_2(\theta) \qquad (3\text{--}46\text{a})$$

where

$$I_1(\theta) = 4I_0 \frac{\sin^2\left(\frac{ka \sin \theta}{2}\right)}{\left(\frac{ka \sin \theta}{2}\right)^2} \qquad (3\text{--}46\text{b})$$

is four times the diffraction intensity from a single slit, and

$$I_2 = \cos^2\left(\frac{kd \sin \theta}{2}\right) \qquad (3\text{--}46\text{c})$$

is the two-slit interference pattern. Note that $\delta_1 - \delta_2 = d \sin \theta$.

What will be the pattern we now see on the viewing screen? Let us remember that all angles θ are small and that, to a good approximation,

$$\sin \theta \approx \theta \approx \tan \theta = \frac{x}{L}$$

Using the relating $k = 2\pi/\lambda$, we then obtain for the intensity distribution on the screen

$$I(x) = I_0 \cdot \frac{\sin^2\left(\frac{\pi ax}{\lambda L}\right)}{\left(\frac{\pi ax}{\lambda L}\right)^2} \cdot \cos^2\left(\frac{\pi dx}{\lambda L}\right)$$

$$= I_0 \, \frac{\sin^2\left(\frac{\pi x}{\alpha}\right)}{\left(\frac{\pi x}{\alpha}\right)^2} \cdot \cos^2\left(\frac{\pi x}{\beta}\right)$$

With our parameters, $a = 0.2$ mm $= 2 \times 10^{-4}$ m and $d = 1.5$ mm $= 1.5 \times 10^{-3}$ m, we have

$$\frac{1}{\alpha} = \frac{a}{\lambda L} = \frac{2 \times 10^{-4} \text{ m}}{7 \times 10^{-7} \text{ m} \times 5 \text{ m}} = 0.57 \times 10^2 \text{ m}^{-1} = \frac{1}{1.75} \text{ cm}^{-1}$$

and

$$\frac{1}{\beta} = \frac{d}{\lambda L} = \frac{1.5 \times 10^{-3} \text{ m}}{7 \times 10^{-7} \text{ m} \times 5 \text{ m}} = 4.3 \times 10^2 \text{ m}^{-1} = \frac{1}{0.233} \text{ cm}^{-1}$$

Thus, we observe a diffraction pattern from a single slit, with its first minimum 1.75 cm away from the center. However, the pattern is chopped up into stripes 2.3 mm wide (Fig. 3–25b). Both the functions $\sin^2 \alpha$ and $\cos^2 \alpha$ have a period of π; thus, $\sin^2\left(\frac{\pi x}{\alpha}\right)$ has the period α and $\cos^2\left(\frac{\pi x}{\beta}\right)$ has the period β. The resulting intensity distribution is shown in Fig. 3–25a.

The two-slit system enables us to measure the wavelength of light, or of any type of wave, but it is not very accurate. For a more precise determination, one uses a very large number of such slits. Such arrangements are called diffraction gratings. They consist basically of a regularly spaced linear array of either absorbing lines or narrow slits. Because of Babinet's principle, the two types

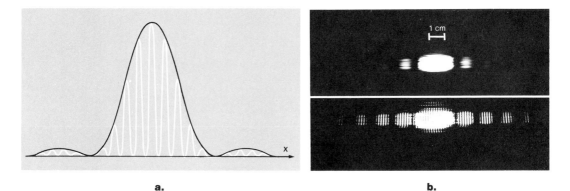

a.   **b.**

Figure 3–25 (a) Interference between two slits: intensity distribution on a faraway screen. (b) Photograph of one slit diffraction pattern (top) and two slit interference (bottom).

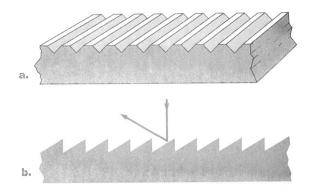

Figure 3–26 Different types of gratings: (a) simple grating, and (b) reflection grating.

are equivalent. The simplest form of grating consists of a glass plate with equally spaced scratches; but many different types of gratings are made for special purposes (Fig. 3–26). High quality gratings have as many as 10,000 grooves/cm or 25,000 grooves/inch. These gratings are extremely difficult to produce, since it is essential that the spacing of the lines be completely regular. Special precision machines have been developed for this purpose. The art of making precision gratings is quite old; the first ones were made by Rowland in 1882. Even today these so-called "Rowland* gratings" are among the best one can find. Such "original" gratings are extremely expensive; however, it is possible today to make "plastic replicas." Some of those are produced by photographic means; only rather coarse gratings can be reproduced this way, because the natural graininess of photographic film imposes a limit on the quality of the gratings. The best replicas are obtained by spraying a plastic film on the original glass grating. After the film has dried, it is carefully peeled off; the surface of the plastic is then an exact replica of the original grating.

The basic principle of all gratings is the same: they produce cylindrical waves originating from a regular array of line sources (Fig. 3–27). If a plane parallel light wave of wavelength λ is normally incident on the grating, all the cylindrical waves will be exactly in phase. Thus, we will obtain maxima in the

---

* Henry Rowland (1848–1901), British physicist; his contributions to physics were mainly in optics and electromagnetism.

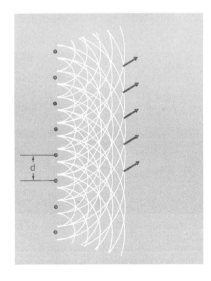

Figure 3–27 The diffraction pattern is obtained by the superposition of cylindrical waves from each line.

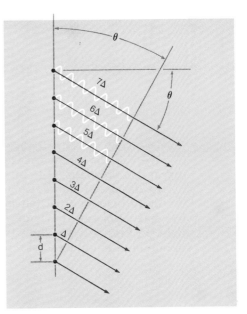

Figure 3-28 Calculating the diffraction pattern maxima of a grating.

intensity distribution $I(\theta)$ when all the individual cylindrical waves interfere constructively with each other, i.e., when they all have the same phase at a faraway point. This occurs for those angles θ for which the difference Δ in the path length from two adjacent sources is equal to a multiple of the wavelength (Fig. 3-28):

$$\Delta = d \sin \theta = n\lambda$$

or

$$\boxed{\sin \theta = \frac{n\lambda}{d}} \qquad\qquad (3\text{-}47)$$

If one knows the angle of reflection and the spacing of the grooves in the grating, one can determine the wavelength λ of the light. Alternatively, one can use a known wavelength and spacing to determine the angle of reflection, as in the following example.

Example 7   What will be the angles under which one sees maxima for light of wavelength $\lambda = 6000$ Å if the diffraction grating has 5000 grooves/cm?

If there are 5000 grooves/cm, the spacing between the grooves is

$$d = \frac{1}{5000 \text{ cm}^{-1}} = 2 \times 10^{-4} \text{ cm} = 2 \times 10^{-6} \text{ m}$$

From Eq. (3-47) we then obtain

$$\sin \theta = \frac{n\lambda}{d} = n \frac{6 \times 10^{-7} \text{ m}}{2 \times 10^{-6} \text{ m}} = 0.3 \, n$$

Thus, there will be several maxima corresponding to

$n = 0$	$\sin \theta = 0.0$	$\theta = 0°$	direct beam
$n = 1$	$\sin \theta = 0.3$	$\theta = 17.5°$	diffraction
$n = 2$	$\sin \theta = 0.6$	$\theta = 36.3°$	maxima
$n = 3$	$\sin \theta = 0.9$	$\theta = 64.2°$	

Since we must have $\sin \theta < 1$, these are all the maxima; $n \geq 4$ cannot occur.

We call the various maxima corresponding to the integer n the *n*th *order diffraction maximum*. In our example, only the first three orders exist.

As discussed before, Eq. (3–47) is also correct if there are only two grooves; the more lines that participate in the diffraction pattern, the narrower will be the maximum and the larger will be the maximum intensity of the diffracted light. In Fig. 3–29 we show the relative intensities of the diffraction patterns if the number of grooves participating is $N = 2$, 5, and 20.

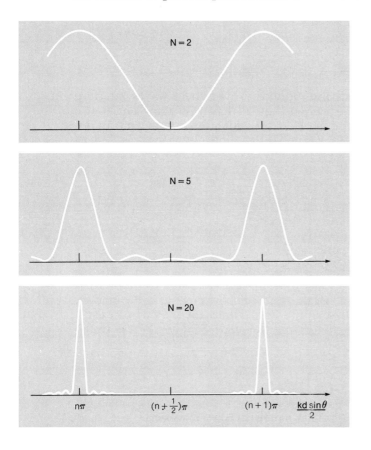

Figure 3–29 Diffraction intensity for a grating having $N = 2$, 5 or 20 lines.

■ The intensity distribution from a diffraction grating with a finite number of grooves can be calculated without too much difficulty. The total wave emitted at an angle θ is, according to Fig. 3–28,

$$u(\theta) = A \{\cos (kx - \omega t) + \cos [k(x + \Delta) - \omega t] + \cos [k(x + 2\Delta) - \omega t] + \ldots\} \quad \text{(3–48a)}$$

where $A \cdot \cos (kx - \omega t)$ is the wave emitted by a single grating and Δ is given by

$$\Delta = d \sin \theta \quad \text{(3–48b)}$$

There are N such cosine terms if the grid has N grooves. We can thus write

$$u(\theta) = A_0 \sum_{n=0}^{N-1} \cos \left[k(x + n\Delta) - \omega t \right] \qquad (3\text{-}48\text{c})$$

where the summation index n goes over all integers $0, 1, \ldots, N-1$. If we had only two or three cosine terms to add, we could use the inverse addition theorems from Appendix C. However, this method would be too cumbersome here, where we have to add a large number of such terms. We use instead a trick which can be used whenever we work with harmonic waves, or with harmonic functions (sines and cosines) in general. Using Moivre's identity, with $i = \sqrt{-1}$,

$$e^{i\alpha} = \cos \alpha + i \sin \alpha$$

which is true for any α, we can write

$$\cos \alpha = \text{Re}(e^{i\alpha})$$

This says that the trigonometric function $\cos \alpha$ is the real part of the complex function $e^{i\alpha}$. As one learns in the elementary theory of complex numbers, the real part of the sum of two numbers is equal to the sum of their real parts:

$$\text{Re}(z_1 + z_2) = \text{Re}(z_1) + \text{Re}(z_2)$$

We can therefore write Eq. (3-44b) in the form

$$u(\theta) = A_0 \text{Re} \left[\sum_{n=0}^{N-1} e^{i[k(x+n\Delta)-\omega t]} \right] \qquad (3\text{-}49)$$

The advantage of writing $u(\theta)$ in the form of Eq. (3-49) lies in the fact that we now can use the identities

$$e^{i(\alpha+\beta)} = e^{i\alpha} \cdot e^{i\beta}$$

and

$$e^{i2\alpha} = (e^{i\alpha})^2; \quad e^{in\alpha} = (e^{i\alpha})^n$$

Using these relations in Eq. (3-49) and taking all common factors in front of the sum yields for $u(\theta)$ from Eq. (3-48c)

$$u(\theta) = A_0 \text{Re} \left[e^{i(kx-\omega t)} \left(\sum_{n=0}^{N-1} e^{ikn\Delta} \right) \right] \qquad (3\text{-}50)$$

The sum in Eq. (3-50) can be written

$$\sum_{n=0}^{N-1} e^{ink\Delta} = \sum_{n=0}^{N-1} (e^{ik\Delta})^n = 1 + e^{ik\Delta} + (e^{ik\Delta})^2 + \ldots + (e^{ik\Delta})^{N-1}$$

We can now use the following formula for adding any geometric series

$$1 + x + x^2 + \ldots + x^{N-1} = \sum_{n=0}^{N-1} x^n = \frac{x^N - 1}{x - 1} \qquad (3\text{-}51\text{a})$$

which is true for any real or complex x. We thus have

$$\sum_{n=0}^{N-1} e^{ink\Delta} = \frac{e^{iNk\Delta} - 1}{e^{ik\Delta} - 1} \qquad (3\text{-}51\text{b})$$

which we can rewrite in the form

$$\sum_{n=0}^{N-1} e^{ink\Delta} = \frac{e^{ik\Delta N/2} \left(e^{iNk\Delta/2} - e^{-iNk\Delta/2} \right)}{e^{ik\Delta/2} \left(e^{ik\Delta/2} - e^{-ik\Delta/2} \right)} \qquad (3\text{-}51\text{c})$$

Since from Moivre's formula one immediately can deduce that

$$\sin \alpha = \frac{e^{i\alpha} - e^{-i\alpha}}{2i}$$

we have for the sum from Eq. (3–51c)

$$\sum_{n=0}^{N-1} e^{ink\Delta} = e^{i(N-1)k\Delta/2} \cdot \frac{\sin (Nk\Delta/2)}{\sin (k\Delta/2)} \tag{3–52}$$

Substituting this into Eq. (3–50), we obtain

$$u(\theta) = A_0 \text{Re}\left[e^{i(kx - \omega t + [N-1]k\Delta/2)} \cdot \frac{\sin (Nk\Delta/2)}{\sin (k\Delta/2)} \right]$$

and since the only complex number in the above expression is the complex exponential, we obtain finally

$$u(\theta) = A_0 \frac{\sin (Nk\Delta/2)}{\sin (k\Delta/2)} \cdot \cos (kx - \omega t + [N - 1]k\Delta/2) \tag{3–53}$$

Of course, throughout our calculation Δ is given by Eq. (3–48b). The intensity for the wave propagating in the direction θ from the diffraction grating will be

$$I = I_0 \frac{\sin^2 \left(\dfrac{Nkd \sin \theta}{2} \right)}{\sin^2 \left(\dfrac{kd \sin \theta}{2} \right)} \tag{3–54a}$$

This is the function which we have plotted against the argument

$$\gamma = \frac{kd \sin \theta}{2} \tag{3–54b}$$

in Fig. 3–29 for various values of N. We note that if the sine function in the denominator vanishes, so does the sine function in the numerator. Indeed, if n is some integer $n = 0$, 1, 2, 3, . . . and

$$\frac{kd \sin \theta}{2} = \gamma = n\pi \tag{3–55}$$

we have

$$I = I_0 \cdot N^2 \tag{3–56a}$$

because

$$\lim_{x \to 0} \frac{\sin \lambda x}{\sin x} = \lim_{x \to 0} \frac{\lambda \cos \lambda x}{\cos x} = \lambda$$

by l'Hôpital's first rule.[°]
   Between the primary maxima, wherever the denominator vanishes, we have $(N - 2)$ maxima and $(N - 1)$ minima of the numerator. There is zero intensity whenever

$$\frac{Nkd \sin \theta}{2} = n\pi \quad \text{but also} \quad \frac{kd \sin \theta}{2} \neq m\pi$$

or whenever n is not a multiple of N. In between, there will be maxima at

$$\frac{Nkd \sin \theta}{2} = n\pi + \frac{\pi}{2}$$

[°] Recall that, if $\lim_{x \to 0} \dfrac{f(x)}{g(x)} = \dfrac{0}{0}$, l'Hôpital's first rule states that $\lim_{x \to 0} \dfrac{f(x)}{g(x)} = \lim_{x \to 0} \dfrac{f'(x)}{g'(x)}$ if $g'(x = 0) \neq 0$.

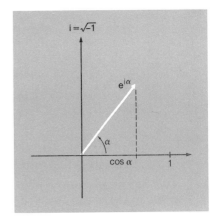

Figure 3–30 The complex number $e^{i\alpha}$ can be considered a phasor.

and since the numerator of Eq. (3–54a) equals unity at such points, we have for the intensity at the secondary maxima

$$I = I_0 \frac{1}{\sin^2\left(\dfrac{kd\,\sin\theta}{2}\right)} \qquad (3\text{–}56b)$$

Because the value of the sine in Eq. (3–56b) is of the order of unity, the intensities of the secondary maxima are indeed much smaller than the intensities of the primary maxima given by Eq. (3–56a). We note in passing that Eq. (3–55) is identical to Eq. (3–47), which we derived using a simple intuitive argument; if we use the identity $k = 2\pi/\lambda$, we have

$$\frac{2\pi}{\lambda} \cdot \frac{d\,\sin\theta}{2} = n\pi$$

which can be written

$$d\,\sin\theta = n\lambda$$

We also note that our method of using the complex exponential is really the same as the method of phasors used in the last chapter. Indeed, the function $e^{i\alpha}$ is a vector of unit length in the complex plane, whose projection on the real axis is equal to $\mathrm{Re}(e^{i\alpha}) = \cos\alpha$ (Fig. 3–30). The advantage of using complex variable theory lies in our ability to add geometric series as we did in Eq. (3–51); but basically the method of phasors and the method of adding complex vectors $Ae^{i\alpha}$ are identical. ■

6. X-RAY DIFFRACTION

Man-made gratings become useless if we want to measure wavelengths which are less than 10 to 100 Å. Such short wavelengths occur not only for electromagnetic rays (X-rays), but also for matter waves (electrons). Luckily, nature has provided us with gratings ideally suited for this purpose, in the form of crystals. A crystalline lattice forms a regular array of scatterers; each atom in the crystal lattice will be the origin of a scattered wave. However, the gratings are no longer one-dimensional, as are macroscopic gratings consisting of parallel lines. Instead, we are dealing with a three-dimensional "grating." Each lattice plane (Fig. 3–31) acts like a mirror; the scattered radiation leaves it at the same angle as that at which the incident wave entered. Maxima will occur whenever all the waves reflected from parallel planes maximally reinforce each other. This happens whenever (Fig. 3–31) the path difference between the waves scattered from neighboring planes is an integer multiple of the wavelength λ:

$$2s = 2d/\sin\theta = n\lambda \qquad n = 1, 2, 3, \ldots$$

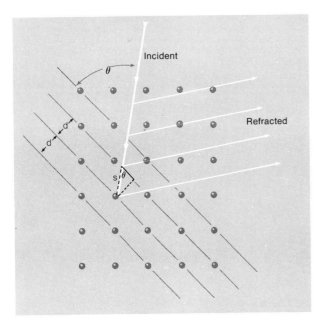

Figure 3–31 Bragg diffraction; reflection from a set of parallel planes.

Diffraction of X-rays by crystals was first predicted by M. von Laue in 1912 and observed by Friedrich and Knipping in the same year. However, it is known as Bragg diffraction or Bragg scattering after W. L. Bragg, who in the twenties and thirties used this effect to study the properties of crystal lattices.

7. FRESNEL DIFFRACTION

Fraunhofer diffraction, in which the incoming wave is a plane wave and the outgoing diffraction pattern is viewed from a large distance, is only a rather special case. It is, however, also the most important case in optics; the very short wavelength of visible light causes just about any macroscopic distance (say 10 cm or more) to be "very far away."

A much more complex type of diffraction, in which neither the original wave source nor the observer (viewing screen, detector, etc.) is very far away, is called Fresnel* diffraction. One of the simpler types of Fresnel diffraction is the interference between two cylindrical waves that we discussed on p. 63 and showed in Fig. 3–8. Instead of the two wave sources we could have, of course, also two narrow slits separated by a distance d and an electromagnetic wave of suitable wavelength (e.g., microwaves of 10 cm wavelength) incident as a plane wave. We then obtain an intensity distribution identical to the one we calculated in Eq. (3–15).

Let us discuss here only one other example of Fresnel diffraction, called the Fresnel lens. A source of spherical waves (sound waves, radio waves of wavelength 3 to 10 cm) is situated at the point P_0 (Fig. 3–32). A receiver (microphone, small antenna) is situated at the point P. In between is a screen which is opaque (i.e., either absorbs or reflects) to the wave we are using. We now punch a small hole in the screen along the axis $P_0 P$. How should we arrange a set of narrow circular slits so as to obtain a maximum intensity at P?

---

* Jean Augustin Fresnel (1788–1827), French engineer and member of the Paris Academy of Sciences. Although his main job was road and bridge construction, he also was first to formulate the mathematical theory of light.

Let us, to simplify our discussion, put the screen exactly halfway between P_0 and P. Huygens' principle tells that the slits should be arranged so that all waves arrive at P exactly in phase. The spherical wave emanating at P_0 reaches each slit; from each point of the slit now emanates a new spherical wave which is in phase with the incident wave. The total phase difference between the point P_0 and P for a wave going through the slit at P_1 is then

$$k(d_1 + d_2) = 2kd$$

because $d_1 = d_2 = d$; the screen is halfway between P_0 and P. We will have a maximum if

$$\cos(2kd - \omega t)$$

has the same value for all rings. Therefore, the phase difference has to be an integer multiple of 2π:

$$k(2d_i) - k(2d_j) = n2\pi \tag{3-57}$$

where n is an integer, and d_i and d_j are the distances from P (or P_0) to two different rings. Since $k = 2\pi/\lambda$, we can write Eq. (3-57) as

$$2d_i - 2d_j = n\lambda$$

for any combination i and j. However, the first "ring" is a hole through the center. Thus,

$$2d_0 = 2L \tag{3-58a}$$

where L is the distance from P_0 to the screen. In consequence, the rings should be arranged so that

$$2d_n = 2L + n\lambda$$

or

$$d_n = L + n\frac{\lambda}{2} \tag{3-58b}$$

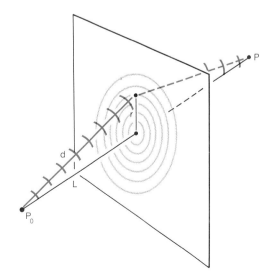

Figure 3-32 Fresnel lens.

But we have

$$d_n = \sqrt{L^2 + r_n^2}$$ (3–59)

where r_n is the radius of the nth ring. We can thus obtain all the ring radii. In order to solve for r_n, we substitute Eq. (3–59) into Eq. (3–58b) and square:

$$L^2 + r_n^2 = \left(L + n\,\frac{\lambda}{2}\right)^2 = L^2 + nL\lambda + \frac{n^2\lambda^2}{4}$$

and this enables us to obtain r_n:

$$r_n = \sqrt{n\lambda L + \frac{n^2\lambda^2}{4}}$$ (3–60)

Let us consider a numerical example, with $\lambda = 10$ cm and $L = 1$ m. Then we have for the radii

$$r_1 = 0.320 \text{ m}$$

$$r_2 = 0.458 \text{ m}$$

$$r_3 = 0.568 \text{ m}$$

.

.

.

Any other arrangement of very narrow rings will lead to a lower total intensity at P.

■   We can also ask the question: How wide should the rings be to obtain the overall maximum intensity? We can find the answer if we consider Fig. 3–33 and ask how large the phase difference ϵ between two waves can be if they are to interfere constructively, i.e., reinforce each other. Obviously, this will be the case if

$$\epsilon < \frac{\pi}{2}$$

because then the second phasor has a positive projection along the direction of the first one. We can make the same statement if we consider Fig. 3–34, the sketch of a wave in

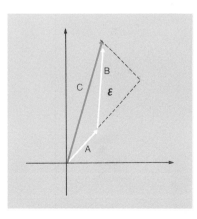

Figure 3–33 The intensity of the sum of two waves is larger than the sum of the intensities if $\epsilon < \pi/2$.

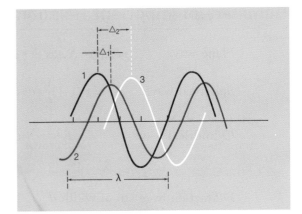

Figure 3-34 Waves 1 and 2 will interfere constructively; waves 1 and 3 interfere destructively.

real space. Two waves reinforce each other if they are shifted relative to each other by less than $\lambda/4$; by this we mean that the resulting intensity is larger than the sum of the individual intensities.

Now we have the clue to solve completely the problem of the Fresnel lens. If we denote the distance for any radius in Fig. 3-32,

$$2d(r) = 2\sqrt{L^2 + r^2}$$

then the interference with the wave passing through the center of the screen will be

constructive for $\qquad 2L \leqslant 2d(r) \leqslant 2L + \dfrac{\lambda}{4}$

destructive for $\qquad 2L + \dfrac{\lambda}{4} \leqslant 2d(r) \leqslant 2L + \dfrac{3}{4}\lambda$

constructive for $\qquad 2L + \dfrac{3}{4}\lambda \leqslant 2d(r) \leqslant 2L + \dfrac{5}{4}\lambda$, etc.

In Fig. 3-35 we show in color the range of $d(r)$ for which a wave with a total path length from P_0 to P of $2d(r)$ will lead to constructive interference. Thus, if we want a maximum intensity at P, the screen should be open wherever $d(r)$ falls in that range and should be closed wherever it does not. Using Eq. (3-59), one can then easily calculate the rings which one should cut into the screen; but we will leave this as an exercise to the reader.

We also stress that we have here treated only one special case of the Fresnel lens, where the source and the detector are equidistant from the screen. This is not necessary; one can construct a Fresnel lens which will give a maximum at any desired distance from the screen. It is also not necessary that the smallest ring be a hole through the center of the opaque screen; one can start with a ring of any radius and then ask for all rings which will reinforce the wave passing through the first ring. ■

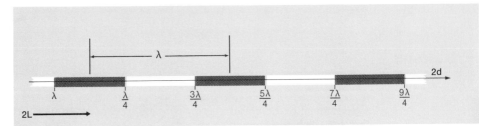

Figure 3-35 Construction of Fresnel lens with rings of optimum width.

Summary of Important Relations

Plane wave $\qquad u(x,t) = A \cos (\mathbf{k} \bullet \mathbf{r} - \omega t)$

Spherical wave $\qquad u(x,t) = \dfrac{A}{r} \cos (kr - \omega t), \quad r = \sqrt{x^2 + y^2 + z^2}$

Cylindrical wave $\qquad u(x,t) = \dfrac{A}{\sqrt{\rho}} \cos (k\rho - \omega t), \quad \rho = \sqrt{x^2 + y^2}$

Diffraction from slit of width d $\qquad I = I_0 \dfrac{\sin^2 \left[\dfrac{kd \sin \theta}{2} \right]}{\left[\dfrac{kd \sin \theta}{2} \right]^2}$

First minimum for circular slit $\qquad \theta_0 = \dfrac{1.22\pi}{kR}$

Diffraction by gratings: maxima at $\quad \sin \theta = \dfrac{\lambda}{d} n, \quad n = 1,2,3, \ldots$

Bragg diffraction $\qquad \sin \theta = \dfrac{\lambda}{2d} n$

Questions

1. How would you experimentally demonstrate Huygens' principle? Babinet's principle?

2. On clear but humid nights one sometimes sees the moon surrounded by a lighted circle of roughly 2 or 3 times the apparent moon diameter. Explain its origin.

3. State more precisely the condition for observing pure Fraunhofer diffraction given on p. 72. Is $L >> d$ really sufficient? Compare also Problem 17.

4. Thin oil films on water or soap bubbles seem to consist of vividly colored stripes or spots which change color if you move your head. Explain the phenomenon.

Problems

1. A large stone is thrown into a lake. The circular waves produced on the surface have a peak-to-valley height of 30 cm, 10 meters away. What will be the wave height at the shore, which is one mile away?

2. A long row of fluorescent lights consists of 48-inch long tubes, each emitting 100

watts of light power. An 8-inch by 11-inch sheet of paper is held 10 feet away with its face toward the lights. What is the total light power hitting the sheet?

°3. A 3-ft. by 3-ft. sheet of paper is held 2 feet away from the row of lights from the previous example, as shown in the figure. Calculate the total light power arriving at the sheet.

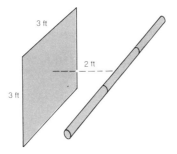

4. Two plane waves of equal wavelength λ and equal intensity cross each other at 45°. Calculate the resulting wave patterns; in particular, find all points where there is no wave motion at all.

5. Two point sources of water waves of wavelength λ are a distance d apart. How large must d be if there are to be exactly 14 directions along which the intensity vanishes?

6. Two loudspeakers, A and B, are emitting sound of wavelength $\lambda = 2$ m; they are separated by a distance $d = 6$ m. Where along the line aa' (see figure) are there minima of sound intensity?

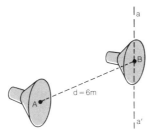

7. Two point sources of water waves of wavelength λ are a distance $d = \lambda$ apart; the two sources are 180° out of phase. Determine the directions of maximum and minimum emission.

8. Two point sources of water waves of wavelength λ are a distance d apart. What are the possible values of d if there is a minimum at $\theta = 45°$, as in the figure? The two sources are in phase.

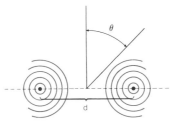

9. Two point sources, A and B, of water waves of wavelength $\lambda = 1$ m are a distance 11.5 m apart. Where on a line 20 meters away from the two sources will there be

minima? Give the distance x from the point P in the figure. The two sources are in phase.

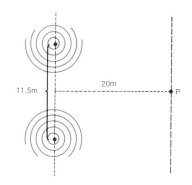

°10. Two point sources, A and B, of sound waves ($\lambda = 1$ m) are distance 11.5 m apart. They are emitting in phase. Find all points on a screen 20 meters away from both sources for which the sound intensity is a minimum. What curve do these points lie on?

11. A laser beam (wavelength $\lambda = 6330$ Å) of 1 mm diameter is aimed at the moon. At least how large will be the size of the beam when it hits the moon? (The distance from earth to moon is 380,000 km.)

12. A deep bay is completely enclosed except for an opening $d = 20$ meters wide. The water waves on the sea outside have a typical wavelength $\lambda = 1$ m and are moving along the normal to the seashore. Mr. Seasick, who wants to fish from a small boat inside the bay, asks you for advice about where the water will be quietest. Where should he position his boat?

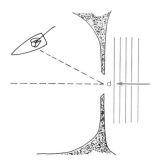

13. A well-insulated room has two open, narrow windows. There is a siren sounding off every half hour, 1000 meters away from the room. Where along the dotted line in the figure should you sit to be least bothered by the noise? The sound of the siren is strongest at a frequency $\nu = 1000$ Hz. The window separation is 1 meter.

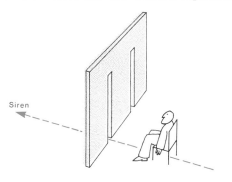

14. Neutrons (mass $= 1.7 \times 10^{-27}$ kg) of 50 MeV kinetic energy are scattered by carbon nuclei, which have a radius of $r \approx 3 \times 10^{-15}$ m. To a good approximation one can consider the carbon nucleus as a black object, which completely absorbs the neutrons. At what angle θ_0 will the first diffraction minimum lie? The typical angular distribution of neutrons scattered from complex nuclei is shown in the figure. (1 MeV $= 1.6 \times 10^{-13}$ Joule.)

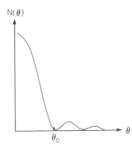

15. Sunlight (average wavelength 5500 Å) is entering a darkened hut through a pinhole of 0.5 mm diameter in the roof. The roof is 2.5 meters (~8 feet) above the floor. How large is the lighted spot on the floor, if the sun is directly overhead?

16. Three narrow slits in a copper screen are separated by a distance $d = 4$ cm. Microwaves of 3 mm wavelength are impinging normally on the screen. Calculate the intensity distribution $I(\theta)$ measured behind the screen.

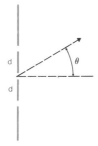

17. A parallel light beam is to be narrowed using a slit, so that the lighted spot on a screen 1 meter away from the slit is narrowest. What size slit would you choose? Estimate the beam spot size you would obtain in this way. Assume a wavelength $\lambda \approx 5000$ m.

18. A diffraction grating has an unknown number N of lines per centimeter. The yellow sodium line ($\lambda = 5900$ Å) shows a diffraction maximum at $\theta = 45°$. What are the possible values for N? How could you, using the same yellow sodium light, quickly decide which value for N is correct?

19. For a certain diffraction grating the second order diffraction maximum for blue light ($\lambda = 4000$ Å) appears at $\theta = 30°$; where will you expect the first order maximum for yellow light ($\lambda = 5900$ Å) to appear?

20. Assume that you want a diffraction grating which will have a maximum for $\lambda = 5000$ Å and for $\lambda = 6000$ Å in the same direction. How many lines per centimeter should the grating have? Are there several possible solutions?

°21. The yellow sodium line consists actually of two lines of wavelengths 5890 Å and 5896 Å. How many total lines should a diffraction grating have (at least) if one wants to be able to separate the two sodium lines? Comment also on what we mean exactly by "how many lines are required." Is this a statement only about the grating or also about the light incident on the grating?

22. In what direction θ will lie the maxima produced by a diffraction grating, if the grating itself is inclined (see figure) by the angle θ_0 relative to the incident wave fronts? Assume that all angles are small: $\theta \approx \sin \theta \approx \tan \theta$.

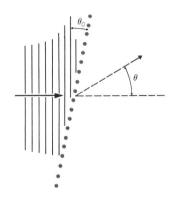

23. Due to a manufacturing fault, a grating with N lines/cm has every 10th line missing. In a perfect grating of N lines/cm, the yellow sodium line has its first diffraction maximum at $\theta = 30°$. (a) Calculate the number N of lines per centimeter. (b) Where will additional maxima appear in the faulty grating? Hint: Use Babinet's principle. The yellow sodium line has $\lambda = 5890$ Å.

24. A crystal is mounted in a goniometer (an apparatus that allows it to be turned around its axis). Using X-rays of wavelength 0.14 Å, you observe maxima when the reading on the goniometer is 12.5°, 14.6°, and 16.7°. What is the spacing between the crystal planes? Note that you do not know how the crystal is oriented in the goniometer.

°25. *Linear Fresnel lens.* A line source S of radiation of wavelength λ is a distance d_1 away from a screen into which slits can be cut. At a point P, a distance d_2 on the other side, is a detector. Assuming all slits to be narrow, where should the slits be cut to achieve maximum intensity at P? Assume that one slit is at $x = 0$, on the line connecting S and P. Assume specifically $\lambda = 3$ m (radio waves), $d_1 = 10$ m, and $d_2 = 20$ m.

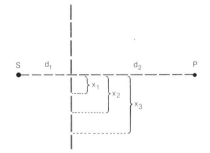

CHAPTER 4

REFLECTION AND REFRACTION

1. SNELL'S LAW

The velocity of a wave is in general a property of the medium in which it travels. Thus, for water waves in shallow water, it depends on the water depth; the sound velocity in a gas depends on the temperature and the molecular composition of the medium. Even the velocity of light is not the same inside a medium as it is in vacuum. In Chapter 1 we said that the refractive index n of a transparent medium is defined as the ratio of the velocity of light in the medium, v, to the velocity of light in vacuum, c:

$$v = \frac{c}{n} \tag{4-1}$$

We should also remember that for many types of waves the velocity, in addition to being dependent on the properties of the medium, also depends on the wavelength of the wave. This is true for water waves, as can be seen in Eq. (1–27). The same is true for light; although the velocity of light *in vacuum* is independent of the wavelength and equal to $c = 3 \times 10^8$ m/sec, the velocity of light *in a medium* and thus the refractive index varies with wavelength. We should note that by "light" we mean not only visible light, but the whole spectrum of electromagnetic waves; radio waves, light, and X-rays are all the same phenomenon. They differ only by their wavelength (see Fig. 4–1), which is macroscopic (a few centimeters to one meter) for radio waves and microscopic (0.001 to 0.1 mm) for infrared waves. Visible light has wavelengths of 4000 to 7000 Å = 0.0004 to 0.0007 mm. Shorter wavelengths are called ultraviolet, while we call the same radiation X-rays when the wavelength is as short as 0.1 to 10 Å.

We will now study what happens to waves when they arrive at the *interface* between two media with different wave velocity. If there are two homogeneous media, there exists a surface between the two where the wave velocity changes abruptly; this we call the interface. For an experimental study we can use either water waves or light waves. In water waves we can vary the wave speed by changing the water depth; for light waves we can find substances with different refractive indices. We use the two types of waves because we can easily see water waves or light waves, but anything we say must be true also for sound waves in solids or gases, or for any other type of waves. Experimentally, we observe that if a plane wave arrives at such an interface, some of the wave will be reflected back into the original medium, but a plane wave will also appear in

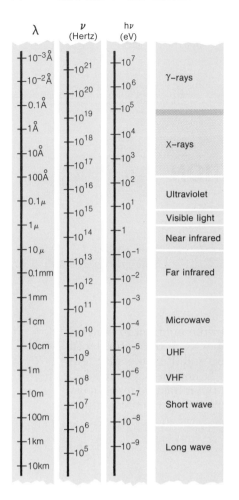

Figure 4-1 The range of wavelengths, frequencies and photon energies of electromagnetic waves.

the second medium; however, it travels in a different direction than does the incident wave. We will call the three waves the incident, the reflected, and the refracted waves, respectively (Fig. 4–2a).

Because of energy conservation, we always have

$$I \text{ (incident)} = I \text{ (reflected)} + I \text{ (refracted)} \qquad (4\text{–}2)$$

i.e., the incident intensity is either reflected or refracted, but cannot disappear. Eq. (4–2) is not sufficient to determine the reflected and refracted intensities separately; we would also have to know how the outgoing intensity is shared between the reflected and refracted waves. This sharing is not the same for waves of different types; it depends on the detailed boundary conditions on the interface between the two media. However, the direction in which the reflected and the refracted waves propagate can be determined from Huygens' principle. Let us consider first the reflected wave. We can consider it to be built up from spherical waves originating on all points of the interface. At each point of the interface these spherical waves have to be in phase with the incident wave *for all times*. This cannot happen unless the reflected wave has the same frequency ν as the incident wave (Fig. 4–2b). Since the reflected wave travels in the same medium as the incident wave, and thus has the same wave velocity, it will have the same wavelength

$$\lambda_i = \lambda_r \qquad\qquad (4\text{–}3\text{a})$$

and the same absolute value for the wave vector

$$|\mathbf{k}_i| = |\mathbf{k}_r| = \frac{2\pi}{\lambda} = \frac{2\pi\nu}{\upsilon} \qquad\qquad (4\text{–}3\text{b})$$

If we consider the lines drawn in Fig. 4–3 as the wave crests (maxima) we see that the indicated distance along the incident wave is

$$BB' = \lambda$$

On the other hand, the wavelength of the reflected wave is also λ; thus,

$$A'A = \lambda = BB'$$

But then we also have

$$\sin \theta_i = \frac{BB'}{AB} = \frac{AA'}{AB} = \sin \theta_r$$

Therefore, the two angles θ_i and θ_r must be equal. The reflected wave fronts

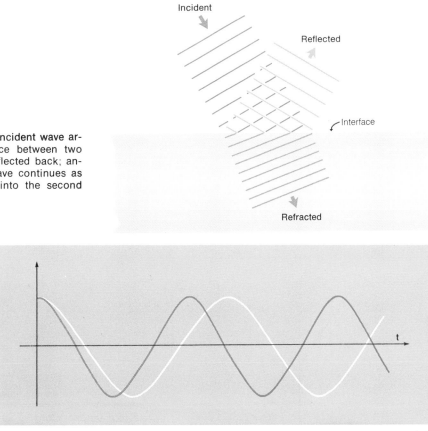

Figure 4–2 a. The incident wave arriving at an interface between two media is partially reflected back; another part of the wave continues as the refracted wave into the second medium.

Figure 4–2 b. Two waves of different frequencies cannot stay in phase over an extended period of time.

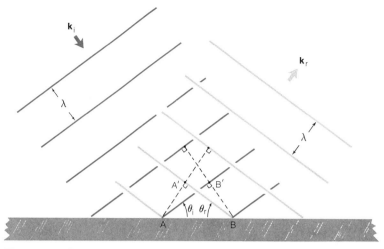

Figure 4-3 Calculation of angle of reflection.

form the same angle with the interface as the incident wave. We usually state this law in the following way: The wave vectors of the incident wave and of the reflected wave form the same angle θ with the normal to the interface between two media. As can be seen from Fig. 4–4, the angle θ between the wave vector (normal to the wave fronts) and the normal to the interface is identical to the angle θ between the wave front and the interface. We can state the law of reflection in the form of an equation. If $\mathbf{n}$ is a unit vector perpendicular to the interface, then

$$(\mathbf{k}_i \cdot \mathbf{n}) = -(\mathbf{k}_r \cdot \mathbf{n}) \tag{4-4}$$

The minus sign in Eq. (4–4) expresses the fact that θ_i is the angle between $\mathbf{k}_i$ and the down-pointing normal in Fig. 4–4, while θ_r is the angle between $-\mathbf{k}_r$ and the down-pointing normal. Eq. (4–4) is then correct if we choose the same direction (down or up) for both scalar products.

We can treat the refracted wave in a similar way. Again because of Huygens' principle, the refracted wave can be considered a superposition of spherical waves originating everywhere on the surface. However, while its frequency has to be the same as that of the incident wave,

$$\nu_2 = \nu_1 = \nu$$

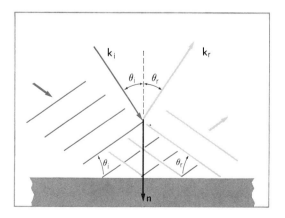

Figure 4-4 The angles between wave fronts and interface are the same as the angles between wave vectors and the normal to the interface.

the wavelength will be different because of the different wave velocities in the two media. We have

$$\lambda_2 = \frac{v_2}{\nu} \tag{4-5a}$$

which cannot be the same as

$$\lambda_1 = \frac{v_1}{\nu} \tag{4-5b}$$

Instead, we have for the ratio of wavelengths

$$\frac{\lambda_2}{\lambda_1} = \frac{v_2}{v_1} \tag{4-6}$$

Let us now consider the lines in Fig. 4–5 as wave crests of the harmonic wave. Then the distance AB is equal to two wavelengths in medium 1:

$$AB = 2\lambda_1$$

Similarly, we have

$$CD = 2\lambda_2$$

Thus, because of Eq. (4–6), the two distances have a well-defined ratio

$$\frac{AB}{CD} = \frac{\lambda_1}{\lambda_2} = \frac{v_1}{v_2} \tag{4-7}$$

But we have another connection between the two distances. From Fig. 4–5 we can read

$$AB = CB \sin \theta_1 \tag{4-8a}$$

and

$$CD = CB \sin \theta_2 \tag{4-8b}$$

Dividing Eq. (4–8a) by Eq. (4–8b), we then obtain

$$\frac{AB}{CD} = \frac{\sin \theta_1}{\sin \theta_2}$$

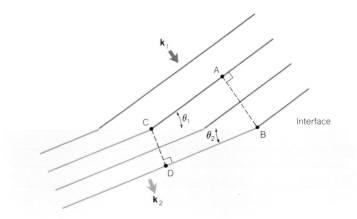

Figure 4–5 Derivation of Snell's law.

which because of Eq. (4–7) can be written

$$\frac{\sin \theta_1}{\sin \theta_2} = \frac{v_1}{v_2} \qquad (4\text{–}8)$$

Note that again we can consider θ_1 and θ_2 as either the angles between the wave fronts and the interface or as angles between the wave vectors (directions of propagation) and the normal to the interface (Fig. 4–6). Eq. (4–8) is known as *Snell's* [*] *law*. In optics, we define the refractive index n of a medium as the ratio between the velocity of light in vacuum, c, and the velocity of light in the medium, as in Eq. (4–1):

$$n = \frac{c}{v}$$

Thus, for light we can also write Snell's law as

$$\frac{\sin \theta_1}{\sin \theta_2} = \frac{c/n_1}{c/n_2}$$

or

$$n_1 \sin \theta_1 = n_2 \sin \theta_2 \qquad (4\text{–}9)$$

If the wave velocity in the second medium is lower, the wave direction will move toward the normal to the interface. (Note that smaller velocity for light means larger refractive index.) For air, we can mostly use for refractive index the value $n = 1$; actually, the exact value at normal density is

$$n_{\text{air}} = 1.00028$$

---

[*] Willebrord Snell van Royen (1591–1626), Dutch mathematician and scientist, was professor of mathematics and Hebrew in Leyden, 1601–1615. His contributions to optics were never published; we know of them mainly through Huygens' works.

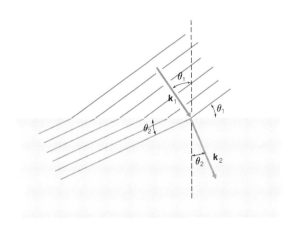

Figure 4–6 The wave vectors and wave fronts of the incident and refracted wave.

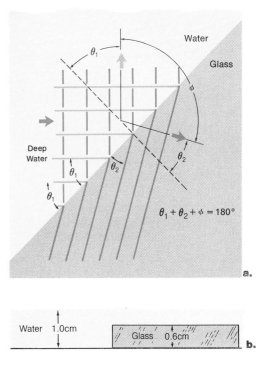

Figure 4–7   Example 1: a. Wave fronts. **b.** The physical arrangement. The surface waves above the glass plate have a lower velocity, because the water is shallower there.

Example 1   In a flat shallow tank the water depth is 1 cm. A 0.6 cm thick glass plate covers part of the bottom of the tank; its edge is inclined at 45° with respect to the direction of an incident water wave. What will be the angle between the reflected wave and the refracted wave?

As we had seen in Eq. (1–29), for shallow water the wave velocity depends on water depth

$$v = \sqrt{gh}$$

The depth is $h_1 = 1$ cm where there is no glass plate, while above the glass plate the water depth is

$$h_2 = 1 \text{ cm} - 0.6 \text{ cm} = 0.4 \text{ cm}$$

The ratio of wave velocities is

$$\frac{v_1}{v_2} = \sqrt{\frac{h_1}{h_2}} = \sqrt{2.5} = 1.58$$

From Snell's law we then have

$$\frac{\sin \theta_1}{\sin \theta_2} = \frac{v_1}{v_2} = 1.58 \qquad (4\text{–}10\text{a})$$

We are also told that $\theta_1 = 45° = \pi/4$. However we are asked to calculate not the angle θ_2 between the normal and the refracted wave, but the angle between the reflected and refracted wave directions. From Fig. 4–7a we see that

$$\phi = 180° - \theta_1 - \theta_2 \qquad (4\text{–}10\text{b})$$

Since $\sin (45°) = \sqrt{2}/2 = 0.707$ we obtain from Eq. (4–10a)

$$\sin \theta_2 = \frac{\sin \theta_1}{1.58} = 0.447$$

and from a trigonometric table we read off

$$\theta_2 = 26.6°$$

In consequence, we have

$$\phi = +180° - 45° - 26.6° = 108.4°$$

Example 2   A piece of glass has a refractive index $n = 1.6$. What should be the angle of an incident light beam, if the reflected and the refracted beam are to form an angle of 90° relative to each other? We have drawn in Fig. 4–8 the wave vectors of the incident, reflected and refracted waves, as well as all relevant angles.

Because of Eq. (4–4) we have

$$\theta_3 = \theta_1$$

and because of Snell's law [Eq. (4–9)] we have

$$\frac{\sin \theta_1}{\sin \theta_2} = \frac{v_1}{v_2} = n$$

Since $\mathbf{k}_3$ and $\mathbf{k}_2$ are to be perpendicular to each other, we can read from Fig. 4–8

$$180° = \theta_3 + 90° + \theta_2$$

or, since $\theta_3 = \theta_1$,

$$\theta_1 + \theta_2 = 90°$$

Since $\sin (90° - \alpha) = \cos \alpha$ for any α, we have

$$n = \frac{\sin \theta_1}{\sin \theta_2} = \frac{\sin \theta_1}{\sin (90° - \theta_1)} = \frac{\sin \theta_1}{\cos \theta_1}$$

or

$$\tan \theta_1 = n = 1.6$$

and from a table of tangents we find

$$\theta_1 = 58°$$

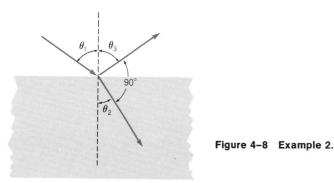

Figure 4–8   Example 2.

2. TOTAL REFLECTION

As we said earlier, there exist no general laws applicable to all waves and capable of predicting how the incident wave intensity is shared between the reflected and the refracted waves. We are familiar with mirrors which reflect all (or nearly all) of the incident light; water waves will also be completely re-

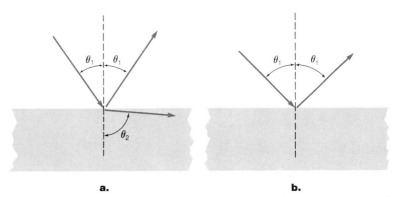

Figure 4-9 An incident wave at slightly smaller and slightly larger angles than the minimum angle required for total reflection.

flected by a vertical wall. These are special cases from which one cannot construct a general theory.

However, we can achieve total reflection under certain conditions for any type of wave if we let it pass from a medium of lower wave velocity to one of higher velocity. From Snell's law

$$\frac{\sin \theta_1}{\sin \theta_2} = \frac{v_1}{v_2}$$

we conclude that if $v_2 > v_1$, then

$$\sin \theta_2 = \frac{v_2}{v_1} \sin \theta_1 > \sin \theta_1$$

However, the sine of any angle cannot be larger than unity. If

$$\frac{v_2}{v_1} \sin \theta_1 = 1$$

then the refracted wave is parallel to the interface (Fig. 4–9a). Any further increase of incident angle, so that

$$\boxed{\frac{v_2}{v_1} \sin \theta_1 > 1} \qquad\qquad (4\text{--}11)$$

has no solution for θ_2; there exists no angle θ_2 for which Snell's law is obeyed. Thus, no refracted wave can exist and the total intensity is reflected back (Fig. 4–9b). We call Eq. (4–11) the condition of total reflection. In optics, where one usually uses the refractive index, one can write the condition of total reflection as

$$\boxed{\frac{n_1}{n_2} \sin \theta_1 > 1} \qquad\qquad (4\text{--}12)$$

Example 3   A glass prism can be used to reflect a light beam back into the original direction, if the condition for total reflection is satisfied at both points P_1

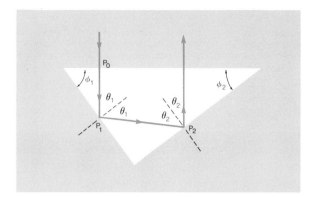

Figure 4-10   Example 3.

and P_2 (Fig. 4–10). What should be the refractive index of the glass, and what is the range of angles ϕ_1 and ϕ_2 for which the prism will work in such a manner?

We assume that the light is normally incident to the glass surface at P_0; however, we are allowed to use an asymmetric prism for which $\phi_1 \neq \phi_2$. From Fig. 4–10 we can see that $\phi_1 = \theta_1$ and $\phi_2 = \theta_2$. Since the beam should be reflected into the original direction, we need

$$2\theta_1 + 2\theta_2 = 180°$$

and thus we need for the prism

$$\phi_1 + \phi_2 = 90° \tag{4–13}$$

We also require that the condition of total reflection be satisfied both at P_1 and P_2. Since the refractive index of air is $n_{air} = 1$ we require not only

$$\sin \phi_1 > \frac{1}{n} \tag{4–14a}$$

but also

$$\sin \phi_2 > \frac{1}{n} \tag{4–14b}$$

We always assume that $\phi_1 > \phi_2$, since in Fig. 4–10 it is arbitrary whether the light follows the path in the direction of the arrows or against them. However, if $\phi_1 > \phi_2$ then also $\sin \phi_1 > \sin \phi_2$ (both ϕ_1 and ϕ_2 are less than 90°). Thus, if Eq. (4–14b) is satisfied, so is Eq. (4–14a). Because of Eq. (4–13) we can also write, instead of Eq. (4–14b),

$$\cos \phi_1 > \frac{1}{n}$$

Since we assume $\phi_1 > \phi_2$, because of Eq. (4–13) we have $\phi_2 < 45°$. The lowest value of n for which all equations are satisfied is then

$$n = \frac{1}{\sin 45°} = \sqrt{2} = 1.41 \tag{4–15}$$

which implies $\phi_1 = \phi_2 = 45°$. If n is larger than $\sqrt{2}$, a range of ϕ_2—the smaller of the two angles—is possible:

$$\sin^{-1}\left(\frac{1}{n}\right) < \phi_2 < 45°$$

and, of course, we must always have $\phi_1 = 90° - \phi_2$.

Example 4   A diver is lying on his back 5 meters under a quiet water surface. If he looks straight up, he can see the sky. Looking sidewards, he will see the
reflection of the bottom of the pool.
  a. How far must a pebble on the bottom be from his eyes for him to see its
     reflection?
  b. How much of the sky does he see?
The refractive index of water is $n = 1.33$.

We can answer the second question first: The diver will see the whole sky!
But he will see it compressed into a cone of angle θ (Fig. 4–11). Thus, he will see
a bird, which in reality is flying just above the water at A; however, the bird will
seem to him to be at B.
  We can also calculate the angle θ—it is the minimum angle of total reflection.
Thus

$$\sin \theta = \frac{1}{n} = \frac{1}{1.33} = 0.752$$

or

$$\theta = 48.8°$$

In order to see the reflection of a pebble on the bottom, he has to view it at
an angle larger than θ. Thus, he will not see a pebble at P_1, because light is only
partially reflected at P_1' and the reflected intensity is much less than the direct sunlight coming in from the sky. On the other hand, a pebble at P_2 will be clearly
visible to the diver. There is no light coming from the sky from the direction of
P_2', and the light from the pebble is totally reflected from the surface. The distance
$P_0 P_2$ is given by

$$P_0 P_2 = 2d \tan \theta'$$

The angle θ' must be larger than the angle of the cone of visibility. Thus

$$P_0 P_2 > 2d \tan \theta$$

where

$$\tan \theta = \frac{\sin \theta}{\cos \theta} = \frac{1/n}{\sqrt{1 - (1/n^2)}} = \frac{1}{\sqrt{n^2 - 1}}$$

Finally, since the depth of the pool is $d = 5$ m,

$$P_0 P_2 > \frac{2d}{\sqrt{n^2 - 1}} = \frac{10 \text{ m}}{\sqrt{1.77 - 1}} = 11.4 \text{ meters}$$

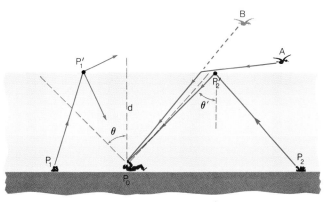

Figure 4–11 **Example 4:** A diver at the bottom of a pool.

3. FERMAT'S PRINCIPLE

We are all familiar with light beams or light rays, such as a laser beam or the light we see coming into a dark room through a keyhole from the neighboring lighted room. We can define a ray as a plane wave whose wave fronts are of finite size (Fig. 4–12). We can imagine the ray being formed by a circular hole in a screen. Of course, the diameter of the hole has to be sufficiently large so that diffraction effects can be neglected. Since diffraction effects are always present, the concept of a ray and of the direction in which it propagates is always only of approximate validity. However, it is a very useful concept in optics, where the wavelength is small compared to the size of everyday objects. It is much less useful for sound in air, for example, where the wavelength is typically of the order of 0.1 to 10 meters. According to Eq. (3–37), the first diffraction minimum occurs at

$$\theta \approx \frac{\lambda}{d} \qquad\qquad (4\text{–}16)$$

where d is the "diameter of the light beam." If we want the direction of the beam defined to 1°, we need a diameter of

$$d > \frac{\lambda}{\theta} = \frac{\lambda}{0.0175} \approx 57\lambda$$

For visible light with $\lambda = 6000$ Å, this implies $d > 3.4 \times 10^{-4}$ m $= 0.34$ mm, a quite small value. For sound waves of wavelength 1 meter, however, we have $d \gtrsim 60$ m. Thus, it is nonsense to talk about "sound rays" propagating in a room – but quite reasonable if we discuss sound being reflected from a large mountain cliff when we study the echo effect.

In the "ray approximation" – also called the "geometric approximation" – we can then consider a point source of a spherical wave as a source of rays emanating in all directions, while a plane wave would consist of parallel rays (Fig. 4–13). In a homogeneous medium, where the wave velocity is everywhere the same, a ray will propagate along a straight line. Since the straight line is the shortest connection between two points, we can also say that in a homogeneous medium light rays travel between two points P_1 and P_2 along a line requiring the minimum transit time

$$\Delta t = \frac{P_1 P_2}{v} \qquad\qquad (4\text{–}17)$$

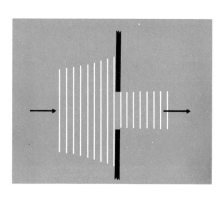

Figure 4–12 A light ray can be produced by a small hole; however, the hole has to be much larger than the wavelength to minimize diffraction.

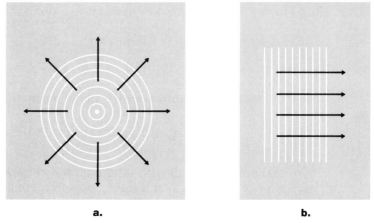

Figure 4-13 A spherical or plane wave can be considered a beam of diverging or parallel rays, as long as diffraction effects can be neglected.

where v is the wave velocity. *Fermat's° principle* is a generalization of this rather trivial statement. It asserts that *any light ray will travel between two end points along a path which is shorter than any other nearby equivalent path.*

We will first show that Fermat's principle is satisfied in refraction. When a ray passes from the point P_1 to the point P_2 (Fig. 4-14) so that it has to cross an interface between two media, the path has to obey Snell's law

$$\frac{\sin \theta_1}{\sin \theta_2} = \frac{v_1}{v_2}$$

We have to show that Snell's law indeed leads to the shortest transit time between the two points P_1 and P_2. The total time the ray needs to travel is

$$\Delta t = \frac{P_1 A}{v_1} + \frac{AP_2}{v_2} \tag{4-18}$$

Let us for the moment assume that θ_1 and θ_2 do not necessarily obey Snell's

---

°Pierre de Fermat (1601–1665), French mathematician and lawyer; founder of the theory of numbers. All his discoveries were communicated in letters to friends, frequently without detailed proofs.

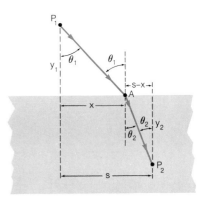

Figure 4-14 Fermat's principle for a refracted ray.

law. However, using the coordinate system defined in Fig. 4–14, we have

$$P_1A = \sqrt{x^2 + y_1^2}$$

and

$$P_2A = \sqrt{(s - x)^2 + y_2^2}$$

Thus, the total time required for the ray to go from P_1 to P_2 is

$$\Delta t = \frac{\sqrt{x^2 + y_1^2}}{v_1} + \frac{\sqrt{(s - x)^2 + y_2^2}}{v_2} \tag{4-19}$$

If we search for the path with minimum travel time, we can vary only the distance x, since the other quantities, y_1, y_2, and s, are given by the positions of the two end points P_1 and P_2. Thus, we search for the value x for which Δt becomes a minimum. We know that at that point

$$\frac{d(\Delta t)}{dx} = 0$$

Differentiating Eq. (4–19) explicitly, we obtain

$$\frac{1}{v_1} \frac{x}{\sqrt{x^2 + y_1^2}} - \frac{1}{v_2} \frac{(s - x)}{\sqrt{(s - x)^2 + y_2^2}} = 0 \tag{4-20}$$

But from Fig. 4–14 we can also deduce that

$$\sin \theta_1 = \frac{x}{\sqrt{x^2 + y_1^2}}$$

and

$$\sin \theta_2 = \frac{(s - x)}{\sqrt{(s - x)^2 + y_2^2}}$$

The shortest transit time will occur exactly if Snell's law is obeyed:

$$\frac{1}{v_1} \sin \theta_1 = \frac{1}{v_2} \sin \theta_2$$

■   We can also show that Fermat's principle is true for reflected rays. Let us consider (Fig. 4–15) two points P_1 and P_2 which lie on the same side of an interface between two media (or alternately, both lie in front of a mirror). In what sense can we consider the

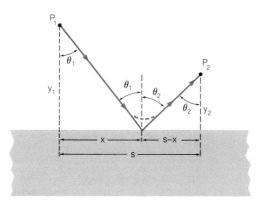

Figure 4–15 Fermat's principle for a reflected ray.

light ray reflected from the interface the "shortest path"? Obviously, the straight line between the two points P_1 and P_2 is shorter than the reflected ray. The dotted path is also obviously shorter. However, let us consider as "equivalent" only paths which go from P_1 to a point A on the interface and from there to P_2. Then, since the wave velocity is everywhere the same we have

$$\Delta t = \frac{P_1A + AP_2}{v}$$

But again from geometry we see that

$$P_1A = \sqrt{x^2 + y_1^2}$$

and

$$AP_2 = \sqrt{(s - x)^2 + y_2^2}$$

and thus the total time difference can be written as

$$\Delta t = \frac{1}{v}[\sqrt{x^2 + y_1^2} + \sqrt{(s - x)^2 + y_2^2}]$$

The minimum will occur when

$$\frac{d(\Delta t)}{dx} = \frac{1}{v}\frac{d}{dx}[\sqrt{x^2 + y_1^2} + \sqrt{(s - x)^2 + y_2^2}] = 0$$

or when

$$\frac{x}{\sqrt{x^2 + y_1^2}} - \frac{s - x}{\sqrt{(s - x)^2 + y_2^2}} = 0$$

However, from Fig. 4–15 we can also see that

$$\sin \theta_1 = \frac{x}{\sqrt{x^2 + y_1^2}}$$

and

$$\sin \theta_2 = \frac{s - x}{\sqrt{(s - x)^2 + y_2^2}}$$

Thus, the transit time will be minimal if

$$\sin \theta_1 = \sin \theta_2$$

or

$$\theta_1 = \theta_2$$

which is exactly the law of reflection.

We stress that in using Fermat's principle one has to be careful to use the correct meaning of "equivalent neighboring path." We can allow only paths which are nearly the same as the real path, and also which have the same basic characteristics, e.g., touch the same interface or mirror surface. ◧

4. LENSES

The art of influencing light rays by lenses is the heart of the usefulness of geometrical optics. We are all familiar with magnifying glasses; many of us wear eyeglasses to correct defects of the eye; the eye itself has a built-in lens made of transparent tissue.

Within the limits of this text, we are unable to discuss all details of lens theory; but we can present a somewhat simplified theory which yields all the main results of the more detailed theory. Let us first consider a glass prism of angle α, as shown in Fig. 4–16. We want to calculate the angle γ by which the

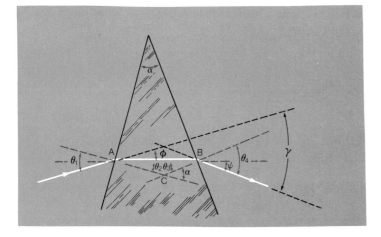

Figure 4-16 Calculating the bending angle γ of a ray passing through a prism.

incident light ray is bent by the prism. However, we do this only for the special case in which both the angle α and the angle θ_1 are sufficiently small that we can use the approximation

$$\sin \theta \approx \theta$$

We replace sines of angles by their arguments. From Fig. 4–16 we can read off a few simple relations. We call θ_1 and θ_2 the angles of the incident and refracted light ray relative to the normal to the glass surface at the point A; θ_3 and θ_4 are the same angles relative to the normal at B. We see that

$$\gamma = \phi + \psi = (\theta_1 - \theta_2) + (\theta_4 - \theta_3) \tag{4-21}$$

which is *an exact relation*, independent of the size of the angles. Another exact relation is given by the sum of the three angles in the triangle ABC. We have

$$\theta_2 + \theta_3 + (180° - \alpha) = 180°$$

or

$$\alpha = \theta_2 + \theta_3 \tag{4-22}$$

(The simple geometric proof that the two angles marked α are indeed equal is left to the reader.) The two equations (4–21) and (4–22) are by themselves not sufficient to determine γ; we also need to use Snell's law, which relates θ_1 and θ_2 and also θ_3 and θ_4:

$$\frac{\sin \theta_1}{\sin \theta_2} = n = \frac{\sin \theta_3}{\sin \theta_4} \tag{4-23}$$

If we now use the approximation of replacing all sines by their arguments, we have

$$\theta_1 \approx n\theta_2 \quad \text{and} \quad \theta_4 \approx n\theta_3$$

and using Eq. (4–22) we then obtain

$$\theta_1 + \theta_4 \approx n(\theta_2 + \theta_3) = n\alpha \tag{4-24}$$

Now we are ready to calculate the angle γ from Eq. (4–21):

$$\gamma = \theta_1 + \theta_4 - (\theta_2 + \theta_3) \approx (n - 1)\alpha \tag{4-25}$$

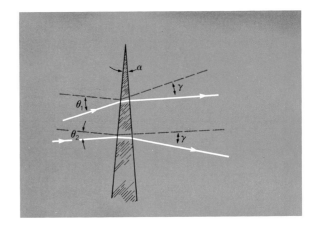

Figure 4-17 If the prism angle α and the angle of incidence θ are both small, then the bending angle is independent of θ.

and we see that the bending angle γ in our approximation is independent of the angle of incidence θ_1 (Fig. 4–17). This is a very important result which will form the basis of our discussion of lenses.

We now want to design a converging lens which has the property that all light rays incident parallel to the axis converge together at a focus F a distance f away from the center of the lens. From Fig. 4–18, we see that

$$\frac{x}{f} = \tan \gamma \tag{4-26a}$$

and if γ is again a small angle we have, using Eq. (4–25) as a requirement for the lens,

$$\alpha = \frac{1}{n-1}\frac{x}{f} \tag{4-26b}$$

Of course, x is really the distance from the axis of the lens. We have drawn in Fig. 4–18 only a cross section of the lens; the reader should imagine the whole figure rotated around the lens axis. By itself, Eq. (4–26) does not yet give us the contour of the lens; but we can now easily calculate it. Let us first assume that we want a symmetrical lens as shown in Fig. 4–19. Since $\alpha_1 = \alpha/2$, and $\tan \alpha_1 \approx \alpha_1$

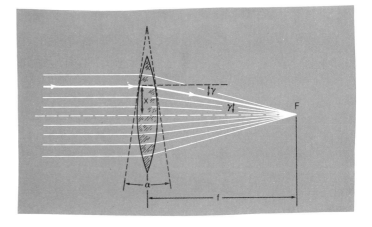

Figure 4-18 In a focusing lens, the bending angle γ is proportional to the distance x from the axis.

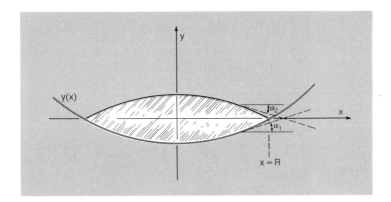

Figure 4-19 Calculating the lens shape; the curve *y(x)* is a parabola.

is the derivative of $y(x)$, we have

$$\alpha_1 = \frac{dy}{dx} = \frac{1}{(n-1)} \cdot \frac{x}{2f}$$

From this we obtain

$$y(x) = \frac{1}{(n-1)} \cdot \frac{x^2}{4f} + \text{constant}$$

as the contour of the lens. If we want the radius of the lens to be R, and if we want sharp corners (not a good choice—the lens will chip easily!), then of course $y(R) = 0$ and thus we have

$$y(x) = \frac{1}{(n-1)} \frac{x^2 - R^2}{4f} \tag{4-27}$$

Of course, there is no need to choose a symmetrical lens; another extreme would be to have one face of the lens flat. It is left to the reader to show that then the other face should lie on the curve (Fig. 4–20)

$$y(x) = \frac{1}{n-1} \frac{x^2 - R^2}{2f} \tag{4-28}$$

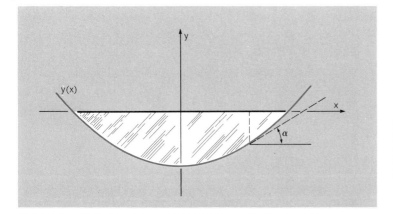

Figure 4-20 A lens does not have to be symmetrical; the curve *y(x)* is still a parabola, although a different one than in Fig. 4–19.

In either case, the surfaces are paraboloids—their cross section is a parabola. Eqs. (4–27) and (4–28) are only approximations; we are assuming that the angle α in Eq. (4–26) is always small. We can translate this, using Eq. (4–27) or (4–28), into the condition

$$f >> R \qquad (4\text{–}28a)$$

The focal length should be large compared to the size of the lens.

■ We can—instead of the paraboloidal approximation to the lens surfaces—also consider the surfaces as portions of two spheres. Indeed, consider the equation for a circle of radius r

$$y = \sqrt{r^2 - x^2}$$

If x is small ($x << r$), then we can expand the square root as

$$y = r\sqrt{1 - \frac{x^2}{r^2}} \approx r\left(1 - \frac{x^2}{2r^2}\right) = r - \frac{x^2}{2r} \qquad (4\text{–}29)$$

Thus, a circle is an approximation to a parabola—and a sphere is an approximation to an axially symmetric paraboloid. If we have two circles of radii r_1 and r_2 as shown in Fig. 4–21, then the angle α is equal to

$$\alpha = \alpha_1 + \alpha_2 = \frac{dy_1}{dx} - \frac{dy_2}{dx}$$

Note that the lower circle has its apex a distance d below the x axis and thus has its center a distance $r - d$ above the x axis. Therefore, it is given by the equation

$$(r_1 - d) - y_1 = \sqrt{r_1^2 - x^2} \approx r_1 - \frac{x^2}{2r_1}$$

or

$$y_1 \approx -d + \frac{x^2}{2r_1} \qquad (4\text{–}30a)$$

while the other circle is given by the equation

$$r_2 - d + y_1 = \sqrt{r_2^2 - x^2} \approx r_2 - \frac{x^2}{2r_2}$$

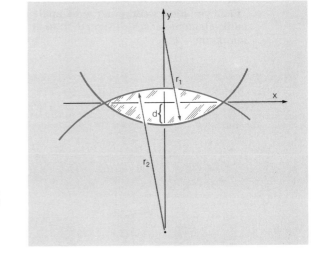

Figure 4–21 Instead of paraboloid surfaces, one can use spherical surfaces, which are easier to grind.

or

$$y_2 \approx d - \frac{x^2}{2r_2}$$ (4–30b)

From this we see that

$$\alpha = \frac{dy_1}{dx} - \frac{dy_2}{dx} \approx \frac{x}{r_1} + \frac{x}{r_2} = x\left(\frac{1}{r_1} + \frac{1}{r_2}\right)$$

and comparing this with Eq. (4–26b) we obtain the relation

$$\boxed{\frac{1}{r_1} + \frac{1}{r_2} = \frac{1}{n-1}\frac{1}{f}}$$ (4–31)

This is called the lens-maker equation. It is a reformulation of Eq. (4–26b) if we approximate the lens surfaces by spheres instead of paraboloids (Fig. 4–22). The equivalent approximation for paraboloids can be read off from Eqs. (4–30a) and (4–30b), which are exact if we consider the surfaces paraboloids. Since the focus of the parabola

$$y = \frac{ax^2}{2}$$

is a distance $d = 1/(2a)$ away from the apex, we can instead of Eq. (4–31) also write

$$\boxed{\frac{1}{d_1} + \frac{1}{d_2} = \left(\frac{2}{n-1}\right)\frac{1}{f}}$$ (4–32)

and to within the approximations used, Eq. (4–32) is completely equivalent to Eq. (4–31).

We stress that the distinction between lenses with spherical or parabolic surfaces is to us purely mathematical. We assume that the lens is small compared to its focal length; but as Eqs. (4–29) and (4–30a) show, this implies that we use an approximation in which we consider a parabola and a circle to be the same curve—we are limiting ourselves to a region near the apex where this is (approximately) true. Real lenses usually have spherical surfaces, because such a surface is easier to grind. However, it is essential that the incoming rays form small angles with the lens axis if a sharp focus is required. "Fisheye" lenses and other wide-angle lenses sometimes found on cameras are always multi-element lenses designed so as to keep the bending angles at each surface to a minimum. ∎

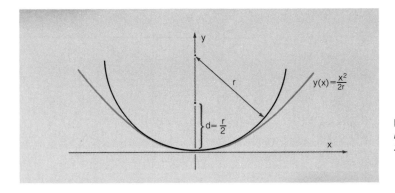

Figure 4–22 A circle of radius r approximates a parabola $y = x^2/2r$.

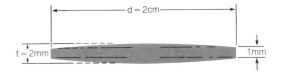

Figure 4-23 Example 5: Calculating the maximum thickness of a lens.

Example 5   What must be the thickness of a flat glass plate of refractive index $n = 1.5$ from which one is supposed to grind a symmetric converging lens of diameter $d = 2$ cm and focal length $f = 10$ cm? At the edge the lens should be at least 1 mm thick to facilitate mounting and prevent chipping (Fig. 4-23).

If the lens had to have zero thickness at its edge, we could directly use Eq. (4-27). We would have for the thickness

$$t_1 = 2|y(0)| = \frac{1}{n-1}\frac{R^2}{2f}$$

where $R = d/2$ is the radius of the lens. However, at the edge the lens has to have the thickness $t_2 = 1$ mm; thus, we have for the thickness at the center

$$t = t_1 + t_2 = \frac{1}{n-1}\frac{d^2}{8f} + t_2$$

and we obtain a minimum glass thickness

$$t = \frac{1}{0.5}\frac{4 \text{ cm}^2}{8 \times 10 \text{ cm}} + 0.1 \text{ cm} = 0.2 \text{ cm} = 2 \text{ mm}$$

Let us now discuss what happens to a parallel light beam incident on the lens, if the beam is not parallel to the lens axis, but forms an angle θ with the axis (Fig. 4-24). If the lens has the focal length f, each light beam will be bent by the angle given by Eq. (4-26a)

$$\gamma(x) \approx \frac{x}{f}$$

Thus, after passing through the lens, the beam is again converging to a point

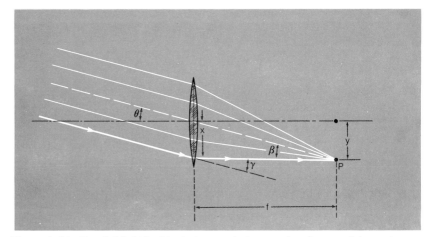

Figure 4-24 Parallel light incident at an angle θ comes to a focus a distance f away from the lens, but not on the axis of a lens.

(in the small angle approximation!), because

$$\beta = \gamma = \frac{x}{f}$$

The point P will still be a distance f away from the lens; we say that it lies in the *focal plane*. Its distance from the axis will be

$$y = f \cdot \tan \theta \approx f \cdot \theta \qquad (4\text{--}33)$$

We can now also discuss more complicated situations; let us consider a point source S of light a distance s away from the lens axis, but also a distance $a > f$ away from the lens, as shown in Fig. 4–25. Since we are consequently assuming that all angles are small, we have to assume $s \ll a$. We want to answer the question of whether—and if so, where—the light rays passing through the lens again converge to a point. We can study this by considering the point light source S as the origin of light rays going in all directions. Of course, some rays will miss the lens completely; we are interested only in those which are bent by the angle

$$\gamma \approx \frac{x}{f} \qquad (4\text{--}34)$$

by the lens material.

We know that the central ray emanating from the source S and going through the center of the lens at $x = 0$ continues undisturbed, because the bending angle $\gamma = 0$. Consider now a ray forming an angle α with the central ray; it will arrive at the lens at

$$x \approx \alpha a \qquad (4\text{--}35a)$$

and will be bent toward the central ray. The two will intersect a distance b from the lens, where

$$b \approx \frac{x}{\beta} \qquad (4\text{--}35b)$$

The bending angle γ is equal to $\gamma = \alpha + \beta$, as can be seen from Fig. 4–25. Using

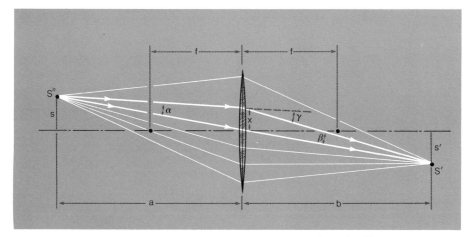

Figure 4–25   Light from a point source S converges again at the point S'.

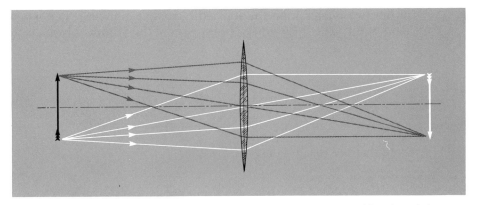

Figure 4–26 Imaging of an object; each point of the object can be considered a point source of light.

Eq. (4–34), we can write instead of Eq. (4–35b)

$$x = \beta b = (\gamma - \alpha)b = \left(\frac{x}{f} - \alpha\right)b$$

and inserting α from Eq. (4–35a) we obtain

$$x = \left(\frac{x}{f} - \frac{x}{a}\right)b$$

or

$$\boxed{\frac{1}{a} + \frac{1}{b} = \frac{1}{f}} \qquad (4\text{–}36)$$

Eq. (4–36) shows that all rays emanating from the source and being bent by the lens intersect the central ray at the same point S'. Thus, we obtain an image of the source S at S'; if we view the light coming from S, it is as if the light source was really located at S'.

The distance between the axis and the image has also a simple relation to the distance of the source from the axis; we have

$$\frac{s'}{b} = \frac{s}{a}$$

or

$$\boxed{s' = \frac{b}{a} s} \qquad (4\text{–}37)$$

If instead of a single light source S we have a lighted object (Fig. 4–26), every point of the object can be considered a point source of light. Every one of these points in the source plane will have an image in the image plane; the distances of the two planes from the lens still satisfy Eq. (4–36). The *magnification*, the ratio of the size of the image to the size of the original object, is then according to Eq. (4–37)

$$M = \frac{s'}{s} = \frac{b}{a}$$

(4–38)

Note that the image is inverted!

Example 6   How far should the object be from the lens for the magnification to be larger than unity, i.e., if the image is to appear larger than the original object?

From Eq. (4–38) we see that if $M > 1$ then

$$b > a$$

But then

$$\frac{1}{a} > \frac{1}{b}$$

and thus

$$\frac{1}{f} = \frac{1}{a} + \frac{1}{b} < \frac{1}{a} + \frac{1}{a} = \frac{2}{a}$$

But if

$$\frac{1}{f} < \frac{2}{a}$$

then

$$a < 2f$$

Thus, the condition for a magnified image is $a < 2f$.

There exists a very simple geometric construction of the image plane and image magnification. We only have to draw those rays, called the *principal rays*, whose directions are easily predicted (Fig. 4–27):

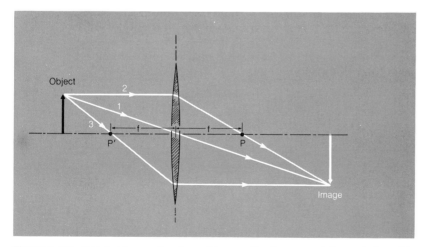

Figure 4–27 Geometric construction of image: the three special rays 1, 2, and 3 are easy to draw.

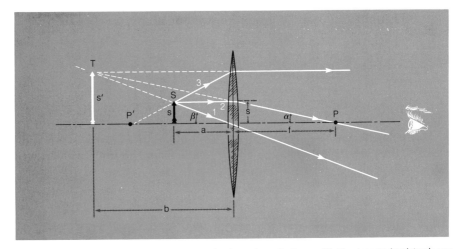

Figure 4–28 If the object is closer to the lens than its focus P', the image is virtual; we have a "magnifying glass."

1. A ray passing through the center of the lens continues undisturbed.

2. A ray which is parallel to the axis before the lens passes through the focus—it crosses the axis a distance f after the lens.

3. A ray crossing the axis a distance f before the lens continues parallel to the axis after the lens.

Thus, the lens actually has two foci P and P'; the relations between image and source are completely symmetrical. If we put an object where the image was, its image will appear at exactly the spot where the original object was placed.

Eq. (4–36) seemingly breaks down if $a < f$, i.e., the object is closer to the lens than its focal distance f. We then have

$$\frac{1}{b} = \frac{1}{f} - \frac{1}{a} < 0$$

and any real distance, of course, has to be positive. However, using the graphical construction method, we can easily discuss this situation (Fig. 4–28). We imagine the tip of the solid arrow to be the point light source S. If we draw the ray 1 going through the lens center and the ray 2 which leaves the source parallel to the axis, we find that after the lens they diverge—there is no real image of the light source S. The third ray, which—extended backward—can be thought to originate at the focus P', will continue parallel to the axis; again there is no convergence. However, to a person viewing the object through the lens the light will seem to originate from the point T; thus, while the lens produces no real image, there exists a *virtual image*. We can also calculate the position of this virtual image. From Fig. 4–28 we see that

$$\tan \alpha = \frac{s}{f} = \frac{s'}{b + f}$$

but also

$$\tan \beta = \frac{s}{a} = \frac{s'}{b}$$

Eliminating s and s' from the two equations, we find that

$$\frac{a}{f} = \frac{b}{b+f}$$

or

$$\frac{1}{a} = \frac{1}{f} + \frac{1}{b} \tag{4-39}$$

This is very similar to Eq. (4-36); indeed, if we set $b' = -b$ and define this to be the (negative!) distance of the image from the lens, we can write

$$\frac{1}{b'} + \frac{1}{a} = \frac{1}{f}$$

which is identical to Eq. (4-36). Thus, if the image is a virtual image on the same side of the lens as the source, the image distance b' is to be considered negative; while if the image is real and on the opposite side of the lens from the source, we consider the image distance positive. Then Eq. (4-36) is always valid.

Example 7   A magnifying glass is held so that a bug is three-fourths of the focal distance away from the lens. What will be the magnification of the bug as seen through the glass? The focal length of the lens is $f = 2.5$ cm.

We can directly use Eq. (4-39) to determine the image distance from the lens:

$$\frac{1}{b} = \frac{1}{a} - \frac{1}{f} = \frac{1}{(3/4)f} - \frac{1}{f} = \frac{1}{3f}$$

or $b = 3f$. From Fig. 4-29a we can also read off

$$\tan \alpha = \frac{s}{a} = \frac{s'}{b}$$

and thus the magnification is

$$M = \frac{s'}{s} = \frac{b}{a}$$

and since $a = \frac{3}{4}f$ and $b = 3f$, we have

$$M = \frac{3f}{(3/4)f} = 4$$

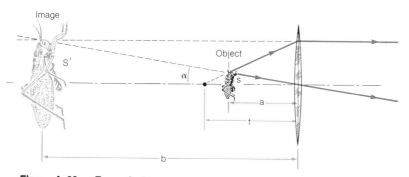

Figure 4-29a   Example 7: Viewing a bug through a magnifying glass.

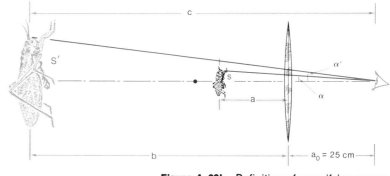

Figure 4–29b Definition of magnifying power.

Thus the bug will be magnified four-fold.

However, our result is somewhat misleading. While the virtual image is larger, it is also further away from the viewer, who is situated to the right of the lens in Fig. 4–29a. In the limit where a is nearly equal to the focal distance f, we have $b \rightarrow \infty$; the virtual image, while becoming infinitely large, also moves infinitely far away. Therefore, one usually rates magnifying glasses by their *magnifying power*, the definition of which is illustrated in Fig. 4–29b. One imagines the observer's eye to be a standard distance $a_0 = 25$ cm (≈ 10 inches) away from the object—in our case, the live bug. The magnifying power is then defined as the ratio of the opening angles α' and α under which the eye sees the virtual image and the object, if the magnifying lens is inserted so that the object is nearly at its focus. Thus, from Fig. 4–29b we have

$$\text{Magnifying power} = \text{M.P.} = \frac{\alpha'}{\alpha} = \frac{s'/c}{s/a_0}$$

because for small angles we can replace the angles by their tangents. However, $c = b + (a_0 - a) \approx b$ because the object is nearly at the focus; thus, $a \approx f$ and $b >> (a_0 - a)$. Therefore,

$$\text{M.P.} \approx \frac{s'/b}{s/a_0}$$

and since $s'/b = s/a$ we can write

$$\text{M.P.} \approx \frac{a_0}{a} \approx \frac{a_0}{f} = \frac{25 \text{ cm}}{f \text{(cm)}}$$

because $a \approx f$. Thus, in our case, where $f = 2.5$ cm, the lens has a magnifying power of 10. This means that if the bug is very near to the focus, its image will be seen under a ten times larger opening angle.

The converging lens is always thickest at the center. Another type of lens, the diverging lens, is thinnest at the center and thicker away from it. It bends light rays away from the axis—the bending angle is again proportional to the distance from the axis of the lens (Fig. 4–30)

$$\gamma = \frac{x}{f} \tag{4–40}$$

If parallel light is incident on the lens, it is defocused; to a viewer who is looking through the lens the light seems to be coming from the focus P a distance f away (Fig. 4–31). If, on the other hand, we have light converging to a point, then interposing the lens a distance f from this point produces parallel light rays. Without proof, we will just mention that if one approximates the lens surfaces by spheres, their radii again obey the lens-maker equation (4–32); to prove it, we would simply repeat the derivation of Eq. (4–32), but with the prism angle $\alpha(x)$ being as shown in Fig. 4–30. We can also use the same type of geometrical construction of the image as we used for the converging lens; the three principal light rays whose direction can be trivially drawn after the lens are again the ray parallel to the axis, the ray which goes through the center of the lens, and the ray which in the absence of the lens would go through its focus P' (Fig. 4–32).

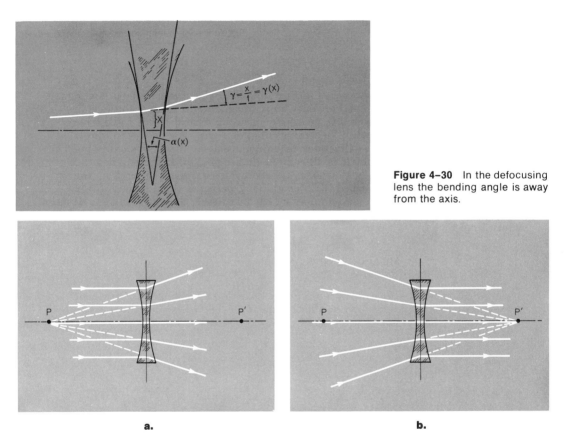

Figure 4–30 In the defocusing lens the bending angle is away from the axis.

Figure 4–31 Defocusing lens: parallel incident rays seem to come from the focus (a); rays converging to the focus (b) leave the lens parallel.

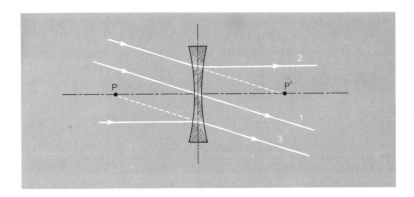

Figure 4–32 Geometric construction of special rays going through axis at the lens (1), converging toward the focus (2), or incident parallel to the axis (3).

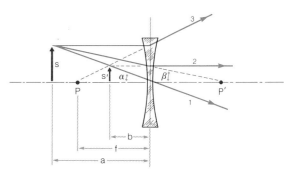

Figure 4–33 **Example 8:** Calculating the object distance *b* and demagnification *s'*/*s* using the geometric construction method.

Example 8   Calculate the image distance *b* for a diverging lens of focal length *f*, if the object is a distance *a* away. Also calculate the magnification and decide whether the image is real or virtual.

We use the graphical construction method to obtain the desired results (Fig. 4–33). We see immediately that the image is always virtual—the light rays coming from a source point *S* always diverge after the lens. We also obtain easily the relations

$$\tan \alpha = \frac{s}{a} = \frac{s'}{b} \qquad \text{from ray 1}$$

and

$$\tan \beta = \frac{s}{a+f} = \frac{s'}{f} \qquad \text{from ray 2}$$

Thus, we have

$$\frac{a+f}{a} = \frac{f}{b}$$

or

$$\frac{1}{b} = \frac{1}{a} + \frac{1}{f} \qquad (4\text{–}41)$$

Since $1/b$ is always larger than $1/a$, we conclude that we always have $b < a$; the virtual image is always closer to the lens than is the source. It is also always smaller than the original object; the magnification is

$$M = \frac{s'}{s} = \frac{b}{a} < 1$$

Inserting *b* from Eq. (4–41), we obtain

$$M = \frac{af}{a+f} \cdot \frac{1}{a} = \frac{f}{a+f}$$

Thus, the greater the object-lens distance *a*, the smaller the image.

5.   LIMITS OF GEOMETRICAL OPTICS

When we discussed light rays and used this concept in deriving relations for lenses, we disregarded the fact that light is really a wave. We completely ignored the fact that light, just as any other wave, is diffracted whenever it goes through a finite opening. But a lens is just such a finite opening; since any light which misses the lens is not refracted by it, we can always imagine the lens surrounded by an absorbing screen. Thus, we have a circular hole which will diffract the incident light. As we said earlier [Eq. (3–36)], such a hole will lead to a diffraction pattern (Fig. 4–34) whose central maximum will have a width[*] in radians of

$$\theta_0 \approx \pm 0.6 \frac{\lambda}{R} \tag{4-42}$$

Let us first consider the case where we want to focus a parallel incident light beam to a point; this would be the case in an astronomical telescope.[†] Light coming from a single star produces a spot of finite size. If light coming from a second star arrives at the lens at an angle different by less than θ_0 from the first, then the two spots will overlap; no separation of the light coming from the two stars is possible. Thus, two stars have to be separated by an angle

$$\theta > \theta_0 = 0.6 \frac{\lambda}{R} \tag{4-43}$$

if the telescope is to resolve them. Eq. (4–43) is called *Rayleigh's criterion*. Of course, it is somewhat arbitrary. The larger the separation angle θ — up to a few times λ/R — the better will be the separation between the two stars. However, Rayleigh's criterion is a good measure of the minimum separation between two stars for which one can still decide whether there are one or two sources of light. To overcome this limitation by the wave nature of light, very large telescopes have been built to improve the resolution; at present, the largest one (at Mount Palomar, California) has a diameter of 200 inches. For radio telescopes, where the wavelength λ is much larger, very large telescopes have been built; the

---

[*] We define the width as the angular difference between the central maximum and the first minimum.

[†] Real astronomical telescopes use focusing mirrors instead of lenses, because there is less light loss in a mirror than in a lens (the glass surface of the lens reflects some of the incident light!) and also because it is much easier to produce large mirrors than large lenses. For the discussion of diffraction, this is unimportant.

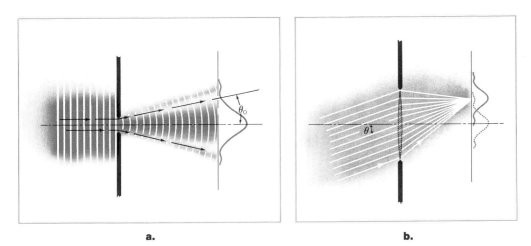

a.   **b.**

Figure 4–34   A finite opening always leads to diffraction (a); parallel incident light will not come to a sharp focus — diffraction produces a finite size spot even for the "perfect" lens.

Figure 4–35   Aerial view of the improved Arecibo Observatory 1000-ft telescope at Arecibo, Puerto Rico. The observatory is operated by the National Ionosphere Center from Cornell University under contract with the National Science Foundation. Photograph courtesy of Professor P. Drake, NIC, Cornell University.

largest radio telescope (at Arecibo, P.R., shown in Fig. 4–35), operating at a wavelength of 21 cm, has a diameter of 1000 ft (305 m).

The wave nature of light not only imposes a limit on the resolution of telescopes, but it also limits our ability to see very small objects, such as in a microscope. Let us somewhat simplify our discussion by assuming the microscope objective to consist of a single focusing lens. We want to produce a real image of as large a magnification as possible. Thus, we put the object very close to the lens focus:

$$a = f + \epsilon$$

where ϵ is assumed to be small (Fig. 4–36). The image distance is then given by Eq. (4–36):

$$\frac{1}{b} = \frac{1}{f} - \frac{1}{a} = \frac{1}{f} - \frac{1}{f + \epsilon}$$

or

$$b = \frac{f(f + \epsilon)}{\epsilon}$$

The magnification is

$$M = \frac{s'}{s} = \frac{b}{a} = \frac{f}{\epsilon} \gg 1 \qquad (4\text{--}44)$$

if $\epsilon \ll f$. However, we see that the typical angle which the light forms with the

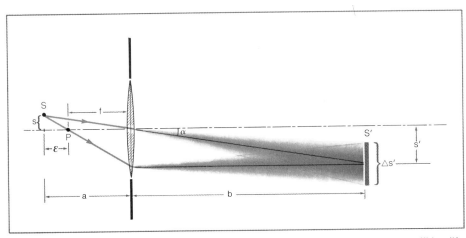

Figure 4-36 Diffraction limit of microscope: The rays leaving the lens are "smeared" by diffraction. A point source S has an image S' of size $\Delta s'$. If $\Delta s' \approx s'$, an object of size s can no longer be resolved.

axis after the lens is very small; typically we have

$$\alpha \approx \frac{s'}{b} = \frac{s}{a} \tag{4-45}$$

The limit will occur if α has a value of less than θ_0 given by Eq. (4-43); i.e., if

$$\alpha \lesssim 0.6 \; \frac{\lambda}{R}$$

If we introduce, instead of the radius R, the lens diameter $D = 2R$, we see that because of Eq. (4-45) we cannot resolve an object whose size s is

$$s \lesssim 1.2 \frac{\lambda a}{D}$$

Since, in addition, the object distance a is nearly equal to the focal distance f, we have as a limit to the size of objects visible in a microscope

$$s = 1.2 \frac{\lambda f}{D} \approx \lambda \frac{f}{D} \tag{4-46}$$

Throughout our discussion we have always assumed that the focal distance is large compared to the lens size

$$f \gg D$$

which was necessary in order for us to consider all angles to be small. However, one can build focusing lenses which have

$$\frac{f}{D} \approx 1$$

Although our derivation of Eq. (4-46) is then no longer quite correct, the result is still valid. Thus, in a microscope one can see objects of size down to the wavelength of light, or to about 0.5 $\mu = 5000$ Å. Smaller objects can be seen with the electron microscope; there, the ultimate limitation is given by the wave nature of matter. However, the de Broglie wavelength

$$\lambda = \frac{h}{mv} = \frac{h}{p} \tag{4-47}$$

can be made as small as 0.1 Å; although the focal length of electron microscopes is much larger than the diameter of the electron beams (which replaces the size D of the lens), it is nevertheless today possible to resolve structures as small as 10 Å with an electron microscope.

Summary of Important Relations

Snell's law $\qquad \dfrac{\sin \theta_1}{\sin \theta_2} = \dfrac{v_1}{v_2} = \dfrac{n_2}{n_1}$

Total internal reflection for $\quad \sin \theta_1 > \dfrac{n_2}{n_1}$

Fermat's principle $\qquad \displaystyle\int dt = \int \frac{ds}{v} = \text{minimum}$

Lensmakers equation

for sperical lenses $\qquad \dfrac{1}{r_1} + \dfrac{1}{r_2} = \dfrac{1}{n-1}\dfrac{1}{f}; \quad r_1, r_2 \text{ radii}$

for paraboloidal lenses $\quad \dfrac{1}{d_1} + \dfrac{1}{d_2} = \dfrac{2}{n-1}\dfrac{1}{f}; \quad \text{parabolas have form } y = \dfrac{x^2}{4d}$

For lens of focal length f, the image distance b is related to the object distance a by

$$\frac{1}{a} + \frac{1}{b} = \frac{1}{f}$$

For virtual images, one has to set b negative: $b < 0$. The magnification is b/a. For diverging lenses, only virtual images exist (b always < 0), because also $f < 0$. If one defines all quantities positive, then one has for diverging lenses

$$\frac{1}{a} - \frac{1}{|b|} = -\frac{1}{|f|}$$

Limits of geometrical optics

for telescopes $\qquad \theta \geq \theta_0 = 0.6\dfrac{\lambda}{R}$

for microscopes $\quad s \geq s_0 = 1.2\dfrac{\lambda f}{D}$

Questions

1. The refractive index of air depends on its density. Explain the following phenomena:

(a) While driving on a sunny day, the road in front of you frequently seems to be a mirror.

(b) When you look at somebody across a log fire, his features will seem to be wavering.

2. Optical fibers consist of an array of many thin polished flexible glass or plastic "wires" strung next to each other. Explain why one can see around corners using this approach (see figure).

3. Prescription glasses for myopia (short-sightedness) are usually shaped as shown in the figure. Is this a focusing or defocusing lens? Why is this particular shape being used?

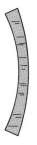

4. Would a helium balloon floating in air act as a focusing lens or as a defocusing lens for sound waves?

Problems

1. A laser beam enters a glass plate (thickness $d = 0.5$ cm, refractive index $n = 1.7$) at an angle of 60° relative to the normal. Behind the glass is a pinhole aligned with the beam. By how much does one have to move the pinhole if the glass is removed?

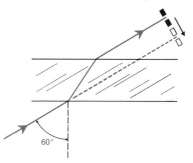

2. A narrow light beam falls on a lucite plate (thickness $d = 1/2$ inch, refractive index $n = 1.49$) at an angle of 45°, a distance x from one end. What is the minimum value of x if the beam is to leave through the other face of the plate?

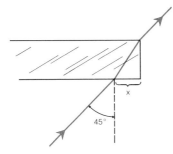

3. Light is impinging at the end of a lucite plate ($n = 1.49$). How large can the angle of incidence θ be, if the light is to leave the plate on the other end of the plate after having undergone total reflection several times?

4. A glass prism has the angle $\alpha = 60°$; its refractive index is $n = 1.7$. Under what angle ϕ should a beam of light arrive at the prism, as in the figure, if it is to be bent by 2ϕ?

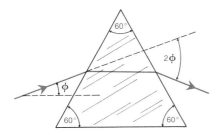

5. Light enters a glass prism of refractive index $n = 1.5$ normally to one surface. What should be the angle α (see figure) if the light beam is to make three total reflections inside the glass before leaving?

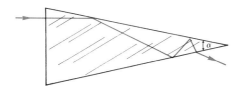

6. A fish swimming 30 cm under the water surface ($n = 1.34$) is viewed from directly above (see figure). How far below the surface does it seem to be? Hint: Assume the angles θ and θ' to be small so that $\theta \approx \sin \theta \approx \tan \theta$.

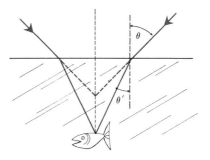

7. Light is impinging from vacuum on a material of refractive index n. For what value of n will the reflected and refracted waves be moving perpendicularly to each other if the angle of incidence θ obeys the relation $\sin \theta = 1/n$?

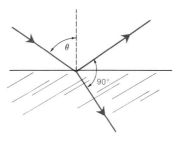

8. A light beam is passing through a series of transparent parallel plates of varying refractive indices. Show that, whatever the angle of incidence, the beam will, after passage through all plates, continue in the same direction it had originally.

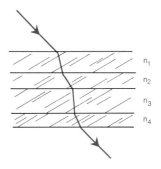

9. Two glass prisms are held together as shown in the figure on the next page. Their refractive indices are $n_1 = 1.4$ and $n_2 = 1.6$. If a light beam enters normally the first prism, by how much will its direction have changed upon leaving the second prism?

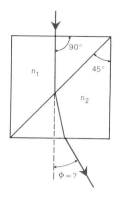

10. The refractive index of air is $n = 1.00028$. If a star is exactly $45°$ from the zenith (vertical), where should you point your telescope to see it in the center of your viewing area?

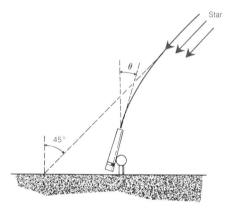

11. A focusing lens is made of glass of refractive index $n = 1.65$. Its thickness at the center is 5 mm, and its diameter is 10 cm. It has sharp edges. Determine its focal length.

12. A defocusing lens is to have a focal length of 30 cm and a diameter of 7.5 cm (≈ 3 inches). If the minimum glass thickness of the lens is to be 3 mm, how thick will the lens be at its edges? The refractive index of the glass used is $n = 1.5$.

13. A lens made of quartz glass ($n_1 = 1.56$) has a focal length in air of 15 cm. What will be its focal length if it is immersed in water ($n_2 = 1.34$)?

14. Two focusing lenses of long focal lengths f_1 and f_2 are right next to each other. What is the effective focal length of the pair?

15. A parallel laser beam has a diameter $d_1 = 1$ mm. Using three identical focusing lenses of focal length $f = 20$ cm, you are to convert this beam into a parallel beam of diameter $d_2 = 2$ cm. How would you space the three lenses? Could you achieve the same result using only two of the three lenses? Hint: Use the method of principal rays.

16. Two identical focusing lenses are spaced by twice their focal length f. If an object is placed a distance a before the first lens, where will the final image be? When will it be a virtual image and when will it be real?

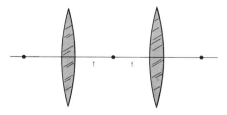

17. A focusing lens of focal length 10 cm and a defocusing lens of focal length 5 cm are 15 cm apart. An object is placed 15 cm before the focusing lens. Where will the image be? Will it be real or virtual, and what is the magnification?

18. An object is a distance d before the focus of a focusing lens of focal length f; its image is a distance d' behind the second focus. Show that $dd' = f^2$.

19. The simplest compound microscope consists of two focusing lenses. The objective produces a real image, which is viewed by the eyepiece used as a magnifying glass. If the objective magnifies 50 times and the eyepiece 10 times, how far apart should the two be spaced in terms of their focal lengths f_1 (objective) and f_2 (eyepiece)?

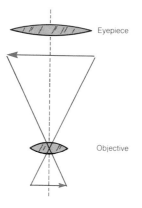

20. A concave mirror has a spherical surface. Show that light rays, which are near the axis aa' and parallel to it, converge to a point after reflection. What is the distance d of the point from the mirror if the radius of curvature is R?

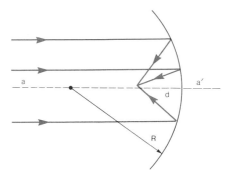

21. A lens of focal length f is right next to a plane mirror. How far should an object be if its image is to coincide with the object?

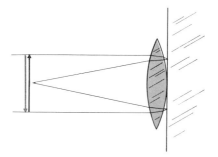

22. What is the minimum size of an object one can see on the moon, using a 24-inch diameter telescope? The distance from moon to earth is 380,000 km. The actual resolution is worse because of the refractive index fluctuations in the atmosphere.

23. A peeping tom is looking into a room through a pinhole of 1 mm diameter. Can he see the eye pupils of a person who is standing 10 feet away from the pinhole?

24. What angular resolution can one achieve with the Arecibo radio telescope (diameter 1000 feet) if it is receiving at a wavelength of 21 cm? How does this compare with the resolution of the 200 inch diameter optical telescope at Mount Palomar? Note that 21 cm is a wavelength emitted strongly by hydrogen molecules in intergalactic space; thus, it is frequently used in radio astronomy.

CHAPTER 5

COULOMB'S LAW

1. INTRODUCTION

The gravitational forces between any two massive bodies are responsible for many everyday phenomena. Because of them, the earth stays near the sun; but they also enable us to sit on chairs and to use a water glass when drinking. Anybody who has seen on television how astronauts have to use special devices to perform everyday tasks can appreciate how important gravity is to our well-being.

There exists another fundamental class of forces, which we call *electromagnetic forces*. If these forces were not present, matter as we know it would not exist. All "atomic" properties of matter—and we include here the reasons why iron is a solid, while nitrogen is a gas at room temperature—are really electromagnetic phenomena. The atoms consist of electrons and atomic nuclei, which interact by electric forces; a crystal of salt—or a microcrystal of iron—has its regular shape because of electric forces between individual atoms.

Both electrons and atomic nuclei have mass, and thus attract each other by gravitational forces. But they also possess another property—an electric charge. The electric charge of two objects causes them to interact by electromagnetic forces, just as their mass causes them to interact by gravitational forces. Electromagnetic forces are, however, much stronger than gravitational forces. When we discuss atomic phenomena, gravitational forces are completely negligible. If all electromagnetic forces between the individual constituents were attractive, all matter would rapidly collapse. But in reality there are two types of charges; objects having an equal type of charge repel each other, and objects having unequal types of charge attract each other. This interplay of attraction and repulsion accounts largely for the finite size of material objects.

However, as we study these atomic phenomena, we will discover that our ordinary everyday intuition, which was so helpful in treating classical mechanics, begins to fail us. The picture of an atom evoked in a layman's mind is of small billiard balls (electrons) moving around a larger central ball, the atomic nucleus. He thinks of the atom as of a miniature solar system. While this picture is not completely false, it is largely misleading. Classical mechanics, which enables us to calculate motions of planets (or helps us understand what a player does at a pool table), fails at the atomic level. The electron is no more a small massive point which travels around the nucleus—it is spread out over the whole atom. We cannot say that the electron is "now here and later there"; it is everywhere at once, represented by a "probability wave."

Quantum mechanics, which replaces classical mechanics on the atomic scale, is often the great stumbling block for the student who tries to understand physics. We learn from early childhood on that things are at definite places, and at least in principle we can always tell where they are—we look at them, feel them, and so forth. It is important to note that we cannot see an individual electron or atom; thus, we cannot observe it as we can observe the planets. We may discover that it arrived somewhere (a click in a counter, a flash on a fluorescent screen), but we will never be able to follow it accurately (knowing its position to within a few atomic diameters) over any period of time. Thus, there is really no reason why electrons should be describable in the same way as we describe macroscopic objects.

Electromagnetic forces can also be noticed on a "macroscopic" scale, the scale of our daily experience. We are all familiar with magnets; we use electric fields which force electrons to move in a light bulb and thus heat it until it emits light. These macroscopic effects are all collective phenomena caused by the individual atoms in a piece of matter. Thus, a permanent magnet is an assembly of individual atomic magnets called magnetic dipoles, most of which are aligned with each other. The electric currents which we use everywhere in life are due to the motion of electrons in the wires.

We call *electric forces* those forces which act between charged bodies when they are at rest. If the charges are moving, they will interact in addition by *magnetic* forces. However, the compound word "electromagnetic" which we have used earlier expresses the fact that they are really inseparable. We learned in classical mechanics that all inertial reference frames are equivalent; and by inertial reference frames we defined all frames which move relative to one another with *constant velocity*. Thus, when a particle is at rest in one frame—and is being acted upon only by electric forces—it will be in motion in another reference frame and will feel the pull of magnetic forces in addition to electric forces.

This may seem surprising and contrary to everyday experience. There is no visible motion in a permanent magnet; also, usually no electric fields can be found in an electromagnet or around a current-carrying wire, while magnetic forces will act on a compass needle in both cases. But the magnetic field in an electromagnet is due to the motion of the electrons in the copper wiring; and in a permanent magnet the field is produced by the rotation (spin) of the electrons in the iron. Also, if we were to observe the electromagnet from a reference frame moving past it at high speed, we would measure an electric field between its pole pieces. Thus, while we are studying macroscopic electric or magnetic phenomena, we should always keep in mind these two points:

(a) What we measure is the cooperative effect of many electrons and atomic nuclei—we see only the average effect of an extremely complex situation.

(b) What electromagnetic field we measure may well depend on whether we ourselves are moving or not. Sometimes the situation becomes much simpler if we consider it from a moving reference frame.

In this chapter we will limit ourselves to discussing forces between charges which are at rest or are moving only very slowly, so that magnetic effects can be neglected.

2. COULOMB'S LAW

We have just introduced a new concept, the electric charge. We do not know "what" charge is, any more than we know what mass is. All that we know is that any two objects which have charges will act upon each other by an electric force, in the same way that any two objects which have mass will act upon each other by a gravitational force. But there is one difference: while all massive objects

attract each other, the force between charged objects can be attractive or repulsive.

We can demonstrate this by using a very crude setup, such as that shown in
Fig. 5–1. We suspend two small metallic spheres on nylon threads. To minimize
their mass, we choose hollow spheres. If we now take a glass rod, rub it, and then
touch both spheres with the rod, they will move apart; *two similar charges repel
each other*. But if we touch one sphere with the glass rod and the second one
with a similarly rubbed lucite rod, they will move together; *dissimilar charges
attract each other*. If we repeat this experiment by trying to charge them any
possible way, we will always end up with either the "glass" charge or the
"lucite" charge. We can also show that the lucite charge can be canceled by the
glass charge: The two types of charge can be added algebraically, if we assign
a positive sign to one type and a negative sign to the other. We call the "glass"
charge positive and the "lucite" charge negative. This definition of the sign of a
charge was first given by Benjamin Franklin.

Our crude experiment does not allow us to measure accurately how the
force varies with the distance between the two spheres; nor does it enable us to
show how the force depends on the amount of charge deposited on each sphere.
This was done by Coulomb in 1785, who showed that the force between two
point charges has the magnitude

$$|\mathbf{F}| = \text{constant} \times \frac{Q_1 Q_2}{r^2} \qquad (5\text{--}1)$$

Here Q_1 and Q_2 are the two charges and r is the distance between them. The
force is repulsive if the two charges are of equal sign, and attractive if they are of
opposite sign. Note that our law is true only for point charges; by this we mean
that the size of the spheres has to be very small compared to their separation.
Note also that whether the force is attractive or repulsive, it always acts along
the line connecting the two (point) charges, just as the (always attractive) gravitational force between two mass points always acts along the line connecting
them.

As we said earlier, all macroscopic electromagnetic phenomena are due to
the cooperative effect of the atomic constituents, the electrons and atomic nuclei.
Let us assume that our metallic spheres are made of copper. A piece of copper
is made of small crystals; each crystal consists of a regular periodic arrangement
of copper nuclei. Each nucleus is surrounded by 29 electrons; we call the
assembly of one nucleus and 29 electrons the copper atom. These electrons all
have the same negative charge $(-e)$. Because the copper nucleus has the charge
$+29e$, the copper atom as a whole carries no charge, and neither does the copper
sphere as a whole. Of the 29 electrons, 28 are tightly bound to the nucleus and
cannot move around freely; the 29th is so loosely tied that it can move freely

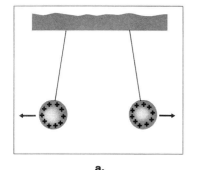

a.

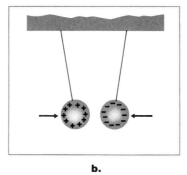

b.

Figure 5–1   Equal charges repel each other; opposite charges attract each other.

around the whole copper sphere. Because of this "free" electron, we call copper *a good conductor of electricity.*

The electrons in the glass rod, on the other hand, cannot move freely. They are too solidly bound by the nuclei of the various atoms (Na, O, Si, etc.) which make up the glass rod. Therefore, glass is an insulator. But by rubbing the glass rod, we remove some of the electrons from the atoms which are close to the surface: We "give it a positive charge." When we later touch the copper sphere with the glass rod, some of the free electrons in the copper are captured by the atoms in the glass rod. After sphere and rod separate, the copper sphere has too few electrons — it is positively charged.

If a lucite rod is rubbed in the same way, electrons from atoms of the cloth are transferred to the rod. When the rod then touches the copper spheres, some of the electrons are transferred to the sphere — the sphere becomes negatively charged. Similar phenomena can be observed on dry days: if you walk on a nylon carpet and then touch a metallic object such as a door knob, you will feel an electric shock. Sometimes even a small spark will jump between your finger and the doorknob. This happens because while you are walking, your feet rubbing against the carpet were continuously charging your body. When you touch the knob, the charge flows from you into the knob.

Note that such electrostatic charging can happen only by rubbing two good insulators on each other. In good conductors, electrons move around freely; thus, even if a few were transferred from one conductor to another by rubbing, they would immediately flow back, since they are attracted by the positive nuclear charge to which they were weakly attached at the beginning. Therefore, no macroscopic charges can be produced by simply rubbing two conductors on each other.

3. UNITS

When we wrote down Coulomb's law [Eq. (5–1)] we used an undefined constant. We had to do it because we had not specified the units in which we are to measure the charges Q_1 and Q_2. Since we now enter a new domain with a confusing array of new units, we should first discuss again the whole problem of units from a deeper point of view. In mechanics we learned that there exist three basic units: those of length, mass, and time. This is correct only because we have been conditioned to accept this from childhood. We say, "See you in an hour," we buy a pound of ham (or a kilogram of sugar), and we say, "It is three miles to school." Everybody knows what we mean when we say that. Our intuition is a bit shaky only when we talk about mass. A pound is a unit of weight, and a kilogram is a unit of mass, but in everyday language we use both weight and mass interchangeably. Because of the constancy of g, the gravitational acceleration on the surface of the earth, we "know" that mass and weight have the same effect. Thus, it is rather difficult for many people to understand that an astronaut in a satellite has no weight, but the same inertia (mass) as on earth. In the twenty-first century, when schoolchildren will make regular weekend trips to the moon (where the gravitational attraction is only one-sixth of its earth value), the confusion between the two concepts of mass and weight will probably no longer exist.

We also saw in mechanics many quantities, such as velocity, acceleration, and force, which could not be measured in one of the three basic units. Nevertheless, there was no doubt about how to define their units. Since velocity is the distance traveled divided by the time required for the travel, its unit must be 1 m/sec. Similarly, the units of acceleration, force, torque, or angular momentum can be deduced from the basic definitions of the respective quantities.

When we want to define the unit of electric charge, our task is a little more difficult. One possible solution is to define the unit of charge as that amount of charge which repels another equal charge by the unit of force, if the two are a unit of distance apart. Expressed as an equation, we write Coulomb's law in the form

$$|\mathbf{F}| = 1 \frac{Q_1 Q_2}{r^2} \tag{5-2}$$

This definition of the unit of charge is used in the "electrostatic system." Specifically, the electrostatic CGS system, which uses the centimeter, gram, and second as the basic units, is used by most theoretical physicists. The unit of force in this system is

$$1 \text{ g}\frac{\text{cm}}{\text{sec}^2} = 1 \text{ dyne}$$

and the electrostatic unit of charge is defined as the charge which repels another equal charge by the force of 1 dyne if the two charges are 1 cm apart.

However, this system of units is awkward because it leads to a strange unit of charge. We can write Eq. (5-2) in the form

$$1 \text{ g}\frac{\text{cm}}{\text{sec}^2} = \frac{(\text{unit of charge})^2}{(1 \text{ cm})^2} \tag{5-3a}$$

or

$$1 \text{ electrostatic unit} = 1 \text{ statcoulomb} = 1 \frac{(\text{g})^{1/2}(\text{cm})^{3/2}}{\text{sec}} \tag{5-3b}$$

As we see, defining the unit of charge using only the mechanical units of length, mass, and time leads to fractional exponents in the units for quantities connected with electromagnetic phenomena.

In this book we will use a different system of units, which is preferred by most engineers. In this system one defines a unit of charge independently of Coulomb's law. We will have to postpone until later a discussion of how this unit is defined in terms of easily measured quantities. Temporarily we will define our unit, the coulomb,[*] by defining the charge of the electron to be equal to

$$-1e = -1.602 \times 10^{-19} \text{ coulomb} \tag{5-4}$$

Since the charge of the electron is measurable, at least in principle, we have a unique definition of the coulomb.

We should also specify here that we will use almost exclusively the mechanical units of 1 meter, 1 kilogram, and 1 second. This system of units, called the MKSC system (C for coulomb) is the system of units used by most engineers —and also by physicists who design or use electrical measuring equipment.

Since we have defined the unit of charge independently of Coulomb's law, we no longer write it in the form of Eq. (5-2). Instead we have to write

$$\boxed{|\mathbf{F}| = K_0 \frac{Q_1 Q_2}{r^2}} \tag{5-5a}$$

---

[*] We will use the abbreviation 1 coulomb = 1 coul. This is different from the "standard abbreviation" 1 coulomb = 1 C, which is endorsed by the IUPAP (International Union for Pure and Applied Physics). However, we want to reserve the symbol C for the capacitance, which will be introduced in Chapter 7.

and the proportionality constant is now a measurable quantity. Since the unit of force is the newton, the dimension of K_0 is

$$[K_0] = \text{newton-m}^2/(\text{coulomb})^2$$

The experimental value of K_0 is

$$K_0 = 8.9875 \times 10^9 \frac{\text{newton-m}^2}{\text{coul}^2} \approx 9.0 \times 10^9 \frac{\text{N-m}^2}{\text{coul}^2} \tag{5-5b}$$

Instead of Eq. (5-5a), it is usual to write Coulomb's law in a slightly different form

$$|\mathbf{F}| = \frac{1}{4\pi\epsilon_0} \frac{Q_1 Q_2}{r^2} \tag{5-6a}$$

where ϵ_0, called the *electrical permittivity of the vacuum*, has the value

$$\epsilon_0 = \frac{1}{4\pi K_0} = 8.854 \times 10^{-12} \frac{\text{coul}^2}{\text{N-m}^2} \tag{5-6b}$$

We stress that the only reason we write Eq. (5-6a) is because this is the traditional way to write Coulomb's law. It would seem more reasonable to use the constant K_0 in Coulomb's law; the factor $1/4\pi$ in Eq. (5-6a) seems at first glance to be downright silly. However, as we will see later, in many equations of electrostatics the expression $4\pi K_0 = 1/\epsilon_0$ appears. Once one realizes this, one has the rather arbitrary choice of whether one wants to carry around the factor of 4π. We follow the voice of tradition and write Coulomb's law in the form of Eq. (5-6a).

Example 1   The copper atom consists of a copper nucleus surrounded by 29 electrons. The atomic weight of copper is 63.5 g/mol. Let us now take two pieces of copper, each weighing 10 g (labeled (a) and (b) in Fig. 5-2); and let us transfer from (a) to (b) one electron for every 1000 atoms in piece (a). What will be the force between (a) and (b), if they are 10 cm apart?
    We first have to calculate how much charge we have transferred from (a) to (b). The number of atoms in a gram-mol (for copper, 63.5 g) is 6×10^{23}; we will have transferred

$$6 \times 10^{23} \frac{\text{atoms}}{\text{g-mol}} \times \frac{10 \text{ g}}{63.5 \frac{\text{g}}{\text{g-mol}}} \times \frac{1}{1000} = 9.45 \times 10^{19} \text{ electrons}$$

Since the charge of each electron is $-1e = -1.6 \times 10^{-19}$ coulomb, we have transferred the charge

$$9.45 \times 10^{19} \times (-1.6) \times 10^{-19} \cong -15 \text{ coulombs}$$

Figure 5-2 Example 1.

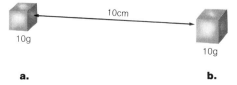

10cm

10g

10g

a.

b.

Since both pieces of copper were originally neutral, block (a) will now carry a charge of $+15$ coulombs, while block (b) carries a charge of -15 coul. The attractive force between the two now would have the magnitude

$$F = K_0 \frac{Q_1 Q_2}{r^2} = 9.0 \times 10^9 \frac{\text{N-m}^2}{(\text{coul})^2} \cdot \frac{(15 \text{ coul})^2}{(0.1 \text{ m})^2} = 2.03 \times 10^{14} \text{ newtons} \qquad (5\text{--}6)$$

How big is this force? An ocean liner has a mass of about 70,000 tons $= 7 \times 10^7$ kg; thus, it will weigh about 7×10^8 newtons. We could lift 300,000 such ocean liners, if we could somehow attach them to one of the little cubes. Actually, we could never deposit 15 coulombs on such a small object; the forces between the charges on the same block would immediately pulverize the block. When we charge an object, we add or take away electrons only from atoms within a few atomic layers of the surface; but even here we disturb at most only one surface atom in a thousand.

4. THE ELECTRIC FIELD

In each atom there are several electrons and one atomic nucleus; all these building blocks of matter have electric charges. Thus, in any piece of matter, which consists of a large number of atoms, there exist also huge numbers of individual point charges. Let us consider a single point particle of charge q; we will for the moment assume q to be positive. In order to distinguish the one particle from all the others, we will call it the *test particle*.

Our test particle will be attracted by all nearby negative charges and repelled by all nearby positive charges, as illustrated in Fig. 5–3. The net result is that a force $\mathbf{F}$ acts on the particle, which is the resultant (sum) of all the individual forces. However, Coulomb's law states that the force between our particle and any single point charge Q is proportional to the charge q:

$$|\mathbf{F} \text{ (due to } Q)| = K_0 \frac{qQ}{r^2}$$

where r is the distance from the test particle to the point charge Q.

If there exists more than one such point charge Q, the force with which any one of them is acting on the test particle is proportional to q. Thus, the net force on our test particle will also be proportional to its charge q:

$$\boxed{\mathbf{F} = q\,\mathbf{E}} \qquad (5\text{--}7a)$$

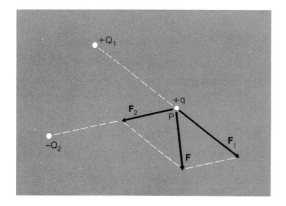

Figure 5–3 The force on a test charge q is the sum of the forces due to all other present charges. Each component and thus also the sum $\mathbf{F}$ is proportional to q.

Let us now use another test particle of negative charge $(-q')$. Of all the nearby changes, it will now be repelled by all negative charges and attracted by all positive charges. The net force on the second test particle is then

$$\mathbf{F}' = (-q')\mathbf{E} \tag{5-7b}$$

It is important to note that $\mathbf{E}$ has the identical value—magnitude and direction! —for both particles if they are placed (one after the other) at the same position in space. We call $\mathbf{E}$ the *electric field*. We can thus define the electric field at any point P as the force acting on a charged test particle at P divided by the charge of the test particle. Since the value of the electric field does not depend on the test particle charge, we can define $\mathbf{E}$ independently of whether there actually is a test charge at P or not.

We said that in matter all charged objects—electrons and atomic nuclei— are point charges.* Let us therefore first calculate the electric field of a point charge of charge Q. The force between Q and a test charge q a distance r away has the magnitude

$$|\mathbf{F}| = K_0 \frac{qQ}{r^2}$$

and thus the electric field has the magnitude

$$|\mathbf{E}| = \frac{|\mathbf{F}|}{q} = K_0 \frac{Q}{r^2} \tag{5-8}$$

The direction of $\mathbf{E}$, according to Eq. (5-7a), is the same as the direction of the force $\mathbf{F}$, if the test charge q is positive. Thus, $\mathbf{E}$ will be a vector pointing away from the charge Q if Q is positive. If Q is negative, $\mathbf{E}$ will point toward the charge Q. If we define the unit vector

$$\hat{\mathbf{u}}_r = \frac{\mathbf{r}}{r} \tag{5-9}$$

which points from the charge Q toward the point at which the electric field is to be measured, we can write an equation for $\mathbf{E}$ which is correct in magnitude and direction:

$$\mathbf{E} = K_0 \frac{Q}{r^2}\hat{\mathbf{u}}_r = \frac{1}{4\pi\epsilon_0}\frac{Q}{r^2}\hat{\mathbf{u}}_r \tag{5-10}$$

If Q is positive, $\mathbf{E}$ as given in Eq. (5-10) has the same direction as $\hat{\mathbf{u}}_r$; while if Q is negative, $\mathbf{E}$ points in the opposite direction. Both situations are shown in Fig. 5-4.

The unit in which we are to measure the electric field can be obtained from the basic definition, Eq. (5-7a). We have

$$[\mathbf{F}] = [q] \cdot [\mathbf{E}]$$

---

* Actually, this is not quite true. To the best of our knowledge, electrons are really point charges; experimentally one knows that the size of the electron must be less than $\sim 10^{-16}$ m. Atomic nuclei, on the other hand, have a measurable size. Their diameter ranges from 1.6×10^{-15} m for the hydrogen nucleus to 1.7×10^{-14} m for the uranium nucleus. However, the typical size of the atom is 1 to 3 Å $= 1$ to 3×10^{-10} m; thus, on the atomic scale, even the uranium nucleus is practically a point particle.

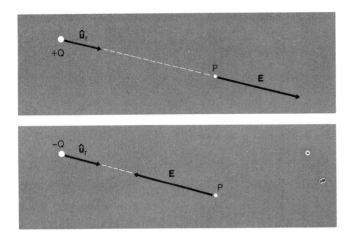

Figure 5–4 The electric field of a positive point charge points away from the charge; the field of a negative charge points toward the charge.

or

$$\text{newton} = \text{coulomb} \cdot (\text{Electric field unit})$$

and therefore we measure the electric field in units of

$$1\,\frac{\text{N}}{\text{coul}} = 1\,\frac{\text{kg-m}}{\text{sec}^2\text{-coul}}$$

because $1\,\text{N} = 1\,\text{kg-m/sec}^2$.

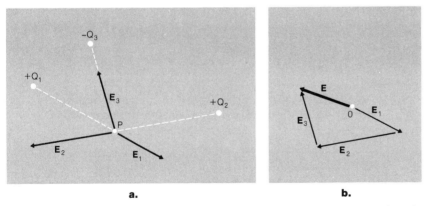

| a. | b. |

Figure 5–5 The electric field due to several point charges can be obtained as the vectorial sum of the electric fields of each point charge.

If there are several charges Q_1, Q_2, . . . near our test charge q, each will either attract or repel the test charge. The net force is the vectorial sum of all forces acting on q. Thus, the net electric field at the point P of the test charge (Fig. 5–5) can be obtained by adding vectorially the electric fields due to the individual charges Q_i. This is illustrated in the following examples:

Example 2   Three equal positive charges $+Q$ are arranged at the corners of an equilateral triangle of sides a, as shown in Fig. 5–6. What is the electric field at the center of the triangle?

We have numbered the three charges Q_1, Q_2, and Q_3 so that we can discuss

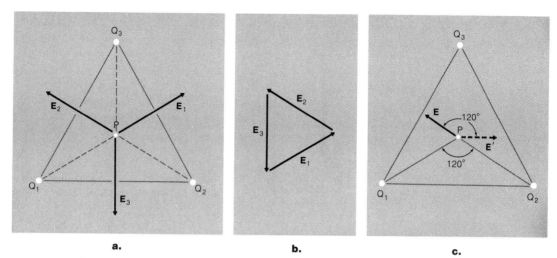

Figure 5-6 Example 2. (a) The three contributions to the electric field. (b) Vectorial addition shows that the sum vanishes. (c) If the field is **E**, rotating the figure clockwise by 120° rotates **E** into **E'**, but the charges are back in their positions. Thus, by symmetry **E** = 0.

them separately. We can calculate the electric field at the center point P in two ways.

a. The explicit way: Each of the three charges gives rise to an electric field pointing away from it and having the magnitude

$$|\mathbf{E}| = \frac{1}{4\pi\epsilon_0}\frac{Q}{d^2}$$

where d is the distance from one corner to the center of the triangle. The total electric field can be obtained as the sum of the three fields of each point charge:

$$\mathbf{E} = \mathbf{E}_1 + \mathbf{E}_2 + \mathbf{E}_3$$

The three fields $\mathbf{E}_1$, $\mathbf{E}_2$, and $\mathbf{E}_3$ have equal magnitude. Adding them graphically (Fig. 5–6b), we note that the three vectors themselves form an equilateral triangle. Thus, their sum is equal to zero; the field at the point **P** vanishes.

b. Using symmetry: Let us assume that the electric field has the magnitude and direction as indicated in Fig. 5–6c. Then if we rotate the figure, the direction of the electric field rotates with it. Rotating by 120° clockwise, we obtain a field pointing in the direction shown by the dashed arrow. But the three charges after the rotation are again at the corners of the original triangle! Since all three charges are equal, we have the same situation we had before rotating. Thus, we conclude that if the electric field points in the direction of the solid arrow, it also has to point in the direction of the dashed arrow. This is possible only if the *magnitude* of the electric field vanishes; then its direction is undefined and there is no paradox.

Example 3   Calculate the force on any of the three charges Q in the previous example.

We will calculate the force acting on Q_3 in Fig. 5–6a; since all charges are equal, we can obtain the forces on the other two charges by rotating Fig. 5–7a clockwise or counterclockwise by 120°. First we calculate the electric field at Q_3 due to Q_1 and Q_2. It is important to remember that Q_3 is now our "test particle"; thus, its own electric field has to be excluded from the calculation. The two fields $\mathbf{E}_1$ due to

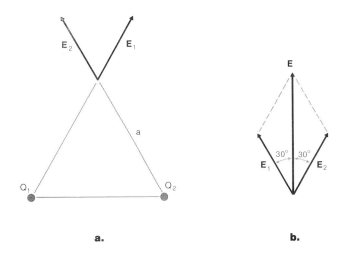

a.

b.

Figure 5–7 Example 3: Calculating the force on any of the three charges from Fig. 5–6.

Q_1 and $\mathbf{E}_2$ due to Q_2 have the same magnitude:

$$|\mathbf{E}_1| = |\mathbf{E}_2| = \frac{1}{4\pi\epsilon_0}\frac{Q}{a^2}$$

but they have different directions! $\mathbf{E}_1$ points away from Q_1, while $\mathbf{E}_2$ points away from Q_2. In order to add them, we have redrawn the two vectors in Fig. 5–7b. The sum $\mathbf{E} = \mathbf{E}_1 + \mathbf{E}_2$ points straight up and has the magnitude

$$|\mathbf{E}| = 2\,|\mathbf{E}_1|\,\cos 30°$$

Since $\cos 30° = \sqrt{3}/2$, we have

$$|\mathbf{E}| = \frac{\sqrt{3}}{4\pi\epsilon_0}\frac{Q}{a^2}$$

Because $Q_3 = Q$ is positive, the force on Q_3 will thus also point up; its magnitude is

$$|\mathbf{F}| = Q\,|\mathbf{E}| = \frac{\sqrt{3}}{4\pi\epsilon_0}\frac{Q^2}{a^2}$$

The two examples show us how to calculate the electric field at any point due to an arbitrary distribution of point charges $Q_1, Q_2, \ldots Q_N$, which do not have to be equal (as they were in our examples). Calculate the electric field $\mathbf{E}_i$ due to each individual point charge using Eq. (5–10); the total electric field is then the sum of all these partial fields

$$\mathbf{E} = \mathbf{E}_1 + \mathbf{E}_2 + \ldots + \mathbf{E}_N \tag{5–11}$$

If we want to calculate the force on any of the charges (e.g., on Q_1), we have to consider it to be a test charge and calculate the electric field $\mathbf{E}'$ which would exist at its position if Q_1 was removed and all the other charges did not move. Then we multiply the resulting electric field by the charge Q_1:

$$\mathbf{F}\ (\text{acting on } Q_1) = Q_1\mathbf{E}' = Q_1(\mathbf{E}_2 + \mathbf{E}_3 + \mathbf{E}_4 + \ldots \mathbf{E}_N) \tag{5–12}$$

However, the method indicated here is no longer useful if the number of point charges becomes very large. There are 8.5×10^{28} atomic nuclei and 29 times as many electrons per cubic meter of copper; thus, even a copper sliver of 1 mm³ volume will still have some 2.5×10^{21} point charges in it. Obviously, using Eq. (5–11) is a hopeless undertaking.

Instead, we can define a charge density, the total charge ΔQ in a small volume $\Delta \tau$ divided by the volume itself:

$$\rho = \frac{\Delta Q}{\Delta \tau} \qquad (5–13a)$$

We now want to calculate electric fields on the macroscopic scale (e.g., near a charged metallic object). The volume element $\Delta \tau$ which is used in the definition Eq. (5–13a) of the density then should be small – but not too small. It should be small compared to the typical size of the metallic object, but still large enough so that there are many atoms inside; only then can we neglect the inherent "graininess" of the electric charge in matter. Although we frequently will write

$$\rho = \frac{dQ}{d\tau} \qquad (5–13b)$$

we should never forget that the differential quantities dQ and $d\tau$ are not infinitesimally small – one is not allowed to take the limit $d\tau \to 0$. If we want to calculate the electric field near a charged macroscopic object (Fig. 5–8), we subdivide its volume into volume elements $d\tau$. The volume element at the point P has the charge

$$dQ_P = \rho_P d\tau_P$$

and thus contributes to the electric field at the point P' the vectorial element $d\mathbf{E}$:

$$d\mathbf{E}_{P'} = \frac{1}{4\pi\epsilon_0} \frac{dQ_P}{r^2_{PP'}} \hat{\mathbf{u}}_{PP'} \qquad (5–14a)$$

which we can write also in the form

$$d\mathbf{E} = \frac{1}{4\pi\epsilon_0} \frac{\rho_P \hat{\mathbf{u}}_{PP'}}{r^2_{PP'}} d\tau_P \qquad (5–14b)$$

The total electric field $\mathbf{E}$ at the point P' then is obtained by summing over all contributions $d\mathbf{E}$. Since in our approximation the charge distribution is continuous, we have to write the sum as an integral over the whole region where

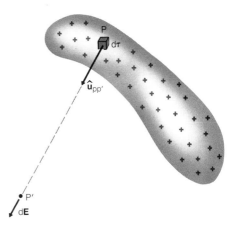

Figure 5–8 The electric field of a continuous charge distribution can be obtained by summing (integrating) over all volume elements $d\tau$.

there are charges:

$$\mathbf{E} = \int d\mathbf{E} = \frac{1}{4\pi\epsilon_0} \int \frac{\rho_P \hat{\mathbf{u}}_{PP'}}{r^2_{PP'}} d\tau_P \qquad (5\text{–}14c)$$

The meaning of Eq. (5–14c) is rather obvious – it is the exact analogy to Eq. (5–11). However, the practical calculation can be much more difficult; instead of a sum we have to calculate an integral. This should not frighten us; we can always calculate the integral numerically on a computer if we are unable to perform the integration analytically.

Example 4   Calculate the electric field a distance a away from the center of a thin, homogeneously charged rod of length $2a$. The total charge on the rod is Q. For numerical purposes, choose $Q = 10^{-9}$ coul and $a = 10$ cm.

We subdivide the rod into little elements of length dx, as shown in Fig. 5–9. If we define the linear charge density λ, the charge per unit length on the rod, as

$$\lambda = \frac{Q}{2a} \qquad (5\text{–}15)$$

then the charge on the element of length dx of the rod is

$$dQ = \lambda \, dx$$

The contribution $d\mathbf{E}$ at P which is due to the charge dQ will point away from the element of length dx. However, there is another element dx' on the other side of the rod. The two will combine to give a resultant pointing straight down. Thus, we know that the electric field will point down in Fig. 5–9, away from the center of the rod. So all we have to calculate is

$$dE_\perp = |d\mathbf{E}| \cos \alpha \qquad (5\text{–}16a)$$

We also read off from Fig. 5–9 that $\cos \alpha = a/r$ or $r = a/\cos \alpha$. The magnitude of $|d\mathbf{E}|$ is

$$|d\mathbf{E}| = \frac{1}{4\pi\epsilon_0} \frac{dQ}{r^2} = \frac{1}{4\pi\epsilon_0} \frac{\lambda \, dx}{(a/\cos \alpha)^2}$$

and we have for the vertical component

$$dE_\perp = \frac{\lambda}{4\pi\epsilon_0} \frac{\cos^3 \alpha}{a^2} \, dx \qquad (5\text{–}16b)$$

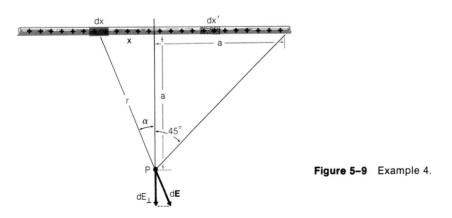

Figure 5–9   Example 4.

However, α depends on x; we have

$$\frac{x}{a} = \tan \alpha$$

and thus

$$dx = a\, d(\tan \alpha) = \frac{a}{\cos^2 \alpha}\, d\alpha$$

Inserting this into Eq. (5–16b), we obtain

$$dE_\perp = \frac{\lambda}{4\pi\epsilon_0 a^2} \cos^3 \alpha \frac{a}{\cos^2 \alpha}\, d\alpha$$

or

$$dE_\perp = \frac{\lambda}{4\pi\epsilon_0 a} \cos \alpha\, d\alpha$$

The integral over α goes from $\alpha = -45°$ to $\alpha = +45°$; thus,

$$|\mathbf{E}| = E_\perp = \frac{\lambda}{4\pi\epsilon_0 a} \int_{-\pi/4}^{\pi/4} \cos \alpha\, d\alpha$$

and the integral can be performed explicitly:

$$\int_{-\pi/4}^{\pi/4} \cos \alpha\, d\alpha = \sin \alpha \Big|_{-\pi/4}^{\pi/4} = \frac{\sqrt{2}}{2} - \left(-\frac{\sqrt{2}}{2}\right) = \sqrt{2}$$

Using the definition $\lambda = Q/2a$, we thus find that the electric field has the magnitude

$$|\mathbf{E}| = \frac{(Q/2a)}{4\pi\epsilon_0 a} \sqrt{2} = \frac{Q}{4\pi\epsilon_0 a^2} \frac{1}{\sqrt{2}} \tag{5–17}$$

and points away from the center of the rod. Since $K_0 = \dfrac{1}{4\pi\epsilon_0} = 9 \times 10^9$ N-m^2/(coul)^2, we obtain numerically

$$|\mathbf{E}| = 9 \times 10^9 \frac{\text{N-m}^2}{(\text{coul})^2} \frac{10^{-9}\,\text{coul}}{(0.1\,\text{m})^2} \times 0.707 = 6.36 \times 10^2 \frac{\text{N}}{\text{coul}}$$

Example 5   How would one calculate the same result as in Example 4 using a computer?

We will only sketch the method without going into details. The procedure consists of dividing the rod into N (e.g., $N = 100$) little segments and calculating the electric field due to each segment. Each segment will carry the charge Q/N. However, in the FORTRAN computer language we cannot add vectors directly. Instead, we use an array [defined by the statement DIMENSION E(2)], and define $E(1)$ to be the component parallel to the rod

$$E(1) = |\mathbf{E}| \sin \alpha$$

while $E(2)$ is the perpendicular component

$$E(2) = |\mathbf{E}| \cos \alpha$$

We now calculate both components, using for simplicity a special value for Q and a such that

$$\frac{1}{4\pi\epsilon_0} \frac{Q}{a^2} = 1$$

We then obtain, if we carry out our DO loops correctly, the result

$$E(1) = 0$$

$$E(2) = 0.707$$

How do we know that Eq. (5–17) is correct—that the factor with which we have to multiply is indeed equal to

$$\frac{1}{4\pi\epsilon_0}\frac{Q}{a^2}?$$

Since every contribution to the sum is proportional to $1/4\pi\epsilon_0$ and to the charge Q, the total sum is also proportional to each of these quantities. On the other hand, if we double the size a, each element will now also be twice as far away; because of Coulomb's law, this implies that the contribution by each segment will be four times smaller. Thus, the contribution of each segment is proportional to $1/a^2$, and therefore the sum is also proportional to $1/a^2$.

Summary of Important Relations

Force between two point charges (Coulomb's law)

$$F = \frac{1}{4\pi\epsilon_0}\frac{Q_1Q_2}{r^2}$$

$$\frac{1}{4\pi\epsilon_0} = K_0 = 9.0 \times 10^9 \frac{\text{N-m}^2}{\text{coul}^2}$$

Force on charge q in electric field $\mathbf{E}$ $\qquad \mathbf{F} = q\mathbf{E}$

Volume charge density $\qquad \rho = \dfrac{dQ}{d\tau}$

Electric field of charge distribution (see Fig. 5–8)

$$\mathbf{E} = \frac{1}{4\pi\epsilon_0}\int \frac{\rho_P\hat{\mathbf{u}}_{PP'}}{r^2_{PP'}}d\tau_{P'} \qquad \text{continuous charge distribution}$$

$$\mathbf{E} = \frac{1}{4\pi\epsilon_0}\sum_i \frac{Q_i}{r_i^2}\hat{\mathbf{u}}_i \qquad \begin{array}{l}\text{discrete charge distribution} \\ (r_i \text{ distance to } Q_i; \hat{\mathbf{u}}_i \text{ unit} \\ \text{vector from charge } Q_i)\end{array}$$

Questions

1. As we learned in mechanics, the gravitational acceleration at the surface of the earth is $g = 9.8$ m/sec². Estimate, using examples from daily experience, by how much g would have to increase in order to:

   (i) make animal life as we know it impossible.

   (ii) quickly make loose ground (sand) as solid as rock.

   (iii) make solids collapse.

2. A metallic (conducting) object carrying no net charge will be attracted by a charged rod. Explain why.

3. Electrostatic demonstrations usually fail on humid days or toward the end of a lecture (if the classroom is crowded). Explain why.

4. It is very difficult to do quantitative electrostatic demonstrations because the forces between charged objects are quite small. But we said that electrostatic forces are very strong. Explain the "paradox." (See also Problem 9.)

5. A philosophy major tells you: "Only the force $\mathbf{F} = q\mathbf{E}$ on a test charge is real, but the electric field $\mathbf{E}$ itself is an abstract concept which has no real meaning." What is your answer?

Problems

1. Calculate the conversion factor between the electrostatic CGS unit of charge and the coulomb:

$$1 \text{ statcoulomb} = (?) \text{ coulomb}$$

2. How far apart would two charges of $+1$ coulomb each have to be to exert a force of 1000 newtons on each other (1000 newton $\approx$ weight of 200 lbs. mass)?

3. Calculate the ratio between the electrostatic attractive force and the gravitational attractive force between a proton (charge $+1e$, mass $M = 1.67 \times 10^{-27}$ kg) and the electron (charge $-1e$, mass $m = 0.91 \times 10^{-30}$ kg) if they are separated by the distance $1 \text{ Å} = 10^{-10}$ m. Does the distance matter?

4. The diameter of the proton is $d = 1.6 \times 10^{-15}$ m. What would be the force between two charges of $e/2$ a distance d apart?

5. The hydrogen atom has a radius $a_0 = 0.53$ Å. What is the force between the proton and electron at that distance?

6. Let us assume that the charge of the copper nucleus is not exactly equal to $+29e$ ($-e =$ charge of electron $= -1.602 \times 10^{-19}$ coul), but instead equals $29e(1 + \epsilon)$ with $\epsilon = 10^{-10}$. Calculate under this assumption the force between two copper blocks, each of 10 g mass and $r = 10$ cm apart, if both blocks are "neutral," i.e., there are exactly 29 electrons for each copper nucleus.

7. Two electrons are initially (at $t = 0$) 1 Å apart at rest. Calculate the acceleration of each and determine (or estimate) how long it will take until they are 10 Å apart.

8. Two objects, each carrying the unknown charge, Q, repel each other with a force $F = 10$ newtons, if they are 1 m apart. How large is Q?

9. On clear days there is, just above the ground, an electric field of about 100 newtons/coulomb pointing downward. Calculate the acceleration of an electron in this field. Will the gravitational attraction of the earth affect your result?

10. A spark discharge will develop in dry air if the electric field is larger than 2×10^4 newtons/coulomb. A piece of copper of mass $m = 1$ g is charged so that the electric field at $r = 10$ cm away has the above value and is pointing away from the copper. What fraction of the copper atoms have had one electron removed?

11. Two conducting balls, each of mass $m = 0.3$ g, are hung on silk threads so that they touch. Each one is given the same unknown charge Q. One then finds that each

thread forms the angle $\alpha = 45°$ with the vertical. What is the value of the charge Q, if the thread length is $L = 30$ cm?

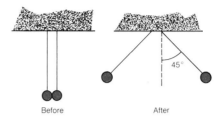

Before   After

12. Four charges are arranged, as shown in the figure, on the corners of a square. Calculate the electric field at the center P of the square.

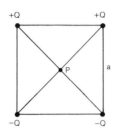

13. Calculate the force acting on any of the four charges of Problem 12 due to the presence of the other three charges.

°14. Four charges $+Q$ are located at the corners of a square. How many points are there along the two dotted lines in the figure where the electric field vanishes?

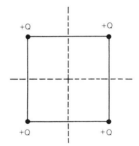

15. In an ionic lattice (e.g., a crystal of NaCl), positive ions (Na$^+$) alternate with negative ions (Cl$^-$). Here we discuss a simpler linear chain of such ions. Calculate to 1% accuracy the electric field at the point P a distance $a = 2$ Å away from an infinitely long linear chain of charges $\pm e$ which are spaced by the same distance a apart. The point P is opposite one of the positive charges.

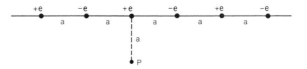

16. Five charges, each of value $+Q$, are spaced along a line at equal distances a. Calculate the force on any one of the charges. Be careful to specify the direction!

17. Three charges $(+Q)$, $(+Q)$ and $(-2Q)$ are located at the corners of an equilateral triangle of side a. Calculate the electric field (magnitude and direction!) at the three halfway points P_1, P_2, and P_3.

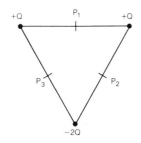

18. Three sides of a square each carry, homogeneously distributed over them, the charge of $+Q$ (the total charge is $+3Q$). Calculate the electric field at the center.

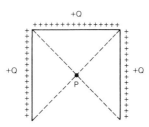

19. A thin rod of length a carries the charge $+Q$ homogeneously distributed over it. Calculate the electric field a distance x away from the end of the rod along the direction of the rod.

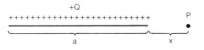

20. The charge $+Q$ is homogeneously distributed over a thin ring of radius R. Calculate — and plot against z — the electric field at an arbitrary point P on the axis through the ring and a distance z away from the ring's center.

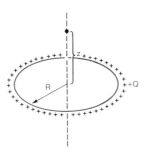

21. The charge $+Q$ is homogeneously distributed over a half ring of radius R. Calculate the electric field at the center P.

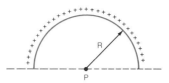

°22. A thin circular sheet of radius R carries homogeneously distributed over it the charge $+Q$. What will be the electric field at a point P a distance z above the center of the sheet? Show also that for $z \gg R$ your result is nearly the same as the field of a point charge Q at the center of the sheet.

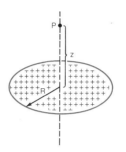

CHAPTER 6

ELECTROSTATIC POTENTIAL

1. POTENTIAL ENERGY AND ELECTROSTATIC POTENTIAL

Whenever a charge moves in the presence of other charges, it is subjected to a force while it changes position. As one learns in mechanics, this implies that work is done on the system.

Let us assume that we are using a test charge of positive charge q to measure the electric field. The force acting on the test charge is

$$\mathbf{F} = q\mathbf{E}$$

As we slowly move the test charge, either we have to supply work (if we move against the electric field) or we can gain work (by moving in the direction of the electric field). In mechanics, we learn that if the force is constant, then the work done by the test charge as it moves along a displacement vector $\mathbf{l}$ is equal to (Fig. 6-1)

$$W = (\mathbf{F} \cdot \mathbf{l}) = |\mathbf{F}||\mathbf{l}| \cos \alpha \qquad (6\text{-}1a)$$

The work done *by the charge* equals the scalar product of the force and the displacement. We can extract the energy—e.g., by braking the test charge along its path. If the test particle is not restrained (i.e., if we release it), it will accelerate along the path and will have a larger kinetic energy after it has traversed the path:

$$\left(\frac{mv^2}{2}\right)_{\text{end}} = \left(\frac{mv^2}{2}\right)_{\text{initial}} + W \qquad (6\text{-}1b)$$

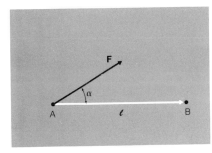

Figure 6-1 Calculating the work of a force if the force is constant.

Of course, if the angle α is larger than 90°, the work W becomes negative; we have to supply work. If the force is not constant but changes along the path as shown in Fig. 6-2, we have to subdivide the path into small elements $\Delta\,\mathbf{l}_i$, calculate the work done by the charge while moving along $\Delta\,\mathbf{l}_i$, and finally sum all the contributions to the work

$$W(A \rightarrow B \text{ along } C) \approx \sum_i (\mathbf{F}_i \bullet \Delta\,\mathbf{l}_i) \qquad (6\text{-}2a)$$

or if we write out the scalar products

$$W(A \rightarrow B \text{ along } C) \approx \sum_i |\mathbf{F}_i|\,|\Delta\,\mathbf{l}_i|\cos\alpha_i \qquad (6\text{-}2b)$$

The expressions (6-2a) and (6-2b) are only approximations; the force $\mathbf{F}$ may change along even a very small finite displacement $\Delta\,\mathbf{l}$. In order to obtain an exact result, we have to go to the limit where all the displacements $\Delta\,\mathbf{l}_i$ become infinitesimal. The sum then has to be replaced by an integral

$$W(A \rightarrow B \text{ along } C) = \int_C (\mathbf{F} \bullet d\,\mathbf{l}) \qquad (6\text{-}2c)$$

called the line integral of $\mathbf{F}$ along C. Eqs. (6-1) and (6-2) are completely general and true for any force. Here we are discussing electric forces on the test charge q. Since the force is always proportional to the charge q, the work done by the force is also proportional to q:

$$W(A \rightarrow B \text{ along } C) = qV(A \rightarrow B \text{ along } C) \qquad (6\text{-}3)$$

In mechanics we defined a *conservative force field* as one for which the work done while moving from the point A to the point B is independent of the path one chooses. Thus, for a conservative force one has (compare Fig. 6-3):

$$W(A \rightarrow B \text{ along } C) = W(A \rightarrow B \text{ along } C')$$

In mechanics, we also introduced the concept of potential energy by showing that for conservative forces there exist a function $U(x,y,z)$ defined at any point so that

$$W(A \rightarrow B \text{ along any curve}) = U(A) - U(B) \qquad (6\text{-}4)$$

We call $U(A)$ the potential energy at the point A. Note that Eq. (6-1b) then re-

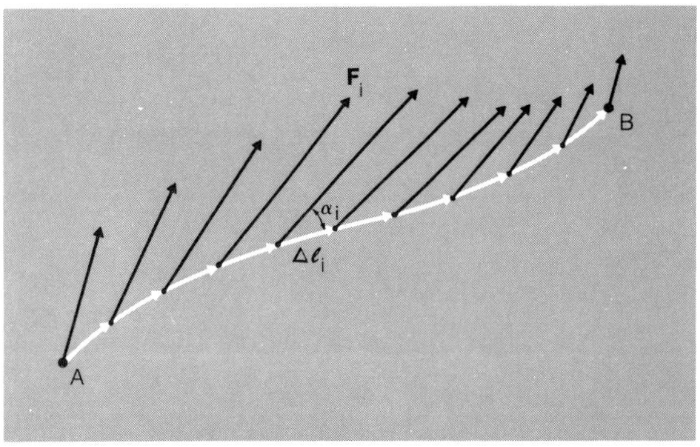

Figure 6-2 Calculating the work of a force if the force changes along the path taken.

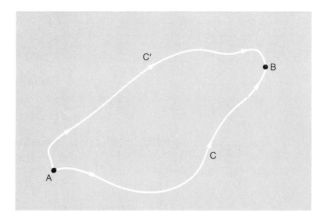

Figure 6–3 For conservative forces, the work done by the force is independent of the path taken.

duces to the energy conservation law

$$\left(\frac{mv^2}{2}\right)_B - \left(\frac{mv^2}{2}\right)_A = U(A) - U(B)$$

or

$$\left(\frac{mv^2}{2} + U\right)_B = \left(\frac{mv^2}{2} + U\right)_A = E = \text{total energy}$$

We will show in the next section that electrostatic forces are indeed conservative; for the moment, we assume it to be true. We thus conclude that our test charge q indeed has everywhere a potential energy. But because the force acting on q at any point is proportional to the charge q, the potential energy will also be proportional to the charge. At any point A we have

$$\boxed{U(A) = qV(A)} \qquad\qquad (6\text{–}5)$$

We call $V(A)$ the *electrostatic potential* at the point A. Thus, the electrostatic potential in the presence of some charge distribution is defined as the potential energy per unit charge of a positive test charge. Note that the electrostatic energy is *not equal* to the potential energy of any of the individual charges which make up the electric field; we first have to assemble the whole charge distribution, and only then do we introduce our test charge q.

   Example 1   In a region of space the electric field is constant as shown in Fig. 6–4. What is the difference in electrostatic potential between the points A and B? What is the difference in the potential energy for a negative charge $(-Q)$?

   The electrostatic potential is defined as the potential energy of a positive test charge. The force on such a test charge is proportional to the electric field:

$$\mathbf{F} = q\mathbf{E}$$

The work done by the test charge is

$$W(A \rightarrow B) = |\mathbf{F}|\, a \cos \alpha$$

and since $\cos (45°) = 1/\sqrt{2}$, we have

$$W(A \rightarrow B) = q\frac{|\mathbf{E}|\, a}{\sqrt{2}}$$

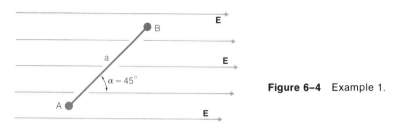

Figure 6–4 Example 1.

According to Eq. (6–4), this is equal to the difference in potential energy of the test charge

$$W(A - B) = U(A) - U(B)$$

and since the electrostatic potential is equal to the potential energy per unit charge, we have

$$V(A) - V(B) = \frac{U(A) - U(B)}{q}$$

or

$$V(A) - V(B) = \frac{|\mathbf{E}|\, a}{\sqrt{2}}$$

Note that we used a positive test charge q. If we now place a negative charge $(-Q)$ at A, the force on Q will have a direction opposite to $\mathbf{E}$

$$\mathbf{F} = -Q\mathbf{E}$$

When we move the charge $-Q$ from A to B, we move *against* the force field; we therefore have to supply the work. The difference in potential energy is negative:

$$U_{(-Q)}(A) - U_{(-Q)}(B) = \frac{-Q\,|\mathbf{E}|\, a}{\sqrt{2}}$$

One should note that Eq. (6–5) is valid for both positive and negative charges; for positive charges the potential energy has the same sign as the electrostatic energy, while for negative charges the two quantities have opposite signs.

As we had already noted in mechanics (Vol. I, p. 182), the potential energy is defined only up to an arbitrary constant. Combining Eqs. (6–2c) and (6–4), we can write

$$U(A) - U(B) = \int_{A}^{B} (\mathbf{F} \bullet d\,\mathbf{l})$$

or

$$U(B) = -\int_{A}^{B} (\mathbf{F} \bullet d\,\mathbf{l}) + U(A) \qquad (6\text{–}6)$$

If we now decide once and for all on some initial point A_0 – usually a point very far away from the charge distribution – we can set $U(A_0)$ equal to any value we choose. For simplicity, we choose $U(A_0) = 0$. Then the potential energy at any point P is:

$$U(P) = -\int_{A_0}^{P} (\mathbf{F} \bullet d\,\mathbf{l}) \qquad (6\text{–}7)$$

Since we are discussing the potential energy of a positive test charge q, we can deduce that the electrostatic potential is equal to

$$V(P) = -\int_{A_0}^{P} (\mathbf{E} \bullet d\,\mathbf{l}) \tag{6-8}$$

The equivalence of Eqs. (6–7) and (6–8) may be shown as follows: If we divide $U(P)$ by the charge q of our test charge, we obtain according to Eq. (6–5) the electrostatic potential $V(P)$. If, however, we divide the right-hand side of Eq. (6–7) by q, we obtain the right-hand side of Eq. (6–8) because of the fundamental definition of the electric field [Eq. (5–7)]

$$\mathbf{E} = \mathbf{F}/q$$

We measure potential energy in units of Joules; thus, we have to measure the electrostatic potential according to Eq. (6–5) in Joules/coulomb. Since this unit is used very frequently, a new name, the volt (V), has been introduced. We thus have

$$1 \text{ volt} = 1 \text{ V} = 1 \text{ Joule/coulomb} = 1 \text{ J/coul}$$

Since 1 joule = 1 newton-meter, we can also write

$$1 \text{ volt} = 1 \text{ newton-meter/coulomb}$$

We have up to now used the unit newton/coulomb for the electric field. From what we just said, we may write

$$1 \frac{\text{newton}}{\text{coulomb}} = 1 \left.\frac{\text{newton-meter}}{\text{coulomb}}\right/ 1 \text{ meter}$$

$$= \text{volt/meter}$$

Thus, we can call the unit of electric field one volt per meter. A number of com-

TABLE 6–1 TYPICAL ELECTRIC FIELDS AND POTENTIAL DIFFERENCES

Breakdown (spark) in air	$(1 - 3) \times 10^6$ V/m
Electric field in atmosphere on clear day	10^2 V/m
Electric field in thundercloud	$\sim 10^6$ V/m
Electric field in hydrogen atom	$\sim 10^{11}$ V/m
Electric field in innermost shell of lead atom	$\sim 10^{17}$ V/m
Electric field near surface of proton	$\sim 10^{21}$ V/m
Household plug (peak value)	$115\ \sqrt{2}$ V = 160 V
Long distance transmission lines	up to 5×10^5 V
Electrostatic potentials inside atoms	10 V $\left(\begin{array}{c}\text{at position of}\\\text{outermost}\\\text{electrons}\end{array}\right) - 10^5$ V $\left(\begin{array}{c}\text{at position of}\\\text{inside electrons}\\\text{in heavy atoms}\end{array}\right)$
Potential near surface of proton	10^6 V
Highest man-made potential difference relative to earth	$\sim 10^7$ V

mon field strengths and potential differences are listed in Table 6–1 to give an idea of the magnitudes that occur in nature.

The introduction of the volt as a unit leads us to also introduce another unit of energy, the electron-volt (eV). It is defined as the potential energy of a test charge of charge $1\ e = 1.602 \times 10^{-19}$ coul at a point where the electrostatic potential equals one volt. Thus,

$$1\ \text{eV} = 1.602 \times 10^{-19}\ \text{J}$$

The electron-volt unit is particularly suitable for measuring energies on the atomic scale.

2. POTENTIAL OF A POINT CHARGE

As we have shown in Eq. (6–8), the electrostatic potential* at some point P is given by

$$V(P) = -\int_{A_0}^{P} (\mathbf{E} \bullet d\,\mathbf{l})$$

where A_0 is some specified initial point. However, Eq. (6–8) has a meaning only if the integral does not depend on the particular path we choose to go from A_0 to P. Thus, we have to prove that the electrostatic force on a test charge — or the electric field itself — is conservative. We will first prove this result for the electric field of a single point charge Q. As we discussed in Section 5–4, an arbitrary charge distribution can always be considered as being built up from individual point charges. Thus, if the electric field of a single point charge is conservative, then the electric field of any charge configuration will also be conservative.

Let us therefore consider (Fig. 6–5) a segment of some path in the neighborhood of an isolated point charge Q. In order to make the example concrete, we will assume Q to be positive. Then at an arbitrary point P along the path, the electric field points away from Q and has the magnitude

$$|\mathbf{E}| = \frac{1}{4\pi\epsilon_0} \frac{Q}{r^2} \tag{6–9}$$

───────────

* We frequently use the abbreviation "potential" for "electrostatic potential."

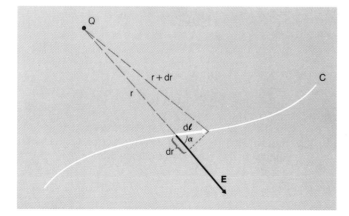

Figure 6–5 The scalar product $(\mathbf{E} \bullet d\,\mathbf{l})$ is equal to $|\mathbf{E}|\ dr$, where dr is the change in distance from the point charge Q.

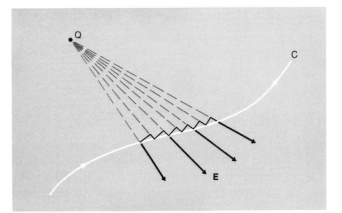

Figure 6-6 We can replace a path by a succession of small steps if we want to calculate $\int \mathbf{E} \cdot d\mathbf{l}$.

The scalar product between the electric field $\mathbf{E}$ and the small path element $d\ell$ equals

$$\mathbf{E} \cdot d\mathbf{l} = |\mathbf{E}|\,|d\mathbf{l}|\cos\alpha$$

From Fig. 6–5 we see that

$$|d\mathbf{l}|\cos\alpha = dr$$

is the change in distance from the charge Q. If we substitute for $|\mathbf{E}|$ the expression (6–9) we have

$$(\mathbf{E} \cdot d\mathbf{l}) = \frac{1}{4\pi\epsilon_0}\frac{Q}{r^2}dr \tag{6–10}$$

Thus, $(\mathbf{E} \cdot d\mathbf{l})$ along each infinitesimal shift $d\ell$ along a curve C is the same as it is along one of the steps of the step-curve shown in Fig. 6–6. We now can perform the integral between any two points A and B, along any path C:

$$\int_A^B \mathbf{E} \cdot d\mathbf{l} = \int_{r_A}^{r_B} \frac{1}{4\pi\epsilon_0}\frac{Q}{r^2}dr \tag{6–11}$$

where r_A and r_B are the distances between the points A and B, respectively, and

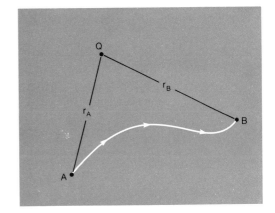

Figure 6-7 The potential difference between two points A and B depends only on the distances r_A and r_B from the point charge Q, but not on the path taken.

the charge Q (Fig. 6–7). Performing the integration, we obtain

$$\int_{r_A}^{r_B} \frac{1}{4\pi\epsilon_0} \frac{Q}{r^2}\, dr = \frac{Q}{4\pi\epsilon_0} \int_{r_A}^{r_B} \frac{dr}{r^2}$$

and since $\int dx/x^2 = -1/x$, we have

$$\int_A^B \mathbf{E} \cdot d\mathbf{l} = \frac{Q}{4\pi\epsilon_0}\left(-\frac{1}{r_B} + \frac{1}{r_A}\right) \tag{6–12}$$

Our result is the same whatever path we integrate along from A to B; it depends only on the end points of the path, i.e., on their distance from the point charge Q. Thus, we have shown that the electric field of a point charge is conservative. We can now define the electrostatic potential of a point charge at an arbitrary point P. Using Eq. (6–8), we have

$$V(P) = -\int_{A_0}^P \mathbf{E} \cdot d\mathbf{l} = \frac{Q}{4\pi\epsilon_0}\left(\frac{1}{r_P} - \frac{1}{r_{A_0}}\right)$$

For simplicity, we choose the initial point A_0 very far away:

$$r_{A_0} \to \infty$$

Then the electrostatic potential at a point P can be written

$$\boxed{V(P) = \frac{1}{4\pi\epsilon_0}\frac{Q}{r}} \tag{6–13}$$

Example 2   The hydrogen atom consists of a hydrogen nucleus (proton) of charge $+1\,e = 1.60 \times 10^{-19}$ coul and an electron of charge $-1\,e$. The radius of the hydrogen atom is $r = 0.53$ Å. What is the electrostatic potential due to the proton at this distance? What is the electric field? What is the potential energy of an electron at this distance?

This example will give us an idea of what are "reasonable" electric fields or electrostatic potentials on the atomic scale. The electrostatic potential is

$$V = \frac{1}{4\pi\epsilon_0}\frac{e}{r} = 9 \times 10^9 \frac{\text{V-m}}{\text{coul}}\frac{1.6 \times 10^{-19}\ \text{coul}}{0.53 \times 10^{-10}\ \text{m}} = 27.2\ \text{volts}$$

By comparing Eq. (6–13) with Eq. (6–9) we see that for a point charge – *but only for a point charge!* – one has

$$|\mathbf{E}| = \frac{V}{r} = \frac{27.2\ \text{volts}}{0.53 \times 10^{-10}\ \text{meters}} = 5.13 \times 10^{11}\ \text{V/m}$$

The electron has a negative charge; since the electrostatic potential is positive, its potential energy is negative:

$$U = (-1\,e)\,V = -27.2\ \text{electron-volts}$$

or expressed in Joules

$$U = -1.602 \times 10^{-19} \times 27.2 = -4.35 \times 10^{-18}\ \text{Joule}$$

We see why it is advantageous to use the electron-volt unit for energies on the atomic scale. No large powers of ten are necessary; in addition, since all atomic charges are small integer multiples of 1 e, conversion between potential energy and electrostatic potential is very easy.

3. POTENTIAL OF CHARGE DISTRIBUTIONS

We have just learned how to calculate the electrostatic potential of a single point charge. We have shown it to be equal to [Eq. (6–13)]

$$V = \frac{1}{4\pi\epsilon_0} \frac{Q}{r}$$

where r is the distance to the charge Q. Let us now assume that we have several point charges $Q_1, Q_2, \ldots Q_N$; each charge can be positive or negative (Fig. 6–8). How do we obtain the electrostatic potential at an arbitrary point P? It is easy to answer this question if we remember that at any point the electric field is the vectorial sum of the electric fields due to the individual point charges:

$$\mathbf{E}_{\text{tot}} = \mathbf{E}(\text{due to } Q_1) + \mathbf{E}(Q_2) + \ldots + \mathbf{E}(Q_N) \tag{6–14}$$

The electrostatic potential at a point P can be calculated by the expression

$$V = -\int_{A_0}^{P} \mathbf{E}_{\text{tot}} \cdot d\mathbf{l} = -\int_{A_0}^{P} \mathbf{E}(Q_1) \cdot d\mathbf{l} - \ldots - \int_{A_0}^{P} \mathbf{E}(Q_N) \cdot d\mathbf{l} \tag{6–15}$$

Thus, it is the sum of N integrals. In each integral, $\mathbf{E}(Q_i)$ is the electric field due to the single point charge Q_i. We can calculate the integral explicitly. If we choose the point A_0 very far away, we obtain, according to Eq. (6–13),

$$-\int_{A_0 \to \infty}^{P} \mathbf{E}(Q_i) \cdot d\mathbf{l} = \frac{1}{4\pi\epsilon_0} \frac{Q_i}{r_i}$$

where r_i is the distance from the point P to the point charge Q_i. The total electrostatic potential is therefore the sum of the electrostatic potentials due to each point charge:

$$V = \sum_{i=1}^{N} V_i = \frac{1}{4\pi\epsilon_0} \sum_{i=1}^{N} \frac{Q_i}{r_i} \tag{6–16}$$

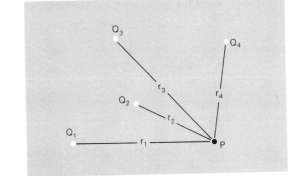

Figure 6–8 To calculate the electrostatic potential due to several point charges $Q_1, Q_2, \ldots$ at the point P, we add the contribution from each point charge.

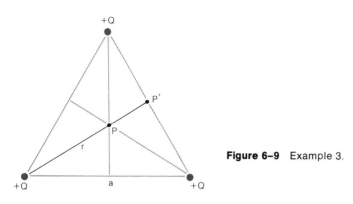

Figure 6-9 Example 3.

Example 3   Calculate the electrostatic potential at the center of a triangle of side a, if at each corner there is the charge $+Q_0$ (Fig. 6–9).

We recognize the situation as being identical to that in Example 2 on p. 150. There we found that the electric field at the center vanishes. But from this we should *not* conclude that the electrostatic potential also vanishes. Indeed, using Eq. (6–16) and taking into account the fact that all three charges are equidistant from P, we obtain

$$V = \sum V_i = 3\,V_i = \frac{3}{4\pi\epsilon_0}\frac{Q}{r}$$

where r is the distance from P to Q_1. From plane geometry we know that

$$r = \overline{QP} = \frac{2}{3}\,\overline{QP'}$$

but by the Pythagorean theorem we obtain

$$\overline{QP'} = \sqrt{a^2 - \left(\frac{a}{2}\right)^2} = \frac{\sqrt{3}}{2}\,a$$

Therefore, we have

$$r = \frac{2}{3}\frac{\sqrt{3}}{2}\,a = \frac{a}{\sqrt{3}}$$

and the electrostatic potential at the center is

$$V(P) = \frac{3\sqrt{3}}{4\pi\epsilon_0}\frac{Q}{a}$$

Example 4   One of the charges in Fig. 6–9 is replaced by the charge $(-Q)$. What is now the electrostatic potential at the center?

Exactly as before, we write

$$V = \sum V_i$$

but now $V_1 = V_2 = -V_3$; thus, $\sum V_i = V_1 + V_2 + V_3 = V_1$ and we have

$$V = \frac{1}{4\pi\epsilon_0}\frac{Q}{r} = \frac{\sqrt{3}}{4\pi\epsilon_0}\frac{Q}{a}$$

Example 5   How much work (energy) was required to assemble the triangle with three equal charges, if initially they were all very far away from each other?

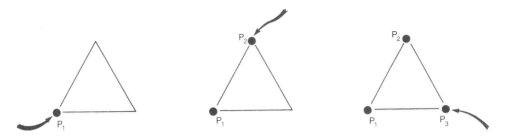

Figure 6–10 Example 5: Calculating the work required to assemble a set of point charges.

At first sight this seems a quite different problem; upon closer inspection we will find that the problem can be trivially reduced to a succession of calculations of the electrostatic potential. First we move one charge into its place (Fig. 6–10a) at P_1. No work is required, since all other charges are very far away — thus there is no force. Then, keeping the first charge fixed, we move the second charge into place at P_2 as shown in Fig. 6–10b. During its motion we can consider it a "test charge." The electrostatic potential due to Q_1 (at P_1) is at the point P_2

$$V(P_2, \text{ due to } Q_1) = \frac{1}{4\pi\epsilon_0} \frac{Q}{a}$$

Bringing the second charge to P_2 thus requires the work

$$W_1 = Q \ V(P_2, \text{ due to } Q_1) = \frac{1}{4\pi\epsilon_0} \frac{Q^2}{a}$$

Once the second charge is in place, we bring in the third charge to P_3 (Fig. 6–10c). The electrostatic potential at P_3 is now

$$V(P_3, \text{ due to } Q_1 \text{ and } Q_2) = 2 \frac{1}{4\pi\epsilon_0} \frac{Q}{a}$$

because there are now two charges $+Q$, each a distance a away from P_3. Thus, the work required to move the third charge into its place is

$$W_2 = Q \cdot V(P_3, \text{ due to } Q_1 \text{ and } Q_2) = 2 \frac{1}{4\pi\epsilon_0} \frac{Q^2}{a}$$

The total work required to assemble the complete triangle is then

$$W = W_1 + W_2 = \frac{3}{4\pi\epsilon_0} \frac{Q^2}{a}$$

The same approach with which we calculated the electrostatic potential for a few charges also can be used if the number of point charges becomes very large. The only modification is that we have to use the concept of charge density as we did on p. 153. Each little volume element (Fig. 6–11) $d\tau$ at a point P' has the charge

$$dQ_{P'} = \rho_{P'} d\tau_{P'}$$

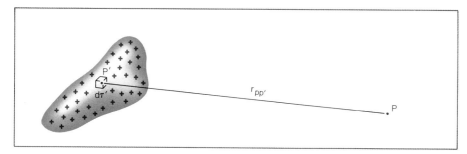

Figure 6–11 Calculating the potential of a continuous charge distribution; we divide space into small volume elements $d\tau'$, calculate the potential due to $\rho d\tau'$ at P, and then add (integrate over) all contributions.

and will at an arbitrary point P give rise to an electrostatic potential

$$dV_P = \frac{1}{4\pi\epsilon_0}\frac{dQ_{P'}}{r_{PP'}} = \frac{1}{4\pi\epsilon_0}\frac{\rho_{P'}}{r_{PP'}}d\tau_{P'}$$

The total electrostatic potential is obtained by adding all contributions; since the summation is over a continuous volume, we have to use the integral

$$V_P = \frac{1}{4\pi\epsilon_0}\int\frac{\rho_{P'}d\tau_{P'}}{r_{PP'}}$$

(6–17)

Example 6   The total charge Q is distributed homogeneously over a thin ring of radius R. What is the electrostatic potential at an arbitrary point on the axis to the ring?

By inspection of Fig. 6–12, we see that any charge element dQ is the same distance r away from the point P:

$$r = \sqrt{d^2 + R^2}$$

Thus, the electrostatic potential at the point P equals

$$V_P = \frac{1}{4\pi\epsilon_0}\int\frac{dQ}{\sqrt{d^2 + R^2}}$$

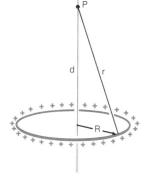

Figure 6–12 Example 6: The potential of a charged ring for a point on the axis.

and since $\int dQ = Q$, we have

$$V_P = \frac{1}{4\pi\epsilon_0} \frac{Q}{\sqrt{d^2 + R^2}}$$

4. THE GRADIENT OF THE POTENTIAL

From Eq. (6–8) it is quite obvious that if we know the electrostatic field at any point in a region of space, we can calculate the electrostatic potential at any point within the region, at least to within an unknown additive constant. Indeed, if we know the electric field everywhere along the curve C_1 (Fig. 6–13), we can calculate for any points P_2 and P_1 the difference between their electrostatic potentials by using Eq. (6–8):

$$V(P_1) - V(P_0) = -\int_{P_1}^{P_2} \mathbf{E} \cdot d\mathbf{l} \tag{6–18}$$
$$\text{along } C_1$$

We will now answer the following two questions:

(a) What statements can one make about the electric field using only the knowledge that there exists an electrostatic potential?

(b) Assuming that we know the value of the electrostatic potential at every point in some region of space, can we then determine the electric field anywhere in the same region of space?

Let us answer first question (a). Since there exists an electrostatic potential, the integral

$$I = \int_{P_0}^{P_1} \mathbf{E} \cdot d\mathbf{l}$$

has to be independent of the path we choose between P_0 and P_1. Thus, if we have two paths C_1 and C_2 as shown in Fig. 6–13, we have

$$\int_{P_0}^{P_1} \mathbf{E} \cdot d\mathbf{l} = \int_{P_0}^{P_1} \mathbf{E} \cdot d\mathbf{l} \tag{6–19}$$
$$\text{along } C_1 \qquad\qquad \text{along } C_2$$

However, let us now go along C_1 from P_0 to P_1 and then back from P_1 to P_0 along

Figure 6–13 The integral $\int \mathbf{E} \cdot d\mathbf{l}$ from P_0 to P_1 is the same whether taken along C_1 or C_2; it follows that the integral $\int \mathbf{E} \cdot d\mathbf{l}$ taken from P_0 to P_1 along C_1 and back to P_0 along C_2 has to vanish.

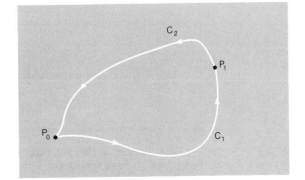

C_2. We then have, by Eq. (6–18),

$$V(P_1) - V(P_0) = -\int_{P_0}^{P_1} \mathbf{E} \cdot d\mathbf{l}$$ (6–20a)

along C_1

but also

$$V(P_0) - V(P_1) = -\int_{P_1}^{P_0} \mathbf{E} \cdot d\mathbf{l}$$ (6–20b)

along C_2

It is important to note that the integration limits in Eq. (6–20b) are reversed; we are moving from P_1 to P_0. If we now add the two equations (6–20a) and (6–20b), we find that the left-hand side vanishes:

$$[V(P_1) - V(P_0)] + [V(P_0) - V(P_1)] = 0$$

Thus, the right-hand has to vanish also, and we have

$$\int_{P_0}^{P_1} \mathbf{E} \cdot d\mathbf{l} + \int_{P_1}^{P_0} \mathbf{E} \cdot d\mathbf{l} = 0$$ (6–21)

along C_1 along C_2

But P_0 and P_1 are completely arbitrary points. The only limitation on the paths C_1 and C_2 is that if one follows the two paths in the sequence given by Eq. (6–21), one ends up at the original point P_0. Thus, C_1 and C_2 form a closed loop and the integral of ($\mathbf{E} \cdot d\mathbf{l}$) over any closed loop has to vanish:

$$\int_{\substack{\text{closed} \\ \text{loop}}} \mathbf{E} \cdot d\mathbf{l} = 0$$ (6–22)

Eq. (6–22) is frequently called *Maxwell's first equation of electrostatics*. We will later learn that it is true only if all charges are at rest or are moving in such a way as to produce a steady-state current. However, for the moment we are discussing only charges at rest.

We will just state here without proof that Eq. (6–22) is the only equation we can deduce from the requirement that there exists an electrostatic potential. However, let us note that it poses a very strong restriction on how we can assume the electric field to behave, because it has to be *true for any closed loop.*

Example 7   Let us assume that the electric field is given in component form

$$\mathbf{E} = E_x \hat{\mathbf{i}} + E_y \hat{\mathbf{j}}$$

We are told that E_x is proportional to y

$$E_x = ay$$

and all we know is that $E_y = 0$ wherever $x = 0$. Can we uniquely determine E_y?

The answer is that we can. To show this, we choose rectangular loops (Fig. 6–14); we go from the origin along the x-axis to A, from there parallel to the y-axis

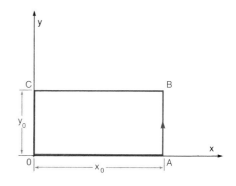

Figure 6–14 Example 7: The loop over which we calculate $\int \mathbf{E} \cdot d\mathbf{l}$.

to B, then to C, and back to O. Thus, the integral can be written as a sum of four terms. Note that along each side of the rectangle only one component contributes to the integral, because

$$d\mathbf{l} = dx\,\hat{\mathbf{i}} + dy\,\hat{\mathbf{j}}$$

is parallel to one of the axes. First, from O to A we have

$$I_1 = \int_0^{x_0} E_x(x, y = 0)\,dx$$

But by assumption $E_x = ay$; thus, $E_x = 0$ for $y = 0$. Thus $I_1 = 0$. From A to B we have

$$I_2 = \int_{y=0}^{y_0} E_y(x = x_0, y)\,dy$$

and as yet we can say nothing about this integral, because it contains the unknown E_y. From B to C we have

$$I_3 = \int_{x=x_0}^{0} E_x(x, y = y_0)\,dx$$

(Note the lower and upper limits!) Since $E_x = ay_0$, we have

$$I_3 = ay_0 \int_{x=x_0}^{0} dx = -ay_0 \int_{x=0}^{x_0} dx = -ax_0 y_0$$

Finally, from C to O we have

$$I_4 = \int_{y=y_0}^{0} E_y(x = 0, y)\,dy$$

and since we are told in the beginning that $E_y = 0$ wherever $x = 0$, we know that $I_4 = 0$. Because of Eq. (6–22) we have

$$I_1 + I_2 + I_3 + I_4 = 0$$

or, since only I_2 and I_3 do not necessarily vanish, we have

$$\int_{y=0}^{y_0} E_y(x = x_0, y)\,dy - ax_0 y_0 = 0$$

which has to be true for any combination of x_0 and y_0. It is easy to see now that the only solution is

$$E_y(x, y) = ax \qquad\qquad (6\text{–}22')$$

Indeed, consider E_y for a certain x_0 as a function of y; then, for two values y_1 and y_2 which are nearly equal $[y_2 = y_1 + \Delta y]$,

$$\int_0^{y_2} f(y)\,dy - \int_0^{y_1} f(y)\,dy = \int_{y_1}^{y_2} f(y)\,dy \approx f(y_1)\,\Delta y$$

But we also have

$$a x_0 y_2 - a x_0 y_1 = a x_0\,\Delta y$$

thus

$$f(y) = a x_0$$

which is equivalent to Eq. (6–22′).

We will now answer the second question posed in the beginning of this section: We will show that if we know the electrostatic potential everywhere in the neighborhood of a point P_0, we can then determine uniquely the electric field at P_0. For this purpose it is sufficient to show that we can determine any component of the electric field. If the point P_0 has the coordinates (x,y,z), we choose a second nearby point with the coordinates $(x + \Delta x, y, z)$ as shown in Fig. 6–15. By our assumption we know the electrostatic potential at both P_0 and P_1, and because of Eq. (6–7) we can write

$$V(P_0) = -\int_{A_0}^{P_0} \mathbf{E} \cdot d\mathbf{l} \qquad (6\text{–}23a)$$

and

$$V(P_1) = -\int_{A_0}^{P_1} \mathbf{E} \cdot d\mathbf{l} \qquad (6\text{–}23b)$$

where A_0 is some specified point and the integrals can be taken along any path. In particular, we can choose the path from A_0 to P_1 such that it goes to P_0 and then along a straight line (parallel to the x-axis) from P_0 to P_1. But then we have

$$V(P_1) = -\int_{A_0}^{P_0} \mathbf{E} \cdot d\mathbf{l} - \int_{P_0}^{P_1} \mathbf{E} \cdot d\mathbf{l} \qquad (6\text{–}24)$$

and because of Eq. (6–23a) this implies that

$$V(P_1) = V(P_0) - \int_{P_0}^{P_1} \mathbf{E} \cdot d\mathbf{l}$$

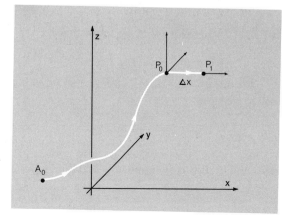

Figure 6–15 A special path chosen to calculate the electric field at P_0 from the electrostatic potential at P_0 and at P_1.

or

$$\Delta V = V(P_1) - V(P_0) = -\int_{P_0}^{P_1} \mathbf{E} \cdot d\mathbf{l} \tag{6-25}$$

However, the path from P_0 to P_1 is parallel to the x-axis; thus the differential vector $d\mathbf{l}$ is

$$d\mathbf{l} = dx\,\hat{\mathbf{i}} + 0\,\hat{\mathbf{j}} + 0\,\hat{\mathbf{k}}$$

and we can write

$$-\int_{P_0}^{P_1} \mathbf{E} \cdot d\mathbf{l} = -\int_{x}^{x+\Delta x} E_x\,dx$$

If P_1 is sufficiently close to P_0, we can then, to any accuracy, claim that

$$-\int_{x}^{x+\Delta x} E_x\,dx \approx -E_x(\text{at }P_0)\,\Delta x$$

and since the coordinates of the two points P_0 and P_1 are (x,y,z) and $(x+\Delta x,y,z)$, we can write Eq. (6–25) in the form

$$\frac{V(x+\Delta x,y,z) - V(x,y,z)}{\Delta x} \approx -E_x(x,y,z) \tag{6-26}$$

In the limit as $\Delta x \to 0$, the near equality becomes an exact equality and we have

$$\lim_{\Delta x \to 0} \frac{V(x+\Delta x,y,z) - V(x,y,z)}{\Delta x} = \frac{\partial V}{\partial x} = -E_x \tag{6-27a}$$

Thus, the x-component of the electric field is equal to the negative of the *partial derivative of* V with respect to x. In the same way, we can derive for the y- and z-components of the electric field

$$E_y = -\frac{\partial V}{\partial y} \tag{6-27b}$$

$$E_z = -\frac{\partial V}{\partial z} \tag{6-27c}$$

If we define as the *gradient* of any scalar function $\phi(x,y,z)$ the vector

$$\boxed{\operatorname{grad} \phi = \frac{\partial \phi}{\partial x}\,\hat{\mathbf{i}} + \frac{\partial \phi}{\partial y}\,\hat{\mathbf{j}} + \frac{\partial \phi}{\partial z}\,\hat{\mathbf{k}}} \tag{6-28}$$

we can rewrite the set of equations (6–27a to c) in the form

$$\boxed{\mathbf{E} = -\operatorname{grad} V} \tag{6-29}$$

■ We can clarify the meaning of the gradient of a function $\phi(x,y,z)$. If ϕ is a sufficiently well-behaved function of x, y, and z, then the equation

$$\phi(x,y,z) = c \tag{6-30}$$

defines a surface in three dimensions; at any point on the surface, ϕ takes on the same value c. Now consider the change of ϕ if we move from a point $P_0(x,y,z)$ to a nearby point $(x + \Delta x, y + \Delta y, z + \Delta z)$. We can write

$$
\begin{aligned}
\phi(x + \Delta x, y + \Delta y, z + \Delta z) - \phi(x,y,z) &= \phi(x + \Delta x, y + \Delta y, z + \Delta z) - \phi(x, y + \Delta y, z + \Delta z) \\
&+ \phi(x, y + \Delta y, z + \Delta z) - \phi(x, y, z + \Delta z) \\
&+ \phi(x, y, z + \Delta z) - \phi(x,y,z)
\end{aligned}
\tag{6-31}
$$

This is an exact identity, because all terms except the first and last on the right-hand side cancel exactly. But now we can consider the limit in which Δx, Δy, and Δz all are very small. Then, the first line in Eq. (6–31) equals

$$
\phi(x + \Delta x, y + \Delta y, z + \Delta z) - \phi(x, y + \Delta y, z + \Delta z) \approx \frac{\partial \phi}{\partial x} \Delta x
$$

and treating the other lines in the same way and writing

$$
\Delta \mathbf{r} = \Delta x\,\hat{\mathbf{i}} + \Delta y\,\hat{\mathbf{j}} + \Delta z\,\hat{\mathbf{k}}
$$

we obtain from Eq. (6–31) with the help of Eq. (6–29)

$$
\Delta \phi \approx \phi(\mathbf{r} + \Delta \mathbf{r}) - \phi(\mathbf{r}) = \operatorname{grad} \phi \bullet \Delta \mathbf{r}
\tag{6-32}
$$

The change of ϕ, if we move to a neighboring point, is equal to the scalar product of the gradient of ϕ and the differential displacement vector. Therefore, if $\Delta \mathbf{r}$ points to any point on the surface defined by Eq. (6–30), then $\Delta \phi = 0$ and thus

$$
\operatorname{grad} \phi \bullet \Delta \mathbf{r} = 0
\tag{6-33}
$$

From this we conclude that $\operatorname{grad} \phi$ is a vector which is perpendicular to the equipotential surface (surface of constant ϕ) defined by Eq. (6–30). This is shown in Fig. 6–16. The magnitude of $\operatorname{grad} \phi$ is equal to the rate of change of ϕ per unit distance, if one goes in the direction along which this rate is maximal:

$$
|\operatorname{grad} \phi| = \frac{d\phi}{d\ell}\bigg|_{\text{maximum}}
\tag{6-34}
$$

In our case the equipotential surfaces are surfaces of constant electrostatic potential. We conclude that the electric field is always perpendicular to these surfaces and that it points in the direction along which V decreases fastest. Thus the electric field, because of the minus sign in Eq. (6–29), is directed from a point of higher V to a point of lower V. ■

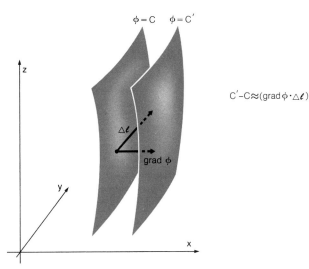

Figure 6–16 Equipotential surfaces for an arbitrary scalar function $\phi(x,y,z)$; grad ϕ everywhere points along the normal to the equipotential surfaces.

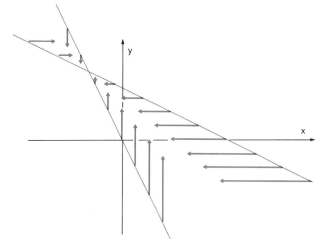

Figure 6–17 Example 8: The lines along which the electric field points in the x- or the y-direction.

Example 8   The electrostatic potential in a region of space is given as

$$V = V_0 \frac{(x^2 + xy + y^2 - ay)}{a^2}$$

Find all points where the electric field vanishes, as well as all points where it is directed along the x-direction or along the y-direction.

It is easy to obtain the components of the electric field

$$E_x = -\frac{\partial V}{\partial x} = -V_0 \frac{(2x + y)}{a^2} \tag{6-35a}$$

and

$$E_y = -\frac{\partial V}{\partial y} = -V_0 \frac{(x + 2y - a)}{a^2} \tag{6-35b}$$

The z-component E_z vanishes everywhere because V is independent of z. The electric field will thus vanish where

$$2x + y = 0 \tag{6-36a}$$

and simultaneously

$$x + 2y - a = 0 \tag{6-36b}$$

There exists only one such solution to both equations:

$$x = -\frac{1}{3}a$$

$$y = +\frac{2}{3}a$$

However, everywhere that Eq. (6–36a) is satisfied we have $E_x = 0$; thus, E_y points in the y-direction. Because of Eq. (6–35b), it will point in the positive y-direction if

$$x + 2y - a < 0$$

(at least if V_0 is positive) and in the negative y-direction if $x + 2y - a > 0$. In the same way, Eq. (6–36b) gives us all points where **E** points in the x-direction, and the sign of E_x can be determined from Eq. (6–35a). In Fig. 6–17 we have sketched the electric field along the two lines. Note that if x and y are measured in units of the length a, then E is given in units of V_0/a.

5. THE DIPOLE

We have learned how to calculate—at least in principle—the electric field and electrostatic potential of any distribution of charges. However, in practical situations such calculations can become very cumbersome. Frequently one is interested only in a rough calculation which gives a correct order of magnitude; at other times one wants to know only the electric field or potential far away from the charge distribution. This will occur frequently when we want to discuss the mutual interactions between molecules or atoms. In a gas, the individual molecules are usually far apart; except during collisions, the details of their charge distributions do not matter. In a liquid, because of the constantly varying orientation and position of each molecule, the situation is usually so complicated that we can calculate only the average values for electric field or potential. Thus, it is frequently useful to substitute for the actual charge distribution another simpler one which will have the same electric field far away.

We call the molecule or atom an *ion* if it has a net electric charge. For example, the sodium ion Na^+ consists of a sodium nucleus of charge $11\,e$ and 10 electrons which each carry the charge $(-1)\,e$. Thus, the Na^+ ion has a net charge $+e$, symbolized by the single $+$ sign in its name. Far away from the ion, the details of the charge distribution of the ion will not matter (Fig. 6–18) and the electric field is given by Coulomb's law

$$\mathbf{E} = \frac{1}{4\pi\epsilon_0} \frac{e}{r^2} \hat{\mathbf{u}}_r$$

Therefore, we can approximate the electric field by the field of a single point charge $(+1)\,e$. The electrostatic potential far away then is also given by Eq. (6–13) (with $Q = 1\,e$) —the potential of a single point charge.

Many molecules have no net charge; we call them neutral or un-ionized. But the detailed charge distribution in general will still produce an electric field. The simplest such arrangement consists of two charges $\pm Q$ which are separated by a distance a, as shown in Fig. 6–19. Such an arrangement is called an electric dipole. Obviously if $a = 0$, the two charges are at the same position and cancel exactly. But the problem before us is to calculate the electric field or potential if the charges are *not exactly* at the same point, but only *nearly* so.

Let us first calculate the electric field of a dipole at a point P along the axis of the dipole. As drawn in Fig. 6–19, the point is nearer to the positive charge. The electric field due to the positive charge $+Q$ points away from the dipole, while the field due to $-Q$ points toward the dipole. Thus, the sum of the two will have the magnitude

$$E = \frac{1}{4\pi\epsilon_0}\left[\frac{Q}{\left(r - \dfrac{a}{2}\right)^2} - \frac{Q}{\left(r + \dfrac{a}{2}\right)^2}\right] \qquad (6\text{--}37a)$$

Figure 6-18 From far away, a charge distribution with a net charge looks like a point charge.

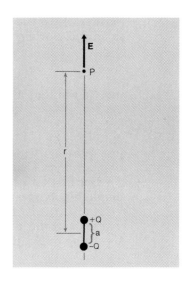

Figure 6-19   The electric field of a dipole on axis.

We can combine the two fractions to read

$$E = \frac{Q}{4\pi\epsilon_0} \frac{\left(r + \frac{a}{2}\right)^2 - \left(r - \frac{a}{2}\right)^2}{\left(r - \frac{a}{2}\right)^2 \left(r + \frac{a}{2}\right)^2}$$

or

$$E = \frac{Q}{4\pi\epsilon_0} \frac{2\,ar}{\left(r^2 - \frac{a^2}{4}\right)^2} \qquad (6\text{-}37\text{b})$$

If we are interested only in the electric field far away, we can neglect the term $a^2/4$ in the denominator because then

$$r^2 >> \frac{a^2}{4}$$

We then obtain for the *far field* of the electric dipole along its axis

$$E \approx \frac{1}{2\pi\epsilon_0} \frac{Qa}{r^3} \qquad (6\text{-}38)$$

We note that the electric field depends only on the product (Qa); if we double both charges, but separate them only by half the distance, the electric field will be the same. The product (Qa) is called the dipole moment of the dipole. More precisely, we call the dipole moment **p** the vector which has the direction pointing from $-Q$ to $+Q$ and the magnitude

$$|\mathbf{p}| = Qa$$

If we introduce the vector a pointing from $-Q$ to $+Q$, we can write the relationship as a vector equation:

$$\boxed{\mathbf{p} = Q\mathbf{a}} \qquad (6\text{-}39)$$

Further, if we use the radius vectors $\mathbf{r}_1$ and $\mathbf{r}_2$ pointing to the charges $\pm Q$ from an arbitrary origin point O (Fig. 6–20), we have

$$\mathbf{a} = \mathbf{r}_1 - \mathbf{r}_2$$

and we can thus also write

$$\mathbf{p} = Q\mathbf{r}_1 + (-Q)\mathbf{r}_2 \qquad (6-40)$$

an expression which we will use later on when we define the dipole moment of an arbitrary charge distribution.

But first let us calculate the electrostatic potential and electric field of the simple dipole at an arbitrary point in space. We introduce for this purpose (Fig. 6–21) a particular coordinate system with the origin halfway between the two charges and the z-axis pointing in the direction of the dipole moment vector. According to Eq. (6–16) the electrostatic potential at an arbitrary point P is

$$V = \frac{1}{4\pi\epsilon_0}\left[\frac{Q}{r_+} - \frac{Q}{r_-}\right] \qquad (6-41)$$

where r_+ and r_- are the distances to the point P from the two charges $+Q$ and $-Q$. Since $+Q$ is at $(x,y,z) = (0,0,a/2)$ and $-Q$ is at $(x,y,z) = (0,0,-a/2)$, we have

$$r_+ = \sqrt{x^2 + y^2 + \left(z - \frac{a}{2}\right)^2} \qquad (6-42a)$$

and

$$r_- = \sqrt{x^2 + y^2 + \left(z + \frac{a}{2}\right)^2} \qquad (6-42b)$$

Again we are interested mainly in the far field, for which

$$r = \sqrt{x^2 + y^2 + z^2} >> a$$

We can now use an approximation which is valid for all sufficiently smooth functions $f(\xi)$ and small changes $\Delta\xi$:

$$f(\xi + \Delta\xi) \approx f(\xi) + \frac{df}{d\xi}\Delta\xi \qquad (6-43)$$

(see Appendix C). Since we have

$$x^2 + y^2 + \left(z \mp \frac{a}{2}\right)^2 = x^2 + y^2 + z^2 \mp az + \frac{a^2}{4}$$

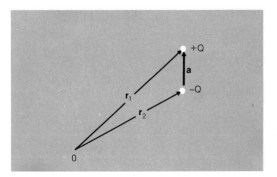

Figure 6–20   Calculation of dipole moment $\mathbf{p} = Q\mathbf{a}$.

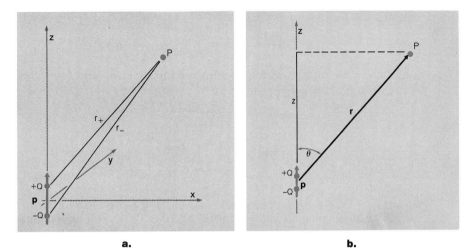

Figure 6-21 Calculating the potential of a dipole at an arbitrary point *P*. **a.** Definition of r_+ and r_-. **b.** The potential depends only on *p*, *r*, and *z*.

we can write

$$r_+^2 = r^2 + \left(\frac{a^2}{4} - az\right) = r^2 + (\Delta r^2)_+$$

Identifying ξ in Eq. (6–43) with r^2 and $\Delta\xi$ with $(\Delta r^2)_+$, we see that we have to calculate for Eq. (6–41) the expression

$$\frac{1}{r_+} = \frac{1}{\sqrt{\xi + \Delta\xi_+}} = f(\xi + \Delta\xi)$$

which according to Eq. (6–43) leads to

$$\frac{1}{r_+} \approx \frac{1}{\sqrt{\xi}} + \frac{-1/2}{\sqrt{\xi^3}} \Delta\xi$$

But $\xi = r^2$ and $\Delta\xi = (a^2/4) - az$; thus

$$\frac{1}{r_+} \approx \frac{1}{r} - \frac{1}{2r^3}\left(\frac{a^2}{4} - az\right)$$

In exactly the same way we arrive at an expression for $1/r_-$:

$$\frac{1}{r_-} \approx \frac{1}{r} - \frac{1}{2r^3}\left(\frac{a^2}{4} + az\right)$$

Therefore, we have

$$\frac{1}{r_+} - \frac{1}{r_-} \approx \left[\frac{1}{r} - \frac{1}{2r^3}\left(\frac{a^2}{4} - az\right)\right] - \left[\frac{1}{r} - \frac{1}{2r^3}\left(\frac{a^2}{4} + az\right)\right]$$

or

$$\frac{1}{r_+} - \frac{1}{r_-} \approx \frac{az}{r^3}$$

Thus, we obtain for the potential of a simple dipole *far away* the expression

$$V = \frac{1}{4\pi\epsilon_0}\frac{Qaz}{r^3} \qquad\qquad (6-44)$$

Using our definition of the dipole moment vector (Eq. 6–39), we see that

$$Qaz = |\mathbf{p}| \; r \cos \theta = (\mathbf{p} \bullet \mathbf{r})$$

and thus we obtain an expression for the electrostatic potential far away from the dipole, which does not depend on any particular coordinate system:

$$V = \frac{1}{4\pi\epsilon_0} \frac{(\mathbf{p} \bullet \mathbf{r})}{r^3} \tag{6-45}$$

Example 9   The water molecule (Fig. 6–22) can be approximated by a dipole of dipole moment $|\mathbf{p}| = 6.2 \times 10^{-30}$ coul-m. What will be the potential energy and the force on an electron which is 5 Å away along the direction of the dipole moment?

Because the radius vector $\mathbf{r}$ is parallel to the dipole moment, we have

$$(\mathbf{p} \bullet \mathbf{r}) = |\mathbf{p}| \; |\mathbf{r}|$$

The potential energy of an electron of charge -1 $e = -1.602 \times 10^{-19}$ coul is then equal to

$$U = -e\,V = -\frac{e}{4\pi\epsilon_0} \frac{pr}{r^3} = -\frac{e}{4\pi\epsilon_0} \frac{p}{r^2} \tag{6-46}$$

We calculate the potential energy directly in units of electron volts. In practice this means setting $e = 1$ in Eq. (6–46) and using MKSC units otherwise. Thus, since $1 \,\text{Å} = 10^{-10}$ meters, we have

$$U = -\frac{1 \times \left(9 \times 10^9 \; \frac{\text{V-m}}{\text{coul}}\right) \times 6.2 \times 10^{-30} \; \text{coul-m}}{(5 \times 10^{-10} \; \text{m})^2} = 2.23 \times 10^{-1} \; \text{eV} = 0.223 \; \text{eV}$$

Is this energy large or small? Since we are talking about an electron—or singly charged negative ion—which is near a water molecule, it may be useful to compare the potential energy to the "typical" thermal temperature kT, where k is Boltzmann's constant and T is the absolute temperature. Numerically we have from Appendix B, Table III, at room temperature

$$kT = 1.38 \times 10^{-23} \; \frac{\text{J}}{\text{°K}} \times 293\text{°K} = 4.04 \times 10^{-21} \; \text{J}$$

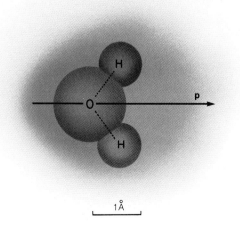

Figure 6–22 The water molecule and the direction of its electric dipole moment.

1Å

or expressed in electron-volts,

$$kT = \frac{4.04 \times 10^{-21} \text{ J}}{1.6 \times 10^{-19} \text{ J/eV}} = 2.5 \times 10^{-2} \text{ eV} \approx \frac{1}{40} \text{ eV}$$

Thus, the potential energy of an electron which is some two molecular distances away has a magnitude roughly nine times larger than the thermal (kinetic and potential) energy of the water molecule. We conclude that there must be large forces between water molecules and the ions of a dissolved salt [e.g., the ions of Na^+ and Cl^- of dissolved sodium chloride (rock salt), NaCl].

We are also asked to calculate the force on the electron; we use Eq. (6–38)

$$|\mathbf{F}| = e \, |\mathbf{E}| = \frac{2e}{4\pi\epsilon_0} \frac{p}{r^3}$$

and thus in units of newtons we have

$$|\mathbf{F}| = 2(1.6 \times 10^{-19} \text{ coul}) \times \left(9 \times 10^9 \frac{\text{V-m}}{\text{coul}}\right) \times \frac{6.2 \times 10^{-30} \text{ coul-m}}{(5 \times 10^{-10} \text{ m})^3}$$
$$= 1.43 \times 10^{-10} \text{ N}$$

To see how large the force is, we calculate how long it would take the electron in such a force to acquire the velocity of 1/100 the velocity of light or 3×10^6 m/sec. Since we have

$$ma = F$$

and the electron mass is 0.91×10^{-30} kg, we have an acceleration of

$$a = \frac{F}{m} = \frac{1.428 \times 10^{-10} \text{ N}}{0.91 \times 10^{-30} \text{ kg}} = 1.57 \times 10^{20} \text{ m/sec}^2$$

The electron thus acquires one percent of the velocity of light in the time

$$\Delta t = \frac{v}{a} = \frac{3 \times 10^6 \text{ m/sec}}{1.57 \times 10^{20} \text{ m/sec}^2} = 1.91 \times 10^{-14} \text{ sec}$$

As we see, forces between water molecules and ions lead to large accelerations.

From Eq. (6–45), which gives us the electrostatic potential of a dipole, we can also calculate the faraway electric field of the dipole by using Eq. (6–29):

$$\mathbf{E} = -\text{grad } V$$

We choose the same coordinate system which we had chosen to calculate the electrostatic potential. Then we can use Eq. (6–44):

$$V = \frac{1}{4\pi\epsilon_0} \frac{pz}{(\sqrt{x^2 + y^2 + z^2})^3} = \frac{1}{4\pi\epsilon_0} \frac{pz}{r^3} \qquad (6\text{–}44')$$

We also note that for any function of $r = \sqrt{x^2 + y^2 + z^2}$ we have

$$\frac{\partial f}{\partial x} = \frac{df}{dr} \frac{\partial r}{\partial x} = \frac{df}{dr} \frac{x}{\sqrt{x^2 + y^2 + z^2}}$$

and since the radius vector $\mathbf{r}$ can be written $\mathbf{r} = x\,\hat{\mathbf{i}} + y\,\hat{\mathbf{j}} + z\,\hat{\mathbf{k}}$ we have

$$\text{grad } f = \frac{\partial f}{\partial x}\hat{\mathbf{i}} + \frac{\partial f}{\partial y}\hat{\mathbf{j}} + \frac{\partial f}{\partial z}\hat{\mathbf{k}} = \frac{df}{dr}\frac{\mathbf{r}}{r}$$

This helps when taking the derivatives in Eq. (6–44′), because V is *nearly* a function of r alone. We can write

$$V = A \cdot z \cdot f(r) = A \cdot z \cdot \frac{1}{r^3}$$

where we have set $A = p/4\pi\epsilon_0$. Thus

$$\frac{\partial V}{\partial x} = Az \frac{\partial f}{\partial x}; \quad \frac{\partial V}{\partial y} = Az \frac{\partial f}{\partial y}; \quad \frac{\partial V}{\partial z} = Af + Az \frac{\partial f}{\partial z}$$

Only the partial derivative with respect to z is different because of the factor z in front of $f(r) = 1/r^3$. Thus we have

$$-E_x = \frac{\partial V}{\partial x} = Az \frac{-3}{r^4} \frac{x}{r} \tag{6–45a}$$

$$-E_y = \frac{\partial V}{\partial y} = Az \frac{-3}{r^4} \frac{y}{r} \tag{6–45b}$$

$$-E_z = \frac{\partial V}{\partial z} = \frac{A}{r^3} - \frac{3Az}{r^4} \frac{z}{r} \tag{6–45c}$$

We now remember that $(\mathbf{p} \bullet \mathbf{r}) = |\mathbf{p}|\, z = pz$ and that $\mathbf{p} = 0\,\hat{\imath} + 0\,\hat{\jmath} + p\,\hat{\mathbf{k}}$. Because $A = p/4\pi\epsilon_0$, we can write

$$E_x = \frac{3}{4\pi\epsilon_0} \frac{(\mathbf{p} \bullet \mathbf{r})}{r^5} x$$

$$E_y = \frac{3}{4\pi\epsilon_0} \frac{(\mathbf{p} \bullet \mathbf{r})}{r^5} y$$

$$E_z = \frac{-1}{4\pi\epsilon_0} \frac{p}{r^3} + \frac{3}{4\pi\epsilon_0} \frac{(\mathbf{p} \bullet \mathbf{r})}{r^5} z$$

and because $\mathbf{p}$ is parallel to the z-axis, we have

$$\boxed{ \mathbf{E} = \frac{1}{4\pi\epsilon_0} \left[-\frac{\mathbf{p}}{r^3} + \frac{3(\mathbf{p} \bullet \mathbf{r})}{r^5} \mathbf{r} \right] } \tag{6–46}$$

Thus, the electric field of a dipole can be considered the sum of two vectors

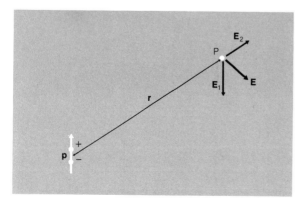

Figure 6-23 The electric field of a dipole at any point P is the sum of two vectors, one parallel (or antiparallel) to $\mathbf{r}$ and the other antiparallel to $\mathbf{p}$.

(Fig. 6–23); one vector, $-\mathbf{p}/4\pi\epsilon_0 r^3$, is always antiparallel to $\mathbf{p}$ and its magnitude depends only on the distance r. The other vector, $\dfrac{3(\mathbf{p}\bullet\mathbf{r})}{4\pi\epsilon_0 r^5}\mathbf{r}$, is parallel or anti-parallel to the radius vector $\mathbf{r}$; its magnitude depends on the distance r, but also on the relative orientation of $\mathbf{p}$ and $\mathbf{r}$.

Example 10   Find all points where the electric field is either parallel or antiparallel to $\mathbf{p}$, and calculate the magnitude of the electric field there.

From Eq. (6–44) we see that $\mathbf{E}$ is a sum of two vectors which, in general, do not point in the same direction. The sum will be collinear (parallel or antiparallel) to $\mathbf{p}$ only if either $\mathbf{r}$ is collinear to $\mathbf{p}$, or the coefficient in front of $\mathbf{r}$ vanishes.

In the first case we have $\mathbf{r}=\pm\lambda\mathbf{p}$ with $\lambda = |\mathbf{r}|/|\mathbf{p}|$. But then

$$\frac{(\mathbf{p}\bullet\mathbf{r})}{r^5}\mathbf{r} = \pm\frac{(\mathbf{p}\bullet\lambda\mathbf{p})(\pm\lambda\mathbf{p})}{r^5}$$

and whether $\mathbf{r}$ is antiparallel or parallel to $\mathbf{p}$, the second term in Eq. (6–44) equals

$$\frac{3(\mathbf{p}\bullet\mathbf{r})}{r^5}\mathbf{r} = \frac{3\lambda^2 p^2\mathbf{p}}{r^5} = \frac{3\mathbf{p}}{r^3}$$

Thus, the electric field has the magnitude

$$|\mathbf{E}| = \frac{1}{4\pi\epsilon_0}\left[-\frac{p}{r^3}+\frac{3p}{r^3}\right] = \frac{2p}{4\pi\epsilon_0 r^3}$$

Note that we have only rederived Eq. (6–38); we already said there that the electric field will point in the direction of $\mathbf{p}$ on either side of the dipole.

The other possibility is $(\mathbf{p}\bullet\mathbf{r}) = 0$; then the coefficient of $\mathbf{r}$ vanishes and we have

$$\mathbf{E} = \frac{-\mathbf{p}}{4\pi\epsilon_0 r^3}$$

i.e., the field is antiparallel to $\mathbf{p}$ everywhere in the direction perpendicular to the dipole axis. In Fig. 6–24 we show the two regions where $\mathbf{E}$ is either parallel or antiparallel to $\mathbf{p}$.

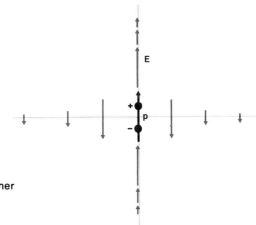

Figure 6–24 Example 10: The regions where $\mathbf{E}$ is either parallel or antiparallel to $\mathbf{p}$.

6. FORCES ON DIPOLES

If an ion finds itself in an electric field $\mathbf{E}$, it will be subject to the force

$$\mathbf{F} = Q\mathbf{E}$$

where Q is the (positive or negative) charge of the ion. We now want to study how an external electric field acts on a dipole. This will again be important when we discuss the behavior of molecules in an electric field. For any distribution of charges $Q_1, Q_2, \ldots$ for which the net charge vanishes (Fig. 6–25), i.e., when

$$Q_{\text{tot}} = \sum Q_i = 0 \qquad (6\text{–}47)$$

we can define the dipole moment in exact analogy to Eq. (6–40). From some arbitrary origin we calculate the position vectors $\mathbf{r}_1, \mathbf{r}_2, \ldots$ to each of the charges. We then define the dipole moment of the charge distribution as

$$\mathbf{p} = Q_1\mathbf{r}_1 + Q_2\mathbf{r}_2 + \ldots = \sum Q_i\mathbf{r}_i \qquad (6\text{–}48)$$

It is easy to prove (see Problem 6–19) that the definition (6–48) is independent of the particular origin, as long as the condition (6–47) is satisfied.

We will calculate the effect of an external field only for the simplest dipole consisting of two charges $\pm Q$; but we stress that our results are valid for an arbitrary dipole, as long as its dipole moment is defined by Eq. (6–48).

Thus, let us first calculate the potential energy and the torque acting on a dipole in a *homogeneous* electric field. The positive charge will be acted upon by a force (Fig. 6–26)

$$\mathbf{F}_1 = +Q\mathbf{E}$$

while the negative charge will feel the force

$$\mathbf{F}_2 = -\mathbf{F}_1$$

Thus, there is no net force on the dipole; however, there *is* a torque whose magnitude is

$$|\mathbf{T}| = |\mathbf{F}_1|\, a = Qd\,\sin\alpha\,|\mathbf{E}|$$

Its direction will be into the paper in Fig. 6–26. Using the vector product notation, since $\mathbf{a} \times \mathbf{b} = |\mathbf{a}|\,|\mathbf{b}|\sin\alpha$ we can write:

$$\mathbf{T} = \mathbf{p} \times \mathbf{E} \qquad (6\text{–}49)$$

where we again use the definition of the dipole moment as a vector of length

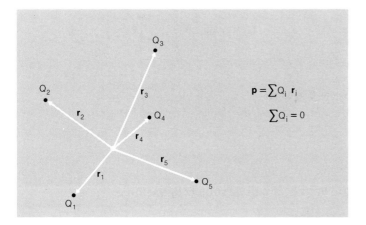

Figure 6–25 Dipole moment of an arbitrary charge distribution whose total charge vanishes.

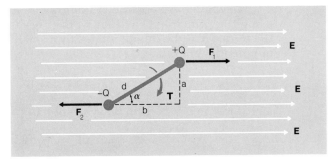

Figure 6–26 Calculating the torque on an electric dipole in a homogeneous electric field.

(Qd) and direction from negative to positive charge. This is equivalent to the definition in Eq. (6–48):

$$\mathbf{p} = \sum Q_i \mathbf{r}_i = Q\mathbf{r}_1 - Q\mathbf{r}_2 = Q(\mathbf{r}_1 - \mathbf{r}_2) \tag{6–50}$$

We can obtain the potential energy just as easily from Fig. 6–26. If both charges were at the same "height," i.e., if $b = 0$, the potential energy would be $U = 0$. However, if $b \neq 0$, then

$$U = QV_+ - QV_- = Q(V_+ - V_-)$$

where V_+ and V_- are the potentials at the positions of the positive and negative charges. However,

$$(V_+ - V_-) = -\int \mathbf{E} \cdot d\mathbf{l} = -|\mathbf{E}|\, b = -|\mathbf{E}|\, d \cos \alpha$$

and we can write the potential energy as

$$\boxed{U = -Qd\, E \cos \alpha = -(\mathbf{p} \cdot \mathbf{E})} \tag{6–51}$$

the scalar product between the dipole moment and the electric field.

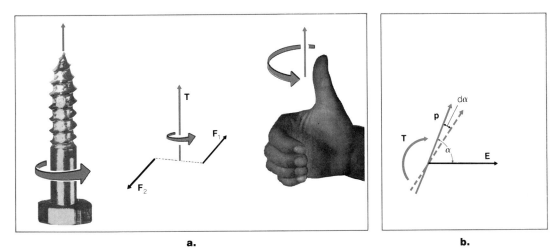

Figure 6–27 a. Definition of torque using right-hand rule. **b.** The work done by the torque $|\mathbf{T}|\, d\alpha$ has to be equal to the change in potential energy $-dU$.

We could have obtained either one of Eqs. (6–49) and (6–51) from the other, using the principle of virtual work (Fig. 6–27b). The work $T\Delta\alpha$, performed by the torque $\mathbf{T}$ if the dipole rotates by a small angle $\Delta\alpha$, has to be equal to the loss of potential energy:

$$|\mathbf{T}|\,\Delta\alpha = -\Delta(|\mathbf{p}|\,|\mathbf{E}|\,\cos\alpha) \tag{6–52}$$

and since $\Delta\cos\alpha = -\sin\alpha\Delta\alpha$, we obtain

$$|\mathbf{T}| = |\mathbf{p}|\,|\mathbf{E}|\,\sin\alpha$$

which is identical to Eq. (6–49).

While the dipole is subject only to a torque in a homogeneous field, there will be a net force acting on it if the field is inhomogeneous, i.e., different at different points in space. Qualitatively we can see this easily by assuming that in Fig. 6–28 the field increases toward one side – let us say toward the right. Then the positive charge will feel a larger force than the negative charge; the two forces no longer cancel exactly and there is a net force on the dipole.

We can calculate the force from the potential energy given in Eq. (6–51). Let us discuss the particular case in which the dipole moment is parallel to the electric field; we will assume both to be pointing in the positive x-direction. Then the net force on the dipole is

$$|\mathbf{F}| = Q\,E_x(x+a) + (-Q)\,E_x(x)$$

where by $E_x\,(x+a)$ we denote the x-component of the electric field at the point $x+a$. If the field changes smoothly, then

$$E_x(x+a) - E_x(x) \approx \frac{dE_x}{dx}a$$

and the force has the magnitude

$$F = Qa\frac{dE_x}{dx} = p\frac{dE_x}{dx} \tag{6–53}$$

and points to the right in Fig. 6–28. We note that, indeed, the work done by this force diminishes the potential energy. If we move the dipole a small distance Δx to the right, we obtain the work

$$W = F\,\Delta x = p\frac{dE_x}{dx}\Delta x$$

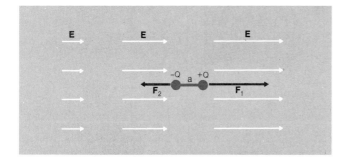

Figure 6–28 In an inhomogeneous electric field there will be a force on the dipole, because the forces on $+Q$ and on $-Q$ no longer exactly cancel.

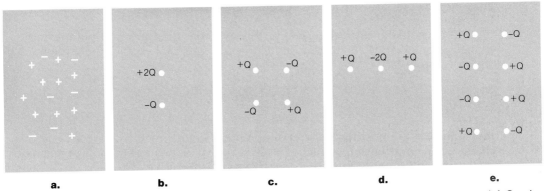

Figure 6-29 Various charge distributions. **a.** Arbitrary. **b.** Monopole and dipole. **c.** and **d.** Quadrupoles. **e.** Example of octupole.

Because of the particular orientation of the dipole, the potential energy is

$$U = -(\mathbf{p} \cdot \mathbf{E}) = -p\,E_x$$

and its change is indeed equal to

$$\Delta U \approx \frac{dU}{dx}\Delta x = -p\frac{dE_x}{dx}\Delta x$$

Thus, the work obtained from the system is compensated by a decrease in potential energy; we could have obtained Eq. (6–53) from the requirement that energy be always conserved.

It is important to note that the force on the dipole depends on the relative orientation of dipole moment and field. If we reverse the dipole in Fig. 6–28, we will also reverse the direction of the force, because now the negative charge is in the stronger field.

Let us summarize what we have learned about dipoles. We have learned to calculate their electric fields and electrostatic potential. We have also studied how they are affected by external electric fields. Returning to Eq. (6–44), we see that in any direction the field of a dipole falls off like the third power of the distance:

$$|\mathbf{E}| \propto \left| \frac{\mathbf{p}}{r^3} - \frac{3(\mathbf{p} \cdot \mathbf{r})}{r^5}\mathbf{r} \right| \propto \frac{p}{r^3}$$

Thus, if we have a complicated charge distribution, such as all the electrons and nuclei of a molecule, then far away the field due to the charge distribution will depend on these rather simple properties of the charge distribution:

(i) If the net charge Q is non-zero, the electric field will be dominated by the total charge:

$$|\mathbf{E}| \sim \frac{Q}{4\pi\epsilon_0 r^2}$$

and the field will be pointing radially out or in, depending on the sign of Q. We say that such a charge distribution is dominated by its monopole term—the *total charge*. The simultaneous presence of a dipole, as shown in Fig. 6–29b, will have no effect at large distances because the dipole has a field which is smaller by a factor $a/r \ll 1$.

(ii) Only if the total charge Q is zero will the field of the dipole term be measurable far away. Its magnitude is given by the total vector dipole moment, which is equal to

$$\mathbf{p} = \sum_i Q_i \mathbf{r}_i,$$

the sum over the position vectors of all charges multiplied by the value of the charge. The contribution of each charge is positive (in the direction of its position vector) if $Q_i > 0$; if $Q_i < 0$, its contribution is negative. The field of a dipole will fall off like $1/r^3$. All un-ionized molecules have no net charge; but many molecules do have an electric dipole moment (e.g., the water molecule H_2O). Thus, many molecules at some distance away have the same electric field as a simple dipole.

(iii) If the charge distribution has no net charge and also its dipole moment vanishes, only then does one have to look in more detail at its charge distribution. The next strongest term (i.e., the term which falls off slowest) is the quadrupole term (a quadrupole is sketched in Fig. 6–29c and d) whose electric field falls off like $1/r^4$. If the quadrupole moment also vanishes, maybe the charge distribution has an octupole moment (an example is shown in Fig. 6–29e), which falls off like $1/r^5$, and so on. The higher the "multipole" (monopole, dipole, quadrupole) order, the faster does the field fall off at large distances, and the more complicated is the detailed angular dependence of the electric field.

Summary of Important Relations

Electrostatic potential (potential energy per unit charge) $V = \dfrac{U}{q}$

relation to electric field $V(P) = -\displaystyle\int_{A_0}^{P} \mathbf{E} \cdot d\mathbf{l}$ along any path

inverse relation $\mathbf{E} = -\text{grad } V = -\dfrac{\partial V}{\partial x}\hat{\mathbf{i}} - \dfrac{\partial V}{\partial y}\hat{\mathbf{j}} - \dfrac{\partial V}{\partial z}\hat{\mathbf{k}}$

Potential of point charge $V = \dfrac{1}{4\pi\epsilon_0}\dfrac{Q}{r};$ r is distance to Q

Potential due to charge distribution

$$V(P) = \frac{1}{4\pi\epsilon_0}\int \frac{\rho_{P'} d\tau}{r_{PP'}} \quad \text{continuous distributions} \atop \text{(see Fig. 6–11)}$$

$$V = \frac{1}{4\pi\epsilon_0}\sum_i \frac{Q_i}{r_i} \quad \text{discrete charge distribution}$$

Potential of dipole $\qquad V = \dfrac{1}{4\pi\epsilon_0}\dfrac{\mathbf{p}\cdot\mathbf{r}}{r^3}$

Electric field of dipole $\qquad \mathbf{E} = \dfrac{1}{4\pi\epsilon_0}\left[\dfrac{-\mathbf{p}}{r^3} + \dfrac{3(\mathbf{p}\cdot\mathbf{r})}{r^5}\mathbf{r}\right]$

Potential energy of dipole
in electric field $\mathbf{E}$ $\qquad U = -\mathbf{p}\cdot\mathbf{E}$

Torque on dipole in field $\mathbf{E}$ $\qquad \mathbf{T} = \mathbf{p}\times\mathbf{E}$

Maxwell's first equation of electrostatics $\qquad \displaystyle\int_{\substack{\text{closed}\\\text{loop}}} \mathbf{E}\cdot d\mathbf{l} = 0$

Problems

1. Calculate the electrostatic potential a distance $d = 0.8 \times 10^{-15}$ m (proton radius) away from a point charge $+e$ ($e = 1.602 \times 10^{-19}$ coul). Also give the potential energy in eV, if a second charge $-e$ is a distance d away from $+e$.

2. The lead nucleus has the charge $+Ze = 82\ e$. The potential energy of one of the innermost electrons of the lead atom is $U \approx -1.7 \times 10^5$ eV. How far away (in Å) is the electron?

3. Six charges $+Q_0$ are arranged at the corners of a hexagon. Calculate the electrostatic potential (i) at the center of the hexagon at P_1; (ii) halfway between two of the charges at P_2.

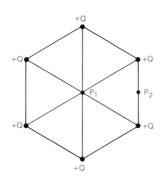

4. Calculate the total work necessary to assemble the six charges $+Q$ on a hexagon of side a (Problem 3).

5. Three charges $+Q$ are arranged at the three corners of a equilateral triangle of side a. How much work is freed when a second triangle with three charges $-Q$ is moved in from very far away, until the two form a hexagon, as shown in the figure on top of p. 194?

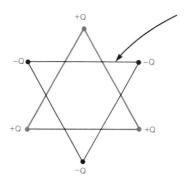

6. N charges $Q_i (1 \le i \le N)$ are arranged so that the separation between Q_i and Q_k is r_{ik}. Show that the potential energy stored in the system is

$$U = \frac{1}{8\pi\epsilon_0} \sum_{\substack{k=1 \\ k \ne i}}^{N} \sum_{i=1}^{N} \frac{Q_i Q_k}{r_{ik}}$$

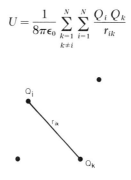

7. Four charges $\pm Q_0$ are arranged at the corners of a square of side a, as shown in the figure. A fifth charge $+Q_0$ is moved from the point P_1 to the point P_2. Calculate the work done by the charge.

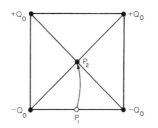

8. A rod of length a has the charge $+Q$ homogeneously distributed over its length. Calculate the electrostatic potential at a point P a distance x away from the rod along its axis.

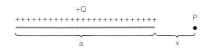

9. The charge $+4Q_0$ is homogeneously distributed over the four edges of a square of side a. Determine the electric field and the electrostatic potential at the center point P.

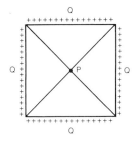

10. In a crystal of NaCl, the separation between the Na$^+$ and Cl$^-$ ions is $d = 2.82$ Å. We consider here the simpler linear chain of ions with separation d. If the chain is very long, calculate to 5% accuracy the work required to remove one of the end charges. Give your answer in eV.

11. Consider a large plane square array of charges $\pm e$ separated by the distance $d = 2.82$ Å. Using a computer, calculate to 5% accuracy the work required to remove (i) one of the corner charges; (ii) one of the charges at the edge, but far away from the corner. Give your answer in eV.

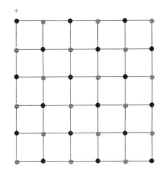

12. Calculate the work required to assemble the charge distribution discussed in Problem 5–16.

13. N equal charges Q are distributed along a line segment; the spacing between neighboring charges is a. Give an approximate formula for the work required to assemble the charge distribution, which is valid for $N \gg 1$.

14. The electric field in a region of space has the form

$$E_x = \frac{E_0}{a}(x - y)$$

$$E_y = \frac{-E_0}{a}(x + y)$$

$$E_z = 0$$

Calculate the electrostatic potential $V(x,y,z)$ and check that Eq. (6–22) is satisfied for any rectangular loop parallel to the coordinate axis.

15. The potential in a region of space is given by

$$V = V_0 \ln\left(\frac{x^2 + y^2}{a^2}\right)$$

Calculate the magnitude of the electric field and the angle it forms with the x-axis at an arbitrary point (x,y). Sketch the electric field distribution.

16. The potential in a region of space is given by

$$V = V_0\left(\frac{x^2 - y^2}{a^2}\right)$$

Find all points where the electric field is pointing away from or toward the z-axis (in two dimensions, away from or toward the origin).

17. Calculate the electrostatic potential due to three charges $(-Q)$, $(-Q)$, and $(+3Q)$ at a distance d away, as shown in the figure. What does your result reduce to for $d >> a$?

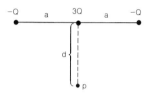

18. Six charges $\pm Q$ are distributed as shown in the figure. Does this charge distribution form a dipole? If so, determine magnitude and direction of the dipole moment.

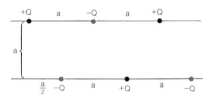

19. Show that the definition [Eq. (6–48)] of the dipole moment of an arbitrary charge distribution is independent of the coordinate origin if and only if Eq. (6–47) is satisfied.

20. Three charges $(+2Q)$, $(-Q)$, and $(-Q)$ are distributed as shown in the figure. (i) Calculate the dipole moment vector. (ii) Calculate the electric field far away on the z-axis.

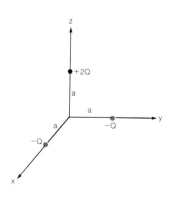

21. The charges $\pm Q_0$ are homogeneously distributed over two touching thin half-rings of radius R. Calculate the direction and magnitude of the dipole moment vector.

22. Find all points—and sketch their positions—for which the electric field of a dipole is perpendicular to the dipole moment direction. Calculate along these points the magnitude of the electric field.

23. Two water molecules (dipole moment $p = 6.2 \times 10^{-30}$ coul-m) are separated by a distance $d = 3$ Å. Estimate (calculate in the dipole approximation) the ratio between their mutual potential energy and kT at room temperature if (i) the two molecules are behind each other; (ii) they are side by side with their dipole moments opposite to each other. Which position is energetically more favored (has lower potential energy)?

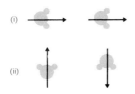

24. Two electric dipoles, each of dipole moment p, are constrained so that they are always perpendicular to each other, and their separation r is always the same. Find the angle θ for which the potential energy is a minimum or a maximum.

25. A dipole (dipole moment $\mathbf{p}$) and a charge $+Q$ are separated by a distance r as shown in the figure. Calculate explicitly (i) the force on the charge due to the presence of the dipole, and (ii) the force on the dipole due to the presence of the charge; also show that Newton's third law (Action = Reaction) is satisfied.

26. The chloroform molecule has the electric dipole moment $p = 3.35 \times 10^{-30}$ coul-m. If chloroform vapor at room temperature is in an electric field of 10^6 V/m, would you expect the molecules to align themselves with the electric field? (Compare the potential energy to kT.)

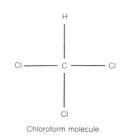

Chloroform molecule

27. One possible form of a quadrupole is shown in the figure. Calculate the electric field and the electrostatic potential at a point P far away along the axis.

CHAPTER 7

GAUSS'S LAW

1. INTRODUCTION

We now know how to calculate the electric field and electrostatic potential of an arbitrary *given* charge distribution. However, frequently we do not know the charge distribution; each charge will move under the effect of the electric field of all other charges, until equilibrium is established.

When we are talking about macroscopic solid objects, there are so many charges present and they are so closely spaced that we can talk about continuous charge distributions. Normally, in any small volume there will be an equal number of positive and negative charges, so that the net average charge is zero. When we say that a macroscopic object is charged, we mean that in a certain region there are more charges of one sign than of the other. In some materials, such as metals, while the atomic nuclei are held firmly in the crystal lattice, some of the electrons can move around freely. Such solids are called conductors. In other solids the electrons are rather tightly bound and cannot move around freely; these solids, such as glass and most plastics, are called insulators. Gases are usually good insulators, unless they are very hot. We will later discuss in detail the conduction of charges through matter and the varied phenomena associated with electrical resistance. Here we are interested only in the *static* properties of materials; and in this sense we can make two basic statements:

(a) We can deposit charge on the *surface* of an insulator, and the charge will stay where we deposited it.

(b) We can deposit charge on a conductor; however, the charge will redistribute itself immediately in such a way that there is no electric field anywhere inside the conductor.

Both statements are intuitively obvious: if the charges cannot move, we have an insulator; but if they can move, they will move around until there is no net force on any of the charges. Let us note that we are tacitly postulating some kind of frictional force on the charges by assuming that they will come "to rest" (whatever that means) at their equilibrium positions. We will later have to study this problem in more detail.

Thus, we need to undertake a further study of the electric field of charge distributions. It is not sufficient to be able to calculate the electric field of a charge distribution at any single point; we therefore will derive—and use in discussing conductors—a general relation between a charge distribution and the electric field on a surface surrounding the charge distribution. This is known as Gauss's law.

2. GAUSS'S LAW

Coulomb's law for a point charge states that the electric field of a point charge falls off like the square of the distance

$$|\mathbf{E}(r)| = \frac{1}{4\pi\epsilon_0}\frac{Q}{r^2}$$

Let us consider a sphere of radius r around the point charge Q (Fig. 7–1). The electric field has the same magnitude everywhere on the sphere's surface and points everywhere perpendicularly to the surface. Further, because of Coulomb's law, we have

$$|\mathbf{E}(r)| \times \text{surface area of sphere} = \frac{1}{4\pi\epsilon_0}\frac{Q}{r^2}4\pi r^2 = \frac{Q}{\epsilon_0}$$

independently of the radius of the sphere. We will now give a generalization of this simple observation, which is valid for any charge distribution and any closed surface.

Let us first define the *flux of the electric field* through a surface. We subdivide the surface into small elements dS. We will assume that the surface is smooth, so that the normal* to the surface is everywhere defined (Fig. 7–2). We can then define for each small element dS the scalar product between the normal to the element and the electric field vector

$$E_n = (\mathbf{E} \bullet \mathbf{n}) = |\mathbf{E}|\cos\alpha \qquad (7\text{–}1)$$

We can multiply this by the surface element area dS and obtain the electric flux through the element dS:

$$d\Phi = E_n\, dS = (\mathbf{E} \bullet \mathbf{n})\, dS \qquad (7\text{–}2)$$

Summing over the finite surface S, we obtain the flux of $\mathbf{E}$ through S:

$$\Phi = \int_{\text{surface } S} d\Phi = \int_{\text{surface } S} (\mathbf{E} \bullet \mathbf{n})\, dS \qquad (7\text{–}3a)$$

---

* By *normal to a surface at a point P* we define a vector of unit length, which is perpendicular to the surface at P.

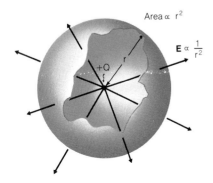

Area $\propto r^2$

$\mathbf{E} \propto \dfrac{1}{r^2}$

$+Q$

Figure 7–1   The electric field of a point charge falls off like $1/r^2$, while the area of a sphere of radius r is $A = 4\pi r^2 \propto r^2$.

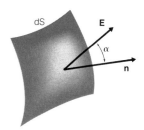

Figure 7–2 Definition of the electric flux through a surface.

We note that Φ depends on the electric field, but also on the particular shape and orientation of the surface S. We can also define the differential vector

$$d\mathbf{S} = \mathbf{n}\,dS \qquad (7\text{–}3b)$$

which has the magnitude of dS and the direction of the normal to the surface. We can then write instead of Eq. (7–3)

$$\Phi = \int_{\text{surface}} (\mathbf{E} \cdot d\mathbf{S}) \qquad (7\text{–}4)$$

Actually, Eqs. (7–3) and (7–4) are somewhat ambiguous, because on any surface at any point one can define two normals $\mathbf{n}_1$ and $\mathbf{n}_2$, one on each side of the surface, as shown in Fig. 7–3. This ambiguity disappears if we use a closed surface, which has a well-defined inside and outside. We then define the normal as always pointing *outward* (Fig. 7–5). In Fig. 7–4 we show an example of a surface for which one cannot define a unique normal.

We now claim—and will prove—that there exists a simple relation between the electric field flux through a closed surface and the total charge inside the surface; the two are proportional to each other:

$$\boxed{\int_{\substack{\text{closed} \\ \text{surface}}} \mathbf{E} \cdot d\mathbf{S} = \frac{1}{\epsilon_0}(Q_{\text{tot}})_{\text{inside}}} \qquad (7\text{–}5)$$

This relation is known as *Gauss's law.* We will derive Gauss's law (Eq. 7–5) first for the electric field of a single point charge Q by showing that for *any*

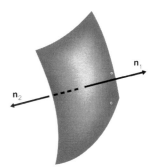

Figure 7–3 There are always two normals to a surface at any point.

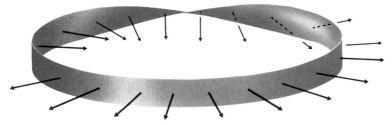

Figure 7-4 A Möbius strip is an example of a surface for which a unique definition of a normal is not possible.

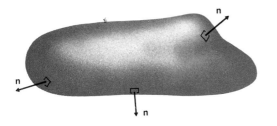

Figure 7-5 For a closed surface we define the normal as always pointing outward.

closed surface S

$$\int_S \mathbf{E} \cdot d\mathbf{S} = \frac{Q}{\epsilon_0} \quad \text{if} \quad Q \text{ is inside}$$

$$\int_S \mathbf{E} \cdot d\mathbf{S} = 0 \quad \text{if} \quad Q \text{ is outside}$$

Thus, let us first consider a surface surrounding a single isolated point charge Q. Figure 7-6 shows a single surface element dS of the surface, while in Fig. 7-7 we have schematically sketched the whole surface. The electric field is

$$\mathbf{E} = \frac{Q}{4\pi\epsilon_0 r^2}\hat{\mathbf{u}}_r \tag{7-6}$$

where $\hat{\mathbf{u}}_r$ is a unit vector pointing away from the point charge. Thus, $\mathbf{E}$ is pointing radially outward. We then have

$$d\Phi = (\mathbf{E} \cdot \mathbf{n})\, dS = \frac{Q}{4\pi\epsilon_0 r^2} dS \cos \alpha \tag{7-7}$$

where $\cos \alpha$ is the angle between $\mathbf{n}$, the normal to dS, and the radius vector $\mathbf{r}$. However,

$$dS \cos \alpha = dS'$$

since (r is very large compared to any linear dimension in dS) dS' is the projection of dS on a plane perpendicular to $\mathbf{r}$, as shown in more detail in Fig. 7-8. Thus,

$$d\Phi = \frac{Q}{4\pi\epsilon_0} \frac{dS'}{r^2} \tag{7-8}$$

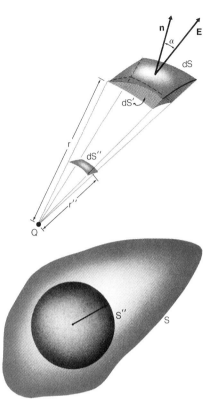

Figure 7-6 For a point charge Q the electric flux is the same through the three surfaces dS, dS', and dS''.

Figure 7-7 For a point charge, the flux through S and S'' is the same.

But for any element dS'' which is an area perpendicular to r bounded by the same four rays, we have

$$\frac{dS''}{r''^2} = \frac{dS'}{r^2} \tag{7-9}$$

If we choose as our element dS'' a surface element of a sphere of radius R, we see that to each differential element dS on the original surface S there corresponds a differential element dS'' on the sphere of radius R so that

$$d\Phi = \frac{Q}{4\pi\epsilon_0}\frac{dS\cos\alpha}{r^2} = \frac{Q}{4\pi\epsilon_0}\frac{dS''}{R^2} \tag{7-10}$$

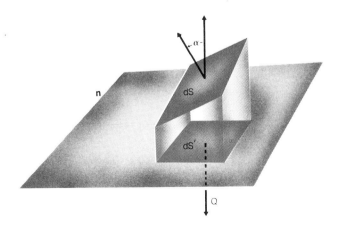

Figure 7-8 The projection dS' of the surface element dS onto a plane has the area $dS' = dS\cos\alpha$.

Thus, the integral over the original surface S is also equal to the integral over the sphere of radius R:

$$\Phi = \int_{\substack{\text{closed} \\ \text{surface } S}} \mathbf{E} \cdot d\mathbf{S} = \frac{Q}{4\pi\epsilon_0} \int_{\substack{\text{sphere of} \\ \text{radius } R}} \frac{dS''}{R^2} \qquad (7\text{–}11)$$

But the integral over the sphere is trivial to perform:

$$\int_{\text{sphere}} \frac{dS''}{R^2} = \frac{1}{R^2} \times (\text{area of sphere}) = \frac{4\pi R^2}{R^2} = 4\pi$$

and so we conclude from Eq. (7–10) that the integral over *any* closed surface *surrounding* the point charge Q is

$$\Phi = \int \mathbf{E} \cdot d\mathbf{S} = \frac{Q}{4\pi\epsilon_0} 4\pi = \frac{Q}{\epsilon_0} \qquad (7\text{–}12)$$

confirming the first part of our proof of Gauss's law.

Let us now choose another closed surface, as shown in Fig. 7–9, so that the point charge Q lies outside the surface. Now, on any straight line from Q either there are two surface elements dS_1 and dS_2, or the line misses the surface altogether. But on the two surface elements the normals point in opposite directions relative to the electric field direction. Thus, while for the surface element dS_2 the differential element (see Fig. 7–10)

$$d\Phi_2 = \frac{Q}{4\pi r_2{}^2} dS_2 \cos\alpha_2$$

is positive, for dS_1 we have

$$d\Phi_1 = \frac{Q}{4\pi r_1{}^2} dS_1 (-\cos\alpha_1)$$

Furthermore, $d\Phi_1$ and $d\Phi_2$ have the same magnitude, as we can show by repeating the argument leading from Eq. (7–7) to Eq. (7–10) for both dS_1 and dS_2. Thus,

$$d\Phi_1 = -d\Phi_2$$

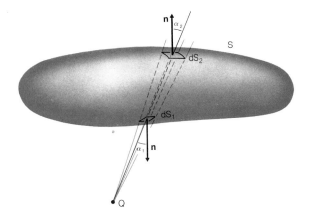

Figure 7–9 If the point source is outside the closed surface S, the fluxes through dS_1 and dS_2 cancel exactly.

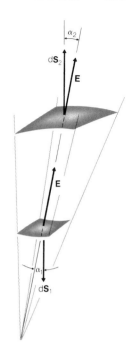

Figure 7-10   The detail view of the two surface elements from Fig. 7-9.

Since for each surface element dS_1 where the normal points against the electric field there exists exactly one corresponding surface element dS_2 where the normal points in the direction of the electric field, the integral over the whole surface also vanishes:

$$\int_{\substack{\text{closed} \\ \text{surface } S}} \mathbf{E} \bullet d\mathbf{S} = 0 \quad \text{if} \quad Q \text{ is outside} \qquad (7\text{--}13)$$

Thus, by proving Eqs. (7–12) and (7–13), we have proved Gauss's law (Eq. 7–5) for a single point charge Q. But we can consider any charge distribution as being built up from individual point charges. If we call the total electric field due to all charges $Q_1 \ldots Q_N$

$$\mathbf{E}\Big(\sum Q_i\Big)$$

then we have

$$\mathbf{E}\Big(\sum Q_i\Big) = \sum \mathbf{E}_i(Q_i)$$

i.e., the electric field of all the charges is the sum of the electric fields due to each charge. If we now calculate the integral $\int \mathbf{E} \bullet d\mathbf{S}$ over some surface, all point charges inside will contribute but none of the charges outside the surface will. Thus,

$$\int_{\substack{\text{closed} \\ \text{surface}}} \mathbf{E} \bullet d\mathbf{S} = \frac{\Big(\sum Q_i\Big)_{\text{inside}}}{\epsilon_0}$$

which is exactly Gauss's law.

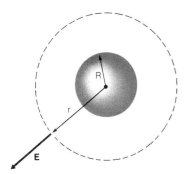

Figure 7-11   Example 1: Electric field outside charged sphere.

We will now show in a few examples how we can use Gauss's law to calculate the electric field of charge distributions of sufficient symmetry. We will see that if we can find surfaces for which we know from symmetry that everywhere on the surface the electric field has the same magnitude and known direction, then we can calculate the electric field anywhere.

Example 1   What is the electric field outside a homogeneously charged sphere of radius R?

Since we have an arrangement with spherical symmetry (any rotation about any axis through the center does not change the physical arrangement), the electric field has to point radially outward and its magnitude can depend only on the distance r from the sphere's center. We can thus choose a particular simple surface, a sphere of radius r, as shown in Fig. 7-11. Everywhere on its surface, the field will have the same magnitude and its direction will be everywhere normal to the surface; thus

$$\int_{\text{sphere}} \mathbf{E} \bullet d\mathbf{S} = |\mathbf{E}(r)|\ 4\pi r^2$$

which is equal to the total charge Q of the sphere divided by ϵ_0:

$$\frac{Q}{\epsilon_0} = |\mathbf{E}|\ 4\pi r^2$$

or

$$|\mathbf{E}| = \frac{Q}{4\pi\epsilon_0}\frac{1}{r^2}$$

(7-14)

As we see, *the field of a charged sphere, at any point outside the sphere, is equal to the field of a point charge located at the center of the sphere.* In particular, it does not matter whether the charge is all on the sphere's surface, or distributed in any *spherically symmetric* way on the inside of the sphere. The radius of the sphere also does not matter, as long as the point at which the field is measured is outside the sphere. But we repeat that it is essential that the charge distribution be spherically symmetric; only then can we argue that the electric field depends only on the distance from the sphere's center.

Example 2   Calculate the electric field inside a homogeneously charged sphere of radius R, as shown in Fig. 7-12.

We have said that the sphere is homogeneously charged, i.e., the charge is evenly distributed over the volume of the sphere. The charge density is

$$\rho = \frac{Q}{\text{Volume}} = \frac{Q}{\dfrac{4\pi}{3}R^3} = \frac{3}{4\pi}\frac{Q}{R^3}$$

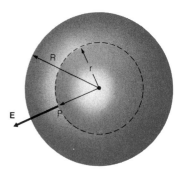

Figure 7-12   Example 2: Electric field inside homogeneously charged sphere.

The field at a point P in Fig. 7–12 will again—because of spherical symmetry—be equal in magnitude to the field at any point the same distance r away from the center. Thus, again we take the integral over a spherical surface of radius r:

$$\int_{\text{sphere}} \mathbf{E} \cdot d\mathbf{S} = |\mathbf{E}(r)|\ 4\pi r^2$$

However, the total charge enclosed by the sphere of radius r is not Q, but

$$\rho \frac{4\pi}{3} r^3 = \frac{Qr^3}{R^3}$$

Gauss's law tells us thus that

$$\frac{Qr^3}{\epsilon_o R^3} = |\mathbf{E}(r)|\ 4\pi r^2$$

or

$$E(r) = \frac{Q}{4\pi\epsilon_0} \frac{r}{R^3} \quad \text{for} \quad r < R \tag{7–15}$$

The field inside the sphere is proportional to the distance from the center. At the surface of the sphere, of course, both Eqs. (7–14) and (7–15) give the same result

$$|\mathbf{E}(r = R)| = \frac{Q}{4\pi\epsilon_0} \frac{1}{R^2}$$

In Fig. 7–13 we show the magnitude of the electric field as a function of the distance from the center of the sphere.

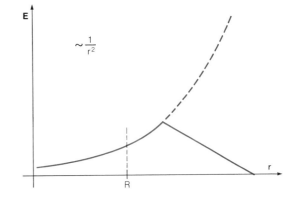

Figure 7-13   Plot of electric field magnitude inside and outside homogeneously charged sphere.

Example 3   Calculate the electric field near an infinitely long homogeneously charged rod with a charge λ per unit length.

Of course, no real charged rod is infinitely long. It will have some real length L and total charge $Q = \lambda L$. However, if we measure the field close to the rod at a distance $r \ll L$, and at a point somewhere near the center of the rod, we can use the approximation that the rod is of infinite length.

Since the rod is infinitely long, we can assume that the field has to point radially away from the rod (for $\lambda > 0$) and that its magnitude depends only on the distance from the rod. Indeed, the distance from the rod is the only distance there is in the problem, since the ends are infinitely far away. Also, any component along the rod $\mathbf{E}'$ or any azimuthal component $\mathbf{E}''$ would change their direction if we turn the rod around by 180° (Fig. 7–14); thus, they have to vanish.

As the surface on which to evaluate Gauss's integral law, we choose a cylinder of radius r and length ℓ. The two ends of the cylinder do not contribute, because the field is parallel to the surface;

$$\int (\mathbf{E} \cdot d\mathbf{S}) = 0 \quad \text{at the ends of the cylinder}$$

Thus, the integral over the cylinder is simply

$$\int (\mathbf{E} \cdot d\mathbf{S}) = |\mathbf{E}(r)| \times (\text{area of cylinder})$$

since the field is everywhere the same and perpendicular to the surface. Thus,

$$\int_{\text{cylinder}} (\mathbf{E} \cdot d\mathbf{S}) = |\mathbf{E}(r)|\, 2\pi r \ell \tag{7–16}$$

This has to be equal to $1/\epsilon_0$ times the total charge *inside* the cylinder; but this is the charge λ per unit length times the cylinder length ℓ. Thus,

$$\frac{\lambda \ell}{\epsilon_0} = 2\pi r \ell\, |\mathbf{E}(r)|$$

or

$$\mathbf{E}(r) = \frac{\lambda}{2\pi\epsilon_0}\frac{1}{r} \tag{7–17}$$

Note that our solution [Eq. (7–17)] fails if the distance r becomes comparable to the (in reality finite) length of the rod; we can no longer argue that the field has to point radially, because turning the rod around leads to a new situation. Also, the field can now depend on the distance from one end of the rod.

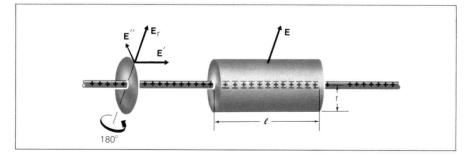

Figure 7-14 Calculating the electric field near a long charged rod.

3. ELECTRIC FLUX LINES

We have by now, starting from Coulomb's law, derived two very fundamental laws about the electric field. In Chapter 6 we found that [see Eq. (6–22)]

(1) $\displaystyle\int_{\substack{\text{closed}\\\text{loop}}} (\mathbf{E} \cdot d\,\mathbf{l}) = 0$ First Maxwell's Equation of Electrostatics   (7–18)

and we have just found that

(2) $\displaystyle\int_{\substack{\text{closed}\\\text{surface}}} (\mathbf{E} \cdot d\mathbf{S}) = \frac{Q_{\text{inside}}}{\epsilon_0}$ Second Maxwell's Equation (Gauss's law)   (7–19)

where Q is the total charge inside the surface. These two laws are both integral laws, and together they are sufficient to determine the electric field of an arbitrary charge distribution. They are equivalent to Coulomb's law.

There exists an intuitive interpretation of the second equation, which explains the name "flux" given to the integral

$$\int (\mathbf{E} \cdot d\mathbf{S})$$

over any surface.

For this purpose, let us consider as an analogy an incompressible liquid, such as water. The volume of liquid which flows through a surface element dS in the small time Δt is, as can be seen from Fig. 7–15, all the liquid in a parallelepiped of base dS and length $|\mathbf{v}| \, \Delta t$. Any molecule which is inside this volume will have gone through the surface element dS before the time Δt has elapsed, because it has moved the distance

$$\Delta\,\mathbf{l} = \mathbf{v}\,\Delta t$$

The volume is equal to

$$d\mathrm{V} = |\Delta\,\mathbf{l}|\,dS \cos\alpha = (\mathbf{v} \bullet \mathbf{n})\,dS\,\Delta t$$

Thus, $(\mathbf{v} \bullet \mathbf{n})\,dS = (\mathbf{v} \bullet d\mathbf{S})$ is the volume of liquid flowing through the surface element dS per unit time. For a finite surface, the total volume of liquid flowing through the surface per unit time is then

$$\int_{\text{surface}} \mathbf{v} \bullet d\mathbf{S}$$

This explains why we called the analogous integral over the electric field the *flux* of the electric field through the surface.

Gauss's law tells us that the total flux of the electric field through a closed

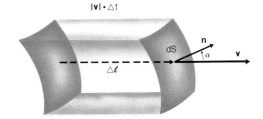

Figure 7–15 In a time interval Δt all the liquid in the volume of height $|\mathbf{v}|\Delta t$ will flow through the surface S.

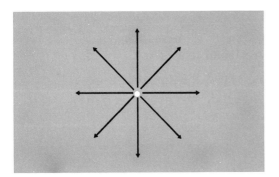

Figure 7-16 A source of electric field lines.

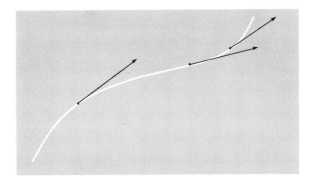

Figure 7-17 At any point the electric field is tangential to the field line.

surface is zero, if there is no charge inside. In the same way, the total flux of a liquid through any closed surface is zero, since as much liquid has to flow out as is flowing in. But even if there is a charge inside, we can use the "liquid flow" analogy: we only have to think of each positive point charge as a "source of liquid" where a certain amount of liquid is constantly created (Fig. 7-16); the liquid then flows away from the source. In the same way, a negative point charge can be considered a "sink," a point where a certain amount of liquid per unit time simply vanishes.

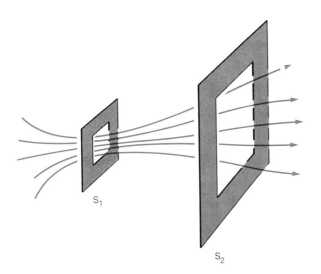

Figure 7-18 More fluid analogy: If the same amount of fluid flows through different areas S, the velocity is indirectly proportional to the area size. We conclude that the electric field magnitude is proportional to the density of the flux lines.

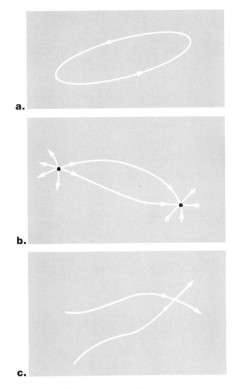

Figure 7–19 Possible and impossible flux line topologies: (a) the lines can never close upon themselves; (b) lines always start at a source (positive charge) and end at a sink (negative charge); (c) lines can never cross.

This analogy helps us to visualize the electric field and enables us to "sketch it." In a flowing liquid we can trace the path of a single particle (molecule) of the liquid (Fig. 7–17). We know that the particle had to start at a positive "source" and end at a negative "sink"—unless, of course, its path closes on itself or the particle moves away towards infinity.

At any point of the path the velocity is tangential to the path. If the particles move away from each other, they will slow down, as can be seen from Fig. 7–18. The same flux has to flow through S_1 as flows through S_2; but since S_2 is larger, the particle velocity has to be smaller. If the flowing liquid has to simulate the electric field, we can also use the first of Maxwell's equations. It says that no particle path can close on itself (Fig. 7–19a). Also, no two particle paths can cross each other, since this would mean that, for example, at the point P, two different directions were possible.

Example 4   Let us apply what we have learned about the idea of electric flux to qualitatively study and sketch the electric field of two point charges $+2Q$ and $-Q$ some distance apart, as shown in Fig. 7–20a (on page 212).

Of course, on a paper we can sketch only in two dimensions; but there is a symmetry axis, so we will sketch only the field lines in a plane. We know that very close to each charge its own effect will dominate; thus, we can draw (Fig. 7–20b) $2n$ lines starting symmetrically at the charge $+Q$, and n lines ending symmetrically at the charge $-Q$. In Fig. 7–20 we chose $n = 5$. We know also that the dashed line starting at $+2Q$ has to end at $-Q$, since the field directly between the charges has to point everywhere from $+2Q$ to $-Q$. This gives us the relative orientation of the two "stars." Next we draw the field far away (Fig. 7–20c). We know that far away the field will be the same as that of a single charge $2Q + (-Q) = Q$. In addition, the field on the symmetry axis to the left of $+2Q$ has to point always to the left (dashed line in Fig. 7–20c). Thus, we again draw n (= 5) lines as if there were a charge $+Q$ sitting in the center. Since paper is expensive, we have not really drawn the distant lines in Fig. 7–20c, but only indicated their direction.

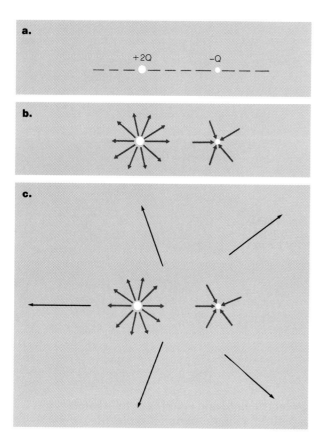

Figure 7-20 Example 4: Sketching the electric flux lines for two charges (+2Q) and (−Q).

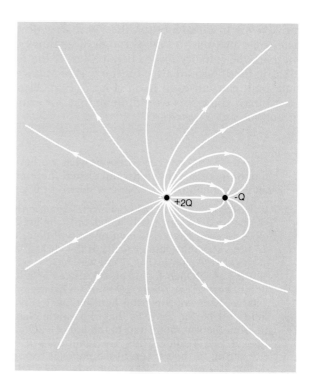

Figure 7-21 Example 4: The final result.

Finally (Fig. 7–21), we simply connect the lines smoothly and have a fairly accurate picture of the electric field anywhere in space. In Fig. 7–21 we have increased the number of lines to get more detail.

4. CONDUCTORS

As we said in the introduction to this chapter, electrons can move around freely in conductors. Thus, for a static charge distribution in a conductor it is necessary and sufficient that there be no electric field anywhere inside the conductor material; then an "inside" electron has no reason to move, because there is no force on it. We will, of course, assume that the conductor is surrounded by insulating material; thus, there can exist an electric field at the conductor surface. From these simple statements we will be able to deduce how charges distribute themselves along a conductor.

Since there is no electric field anywhere inside the conductor, the line integral

$$\int_{P_1}^{P_2} \mathbf{E} \cdot d\mathbf{l} = 0$$

for any two points P_1 and P_2 inside the conductor. The statement follows immediately if the path from P_1 to P_2 lies completely inside the conductor (Fig. 7–22); but since the value of the integral is independent of the path, it will also vanish for paths which lie partially outside the conductor. Thus, the electrostatic potential is the same everywhere inside or on the surface of a conductor, whatever the conductor's shape. In particular, two conductors connected by a metallic (conducting) wire will be at the same potential.

We can also convince ourselves that since there is no electric field, there also can be *no net charge* at any point *inside the conductor*. If there were such a spot, we could enclose it in a closed surface (e.g., a sphere) and Gauss's law would tell us that the integral of the electric field over the surface is

$$\int_S \mathbf{E} \cdot d\mathbf{S} = \frac{Q}{\epsilon_0}$$

where Q is the charge inside the closed surface. But since our surface is completely *inside* the conductor, there can be no electric field anywhere on it. Thus, the charge inside also has to be equal to zero. We have learned that *on a conductor all static charges have to be at its surface.*

How about a hollow conductor, such as a piece of copper with a completely enclosed cavity? Can there be a charge on its inside cavity? The answer depends on whether there is a charge (such as another charged conductor) inside the cavity itself or not. If there is *no charge inside the cavity*, there will also exist *no charge on the surface of the cavity*. We can show this by a simple argument.

Figure 7–22 The potential difference between any two points on or inside the same conductor has to vanish. Also, for any closed surface S we have $\int_S \mathbf{E} \cdot d\mathbf{S} = 0$. Hence one concludes that there are no charges inside a conductor.

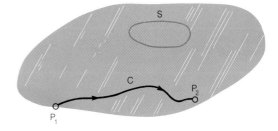

Imagine the hollow cavity in a piece of copper filled with a conducting liquid such as mercury. Now we have a good conductor throughout—there cannot be any charge inside. But this means that there is also no charge on the mercury-copper interface. We can now drain the mercury from the cavity without disturbing any charges—there will still be no charges on the inside cavity surface.

■   We can also prove this in a more mathematical way. Let us take a closed surface S completely inside the conductor, but enclosing the cavity. Since every point of the surface is inside the conductor, the electric field vanishes everywhere on the surface. Thus

$$\int_S \mathbf{E} \cdot d\mathbf{S} = 0$$

which implies that the total charge on the inside surface is zero. Thus, if there is a region of positive charge, there has to be also a region of negative charge on the inside surface (Fig. 7–23). Since the electric field points from positive charge to negative, the line integral along the path from one region to the other is

$$\int_{P_1}^{P_2} \mathbf{E} \cdot d\mathbf{l} > 0$$

$C_1 = \text{path through cavity}$

But we can choose another path C_2 which lies completely inside the copper where the electric field is zero:

$$\int_{P_1}^{P_2} \mathbf{E} \cdot d\mathbf{l} = 0$$

$C_2 = \text{path inside conductor}$

which violates Maxwell's first equation of electrostatics. Therefore, there can be no electric charge anywhere on the inside surface.   ■

This argument fails if there is a charge such as $+Q$ inside the volume of the cavity. We cannot fill the cavity with mercury—the charge $+Q$ would be free to move away. Also, the fact that the integral [Eq. (7–19)] vanishes for a surface S lying completely inside the conductor shows us that now there has to be the charge $-Q$ distributed over the cavity surface. Only then will the total charge inside S be equal to zero, thus satisfying Gauss's law (Fig. 7–24).

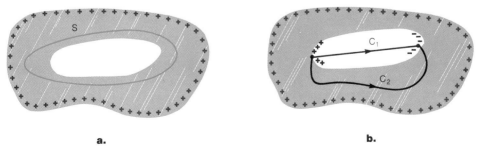

a.   **b.**

Figure 7–23 An empty cavity (with no charge inside the cavity) has no effect on the distribution of the charge in the conductor; all charge is still on the outside surface.

Figure 7–24 If there is a charge $+Q$ inside the cavity in a conductor, there has to be a negative charge on the cavity wall which screens it so that $\mathbf{E}=0$ inside the conductor.

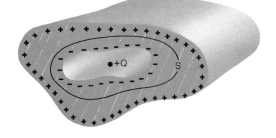

Thus far we have learned that all charges reside on the surface of a conductor. But we still have to learn how the charge distributes itself over the surface. For that purpose let us study the electric field just outside the conductor, as sketched in Fig. 7–25a. We can easily show that there can be no field parallel to the conductor surface. Indeed, we can define a closed loop which goes a distance a from P_1 to P_2 just outside the conductor and returns just inside the conductor. The distances from P_2 to P_3 and from P_4 to P_1 can be kept negligibly short. The electric field inside the conductor vanishes. Thus, the only contribution to the closed-loop integral over the electric field comes from the segment $\overline{P_1 P_2}$. We have, because of Maxwell's first equation,

$$0 = \int_{\substack{\text{whole} \\ \text{loop}}} \mathbf{E} \cdot d\mathbf{l} \approx |\mathbf{E}_t|\, a$$

and therefore we conclude that $|\mathbf{E}_t| = 0$ just outside the conductor surface.

We can also relate the normal component of the electric field to the surface charge density σ. We define as surface charge density the charge per unit area of the surface. Thus, σ is measured in units of coulombs/(meter)^2, in distinction to the spatial density ρ, which has the unit of coulombs/(meter)^3.

We will relate the surface charge density to the electric field just outside the surface using Gauss's law. In Fig. 7–25b we have introduced a small Gaussian surface in the form of a flat cylinder (coin-shaped disk of radius r), one face of which lies inside the conductor and the other outside. Inside the conductor the electric field vanishes; thus, only the outside face of the disk contributes

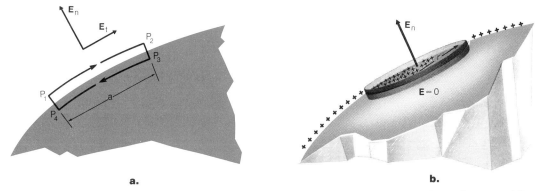

a.   b.

Figure 7–25 (a) There can be no tangential electric field just outside a conductor because of the first Maxwell equation. (b) The normal component of the electric field just outside a conductor is proportional to the surface charge density; to prove this, one applies Gauss's law to a coin-shaped closed surface half inside and half outside the conductor.

to the integral over the closed cylindrical surface:

$$\int_{\text{disk}} \mathbf{E} \bullet d\mathbf{S} = |\mathbf{E}_n| \; \pi r^2$$

where πr^2 is the area of the disk. By Gauss's law this has to be equal to $1/\epsilon_0$ times the charge inside the disk. But this charge is equal to the surface charge density times the disk area:

$$q = \sigma \pi r^2$$

Thus we have from Gauss's law

$$|\mathbf{E}_n| \; \pi r^2 = \frac{\sigma \pi r^2}{\epsilon_0}$$

or

$$\boxed{|\mathbf{E}_n| = \frac{\sigma}{\epsilon_0}} \qquad\qquad (7\text{--}20)$$

The electric field will point away from the conductor if σ is positive, and toward the conductor if σ is negative.

Example 5   A flat metallic sheet of area $A = 4 \text{ ft} \times 8 \text{ ft}$ carries the unknown charge Q. Experimentally one measures that the electric field at the surface has the magnitude $E_n = 100 \text{ V/cm}$. Determine the total charge Q (Fig. 7–26).

The charge will distribute itself over both surfaces of the sheet. The surface charge density can be determined from Eq. (7–20):

$$\sigma = \epsilon_0 \; |\mathbf{E}_n| = 8.85 \times 10^{-12} \frac{\text{coul}}{\text{V-m}} \times 100 \; \frac{\text{V}}{\text{cm}}$$

We note that we do not use consistent units, because the electric field is given in V/cm, while ϵ_0 is in MKSC units. However, $10^2 \text{ V/cm} = 10^4 \text{ V/m}$ and thus

$$\sigma = 8.85 \times 10^{-12} \times 10^4 = 8.85 \times 10^{-8} \frac{\text{coul}}{\text{m}^2}$$

The area is given in square feet; but we know (or can look up in Table II of Appendix A) that $1 \text{ ft} = 0.305 \text{ m}$. Thus

$$A = 32 \text{ ft}^2 = 32 \times (0.305)^2 \text{ m}^2 = 2.97 \text{ m}^2$$

is the area of the sheet. The total charge is distributed on both sides; thus

$$Q = \sigma \times 2A = 8.85 \times 10^{-8} \times 2 \times 2.97 = 5.26 \times 10^{-7} \text{ coul}$$

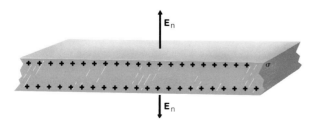

Figure 7–26 Example 5: The electric field outside a charged plate.

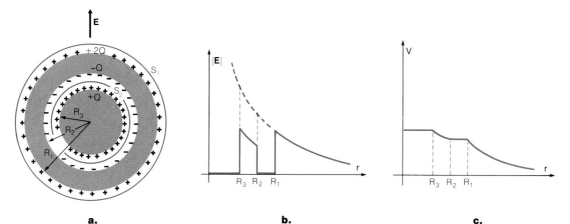

Figure 7-27 Example 6: (a) The two spheres and the Gaussian surfaces S_1 and S_2 used; (b) the radial dependence of the electric field; (c) the radial dependence of the electrostatic potential.

Example 6   A hollow metallic sphere has the outside radius $R_1 = 10$ cm and inside radius $R_2 = 7$ cm. Centered in the hollow inside is a solid metallic sphere of radius $R_3 = 5$ cm (Fig. 7-27). The same charge $Q_0 = 10^{-8}$ coul is deposited on each of the two spheres. Determine how the charges will be distributed over the two bodies and calculate the electrostatic potential at the center of the inner sphere.

In order to calculate the electrostatic potential, we first have to calculate the electric field everywhere. Since we have a situation of complete spherical symmetry, we can use Gauss's law on spherical surfaces for calculating the integral of the electric field.

Let us start from the inside. The electric field will everywhere be pointing radially outward. For $r < R_3$, we are inside a conductor; thus,

$$E_r = 0 \quad \text{for} \quad r < R_3 \tag{7-21}$$

All the charge on the inside sphere will be distributed over its surface at $r = R_3$. Thus, for any spherical surface S_2 for which $R_3 < r < R_2$, we have

$$\int \mathbf{E} \cdot d\mathbf{S} = \frac{Q_0}{\epsilon_0}$$

or

$$E_r(4\pi r^2) = \frac{Q_0}{\epsilon_0}$$

from which we conclude that

$$E_r = \frac{Q}{4\pi\epsilon_0} \frac{1}{r^2} \quad \text{for} \quad R_3 < r < R_2 \tag{7-22}$$

For all distances r for which $R_2 < r < R_1$, we are inside the metal of the outer sphere. There can be no electric field there; thus,

$$E_r = 0 \quad \text{for} \quad R_2 < r < R_1 \tag{7-23}$$

However, Gauss's law applied to a spherical surface lying inside the outer hollow sphere says that

$$\int \mathbf{E} \cdot d\mathbf{S} = \left(\frac{Q}{\epsilon_0}\right)_{\text{inside}}$$

Since the electric field vanishes, the charge inside such a Gaussian surface also has to vanish. There is a charge $+Q_0$ residing on the inner sphere; thus, the charge $-Q_0$ has to reside on the inner surface of the outer hollow sphere. Since the total charge of the outer sphere is $+Q_0$, this implies that $+2Q_0$ is the total charge distributed over the outer surface of the hollow outer sphere.

Outside the hollow sphere we can again use a spherical Gaussian surface of radius $r > R_1$ to obtain

$$\int \mathbf{E} \cdot d\mathbf{S} = E_r(4\pi r^2) = \left(\frac{Q}{\epsilon_0}\right)_{\text{inside}}$$

The total charge inside this Gaussian surface is

$$(Q)_{\text{inside}} = +Q_0 - Q_0 + 2Q_0 = 2Q_0$$

and thus

$$E_r = \frac{2Q_0}{4\pi\epsilon_0 r^2} \quad \text{for} \quad r > R_1 \tag{7–24}$$

In Fig. 7–27b we have sketched the radial dependence of the electric field.

We obtain the electrostatic potential by integrating the electric field starting from a point at infinity. On the surface of the outer hollow sphere we have

$$V(R_1) = -\int_\infty^{R_1} \frac{2Q_0}{4\pi\epsilon_0 r^2} \, dr = -\frac{2Q_0}{4\pi\epsilon_0}\left(-\frac{1}{r}\right)\Big|_\infty^{R_1}$$

or

$$V(R_1) = \frac{2Q_0}{4\pi\epsilon_0 R_1}$$

There is no electric field between $r = R_1$ and $R = R_2$; thus

$$V(R_2) = V(R_1) = \frac{2Q_0}{4\pi\epsilon_0 R_1} \tag{7–25}$$

Next we calculate the electrostatic potential for $r = R_3$. We have

$$V(R_3) - V(R_2) = -\int_{R_2}^{R_3} E_r \, dr$$

and using Eq. (7–22) we obtain

$$V(R_3) - V(R_2) = -\int_{R_2}^{R_3} \frac{Q_0}{4\pi\epsilon_0} \frac{dr}{r^2} = -\frac{Q_0}{4\pi\epsilon_0}\left(-\frac{1}{r}\right)\Big|_{R_2}^{R_3}$$

or

$$V(R_3) - V(R_2) = \frac{Q_0}{4\pi\epsilon_0}\left(\frac{1}{R_3} - \frac{1}{R_2}\right) \tag{7–26}$$

We can use the value of $V(R_2)$ calculated in Eq. (7–25) to obtain

$$V(R_3) = \frac{Q_0}{4\pi\epsilon_0}\left(\frac{1}{R_3} - \frac{1}{R_2} + \frac{2}{R_1}\right)$$

The electrostatic potential at the center of the inner sphere will be the same as on its surface, because there is no electric field for $r < R_3$. Thus,

$$V(r = 0) = \frac{Q_0}{4\pi\epsilon_0}\left(\frac{1}{R_3} - \frac{1}{R_2} + \frac{2}{R_1}\right)$$

Inserting the numerical values we obtain, since $1/4\pi\epsilon_0 = 9 \times 10^9$ V-m/coul:

$$V(r = 0) = 9 \times 10^9 \frac{\text{V-m}}{\text{coul}} \times 10^{-8} \text{ coul} \left(\frac{1}{0.05 \text{ m}} - \frac{1}{0.07 \text{ m}} + \frac{2}{0.1 \text{ m}}\right)$$
$$= 90 \text{ V-m} \times (25.7 \text{ m}^{-1}) = 2.31 \times 10^3 \text{ V} \approx 2300 \text{ V}$$

5. CAPACITORS

Frequently in technology one encounters pairs of conductors which are isolated from each other. Electrical power is distributed in a household through a pair of insulated wires. Electric fields are usually created by using two conductors, between which one imposes a potential difference by joining them to opposite terminals of a battery. Pairs of conductors are also used for short-term storage of electric charges. We call a *capacitor* a pair of insulated conductors, of which one carries the charge $+Q$, while the other carries the charge $-Q$. The two conductors, sketched in Fig. 7–28, are called *plates*.

There will exist an electric field in the space between the two plates of a capacitor. Therefore, there will also exist a potential difference between the two plates:

$$V = -\int_2^1 \mathbf{E} \cdot d\mathbf{l} \tag{7-27}$$

We claim—and will prove—that the potential difference is proportional to the charge $\pm Q$ stored on the plates:

$$\boxed{V = \frac{1}{C} Q} \tag{7-28a}$$

where C is a quantity which depends only on the shape and separation of the two plates, but not on the stored charge $\pm Q$. We call C the *capacitance* of the capacitor; it is a measure of how much charge $\pm Q$ one can store on the plates, given the potential difference V. We can of course also write

$$\boxed{Q = CV} \tag{7-28b}$$

We will now proceed to prove Eq. (7–28). The electric field between the two plates is due to the charges $\pm Q$ stored on each plate. We assume that there are no other nearby charges which could influence the electric field. The charges $\pm Q$ are distributed over the surface of the two conducting plates so that on each surface Eq. (7–20) is everywhere valid:

$$|\mathbf{E}_n| = \frac{\sigma}{\epsilon_0} \tag{7-20}$$

Let us now assume that the charge Q stored on the plates is some unknown function of the potential difference: $Q = Q(V)$. Obviously, if there is no potential

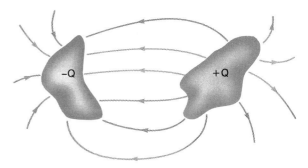

Figure 7–28 A capacitor consists of two conductors with charges $\pm Q$ separated by an insulator.

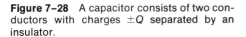

difference, and thus no electric field, then there will be also no charge: $Q(V = 0) = 0$. Now let us assume that for some charge Q_0 the potential difference is $V = V_0$. Thus, we have $Q(V_0) = Q_0$. Next let us increase the electric field everywhere by some factor μ; because of Eq. (7–27), the potential difference will also be multiplied by the same factor μ:

$$V = -\int_2^1 (\mu \mathbf{E}) \cdot d\mathbf{l} = -\mu \int_2^1 \mathbf{E} \cdot d\mathbf{l}$$

However, because of Eq. (7–20) the surface charge density at any point of each plate will be multiplied by the same factor μ:

$$\mu \, |\mathbf{E}_n| = \frac{\mu \sigma}{\epsilon_0}$$

Thus, the total charge, which is obtained by integrating the charge density over the conductors' surface, is also multiplied by the same factor μ and we have

$$Q(\mu V_0) = \mu Q_0$$

But since $V = \mu V_0$, we can also write $\mu = V/V_0$ and then we have

$$Q = \frac{V}{V_0} Q_0$$

By defining $Q_0/V_0 = C$, we obtain

$$Q = CV$$

which is identical to Eq. (7–28). Our proof only expresses the fact that relations (7–27) and (7–20) are both homogeneously linear: the potential difference is proportional to the electric field, and so is the surface charge density. Thus, the potential difference is also proportional to the surface charge density at any point of the plates, and therefore also to the total charge Q.

The unit of capacitance is 1 coulomb/volt. A new name has been introduced for this unit because it is so frequently used. We define

$$1 \text{ F} = 1 \text{ farad} = 1 \frac{\text{coul}}{\text{V}}$$

One farad is a very large unit; smaller units are frequently used:

$$1 \text{ microfarad} = 1 \, \mu\text{F} = 10^{-6} \text{ F}$$

$$1 \text{ nanofarad} = 1 \text{ nF} = 10^{-9} \text{ F}$$

$$1 \text{ picofarad} = 1 \text{ pF} = 10^{-12} \text{ F}$$

While any two conductors are sufficient to make up a capacitor, in practice some simple types are usually used to store charge.

(i) *Plane Parallel Capacitor.* This consists of two parallel plates of area A separated by a small distance d, as shown in Fig. 7–29. If we deposit the charges $\pm Q$ on each plate, all the charge will be on the surfaces facing each other. Thus, on the inside surface of each plate there will be a surface charge density

$$\pm \sigma = \pm \frac{Q}{A}$$

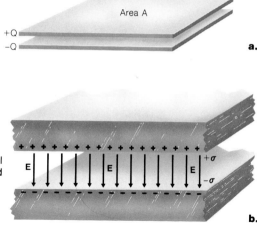

Figure 7–29 The plane parallel capacitor: (a) overall view; (b) electric charge distribution and electric field inside the gap.

where A is the area of the plates. Because of Eq. (7–20), the electric field between the plates then has the magnitude

$$|\mathbf{E}| = \frac{\sigma}{\epsilon_0} = \frac{Q}{\epsilon_0 A} \tag{7–29}$$

and the potential difference between the two plates is

$$V = -\int \mathbf{E} \cdot d\mathbf{l} = |\mathbf{E}|\, d \tag{7–30}$$

because the field is constant. Substituting from Eq. (7–29), we obtain

$$V = \frac{d}{\epsilon_0 A} Q = \frac{1}{C} Q \tag{7–31}$$

Thus, the capacitance of the parallel plate capacitor is

$$\boxed{C = \frac{\epsilon_0 A}{d}} \tag{7–32}$$

The electrostatic potential, because it is the potential energy per unit charge for a positive test charge, is higher at the positively charged plate.

Example 7   A plane parallel capacitor's plates are squares of side a. The distance between the plates is $d = 1$ mm. How large do we have to choose a in order to store the charge ± 1 coul on the plates, if the potential difference is $V = 100$ volts?

If 1 coul is to be stored at a potential difference of 100 volts, we need a capacitance

$$C = \frac{Q}{V} = \frac{1 \text{ coul}}{100 \text{ V}} = 10^{-2} \text{ F}$$

Since the area of the square plate is $A = a^2$, we deduce from Eq. (7–32)

$$\frac{\epsilon_0 a^2}{d} = C$$

or

$$a = \sqrt{\frac{Cd}{\epsilon_0}}$$

We have $d = 1$ mm $= 10^{-3}$ m and $\epsilon_0 = 8.85 \times 10^{-12}$ coul/V-m; thus,

$$a = \sqrt{\frac{10^{-2} \text{ F} \times 10^{-3} \text{ m}}{8.85 \times 10^{-12} \text{ coul/V-m}}} = 1.06 \times 10^3 \text{ m}$$

or about 1 kilometer $\approx$ 2/3 mile. Obviously this is not realistic; 10^{-2} F is a very large capacitance, which we cannot produce with two parallel plates.

Practical capacitors, instead of two plates separated by air, usually consist of two aluminum foils separated by a very thin insulating material, such as oil paper or Mylar foil. Then one can roll the "plates" into a tight little cylinder, as shown in Fig. 7–30; actually the capacitance increases, because we now use both sides of the aluminum foil to store charge. One can thus achieve capacitances of 0.1 to 1 μF in an object 1 inch long and 3/4 inch in diameter.

(ii) *Concentric Cable.* The concentric cable consists of an inner cylindrical conductor of radius r, surrounded by an outer conductor of inside radius R (Fig. 7–31). We have here, in a way, a plane parallel capacitor rolled into a tube. We will assume that the length L of the cable is much larger than its radius, $L >> R$; we can then neglect the effects of the end of the tube. Because of the cylindrical shape, we can no longer assume the electric field to be the same everywhere in the region between the two conductors. In order to determine the capacitance, we imagine a charge $+Q$ to be deposited on the inside conductor and $-Q$ on the outer conductor. Then the charge per unit length on the two surfaces will be

$$\lambda = \frac{Q}{L}$$

If we enclose the inner conductor by a cylindrical surface of radius ρ and length a (Fig. 7–31b), Gauss's law says that

$$\frac{\lambda a}{\epsilon_0} = |\mathbf{E}(\rho)| \, 2\pi\rho a$$

or

$$|\mathbf{E}(\rho)| = \frac{\lambda}{2\pi\epsilon_0} \frac{1}{\rho} \tag{7–33}$$

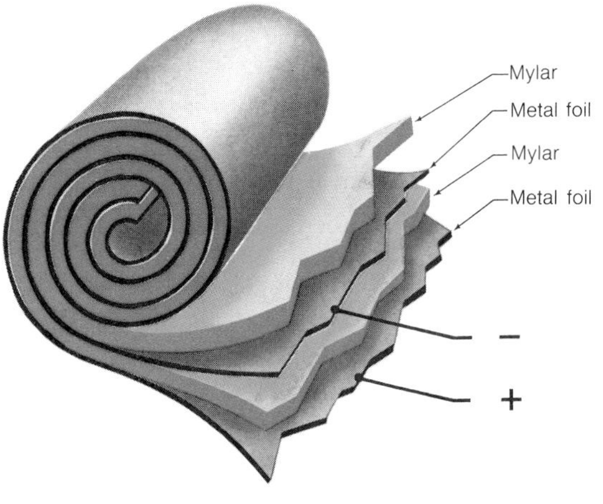

—Mylar
—Metal foil
—Mylar
—Metal foil

−

+

Figure 7–30 Practical capacitors are basically plane-parallel capacitors rolled up.

Figure 7–31 The capacitance of a concentric cable: (a) overall view; (b) the Gaussian surface used to calculate the electric field.

Note that Eq. (7–33) is identical to Eq. (7–17); it describes the electric field in exactly the same situation. The potential difference between the two conductors will be

$$\Delta V = -\int \mathbf{E} \bullet d\mathbf{l} = \frac{\lambda}{2\pi\epsilon_0} \int_r^R \frac{d\rho}{\rho} = \frac{\lambda}{2\pi\epsilon_0}\left[\ln R - \ln r\right]$$

or

$$\Delta V = \frac{Q}{2\pi\epsilon_0 L} \ln\left(\frac{R}{r}\right) = \frac{1}{C}Q$$

From this we can determine the capacitance

$$C = \frac{2\pi\epsilon_0 L}{\ln\left(\dfrac{R}{r}\right)}$$

(7–34)

Should we choose the two conductors nearly touching each other:

$$R = r + d \quad \text{with} \quad d \ll r$$

then we can use the approximation

$$\ln\left(\frac{r+d}{r}\right) = \ln\left(1 + \frac{d}{r}\right) \approx \frac{d}{r}$$

We then obtain for the capacitance from Eq. (7–34):

$$C \approx \frac{2\pi\epsilon_0 L}{\dfrac{d}{r}} = \frac{\epsilon_0 2\pi r L}{d}$$

This is the same as the capacitance of a parallel plate capacitor of area $A = 2\pi r L =$ area of cylinder and spacing d. We see that the geometry does not matter if the distance between the conductors is sufficiently small: a "rolled-up" capacitor has the same capacitance as a "flat" one, if we do not roll it up too tight (the typical radius r or R has to be much larger than the spacing d).

Example 8   Calculate the capacitance of two spheres of radius r separated by a large distance $L \gg r$ (Fig. 7–32).

We assume $L \gg r$ so that we can calculate the electric field due to the charge on each sphere. If the two spheres are close together, the charges $\pm Q$ will no longer be distributed evenly over each sphere; the mutual attraction of opposite charges will lead to a charge distribution like that shown in Fig. 7–33. However, if the two spheres are far from each other, this effect can be neglected. We can then assume that the charges $\pm Q$ are distributed isotropically over each sphere's surface. The positive sphere will produce at a point P an electric field

$$\mathbf{E}_1 = \frac{Q}{4\pi\epsilon_0 d_1{}^2}\hat{\mathbf{u}}_+$$

where $\hat{\mathbf{u}}_+$ (Fig. 7–32) is the unit vector pointing away from the positively charged sphere. The negative sphere will also produce a field at the same point P:

$$\mathbf{E}_2 = \frac{(-Q)}{4\pi\epsilon_0 d_2{}^2}\hat{\mathbf{u}}_-$$

The total field is

$$\mathbf{E} = \mathbf{E}_1 + \mathbf{E}_2$$

In order to calculate the potential difference, we choose a simple path: the straight line going from one sphere to the other. Along this path both electric fields point in the same direction; at any distance x from the positive charge we have (Fig. 7–32)

$$|\mathbf{E}| = \frac{Q}{4\pi\epsilon_0}\left(\frac{1}{x^2} + \frac{1}{(L-x)^2}\right)$$

We have to integrate this from $x = r$ to $x = L - r$; thus,

$$V = \frac{Q}{4\pi\epsilon_0}\int_{x=r}^{L-r}\left(\frac{1}{x^2} + \frac{1}{(L-x)^2}\right)dx$$

The integral can be easily performed and yields

$$\int_r^{L-r}\left(\frac{1}{x^2} + \frac{1}{(L-x)^2}\right)dx = \left(-\frac{1}{x} + \frac{1}{L-x}\right)\Big|_r^{L-r} = -\frac{1}{L-r} + \frac{1}{r} + \frac{1}{r} - \frac{1}{L-r}$$

Thus we obtain

$$V = \frac{Q}{4\pi\epsilon_0}\left(\frac{2}{r} - \frac{2}{L-r}\right)$$

Because we assume $L \gg r$, we can neglect the second term in the bracket

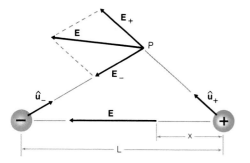

Figure 7-32 Example 8: The electric field of two small charged spheres.

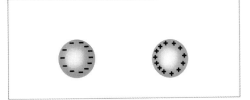

Figure 7-33 If the two spheres are close together, the charge distribution on each will no longer be spherically symmetrical.

$\left(\dfrac{1}{L-r} << \dfrac{1}{r} \right)$ to obtain

$$V = \frac{1}{2\pi\epsilon_0 r} Q$$

Thus the capacitance is

$$C = 2\pi\epsilon_0 r$$

and is independent of the distance L.

6. CAPACITORS IN PARALLEL AND IN SERIES

In electrical engineering practice one frequently connects several capacitors together, either by design or by necessity.[*] Thus, it is necessary that we look briefly at such arrangements.

In Fig. 7-34 we show two such capacitors connected by wires. By our definition the arrangement is a single capacitor; it consists of two conductors which are separated by insulating material. However, in many practical applications we can neglect the surface charge on the wires connecting the two parallel plate capacitors. We then think of this arrangement as *two capacitors in parallel.* Both top plates, because they are connected by a conducting wire, will always be at the same potential; the same can be said of the two bottom plates. Let us now deposit the charge Q on the top plates and $-Q$ on the bottom plates. The charge $-Q$ will distribute itself among the two capacitors. There will be on C_1 the charge

$$Q_1 = C_1 V$$

and on C_2

$$Q_2 = C_2 V$$

---

[*] Because any two conductors separated by an insulator constitute a capacitor, most physical electrical systems contain capacitances, regardless of the designer's intentions. These "stray" capacitances are not always negligible; they can strongly affect the behavior of high-frequency equipment.

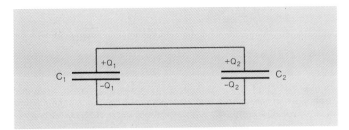

Figure 7-34 Two capacitors in parallel.

where V *is the same* potential difference and C_1 and C_2 are the capacitances of the two capacitors. For the total charge we then have

$$Q = Q_1 + Q_2 = (C_1 + C_2)V$$

and we can say that the capacitance of the two parallel capacitors is the same as that of a simple capacitor of capacitance

$$C = C_1 + C_2 \qquad (7\text{-}35)$$

If we put the two capacitors in parallel, their capacitances add.

Let us now take the same two capacitors, but connect them *in series* as shown in Fig. 7-35a. Now we no longer have a single capacitor by our definition, because there are three conductors mutually insulated from each other. But we still can talk about the capacitance of the circuit. If we deposit the charges $\pm Q$ on the outer conductors (in color in Fig. 7-35), but also make sure that the central conductor (black) has no net charge, then there will be a potential difference V between the two outer conductors which is proportional to Q:

$$Q = CV \qquad (7\text{-}36)$$

The question which arises is: How do we calculate C for such an arrangement — or for a basically identical arrangement shown in Fig. 7-35b? The answer is that the central conductor, while carrying no net charge, will become *polarized*. The charge $-Q$ will appear opposite the positively charged top outer conductor, because there can be no electric field inside the metal of the central conductor. On the other hand, the charge $+Q$ will appear on the top plate of the bottom conductor C_1.

There is a potential difference

$$V_1 = \frac{1}{C_1} Q$$

between the bottom conductor and the (black) central conductor. There is also

a.   **b.**

Figure 7-35 (a) Two capacitors in series: idealized arrangement. (b) Two capacitors in series: in a real arrangement one has to take care that the charges on the two end conductors do not affect each other; otherwise Eq. (7-37) is no longer valid.

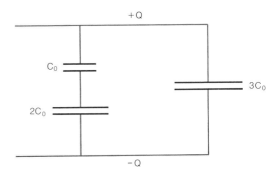

Figure 7-36 Example 9: A circuit made of three capacitors.

a potential difference

$$V_2 = \frac{1}{C_2} Q$$

across the plates of the capacitor C_2. The total potential difference is

$$V = V_1 + V_2 = \left(\frac{1}{C_1} + \frac{1}{C_2} \right) Q$$

Comparing this with Eq. (7–36), we see that the net capacitance C of the circuit obeys the relation

$$\frac{1}{C} = \frac{1}{C_1} + \frac{1}{C_2} \tag{7-37}$$

For two capacitors in series, the inverse capacitances add to yield the inverse of the total capacitance.

Example 9   Three capacitors are arranged as shown in Fig. 7–36. What is the total capacitance of the circuit in units of the capacitance C_0?

The two capacitors to the left are in series and thus equivalent to a single capacitor of capacitance C, where

$$\frac{1}{C} = \frac{1}{1C_0} + \frac{1}{2C_0} = \frac{3}{2C_0}$$

or

$$C = \frac{2}{3} C_0$$

Thus, the circuit is equivalent to two capacitors of capacitance $3C_0$ and $2/3\,C_0$ in parallel. The total capacitance thus is

$$C_{\text{tot}} = 3C_0 + \frac{2}{3} C_0 = \frac{11}{3} C_0 = 3.667 C_0$$

7. ENERGY STORED IN CAPACITORS

Capacitors can be used not only as storage devices for electric charge, but also as storage devices for electrical energy. Consider the capacitor shown in

Fig. 7–37. If we move the charge dQ from the negatively charged plate to the positively charged plate, we have to supply externally the work

$$dW = dQ\,V = dQ\,\frac{Q}{C}$$

However, at the same time we have increased the charges $\pm Q$ to $\pm(Q + dQ)$, because we took the charge dQ away from $-Q$ and added it to $+Q$. We can now consider the whole charge Q successively moved in small increments dQ from one plate to the other. During this process the potential difference V will continuously change, because we have

$$V = \frac{Q}{C}$$

where Q is the charge which has been transmitted before. The total energy stored in the system then is

$$W = \int dW = \int_{Q=0}^{Q=Q} \frac{Q}{C}\,dQ$$

and this yields

$$W = \frac{Q^2}{2C}$$

which, because of Eq. (7–28), we can write in three equivalent forms:

$$\boxed{W = \frac{Q^2}{2C} = \frac{1}{2}QV = \frac{1}{2}CV^2} \qquad (7\text{–}38)$$

Example 10   Calculate the total energy stored in a parallel plate capacitor, if the plates have an area of 100 cm², the distance between them is 1 mm, and the potential between the two plates is 1000 volts.

The total stored energy is equal to

$$E = \frac{1}{C}\frac{Q^2}{2} = \frac{CV^2}{2}$$

where

$$C = \frac{\epsilon_0 A}{d}$$

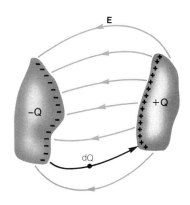

Figure 7–37 If the charge dQ is moved from the negatively charged plate to the positively charged plate, one has to supply the energy $dW = V\,dQ$.

is the capacitance of the capacitor. Numerically,[*]

$$E = \frac{\epsilon_0 A}{d} \frac{V^2}{2} = \frac{(8.85 \times 10^{-12} \text{ coul/V-m}) \times 10^{-2} \text{ m}^2}{10^{-3} \text{ m}} \frac{(10^3 \text{ V})^2}{2} = 4.43 \times 10^{-5} \text{ J}$$

This is quite a small energy; it is equivalent to the potential energy stored in a mass of 4.5 g which is 1 mm above ground. One can see that the simple parallel plate capacitor is not a very efficient energy storage device. For such storage one usually uses large banks of capacitors having a total capacitance of 10^{-3} to 10^{-2} F.

Summary of Important Relations

Gauss's law (Maxwell's second equation)

$$\int_{\substack{\text{closed} \\ \text{surface}}} \mathbf{E} \cdot d\mathbf{S} = \frac{1}{\epsilon_0} [Q]_{\text{inside}}$$

Electric field of homogeneously charged sphere of radius R

$$E = \frac{1}{4\pi\epsilon_0} \frac{Q}{r^2} \quad \text{outside}$$

$$= \frac{1}{4\pi\epsilon_0} \frac{Qr}{R^3} \quad \text{inside}$$

Electric field of long charged rod (linear charge density λ, distance from rod r)

$$E = \frac{\lambda}{2\pi\epsilon_0 r}$$

Electric field at surface of conductor (surface charge density σ) $E = \dfrac{\sigma}{\epsilon_0}$

Capacitor equation $Q = CV$

plane parallel capacitor
A = area, d = separation $C = \dfrac{\epsilon_0 A}{d}$

concentric cable
r, R radii; L length $C = \dfrac{2\pi\epsilon_0 L}{\ln\left(\dfrac{R}{r}\right)}$

two capacitors C_1, C_2 in parallel $C = C_1 + C_2$

in series $\dfrac{1}{C} = \dfrac{1}{C_1} + \dfrac{1}{C_2}$

---

[*] Note that 1 coul $\times$ 1 V = 1 J.

Energy stored in capacitor $W = \dfrac{Q^2}{2C} = \dfrac{1}{2}QV = \dfrac{1}{2}CV^2$

Questions

1. Is the first Maxwell's equation (7–18) usually satisfied for the flow of a liquid? State some specific examples where it is probably satisfied and some where it is not.

2. Sketch the electric field lines for:
   (a) an electric dipole.
   (b) an electric quadrupole as shown in the figure.

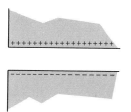

3. Sketch the electric field lines near the edge of a plane parallel capacitor. Remember that $\int \mathbf{E} \cdot d\mathbf{l} = \Delta V$ for *any* path.

4. From Eq. (7–17), calculate the electrostatic potential near an infinitely long rod. Can we normalize the potential so that it vanishes at infinity? If not, why not?

5. A metallic object shaped as in the figure carries the charge $\pm Q$ and is far away from any other charges. Where would you expect the surface charge density to be largest? Smallest? Why?

6. A point charge $+Q$ is near a grounded conductor as shown. Sketch the electric field lines and the surface charge density.

1. Four closed surfaces, S_1 through S_4, are sketched in the figure, together with the charges $+Q$, $+Q$, and $-2Q$. Give the electric field flux through each surface.

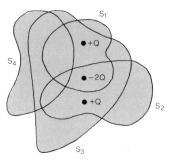

2. A charge $+Q$ is at the center of a cube of edge length a. What is the electric field flux through one of the faces of the cube?

3. A uranium nucleus can be approximated by a sphere of radius $R = 0.8 \times 10^{-14}$ m and charge $+Ze$ with $Z = 92$. What is the electrostatic force on a proton on the uranium nucleus surface? (Note: The force is counteracted by the much stronger attractive nuclear force between the protons and neutrons in the nucleus.)

4. A long rod of radius $r = 1$ cm carries everywhere inside the volume charge density (charge per unit volume) $\rho = 10^{-8}$ coul/m^3. Calculate the electric field *inside* the rod and the difference in electrostatic potential between the center of the rod and its surface.

5. On a clear day, the electric field near the earth's surface is 100 V/m, pointing downward. If the same electric field existed everywhere on the earth's surface, what would be the total charge stored in the earth? (Earth radius $= 6.36 \times 10^6$ m.) Actually, the electric field reverses its sign under cloud cover; thus, the net charge of the earth is certainly much smaller, possibly zero.

°6. Two very long metallic wires, each of radius $r = 2$ mm, are separated by $d = 20$ cm. One carries the linear charge density $+\lambda = 10^{-9}$ coul/m, the other the opposite charge density $-\lambda$. Calculate the potential difference between the two wires. (Hint: Calculate first the electric field and then the electrostatic potential at an arbitrary point P a distance ρ_1 and ρ_2 from each wire.)

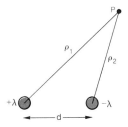

7. Five uncharged spherical metallic shells with radii 1, 2, 3, 4, and 5 cm are all concentric, as shown in the figure on p. 232. The two spheres of radius 2 and 3 cm are connected by a wire, as are the two spheres of radius 4 and 5 cm. A charge $Q = 10^{-10}$ coul is then deposited on the innermost shell.

    (a) What is the charge on each shell?

    (b) Calculate the potential difference between the innermost shell and a point very far away.

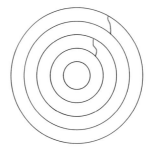

8. A small charged sphere (charge $+Q = 10^{-9}$ coul) is at the center of a hollow sphere; the net charge of the hollow sphere is zero. Outside, a distance $d = 1$ m away from the center, is another small sphere with charge $-Q$. All spheres are made of steel, which is a good conductor. Estimate the electrostatic force on each sphere.

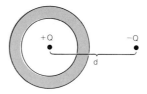

9. An isolated conducting sphere of radius $r = 5$ m is charged to a potential $V = 10^6$ volts.

    (a) What is the charge stored on the sphere?

    (b) What is the potential energy stored in the charge on the sphere?

10. A plane parallel capacitor consists of two metallic plates of area $A = 2$ m² each, separated by $d = 0.5$ mm of air. Assuming that a spark occurs if the electric field is larger than 10^6 V/m, how much charge can be stored on the plates?

11. Calculate the capacitance between two thin metallic spherical shells of radii $R_1 = 10$ cm and $R_2 = 15$ cm.

12. The electrons in an electron beam are traveling with the velocity $v = 10^6$ m horizontally. The beam enters the region between the plates of a parallel plate capacitor very close to the top plate. The plate spacing is $d = 1$ cm, the plate length $a = 10$ cm. How large a potential difference V can one apply between the plates, if the electron beam is to barely miss the bottom plate upon leaving?

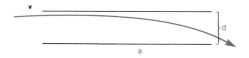

13. We want to estimate the energy stored in thunderclouds. The breakdown voltage in air is $E_0 \sim 2 \times 10^6$ V/m; the typical cloud height is 3 km (~2 miles). If the average electric field between cloud and ground is $1/2\ E_0$, what is the energy stored in a thunderstorm covering an area of 50×50 miles?

14. What are the electrostatic potential and the stored charge on a capacitor of $C = 50\ \mu$F, if the total electrical energy of 1 Joule is stored on the plates?

15. A wire of diameter $d = 1$ mm is stretched along the whole axis of a steel pipe of 2 inches inner diameter and $L = 10$ ft length. Calculate the capacitance between the wire and the steel pipe.

16. A variable capacitor typically consists of two sets of several plates each, which can rotate with respect each other to vary the overlap area. Calculate the maximum capacitance of the variable capacitor shown in the figure, if each plate area is $A = 10$ cm² and the plate spacing is $d = 2$ mm.

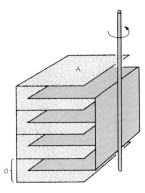

17. Two square aluminum sheets of side $a = 50$ cm are to be made into a plane parallel capacitor of $C = 500$ pF capacitance. What should be their spacing?

18. The plates of a parallel plate capacitor have the area $A = 30$ cm² and spacing $d = 5$ mm. They are charged to a potential difference $V = 1000$ V and then disconnected from the charging device. Afterwards, a copper plate $d' = 3$ **mm thick** is inserted between the capacitor plates – without ever touching them. **What** is now the potential difference V' between the plates?

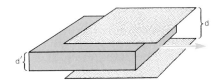

19. You are supposed to construct a bank of capacitors in parallel, able to store the energy of 1000 Joules. From a catalogue you find that electrolytic capacitors of $C = 3000$ μF rated at 15 V maximum voltage between the plates cost $1.00 each. On the other hand, oil-filled capacitors of $C' = 0.5$ μF rated at 7500 V cost $2.60 each. If the cost of wiring (including labor) is $500 independent of the type used, which type of capacitor would you choose? How much would the capacitor bank cost?

20. An unknown capacitor C' is connected in a circuit as shown. The other two capacitors have each a capacitance $C_0 = 1$ μF.

    (a) Between what limits can the overall circuit capacitance C vary, if C' varies between 0 and ∞?

    (b) If the overall capacitance is $C = 4/3$ μF, what is the value of C'?

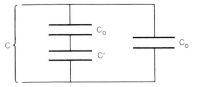

21. You are given 10 capacitors with values 0.1, 0.1, 0.5, 0.5, 1.0, 1.0, 2.0, 2.0, 5.0, and 5.0 μF each. Using at most five of them (the size of your circuit assembly will not allow more to be packed in), combine them to get as close as possible to a capacitance of 2/3 μF.

CHAPTER 8

DIELECTRICS

1. INTRODUCTION

In this chapter we will study the effect of inserting insulating materials into regions of space where there is an electric field. We have said earlier that there are no free charges in an insulator as there are in a conductor; but this does not mean that these materials are unaffected by an electric field. Any piece of matter consists of molecules, and each molecule consists of individual atoms. As we have mentioned before, an atom consists of a central, small, heavy, positively charged atomic nucleus, which is surrounded by negatively charged electrons. The electrons form a "cloud" around the nucleus (Fig. 8–1). If there is an external electric field, the nucleus will experience a force along the field, while the oppositely charged electrons will be acted upon by a force in the opposite direction. The positive and negative charges will move slightly, and the atom acquires a dipole moment; we say that *the material has become polarized.*

Some materials – such as water or acetone – consist of molecules which have permanent dipole moments even in the absence of an electric field. In the absence of an external electric field these dipoles will be oriented randomly, as shown in Fig. 8–2a. The average dipole moment will vanish, because for every dipole there will be another nearby oriented in the opposite direction.

However, these dipoles will reorient themselves in an external field. The

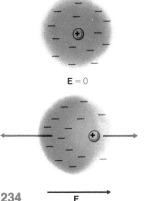

E = 0

E

Figure 8–1 An electric field acts with opposite forces on the central nucleus and on the electrons in an atom. The atom deforms slightly from its symmetric shape and now has a dipole moment.

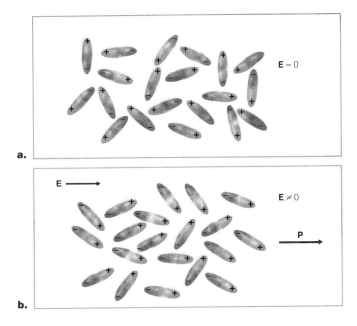

Figure 8-2 Even if each molecule has a permanent dipole moment, the polarization will vanish in the absence of an electric field because the dipoles will be oriented randomly. In the presence of an external electric field they will tend to align themselves with the field, thus generating a nonzero polarization.

potential energy given by Eq. (6–51)

$$U = -(\mathbf{p} \bullet \mathbf{E})$$

is lowest if all dipoles are aligned parallel to the electric field. This alignment will be disturbed by the random thermal motion of the molecules. Thus, thermal equilibrium establishes itself; there will be more dipoles oriented along the field than in the opposite direction. In other words, the material has again become polarized. In general, we call all polarizable materials *dielectrics*, because their behavior is intimately connected to the fact that they are built up from constituents of both charges – the positive nuclei and negative electrons.

Before we discuss these phenomena, we want to discuss more in detail the energy stored in a charge distribution. As we will see, we can associate energy with the electric field itself; this will become even more important later when we discuss time-dependent phenomena. The electric field arriving at our eyes from a distant star carries energy; but the original charge distribution which generated the energy may not even exist any more when we detect the arriving light wave.

2. ENERGY OF AN ELECTRIC FIELD

We have already seen that bringing individual charges close together changes their potential energy. Each charge is subject to a force because of the other charges present, and thus work has to be performed to move the charges around. In the case of a simple capacitor we have already solved the problem; the stored energy is, according to Eq. (7–38),

$$U = \frac{1}{2}CV^2 = \frac{1}{2}\frac{Q^2}{C} = \frac{1}{2}QV \tag{8–1}$$

There exists a relationship between the stored energy and the electric field which we will prove only for the example of the simple plane parallel capacitor.

For this purpose, let us note that the electric field between the capacitor plates has the magnitude

$$|\mathbf{E}| = \frac{V}{d} \tag{8-2}$$

where d is the distance between the plates (Fig. 8–3). Let us now calculate the integral over the whole volume of the capacitor of

$$\rho = \frac{\epsilon_0 |\mathbf{E}|^2}{2} \tag{8-3}$$

Since the electric field is everywhere the same, we obtain for the integral I

$$I = \rho \times \text{volume} = \frac{\epsilon_0 |\mathbf{E}|^2}{2} Ad$$

where A is the area of the plates. Substituting from Eq. (8–2) we obtain

$$I = \frac{\epsilon_0}{2} \left(\frac{V}{d}\right)^2 Ad$$

or

$$I = \frac{\epsilon_0}{2} \frac{A}{d} V^2 \tag{8-4}$$

However, the capacitance of the plane parallel capacitor is $C = \epsilon_0 A/d$; thus we have

$$I = \frac{C}{2} V^2$$

which according to Eq. (8–1) is equal to the energy stored in the charged capacitor. This is only a special case of a very general law. If we have an arbitrary charge distribution, the total energy stored in the system is

$$\boxed{W = \frac{\epsilon_0}{2} \int |\mathbf{E}|^2 \, d\tau} \tag{8-5}$$

where $d\tau$ is a small volume element. The integral is to be taken over all points in space where there is an electric field.

We can also interpret Eq. (8–5) to mean that with an electric field there is associated an *energy density*

$$\boxed{\rho_E = \frac{\epsilon_0}{2} |\mathbf{E}|^2} \tag{8-6}$$

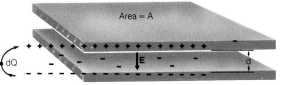

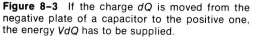

Figure 8–3 If the charge *dQ* is moved from the negative plate of a capacitor to the positive one, the energy *VdQ* has to be supplied.

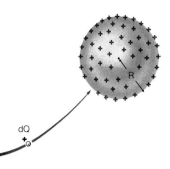

Figure 8-4 Example 1: Energy stored on the charged sphere.

Note that there is a large conceptual difference between talking about an energy density as given by Eq. (8-6) and talking about the energy required to assemble a given charge distribution. In one case we think of the energy distributed over all of space, while in the other we think of the energy localized in the actual charge distribution. In the static case which we are discussing here, there is no large difference between the two points of view; but from Eq. (8-6) we can already anticipate a generalization to time-dependent electric fields. In that case, Eq. (8-6) describes the real situation — the electric field itself carries energy with it. Thus, when a light wave arrives at a detector, we detect the energy which a time-dependent charge distribution emitted some finite time earlier.*

Example 1   A metallic sphere of radius R carries the charge Q (Fig. 8-4). Calculate the energy needed to charge the sphere in small increments and check that this agrees with Eq. (8-5).

The electrostatic potential of a charged sphere is given by Eq. (6-13)

$$V = \frac{1}{4\pi\epsilon_0} \frac{Q}{R}$$

If we want to add the charge dQ, we have to bring it in from infinitely far away (Fig. 8-4); thus, we have to provide the external work

$$dW = V\,dQ = \frac{1}{4\pi\epsilon_0} \frac{Q\,dQ}{R}$$

If we start out with $Q = 0$, and successively charge the sphere, the total work is

$$W = \int dW = \frac{1}{4\pi\epsilon_0 R} \int_{Q=0}^{Q} Q\,dQ$$

or

$$W = \frac{Q^2}{8\pi\epsilon_0 R} \tag{8-7}$$

On the other hand, Eq. (8-5) claims that

$$W = \frac{\epsilon_0}{2} \int |\mathbf{E}|^2\,d\tau \tag{8-8}$$

The electric field vanishes inside the metallic sphere. Outside (i.e., for $r > R$), it has the magnitude

$$|\mathbf{E}| = \frac{1}{4\pi\epsilon_0} \frac{Q}{r^2}$$

---

* The energy in an electromagnetic wave consists not only of the energy of the electric field, but also of the magnetic energy carried by the magnetic field. See Chapter 12.

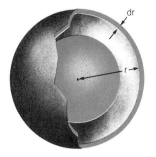

Figure 8-5 Volume element $d\tau$ used to calculate energy stored in electric field.

As volume element $d\tau$ we take a spherical shell of radius r and thickness dr (Fig. 8-5); everywhere inside this shell the electric field has the same magnitude. Its volume is the surface area of the sphere times the thickness of the shell

$$d\tau = 4\pi r^2 \, dr$$

Thus, according to Eq. (8-8) we have to perform the integral

$$W = \frac{\epsilon_0}{2} \int \left(\frac{1}{4\pi\epsilon_0} \frac{Q}{r^2} \right)^2 4\pi r^2 \, dr$$

and we have to include all space outside the sphere; thus, the integral limits are from $r = R$ up to $r = \infty$. Taking all constants before the integral sign and simplifying, we have

$$W = \frac{Q^2}{8\pi\epsilon_0} \int_{r=R}^{\infty} \frac{dr}{r^2} = \frac{Q^2}{8\pi\epsilon_0} \left(-\frac{1}{r} \Big|_{r=R}^{r=\infty} \right)$$

or

$$W = \frac{Q^2}{8\pi\epsilon_0 R}$$

which is indeed the same as Eq. (8-7).

Example 2   Calculate the attractive force between the two plates of a plane parallel capacitor, if the plate area is $A = 0.1 \text{ m}^2$, the separation is $d = 1$ mm, and the applied potential difference is $V = 1000$ volts.

This problem seemingly has little to do with the energy content of the electric field. The two plates attract each other because they carry opposite charge; naively, we would say that we have to calculate the attractive force between any element $+dQ_i$ on one plate and $-dQ_j$ on the other plate, and then sum over all elements. However, this is not necessary. Imagine moving the top plate up by a small distance dx (Fig. 8-6); we have supplied the work

$$dW = \text{force} \times dx$$

This energy has to appear as potential energy in the system; the total energy stored in the system has indeed increased because the volume where there is an electric field has increased. The additional volume is

$$d\tau = \text{area} \times dx = A \, dx$$

and thus the additional energy stored is

$$dW = \frac{\epsilon_0}{2} |\mathbf{E}|^2 \, d\tau = \frac{\epsilon_0}{2} \left(\frac{V}{d} \right)^2 A \, dx$$

Figure 8-6 Example 2: Moving the upper plate increases the region where there is an electric field, thus increasing stored energy. The attractive force is the change in stored energy per unit change in distance.

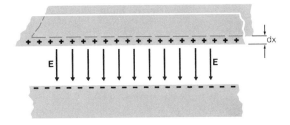

From this we see that the attractive force is equal to

$$F = \frac{\epsilon_0 A}{2d^2} V^2 \qquad (8-9)$$

which in our explicit case yields

$$F = \frac{8.85 \times 10^{-12}}{2} \frac{\text{coul}}{\text{V-m}} \times \frac{0.1 \text{ m}^2}{10^{-6} \text{ m}^2} \times (10^3 \text{ V})^2 = 0.443 \frac{\text{coul-V}}{\text{m}} = 0.443 \frac{\text{J}}{\text{m}} = 0.443 \text{ N}$$

This is roughly the weight of a mass of 45 grams. We could have tried to use another method by saying that the charge $+Q$ on the top plate feels the influence of the electric field $\mathbf{E}$; thus the force acting on it is $\mathbf{F} = Q\mathbf{E}$. *This would be wrong.* If we want to use such an argument, we have to take into account that the surface is, strictly said, not infinitely sharp. The surface charge will be distributed, as shown in Fig. 8-7a, over a small region (a few atomic distances just inside the conductor surface). The electric field also will only gradually reach its full value (Fig. 8-7b). The mean electric field as seen by the surface charge is one-half of the field just outside the surface. Thus, the force is also only one-half of what we naively expected:

$$\boxed{\mathbf{F} = \frac{Q\mathbf{E}}{2}} \qquad (8-10)$$

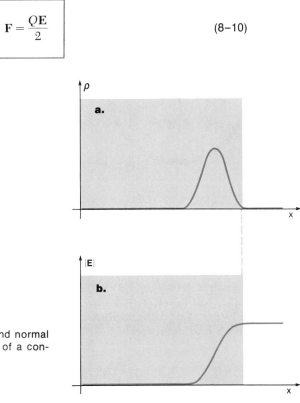

Figure 8-7 Electric charge distribution (**a**) and normal electric field magnitude (**b**) near the surface of a conductor.

Because we have

$$Q = CV = \frac{\epsilon_0 A}{d} V$$

and $|\mathbf{E}| = V/d$, we obtain

$$|\mathbf{F}| = \frac{\epsilon_0 A}{2d^2} V^2$$

which is indeed identical to Eq. (8–9).

■ All *electrostatic voltmeters*, also called *electrometers*, are capacitors of one form or another. The repulsive force between equal charges—and the attractive force between opposite charges—enables us to measure the potential difference between two points.

The simplest such device is the gold leaf electrometer (Fig. 8–8). One of the two capacitor plates is the outer metallic housing, and the other is a metallic rod leading to the inside of the housing. A very thin gold leaf is tied to the rod. If a potential difference exists between housing and rod, there will be a charge on the surface of the rod—and on the gold leaf—while an opposite charge resides on the inside of the housing. The very light gold leaf is repelled by the charge on the rod and attracted by the housing; thus, it will stand out at an angle. By measuring the angle θ we can obtain a crude measurement of the potential difference between housing and rod.

Naturally, if we connect an electrostatic voltmeter to some equipment such as a capacitor, some of the charge will flow to the two electrodes of the voltmeter. This changes the charge distribution and introduces an error in the measurement. To minimize this effect, one tries to keep the intrinsic capacitance of the voltmeter as small as possible (usually 5 to 100 pF). ■

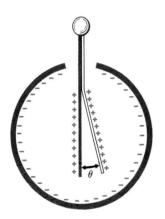

Figure 8-8 Gold leaf electroscope.

3. POLARIZATION

As we said in the introduction to this chapter, insulated materials become polarized if there is an electric field in the medium. First we discuss some experimental evidence for this so far unsubstantiated claim. If we charge up a parallel plate capacitor to a potential difference V_0, and then carefully isolate it, we can measure the potential difference between the two plates using an electrostatic voltmeter. If we now insert an insulating slab of material (e.g., Teflon) between the plates as shown in Fig. 8–9, we find that the potential between the plates had decreased

$$V' \text{ (slab in)} < V_0 \text{ (slab out)}$$

When we remove the slab, the potential reverts to its original value. If we repeat

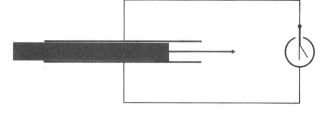

Figure 8-9 Inserting an insulating dielectric between the plates of a capacitor reduces the potential difference between them.

the experiment, varying the size of the capacitor plates, their spacing, or the initially applied potential difference, we always find the same result:

$$\frac{V' \text{ (slab in)}}{V_0 \text{ (slab out)}} = \frac{1}{K} \tag{8-11}$$

where K, called the dielectric constant of the material, is a number larger than unity; its value depends *only on the type of insulator*, but not on the capacitor size or initial charge, as long as all the space between the plates is filled by the insulator. We could choose a cylindrical capacitor or any shape we wanted, and we would always find Eq. (8–11) obeyed as long as we make sure that all regions of space where there is an electric field are filled with the same material.

When we insert a dielectric (insulator) between the plates of a parallel plate capacitor, we do not change the charge on the plates. However, the electric field **E** between the two plates has decreased. If the plate separation is d, then the potential difference is

$$V = |\mathbf{E}|d$$

From Eq. (8–11) we therefore deduce that the electric field in the dielectric slab is less than when the slab is removed:

$$|\mathbf{E}'| \text{ (dielectric in)} = \frac{1}{K} |\mathbf{E}| \text{ (dielectric out)}$$

We can explain our results by assuming that under the influence of an external electric field the individual molecules of the dielectric form dipoles. We define the *polarization* **P** in a material as the dipole moment per unit volume. If there are N molecules in a volume $\mathscr{V}$ (Fig. 8–10), and each has the same dipole moment vector **p**, then the polarization is

$$\mathbf{P} = \frac{N\mathbf{p}}{\mathscr{V}} \tag{8-12}$$

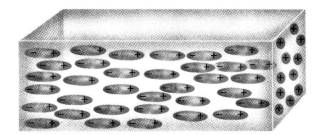

Figure 8-10 Polarized medium: the polarization is defined as the total dipole moment per unit volume.

Note that we assume that all dipoles have not only the same magnitude, but also the same direction. The polarization is measured in units of coul/m², as can be seen from Eq. (8–12):

$$[P] = \frac{[N][p]}{[\mathscr{V}]} = \frac{1 \times \text{coul-m}}{\text{m}^3} = \frac{\text{coul}}{\text{m}^2}$$

The problem before us is to calculate the electric field due to a continuous distribution of dipoles; since the electric field of each dipole (see Fig. 10–34) is quite complex, we cannot do this by adding the fields of all individual dipoles. Instead we use a model. Let us assume (Fig. 8–11) that the dielectric has everywhere a positive charge density ρ_+ and an equal negative charge density $\rho_- = -\rho_+$. In the unpolarized dielectric the charge densities overlap completely; there is no net charge density and thus also no electric field, as shown in Fig. 8–11b for a rectangular slab of material.

The model material can be polarized by moving all the positive charges relative to all the negative charges a small distance δ. We will first assume that the motion is along the direction shown in Fig. 8–11c. The material has now acquired a polarization which is equal to the charge density times the separation of the two charge densities:

$$|\mathbf{P}| = \rho_+\delta \qquad (8\text{–}13)$$

However, no change has really occurred inside the material; at any point inside there is still no net charge density, since the positive and negative charges cancel. A net charge appears only on the two end surfaces; on one side for a thickness δ there is only the positive charge density, while on the other side only the negative charge density remains. Since we assume the distance δ to be very small, we can say that there is a polarization surface charge density

$$\sigma_{\text{pol}} = \pm\rho_+\delta \qquad (8\text{–}14)$$

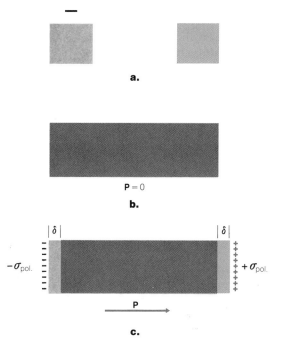

a.

b.   P = 0

c.

$-\sigma_{\text{pol.}}$ $+\sigma_{\text{pol.}}$ P

Figure 8–11 Calculating effect of polarization: We think of the dielectric as composed of equal amounts of positive and negative charges shown by different shading in (**a**). If there is no polarization, the two charges cancel exactly (**b**). The presence of a polarization **P** can be interpreted as moving all positive charges relative to the negative charges by a distance δ (**c**). Inside the two charges still cancel, but a surface charge now exists on both ends.

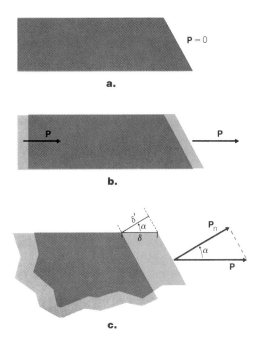

Figure 8-12 Polarization surface charge density. If the polarization is not perpendicular to the surface, the "as if" equivalent surface charge density is smaller by the factor cos α.

By comparison with Eq. (8–13) we see that

$$\sigma_{pol} = \pm |\mathbf{P}| \qquad (8\text{–}15\text{a})$$

The polarization surface charge is positive where the vector $\mathbf{P}$ points toward the surface; it is negative wherever the polarization vector points away from the surface. It is important to note that σ_{pol} is not a real surface charge density; we cannot carry it away by touching a conductor to the surface of the dielectric. It is an "as if" surface charge[°]: the electric field due to the polarization everywhere in the slab is the same as the electric field due to the surface charge density on each end.

Our calculation leading to the result (8–15a) is not sufficiently general, because in the situation shown in Fig. 8–11 the polarization is perpendicular to the end surfaces. Thus, let us assume (Fig. 8–12) that the slab has one end surface such that the normal forms an angle α with the polarization vector. The surface charge density is now smaller than before. As we see from Fig. 8–12c, the thickness of the surface layer is only $\delta' = \delta \cos \alpha$; thus

$$\sigma_{pol} = \rho_+ \delta' = \rho_+ \delta \cos \alpha$$

because the shift of the two charges relative to each other is not perpendicular to the surface. However, the polarization still has the magnitude $\rho_+ \delta$ as given by Eq. (8–13); thus, we conclude that for an arbitrary surface the equivalent surface charge density is

$$\sigma_{pol} = |\mathbf{P}| \cos \alpha \qquad (8\text{–}15\text{b})$$

But $|\mathbf{P}| \cos \alpha$ is the component of the polarization normal to the surface. Thus we

---

[°] We use the term "as if" instead of the usual term "effective charge" to stress that there is no real charge on the surface of a dielectric.

have

$$\boxed{\sigma_{\text{pol}} = P_n} \tag{8-16}$$

Eq. (8–16) says that the electric field due to a homogeneous polarization of a piece of insulator is equivalent to a surface charge density on the surface of the insulator; the surface charge is positive if **P** points out of the material and negative if **P** points into the material.

We still have to show that a polarization in an insulating material correctly explains the observation that the electric field in a capacitor is reduced if we insert a slab of insulating dielectric. Thus, let us return to our original example, the plane parallel capacitor (Fig. 8–13). In the absence of the insulator there would be the electric field $\mathbf{E}_0$ given by Eq. (7–29)

$$|\mathbf{E}_0| = \frac{\sigma}{\epsilon_0} \tag{8-17}$$

where σ is the *real* surface charge on the conductor.

If we insert the dielectric material, it will become polarized. The polarization will have the same direction as the external electric field $\mathbf{E}_0$. We can calculate the effect of the polarization if we replace the material by the "as if" polarization surface charge density. It will *reduce* the net effective surface charge density; the electric field due to the polarization has the magnitude

$$|\Delta\mathbf{E}| = \frac{\sigma_{\text{pol}}}{\epsilon_0} = \frac{|\mathbf{P}|}{\epsilon_0}$$

and points in the direction opposite to **P**. Thus,

$$\Delta\mathbf{E} = -\frac{\mathbf{P}}{\epsilon_0} \tag{8-18}$$

The electric field in the material is then

$$\mathbf{E} = \mathbf{E}_0 + \Delta\mathbf{E} = \mathbf{E}_0 - \frac{\mathbf{P}}{\epsilon_0} \tag{8-19}$$

Experimentally we found

$$\mathbf{E} = \frac{\mathbf{E}_0}{K} \tag{8-20}$$

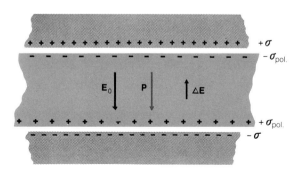

Figure 8–13   A dielectric inside a parallel plane capacitor. The "as if" polarization surface charge reduces the electric field inside the dielectric.

where $K > 1$ is the dielectric constant. Substituting this into Eq. (8–19), we find

$$\frac{\mathbf{E}_0}{K} = \mathbf{E}_0 - \frac{\mathbf{P}}{\epsilon_0}$$

or

$$\frac{\mathbf{P}}{\epsilon_0} = \left(1 - \frac{1}{K}\right)\mathbf{E}_0 = \frac{(K-1)\mathbf{E}_0}{K}$$

which because of Eq. (8–20) we can also write

$$\boxed{\mathbf{P} = (K-1)\epsilon_0\mathbf{E} = \chi\epsilon_0\mathbf{E}} \qquad (8\text{–}21)$$

Thus, we can explain the observed effects if we assume that the polarization $\mathbf{P}$ in the material is proportional to the electric field in the material. This is quite reasonable; the individual dipole moments – and thus the polarization – will tend to align themselves with the electric field. It is also reasonable to assume that the polarization is proportional to the electric field. Actually, Eq. (8–21) is not true for all substances; in some materials the polarization will no longer increase if the electric field becomes very large.

We call χ the *dielectric susceptibility* of the material; it is a measure of how susceptible the material is to being polarized by an electric field. We have the following relation between dielectric constant and susceptibility:

$$K = 1 + \chi \qquad (8\text{–}22)$$

which we can read directly from Eq. (8–21). The quantity

$$\epsilon = K\epsilon_0 \qquad (8\text{–}23)$$

is called the *dielectric permittivity* of the material. Since $K = 1$ for vacuum, one sometimes calls ϵ_0 the dielectric permittivity of the vacuum.

Example 3   A material of dielectric constant $K = 3$ is introduced between the plates of a plane parallel capacitor of spacing $d = 1$ mm and area $A = 0.2$ m^2. Before inserting the material, the potential difference was 1000 volts. Calculate the electric field, the polarization, the surface charge on the conductor, and the "as if" polarization surface charge on the dielectric.

We can obtain the electric field before introduction of the material as

$$E_0 = |\mathbf{E}_0| = \frac{V}{d} = \frac{1000 \text{ V}}{10^{-3} \text{ m}} = 10^6 \frac{\text{V}}{\text{m}}$$

After introducing the material, the electric field is

$$E = |\mathbf{E}| = \frac{E_0}{K} = 3.33 \times 10^5 \frac{\text{V}}{\text{m}}$$

The polarization is

$$|\mathbf{P}| = \chi\epsilon_0 E = 2 \times 8.85 \times 10^{-12} \frac{\text{coul}}{\text{V-m}} \times 3.33 \times 10^5 \frac{\text{V}}{\text{m}} = 5.90 \times 10^{-6} \frac{\text{coul}}{\text{m}^2}$$

The real surface charge density on the conductor is the same whether the dielectric is in or out:

$$\sigma_{\text{real}} = \pm\epsilon_0 \; |\mathbf{E}_0| = \pm 8.85 \times 10^{-6} \frac{\text{coul}}{\text{m}^2}$$

while the "as if" polarization surface charge density which reduces the electric field in the dielectric would be

$$\sigma_{\text{pol}} = \pm \; |\mathbf{P}| = \pm 5.90 \times 10^{-6} \frac{\text{coul}}{\text{m}^2}$$

or $(K - 1)/K = 2/3$ of the real surface charge density.

The total real charge on the capacitor plates is

$$Q_{\text{real}} = A\sigma_{\text{real}} = \pm 0.2 \text{ m}^2 \times 8.85 \times 10^{-6} \frac{\text{coul}}{\text{m}^2} = \pm 1.77 \times 10^{-6} \text{ coul}$$

while the "as if" charge on the dielectric surface would be

$$Q_{\text{pol}} = A\sigma_{\text{pol}} = \pm\frac{2}{3} Q_{\text{real}} = 1.18 \times 10^{-6} \text{ coul}$$

Example 4   Calculate the electric field inside an "infinitely" large flat sheet of a dielectric if the polarization **P** in the sheet is perpendicular to the sheet (Fig. 8–14). There are no external electric fields. The magnitude of **P** is 10^{-6} coul/m².

The reader will ask whether such an arrangement is physically possible. It is; as we will see later, there exist substances (e.g., barium titanate, BaTiO₃) which can exhibit permanent electric polarization even in the absence of external electric fields. Other substances, such as quartz (SiO₂), develop a polarization if external mechanical pressure is applied.

We have not been given a value for the thickness of the sheet; let us call it d. As we will see, the result does not depend on the value of d. Because we assume

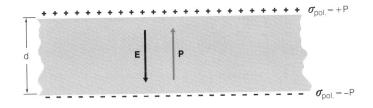

a.

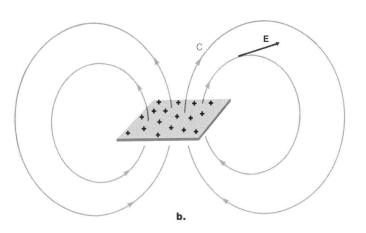

b.

Figure 8–14 Example 4: **a.** The electric field inside the polarized sheet. **b.** The electric field outside of a sheet of finite area.

the sheet to be "infinitely" large, the electric field for reasons of symmetry can point only perpendicularly to the sheet area.

To calculate the electric field, we can replace the slab by the surface charge density $\pm\sigma_{\text{pol}} = |\mathbf{P}|$ on both surfaces, as shown in Fig. 8–14a. Now the charge distribution is identical to that of a parallel plane capacitor of plate spacing d and surface charge density σ_{pol} on the inner surfaces of the plates. The electrical field is thus

$$|\mathbf{E}| = \frac{\sigma_{\text{pol}}}{\epsilon_0} = \frac{|\mathbf{P}|}{\epsilon_0}$$

and points in a direction opposite to the polarization. Thus

$$\mathbf{E} = -\frac{\mathbf{P}}{\epsilon_0}$$

inside the slab. Outside the slab the field vanishes—because the "as if" analogy to the outside of the dielectric slab is the outside of the "as if" capacitor plates. The magnitude of $\mathbf{E}$ is

$$|\mathbf{E}| = \frac{10^{-6}}{8.85 \times 10^{-12}} = 1.13 \times 10^5 \, \frac{\text{V}}{\text{m}}$$

The astute reader (and we assume all our readers to be astute) may note a seeming discrepancy in our treatment when we compare it to what we said earlier about dipoles: Far away the sheet should produce an electric dipole field as sketched in Fig. 8–14b. Since we can never manufacture a sheet which is infinitely large, is our example meaningless or even false? After all, we "showed" that the electric field outside the sheet vanishes. The paradox resolves itself if we note that in our example we were calculating the *near field* of the overall dipole formed by the sheet, which of course necessarily will have a finite area. There is an electric field outside the sheet, but it is much weaker than inside. The integral

$$V = -\int \mathbf{E} \cdot d\mathbf{l}$$

along the path C in Fig. 8–14b is equal to the integral through the sheet—going from top to bottom in Fig. 8–14a. We have

$$|V| = \int_C \mathbf{E} \cdot d\mathbf{l} = |\mathbf{E}|_{\text{inside}} \, d = \frac{1}{\epsilon_0} |\mathbf{P}| d$$

where d is the sheet thickness. As long as the sheet is large, and we discuss the electric field inside the sheet far away from the edges, the path C is much longer than d. Thus the electric field outside is indeed much smaller than inside.

4. ELECTRIC FIELDS IN A DIELECTRIC

Let us now study the electric field just inside and just outside a dielectric. Our discussion also applies also to any abrupt change of dielectric constant on the interface[*] between two materials of dielectric constants K_1 and K_2. We can always imagine the two dielectrics to be separated by a very narrow empty space (Fig. 8–15). In Fig. 8–16 we show the electric fields just inside and outside a dielectric. We assume that there are no free (real) charges directly on the

---

[*] Compare p. 103 for the definition of the interface between two media.

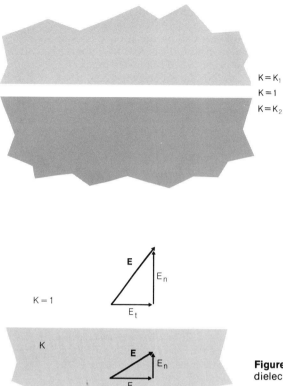

$K = K_1$
$K = 1$
$K = K_2$

Figure 8–15 Two materials of different dielectric constants can always be thought of as having a thin region of vacuum between them.

$K = 1$

E

E_n

E_t

K

E

E_n

E_t

Figure 8–16 The electric fields near the surface inside a dielectric and outside do not have to be the same.

surface. Inside the material, of course, exists the polarization

$$\mathbf{P} = (K - 1)\epsilon_0\mathbf{E}$$

which we will replace by the equivalent "as if" surface charge density

$$\sigma_{\text{pol}} = P_n$$

at the surface. We will use Maxwell's equations of electrostatics (p. 209) to derive relations between the electric fields just inside and just outside the dielectric. First we use the first Maxwell equation, which says that the integral of $\mathbf{E} \bullet d\mathbf{l}$ over any closed loop vanishes. Let us choose two points P_1 and P_2 on the surface separated by a small distance a (**Fig. 8–17**). The loop will consist of going from P_1 to P_2 just outside the surface and back to P_1 just inside the material. We then have

$$\int \mathbf{E} \bullet d\mathbf{l} = \int_{P_1}^{P_2} \mathbf{E} \bullet d\mathbf{l} + \int_{P_2}^{P_1} \mathbf{E} \bullet d\mathbf{l} = 0 \qquad (8\text{–}24)$$

outside inside

But the distance a is very small; thus we can write

$$\int_{P_1}^{P_2} \mathbf{E} \bullet d\mathbf{l} \approx (E_t)_{\text{outside}}\, a \qquad (8\text{–}25a)$$

outside

where $(E_t)_{\text{outside}}$ is the tangential component of the electric field just outside the

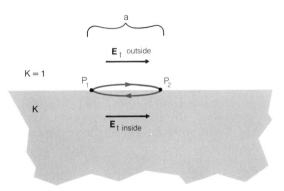

Figure 8-17 The tangential component of the electric field is the same on both sides; the potential difference between P_1 and P_2 is the same whether calculated inside or outside.

material. In the same way, we obtain

$$\int_{P_2}^{P_1} \mathbf{E} \bullet d\mathbf{l} = -\int_{P_1}^{P_2} \mathbf{E} \bullet d\mathbf{l} \approx -(E_t)_{\text{inside}}\, a \qquad (8\text{-}25b)$$
$$\underset{\text{inside}}{} \qquad \underset{\text{inside}}{}$$

Because we are going in the opposite direction inside the material, we have to include a minus sign in Eq. (8–25b). Substituting into Eq. (8–24), we obtain

$$\int \mathbf{E} \bullet d\mathbf{l} \approx (E_t)_{\text{outside}}\, a - (E_t)_{\text{inside}}\, a = 0$$

and thus

$$\boxed{(E_t)_{\text{inside}} = (E_t)_{\text{outside}}} \qquad (8\text{-}26)$$

The tangential components of the electric field are the same inside and outside the material.

The normal components of the electric field, however, will not be the same on both sides of the surface. The "as if" polarization surface charge density will cause a sudden change. To see this, we use the second Maxwell equation, Gauss's law. We replace the effect of the polarization by the surface charge density according to Eq. (8–16):

$$\sigma_{\text{pol}} = P_n$$

and apply Gauss's law over a coin-shaped surface, one face of which lies inside the dielectric and the other outside, as shown in Fig. 8–18. If the area of the

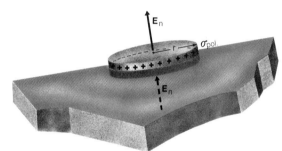

Figure 8-18 The Gaussian surface used to compare the normal components of the electric field at an interface.

circle of the surface is πr^2, we have

$$\int_{\text{coin}} \mathbf{E} \cdot d\mathbf{S} \approx (E_n)_{\text{outside}} \pi r^2 - (E_n)_{\text{inside}} \pi r^2 \qquad (8\text{-}27)$$

Because the side of the cylinder can be kept very small, we can neglect any contribution from them to the integral over the whole closed surface. By Gauss's law, the integral is proportional to the total polarization surface charge inside the coin. But this is the surface charge density times the area of the coin

$$(Q)_{\text{inside}} = \pi r^2 \sigma_{\text{pol}} \qquad (8\text{-}28)$$

Writing out Gauss's law,[*] we thus have

$$[(E_n)_{\text{outside}} - (E_n)_{\text{inside}}] \pi r^2 = \frac{\pi r^2 \sigma_{\text{pol}}}{\epsilon_0}$$

$$(E_n)_{\text{outside}} - (E_n)_{\text{inside}} = \frac{\sigma_{\text{pol}}}{\epsilon_0} = \frac{(P_n)_{\text{inside}}}{\epsilon_0} \qquad (8\text{-}29)$$

However, by the definition of the dielectric constant of the material we know that $P_n = (K - 1)\epsilon_0 E_n$ inside the material. Thus, we have

$$(E_n)_{\text{outside}} = (E_n)_{\text{inside}}\left(1 + \frac{(K - 1)\epsilon_0}{\epsilon_0}\right)$$

or

$$(E_n)_{\text{outside}} = (KE_n)_{\text{inside}} \qquad (8\text{-}30)$$

We can immediately apply Eq. (8–30) to the surface between two media of different dielectric constant; since (KE_n) in both media is equal to the normal component of the electric field in the small empty space between the media, we have

$$\boxed{(KE_n)_{\text{medium 1}} = (KE_n)_{\text{medium 2}}} \qquad (8\text{-}31)$$

Thus, we see that while the tangential components of the electric field are the same on both sides, the normal components are not; instead, it is the normal component of $K\mathbf{E}$ which is the same. We can also write instead of Eq. (8–31)

$$(\epsilon_0 E_n + P_n)_{\text{medium 1}} = (\epsilon_0 E_n + P_n)_{\text{medium 2}} \qquad (8\text{-}32)$$

Example 5   A dielectric slab of dielectric constant $K = 2$ is positioned at 45° relative to an outside homogeneous electric field E_0 as shown in Fig. 8–19; calculate the electric field inside the slab.

We choose a coordinate system so that the slab is parallel to the $(x - y)$ plane and so that E_0 lies in the $(x - z)$ plane. Then there will be no electric field – either inside or outside – in the y direction. We have effectively a two-dimensional problem which is easier to visualize.

The electric field outside can be written

$$\mathbf{E}_0 = \frac{E_0}{\sqrt{2}}(\hat{\mathbf{i}} + \hat{\mathbf{k}})$$

---

[*] Note the similarity between the derivation of Eq. (7–20) and the argument used here. Eq. (8–29) is the analogous result to Eq. (7–20).

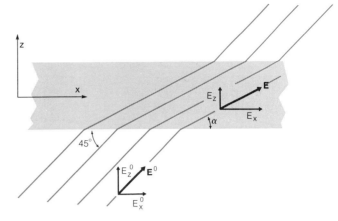

Figure 8–19 Example 5: Electric field inside a slab.

where $\hat{\mathbf{i}}$, $\hat{\mathbf{j}}$, and $\hat{\mathbf{k}}$ are unit length vectors along the x, y, and z directions. Inside the slab, the x-component of the electric field will be the same as outside, but the z-component will be smaller by a factor K:

$$\mathbf{E} = \frac{E_0}{\sqrt{2}}\left(\hat{\mathbf{i}} + \frac{1}{K}\hat{\mathbf{k}}\right) = E_x\hat{\mathbf{i}} + E_z\hat{\mathbf{k}}$$

The magnitude of the electric field will be

$$|\mathbf{E}| = \sqrt{\mathbf{E}^2} = \frac{E_0}{\sqrt{2}}\sqrt{1 + \frac{1}{K^2}} = \frac{E_0}{2}\sqrt{\frac{5}{2}} = 0.79\, E_0$$

since we assumed $K = 2$. The angle α between the electric field inside and the x-axis is given by

$$\tan\alpha = \frac{E_z}{E_x} = \frac{1}{K} = \frac{1}{2}$$

as can be read from Fig. 8–19. From this we obtain $\alpha = 26.6°$.

We can also calculate the energy stored in the electric field inside a dielectric. Let us assume that we have a slab of dielectric material and that the outside field $\mathbf{E}_0$ is perpendicular to the slab. If we remove the slab, the energy density will everywhere be given by Eq. (8–3)

$$\rho_E = \frac{\epsilon_0|\mathbf{E}_0|^2}{2}$$

When we insert the material, it acquires a dipole moment per unit volume $\mathbf{P}$ (Fig. 8–20). We thus have the negative potential energy of the dipoles in the external field E_0, which we have to subtract. The potential energy of a dipole in an electric field is $-\mathbf{p} \cdot \mathbf{E}_0$; we have in a unit volume the total dipole moment $\mathbf{P}$ affected by the external electric field $\mathbf{E}_0$. The energy per unit volume is then

$$\Delta\rho_E = -\mathbf{P} \cdot \mathbf{E}_0$$

and so we obtain inside the material

$$(\rho_E)_{\text{inside}} = \frac{\epsilon_0\mathbf{E}_0^2}{2} - (\mathbf{P} \cdot \mathbf{E}_0) \tag{8–33}$$

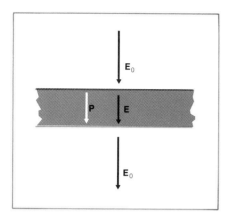

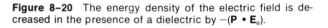

Figure 8-20 The energy density of the electric field is decreased in the presence of a dielectric by $-(\mathbf{P} \cdot \mathbf{E}_0)$.

Inside the slab, the electric field is now no longer $\mathbf{E}_0$, but instead

$$\mathbf{E} = \frac{\mathbf{E}_0}{K}$$

The polarization $\mathbf{P}$ can also be expressed in terms of the electric field $\mathbf{E}$ inside the material:

$$\mathbf{P} = \chi\epsilon_0\mathbf{E} = (K - 1)\epsilon_0\mathbf{E}$$

Substituting this into Eq. (8–33), we obtain for the energy density inside the field

$$(\rho_E)_{\text{inside}} = \frac{\epsilon_0(K\mathbf{E})^2}{2} - (K - 1)\epsilon_0(\mathbf{E} \cdot K\mathbf{E})$$

or

$$(\rho_E)_{\text{inside}} = \frac{K\epsilon_0}{2}\mathbf{E}^2 \tag{8–34}$$

■ We could have also derived Eq. (8–34) by studying the dielectric when inserted between the plates of a capacitor. The total stored energy in a capacitor is

$$W = \frac{Q^2}{2C} = \frac{C}{2}V^2 = \frac{1}{2}QV$$

The capacitance of the empty capacitor is

$$C_0 = \frac{\epsilon_0 A}{d}$$

while the capacitor which is filled by a material of dielectric constant K has the capacitance

$$C = \frac{K\epsilon_0 A}{d}$$

The stored energy for a potential difference V is then

$$W = \frac{K\epsilon_0 A}{2d}V^2 \tag{8–35}$$

Since the electric field has the magnitude $|\mathbf{E}| = V/d$, we can write in Eq. (8–35)

$$W = \frac{K\epsilon_0 A}{2d} \mathbf{E}^2 d^2$$

which can be written as

$$W = \frac{K\epsilon_0 \mathbf{E}^2}{2} Ad = \text{energy density} \times \text{volume} \qquad (8\text{–}36)$$

Thus, the energy density again must be equal to $K\epsilon_0 \mathbf{E}^2/2$; we have derived Eq. (8–34) using the energy stored in a capacitor. ∎

Example 6 A parallel plate capacitor with plate area A and spacing d has its left half filled with a dielectric of dielectric constant K, as shown in Fig. 8–21a. Calculate the capacitance of the system, if a charge $\pm Q$ is deposited on each plate.

Because of the continuity condition [Eq. (8–26)], we know that the electric field has to be the same inside and outside the dielectric. This is shown in Fig. 8–21b. We could have also derived Eq. (8–26) for our special case by noting that the potential difference between the plates

$$V = |\mathbf{E}|d \qquad (8\text{–}37)$$

has to be the same whether we use the electric field inside the dielectric or the field outside it. However, this implies that the real surface charge density *cannot be the same everywhere* on the plates; in the region where there is no dielectric we have from Eq. (8–17)

$$\sigma_1 = \epsilon_0 |\mathbf{E}|$$

while above and below the dielectric we have to include the "as if" polarization charge

$$\sigma_2 - \sigma_{\text{pol}} = \epsilon_0 |\mathbf{E}|$$

Since $\sigma_{\text{pol}} = |\mathbf{P}| = (K-1)\epsilon_0 |\mathbf{E}|$, we have

$$\sigma_2 = K\epsilon_0 |\mathbf{E}|$$

Figure 8–21 Example 6: **a.** Overall view **b.** Detailed view at limit of dielectric

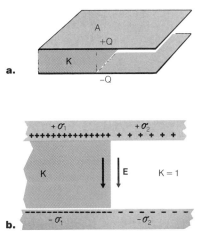

Figure 8–22   Example 6: One can consider the partially filled capacitor as two capacitors in parallel – one empty and one filled with dielectric.

The total charge stored on the capacitor plates is

$$Q = \sigma_1 \frac{A}{2} + \sigma_2 \frac{A}{2}$$

since one-half the area between the plates is filled with dielectric. We thus have

$$Q = \epsilon_0 |\mathbf{E}| \frac{A}{2} + K\epsilon_0 \mathbf{E} \frac{A}{2}$$

or

$$Q = \frac{K+1}{2} \epsilon_0 A |\mathbf{E}|$$

But the potential difference is given by Eq. (8–37); thus, we have

$$Q = \frac{K+1}{2} \epsilon_0 A \frac{V}{d} = CV$$

and we conclude that the capacitance is equal to

$$C = \frac{K+1}{2} \epsilon_0 \frac{A}{d} \tag{8–38}$$

We could also have obtained our result [Eq. (8–38)] by replacing our capacitor by two capacitors connected in parallel. Each capacitor has the area $A/2$; one is filled with the dielectric, while the other is empty. The total capacitance of the two connected capacitors is $C = C_1 + C_2$, and since

$$C_1 = \frac{\epsilon_0}{d} \frac{A}{2} \tag{8–39a}$$

and

$$C_2 = \frac{K\epsilon_0}{d} \frac{A}{2} \tag{8–39b}$$

we could directly obtain Eq. (8–38). However, from Fig. 8–22 it might seem that we are neglecting the condition that the tangential electrical field $\mathbf{E}$ be continuous along the interface between dielectric and air. In Fig. 8–22 the two capacitors are physically separated. There will be end effects in each capacitor, which are neglected in both Eqs. (8–39a) and (8–39b). In our previous discussion we have taken this into account: the electric field is the same on both sides of the interface – this is how we obtained Eq. (8–38).

5.   ATOMIC THEORY

The dielectric constant K is a property of each dielectric; it enables us to calculate macroscopic electric fields in the presence of such materials. Here we are interested in another problem: What is the dielectric constant of a given material? To be more specific, we want to relate the dielectric constant to the atomic or molecular structure of the material. As we said earlier, some molecules have a natural permanent dipole moment. Others, such as all symmetric molecules (N_2, O_2, H_2), have no electric dipole moments in the absence of an external electric field.

Let us discuss first the symmetric molecules. In the presence of an external (i.e., not due to the electrons and nuclei of the molecule itself) electric field $\mathbf{E}_{eff}$ they will acquire a dipole moment, since the force due to the electric field acts in opposite directions on the central nucleus and on the electrons. (See Fig. 8–1.) We can write for the dipole moment

$$\mathbf{p} = \alpha \mathbf{E}_{eff} \qquad (8\text{–}40)$$

where α, the so-called *polarizability*, depends on the electronic structure of the atom.

It is important to note that the effective field $\mathbf{E}_{eff}$ is *not the same* as the average field inside the dielectric. The average field inside a slab of material oriented as shown in Fig. 8–23 has the magnitude

$$\mathbf{E} = \mathbf{E}_0 - \frac{\mathbf{P}}{\epsilon_0} \qquad (8\text{–}41)$$

Here $\mathbf{P}$ is the polarization of the medium itself, i.e.,

$$\mathbf{P} = N\mathbf{p} \qquad (8\text{–}42)$$

where N is the number of all molecules per unit volume. Thus, the molecule under consideration, whose electric dipole moment is expressed in Eq. (8–40), itself contributes to the mean field $\mathbf{E}$. However, the effective field $\mathbf{E}_{eff}$ in Eq. (8–40) is the electric field which would have been at the point of the molecule *if this particular molecule was absent*; its own dipole field does not contribute to $\mathbf{E}_{eff}$. This is illustrated in Fig. 8–23. One can show that[*]

$$\mathbf{E}_{eff} = \mathbf{E} + \frac{\mathbf{P}}{3\epsilon_0} \qquad (8\text{–}43)$$

under the following assumptions: (i) The individual molecules are distributed either randomly in space or in crystals of cubic symmetry, and (ii) neighboring molecules affect each other so weakly that their effect can be neglected. Com-

---

[*] See p. 258 for a derivation of Eq. (8–43).

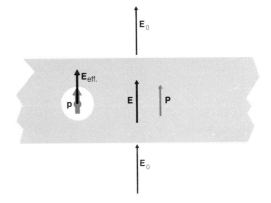

Figure 8–23 The effective field which polarizes a molecule is the field which would exist at the position of the molecule in the molecule's absence.

bining Eqs. (8–40), (8–42), and (8–43) we then obtain

$$\mathbf{P} = N\mathbf{p} = N\alpha\left(\mathbf{E} + \frac{\mathbf{P}}{3\epsilon_0}\right) \tag{8-44}$$

We can now use Eq. (8–21),

$$\mathbf{P} = \chi\epsilon_0\mathbf{E} = (K - 1)\epsilon_0\mathbf{E}$$

which gives us the macroscopic relation between the polarization **P**, the dielectric constant K, and the average electric field **E**. Inserting this into Eq. (8–44) we obtain

$$(K - 1)\epsilon_0\mathbf{E} = N\alpha\left(\mathbf{E} + \frac{K - 1}{3}\mathbf{E}\right) = \frac{N\alpha}{3}(K + 2)\mathbf{E}$$

and since the electric field **E** appears on both sides of the equation, we have obtained a relation between the polarizability α and the dielectric constant K:

$$\boxed{\frac{K - 1}{K + 2} = \frac{N\alpha}{3\epsilon_0}} \tag{8-45}$$

Eq. (8–45) is called the *Clausius-Mosotti* relation, and it gives us the required relation between the atomic property of the polarizability α and the macroscopic property of the dielectric constant K. The unit of the polarizability can be read off from Eq. (8–40); it produces a dipole moment of 1 coul-meter for an electric field of 1 V/m, so its unit is

$$1\frac{\text{coul-m}}{\text{V/m}} = 1\frac{\text{coul-m}^2}{\text{V}}$$

Example 7   The dielectric constant of gaseous nitrogen at 20°C and 1 atmosphere is $K = 1.000548$; the density of the gas is $\rho = 1.17$ kg/m³. Calculate the polarizability of the nitrogen molecule.

Nitrogen forms a diatomic molecule, N_2. The mass number of the nitrogen atom is $A = 14.0$. By the definition of Avogadro's number, there are $N_0 = 6.02 \times 10^{26}$ molecules in one kilogram-mole of nitrogen, or in 2A kg = 28 kg. If we denote by N the number of molecules per unit volume, then the density is

$$\rho = N\frac{2A}{N_0}$$

because the mass of each molecule is $2A/N_0$. From this we obtain

$$N = \frac{N_0\rho}{2A} = \frac{6 \times 10^{26} \text{ molecules/kg-mole} \times 1.17 \text{ kg/m}^3}{28.0 \text{ kg/kg-mole}}$$

$$= 2.51 \times 10^{25} \text{ molecules/m}^3$$

The polarizability α can be written according to Eq. (8–45):

$$\alpha = \frac{3\epsilon_0}{N}\frac{K - 1}{K + 2}$$

and since $\epsilon_0 = 8.85 \times 10^{-12}$ coul/V-m, we have

$$\alpha = \frac{3 \times (8.85 \times 10^{-12} \text{ coul/V-m})}{2.51 \times 10^{25}/\text{m}^3} \times \frac{0.000548}{3}$$

$$= 1.93 \times 10^{-40} \frac{\text{coul-m}^2}{\text{volt}}$$

Here we have replaced $K + 2$ by 3 in the denominator; this is always justifiable if K is very close to unity.

In order to get an idea of the magnitude of the effect, let us calculate also the induced dipole moment in an electric field of $|\mathbf{E}| = 3 \times 10^4$ V/cm $= 3 \times 10^6$ V/m, which is the breakdown voltage in nitrogen gas. We obtain

$$p = \alpha E = (1.93 \times 10^{-40}) \times (3 \times 10^6) = 5.8 \times 10^{-34} \text{ coul-m}$$

This is a very small dipole moment, as we can see if we estimate by how much we would have to move one electron from its normal orbit to achieve this dipole moment. This distance would be

$$d = \frac{p}{e} = \frac{5.8 \times 10^{-34} \text{ coul-m}}{1.6 \times 10^{-19} \text{ coul}} = 3.6 \times 10^{-15} \text{ m} = 3.6 \times 10^{-5} \text{ Å}$$

Thus, the electrons in the nitrogen atom move only by a few parts in one million of the actual size of the molecule (~ 3 Å). Indeed, they move only by a distance comparable to the size of the nitrogen nucleus ($\sim 3 \times 10^{-15}$ m).

Example 8   Calculate the dielectric constant of liquid nitrogen. Nitrogen liquefies at 77.4 °K $= -195.8$°C $= -320$°F; the density of the liquid is $\rho = 8.2 \times 10^2$ kg/m³.

We could use the polarizability of the nitrogen molecules obtained in Example 7. On the other hand, we can derive from the Clausius-Mosotti equation (8–45) that $(K - 1)/(K + 2)$ is proportional to the density of the material. Indeed,

$$\frac{1}{\rho}\left(\frac{K-1}{K+2}\right) = \frac{N}{\rho}\frac{\alpha}{3\epsilon_0} = \frac{N_0}{M}\frac{\alpha}{3\epsilon_0}$$

is independent of the density of the material, as long as the molecular weight M and the polarizability α do not change. Thus we have

$$\left(\frac{1}{\rho}\left(\frac{K-1}{K+2}\right)\right)_{\text{liquid N}_2} = \left(\frac{1}{\rho}\frac{(K-1)}{(K+2)}\right)_{\text{gaseous N}_2}$$

or for the liquid

$$\frac{K-1}{K+2} = \frac{\rho_{\text{liquid}}}{\rho_{\text{gas}}}\left(\frac{K-1}{K+2}\right)_{\text{gas}}$$

Numerically,

$$\frac{K-1}{K+2} = \frac{8.2 \times 10^2 \text{ kg/m}^3}{1.17 \text{ kg/m}^3} \times \frac{0.000548}{3} = 1.28 \times 10^{-1} = 0.128$$

From this we obtain easily

$$K = \frac{1 + (2 \times 0.128)}{1 - 0.128} = 1.45$$

This agrees very well with the experimentally measured value of $K = 1.454$. Note that we have not only calculated the polarizability of the nitrogen atom, but we

have also shown by comparison with experiment that it is independent of temperature, at least in the range between room temperature and the temperature of liquid
air (nitrogen).

■   We will now derive Eq. (8–43), which relates the electric field in the material $\mathbf{E}$ to
the effective field $\mathbf{E}_{eff}$ acting on a single dipole. We have to exclude the contribution of
the dipole itself to the electric field; we will further assume that we can also neglect all
dipoles which are in a small sphere of radius R around the dipole. We thus assume that
$\mathbf{E}_{eff}$ is the electric field which would be measured in the center of a small spherical
cavity of radius R in the dielectric. The effect of the polarization $\mathbf{P}$ (Fig. 8–24) can be
replaced by the effect of the "as if" polarization charge

$$\sigma_{pol} = P_n = P \cos \theta \tag{8–46}$$

on the surface of the cavity. The electric field at the center of the cavity is then the sum
of the electric field $\mathbf{E}$ in the material and the electric field $\mathbf{E}'$ produced by the "as if"
polarization surface charge on the cavity walls. This field can be calculated if we divide
the sphere into rings, as shown in Fig. 8–25. Each ring will have the area

$$dS = 2\pi\rho R \, d\theta$$

and since $\rho = R \sin \theta$, we have

$$dS = 2\pi R^2 \sin \theta \, d\theta \tag{8–47}$$

The "as if" polarization surface charge

$$dQ = \sigma_{pol} \, dS = 2\pi R^2 P \cos \theta \sin \theta \, d\theta \tag{8–48}$$

will be distributed evenly over the surface. Thus, our problem can be reduced to (i)
calculating the electric field of each ring, and (ii) summing over all rings.

We first calculate the field of each ring at the center of the cavity. The distance of
any point on the ring to the center is R. Because of the cylindrical symmetry, we know
that the electric field due to the charge dQ on the ring points along the z-axis, which is
the direction of $\mathbf{P}$ in the material. Thus, the field is smaller by a factor $\cos \theta$ than the field
of a point charge dQ a distance R away:

$$dE_z' = |d\mathbf{E}_0| \cos \theta = \frac{1}{4\pi\epsilon_0} \frac{dQ}{R^2} \cos \theta$$

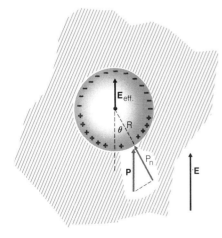

Figure 8–24 To calculate the effective field, we eliminate all
molecules in a small sphere of radius R. The polarization can
then be replaced by the "as if" polarization surface charge.

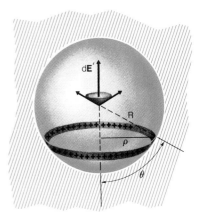

Figure 8-25 The surface charge can be decomposed into individual rings and the electric field of each added.

and using the expression (8–48) for dQ we obtain

$$dE_z' = \frac{P}{2\epsilon_0} \cos^2 \theta \sin \theta \, d\theta$$

The total electric field can be obtained by integrating over θ from $\theta = 0$ to $\theta = \pi$:

$$E_z' = \int dE_z' = \frac{P}{2\epsilon_0} \int_0^\pi \cos^2 \theta \sin \theta \, d\theta$$

Since $d(\cos \theta) = -\sin \theta \, d\theta$, we have

$$E_z' = \frac{P}{2\epsilon_0} \left[-\frac{\cos^3 \theta}{3} \right]_{\theta=0}^{\theta=\pi} = \frac{P}{2\epsilon_0} \left[-\frac{(-1)^3}{3} + \frac{(+1)^3}{3} \right] = \frac{P}{3\epsilon_0} \qquad (8–49)$$

Thus, the difference between the electric field in the center of the cavity and that in the surrounding material is $\mathbf{P}/3\epsilon_0$; from this Eq. (8–43) follows immediately. ■

In some solids which form ionic crystals (Fig. 8–26), there are two contributions to the polarizability α. First, there is the electronic polarizability α_e, which is due to the electron cloud of each ion shifting relative to the nucleus. Second,

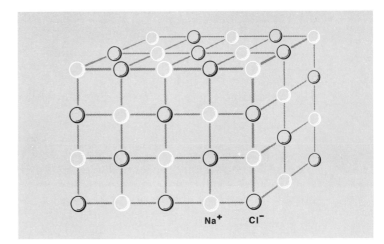

Figure 8-26 Crystal of rock salt (schematic).

Na$^+$ Cl$^-$

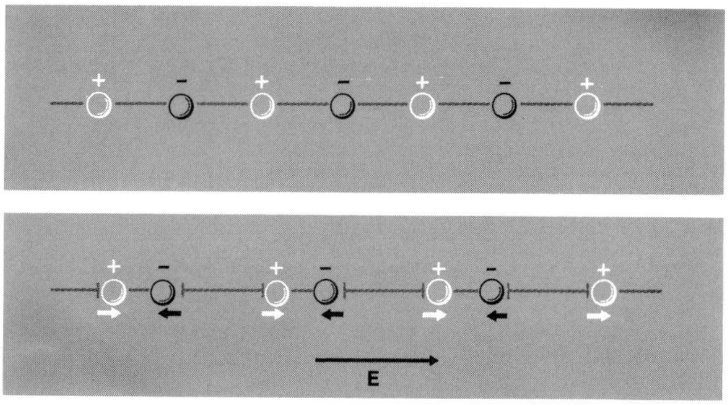

Figure 8–27 Ionic polarization: the individual ions move relative to each other in the crystal.

the ions which form the crystal lattice will also move slightly relative to each other under the influence of an external electric field, as shown in Fig. 8–27; this gives rise to an ionic polarizability α_i. The total polarizability then is

$$\alpha = \alpha_e + \alpha_i \tag{8-50}$$

Both effects are temperature independent.

Let us now turn our attention to those materials in which the individual molecules have a permanent dipole moment. In the absence of an external electric field, the individual dipoles will point in random directions (Fig. 8–2a). There will be no polarization. Even if there is an external local field $\mathbf{E}$, the thermal motion will still prevent the dipoles from aligning themselves *exactly* along the field. Nevertheless, more dipoles will be pointing roughly in the field direction than in the opposite direction (Fig. 8–2b); on the average, the molecules then have a positive component of the dipole moment along the field.

Let us study this in a more quantitative way. In an electric field the individual dipole will have an energy (Fig. 8–28)

$$W = -(\mathbf{p} \cdot \mathbf{E}) = -|\mathbf{p}| \, |\mathbf{E}| \cos \theta \tag{8-51}$$

where θ is the angle between the dipole and the electric field. According to statistical mechanics, the probability that the dipole points in a certain direction is proportional to the Boltzmann factor[*]

$$\mathscr{P}(\theta) = e^{-W/kT} = e^{pE(\cos \theta)/kT} \tag{8-52}$$

At room temperature, the denominator kT is typically much larger than W, so we can write (since $e^x \approx 1 + x$ for $x \ll 1$)

$$\mathscr{P}(\theta) \approx 1 + \frac{pE}{kT} \cos \theta \tag{8-53}$$

The mean dipole moment in the direction of the electric field $\mathbf{E}$ has to be obtained by averaging over all directions

$$<p> = p <\cos \theta> \tag{8-54}$$

We also have to take into account that when the dipole forms the angle θ with the electric field, it still can point in many directions. We can represent any direction by a point on a sphere of unit radius (Fig. 8–29); we then can say that the

---

[*] See Vol. I, p. 448 and p. 650, for a discussion of the Boltzmann factor.

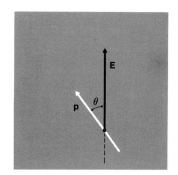

Figure 8-28 Orientational polarization: a dipole will tend to align itself with the electric field; thermal random motion will prevent complete alignment.

probability that the dipole points in the specified direction is

$$C \, \mathscr{P}(\theta) \, dS = C \, \mathscr{P}(\theta) 2\pi \sin \theta \, d\theta \qquad (8\text{-}55)$$

The constant C has to be determined from the fact that the sum of the probabilities over all directions is one—we are certain that the dipole points somewhere:

$$C \int_{\theta = 0}^{\pi} \mathscr{P}(\theta) 2\pi \sin \theta \, d\theta = 1 \qquad (8\text{-}56)$$

The mean value of $\cos \theta$ can then be obtained easily:

$$< \cos \theta > = C \int_{\theta = 0}^{\pi} \cos \theta \, \mathscr{P}(\theta) 2\pi \sin \theta \, d\theta \qquad (8\text{-}57)$$

Using the easily derivable integrals

$$\int_{0}^{\pi} \sin \theta \, d\theta = 2; \qquad \int_{0}^{\pi} \cos \theta \sin \theta \, d\theta = 0; \qquad \int_{0}^{\pi} \cos^2 \theta \sin \theta \, d\theta = \frac{2}{3}$$

we obtain [by substituting the expression (8–53) for $\mathscr{P}(\theta)$ into Eqs. (8–56) and (8–57), after a little algebra]:

$$< \cos \theta > = \frac{pE}{3kT} \qquad (8\text{-}58)$$

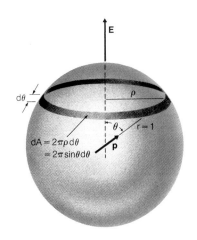

Figure 8-29 The probability that the dipole forms an angle between θ and $\theta + d\theta$ with the electric field is proportional to the area dA available to the unit vector parallel to p.

and thus we have a mean dipole moment in the direction of the field $\mathbf{E}$

$$<\mathbf{p}> = \frac{p^2}{3kT}\mathbf{E} = \alpha_p\mathbf{E} \qquad (8-59)$$

The *orientational polarizability* α_p, which is due to permanent dipole moments of the individual molecules, is inversely proportional to the absolute temperature:

$$\alpha_p = \frac{p^2}{3kT} \qquad (8-60)$$

We stress that Eq. (8–60) is valid only if the individual dipoles are free to rotate, and thus it is necessary that neighboring dipoles have no strong effect on its orientation. However, if such an influence by neighboring dipoles is present, we cannot use even the Clausius-Mosotti equation; for then Eq. (8–43) is no longer valid. This situation occurs, for example, in water; the dipole moment of the water molecule, $p_{H_2O} = 6.2 \times 10^{-30}$ coul-m, is so large that water molecules will make chains, as shown in Fig. 8–30. However, we could use Eq. (8–60) for water vapor or for alcohol containing a small amount of water; then the individual water molecules are sufficiently far away from each other that they no longer influence each other directly.

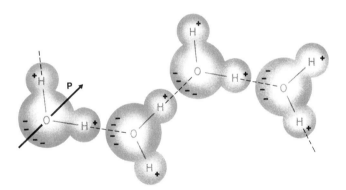

Figure 8–30 Because of their large dipole moment, water molecules tend to form chains. The molecules are so close to each other that the dipole approximation is no longer very good.

Example 9   The dielectric constant of chloroform ($CHCl_3$) at 0°C is $K = 5.4$; its density is $\rho = 1.50 \times 10^3$ kg/m³. Calculate the dipole moment of the chloroform molecule, assuming all of the polarizability to be due to the permanent dipole moment.

We first have to calculate the mass A of a kilogram-mole of chloroform. We have

$$
\begin{array}{llll}
1 \text{ C:} & 1 \times 12.01 & = & 12.0 \text{ kg} \\
1 \text{ H:} & 1 \times 1.008 & = & 1.0 \text{ kg} \\
3 \text{ Cl:} & 3 \times 35.45 & = & \underline{106.4 \text{ kg}} \\
& & & 119.4 \text{ kg}
\end{array}
$$

Thus, $N_0 = 6.02 \times 10^{26}$ molecules have the mass of $M = 119.4$ kg.
The number N of molecules per cubic meter is therefore

$$N = \frac{N_0\rho}{M} = 7.56 \times 10^{27} \text{ molecules/m}^3$$

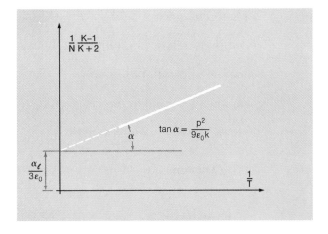

Figure 8–31 If a molecule possesses both an orientational as well as an electronic polarizability, the quantity $\dfrac{1}{N}\dfrac{K-1}{K+2}$ is a linear function of $1/T$. By measuring the intercept and the slope of the straight line, one can determine both contributions separately.

The polarizability is, according to the Clausius-Mosotti equation,

$$\alpha = \frac{3\epsilon_0}{N}\frac{K-1}{K+2} = 2.09 \times 10^{-39}\,\frac{\text{coul-m}^2}{\text{V}}$$

If we assume that the polarizability is completely due to the dipole moment, we have

$$\alpha = \frac{p^2}{3kT}$$

or

$$p = \sqrt{3kT\alpha} = \sqrt{3 \times 1.38 \times 10^{-23}\ \text{J/}^\circ\text{K} \times 273\ ^\circ\text{K} \times 2.09 \times 10^{-39}\,\frac{\text{coul-m}^2}{\text{V}}}$$

$$= \sqrt{23.7 \times 10^{-60}} = 4.86 \times 10^{-30}\ \text{coul-m}$$

Actually, our result is too large by ~25%; nearly one-half of the polarizability of chloroform at room temperature is electronic in nature. However, in order to separate the two, we would have to measure the temperature dependence of the dielectric constant K. According to the Clausius-Mosotti equation, if we plot

$$\frac{1}{N}\frac{K-1}{K+2} = \frac{1}{3\epsilon_0}\left(\alpha_e + \frac{p^2}{3kT}\right)$$

against $1/T$, we obtain a straight line; its intercept (at $1/T = 0$ or $T = \infty$) yields α_e, while the dipole moment can be determined from the slope. This is illustrated in Fig. 8–31.

Summary of Important Relations

Polarization $\mathbf{P} = \dfrac{\text{dipole moment}}{\text{volume}}$

"As if" surface charge density $\sigma_{pol} = P_n$

Relation to electric field $\qquad \mathbf{P} = (K - 1)\epsilon_0 \mathbf{E} = \chi\epsilon_0 \mathbf{E}$

Continuity relations on interface between two media 1 and 2

component parallel to interface $\qquad (E_t)_1 = (E_t)_2$

component normal to interface $\quad (KE_n)_1 = (KE_n)_2$

Energy density of electric field

in vacuum $\qquad\qquad\qquad \rho_E = \dfrac{\epsilon_0 |\mathbf{E}|^2}{2}$

in dielectric $\qquad\qquad\quad \rho_E = \dfrac{K\epsilon_0 |\mathbf{E}|^2}{2}$

Polarizability α $\qquad\qquad\qquad \mathbf{P} = \alpha\mathbf{E}_{eff}$

Clausius-Mosotti relation $\qquad \dfrac{K - 1}{K + 2} = \dfrac{N}{3\epsilon_0}\alpha$

where $\qquad\qquad \alpha = \alpha(\text{electronic}) + \alpha(\text{ionic}) + \dfrac{p^2}{3kT}$

and p is the permanent dipole moment of the molecule.

Questions

1. Could we define a "gravitational field energy density" just as we defined the energy density of the electric field?

2. Assume a configuration with arbitrarily shaped conductors and/or point charges and the rest of space filled by an insulator of dielectric constant K. Show that in such a configuration one can replace ϵ_0 by $\epsilon = K\epsilon_0$, and that thereafter one can ignore the presence of the dielectric, replacing it by vacuum.

3. One sometimes says that "One can think of a conductor in electrostatics as just a dielectric of $K \to \infty$." Discuss in what sense this statement is correct; point out the physical difference between conductor and dielectric.

4. Will an insulating dielectric object be subject to a force in a homogeneous electric field? A nonhomogeneous field?

Problems

1. Calculate the energy stored in a cylindrical (air-filled) capacitor (inner radius r, outer radius R, length L) (a) using Eq. (7–38) (b) using Eq. (8–5) and check that the results agree.

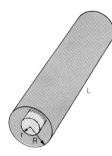

2. One of the problems of using underground cables – proposed to eliminate ugly over-head transmission lines – for long-distance transmission of electric power is their large capacitance. Consider a 100-mile-long cable with inner conductor diameter $d_1 = 2.5$ cm (≈ 1 inch) and outer shield diameter 5 cm; the insulating material has a dielectric constant $K = 2.2$.

   (a) Calculate the total capacitance.

   (b) Calculate the energy stored in the cable if the potential difference be-tween wire and shield is 100,000 volts.

3. We can approximate the electric field of the proton by that of a homogeneously charged sphere of radius $R = 0.8 \times 10^{-15}$ m and total charge $+e$. Estimate the total energy stored in the electric field. What fraction of the total proton mass would this energy correspond to if one uses Einstein's relation, $E = mc^2$?

4. Two identical capacitors are connected in parallel, and a charge $\pm 2Q_0$ is deposited on the plates of the pair. Afterward, one of the capacitors is filled with insulating material of dielectric constant $K = 2.5$. Calculate the charge on the plates of each capacitor.

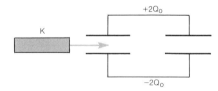

5. Two sheets of aluminum of area $A = 1000$ cm² are separated by a sheet of paper 0.2 mm thick. What is the maximum charge $\pm Q$ which can be stored on this capacitor? Use Table II-a of Appendix B. The paper sheet will withstand an electric field of 3×10^5 V/cm.

6. In a drawer you find several identical capacitors of unknown capacitance. Cutting one open, you find that it consists of two aluminum strips of width 2.0 cm and length 2.5 m separated by a 0.05 mm thick acetate sheet (dielectric constant $K = 3.3$). The assembly was rolled into a tight cylinder. What is the capacitance of each capacitor in the drawer?

7. The nitrogen nucleus has a charge $+Ze$ with $Z = 7$. Assume that the nucleus of each atom moves by 0.001 Å relative to its electrons. Calculate the resulting polarization in nitrogen gas of density $\rho = 1.30$ kg/m³ (atomic weight of nitrogen $A = 14.0$).

8. A plane parallel capacitor (area $A = 250$ cm², plate separation $d = 0.5$ mm) is filled by an insulator of dielectric constant $K = 2.5$. The potential difference between the plates is 1000 V. What is the electrostatic force on each plate? (Hint: Think of the plates separated from the dielectric by a thin layer of air.)

9. A plane parallel capacitor is made of two square plates of edge $a = 10$ cm; the plates are spaced by putting between them four Teflon disks ($K = 2.2$) of diameter $D = 2$ cm and thickness $d = 0.5$ mm. Calculate the fractional change in capacitance if one of the Teflon disks falls out.

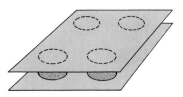

10. A plane parallel capacitor of area A and plate separation d is half filled by an insulating slab of dielectric constant K. The charge $\pm Q$ resides on each plate. Calculate the change in stored energy if the slab is moved inward by a small distance dx. What do you conclude from this? What is the force on the insulating slab? The capacitor plates are square.

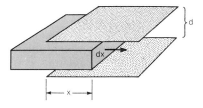

11. A metallic sphere of radius $r_0 = 1$ cm carries the charge $Q = -10^{-9}$ coul. It is surrounded by insulating material (dielectric constant $K = 3.0$) out to a distance $R = 5$ cm. Calculate the electrostatic potential difference between the sphere and a point at infinity (very far away).

12. Outside the flat surface of a dielectric ($K = 3.0$), one measures a dielectric field of magnitude $|\mathbf{E}| = 3 \times 10^5$ V/m; the field forms the angle $\alpha = 30°$ with the normal to the surface. Calculate the electric field and the polarization in the dielectric; state their magnitudes and the angle they form with the normal to the surface.

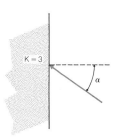

13. A sheet (area $A = 100$ cm², thickness $d = 2$ mm) of a dielectric ($K = 3.0$) is inserted into a homogeneous electric field $\mathbf{E}_0$ (magnitude $E_0 = 10^4$ V/m) which forms an angle of 45° with the **sheet**. Calculate the total *induced dipole moment* of the sheet.

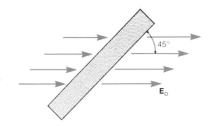

*14. A sphere of radius r is filled with material with the total polarization $\mathbf{P}$ pointing upward. What is the "as if" surface charge density σ_{pol} at an arbitrary point P on the surface? What is the total "as if" charge on the top half? The bottom half?

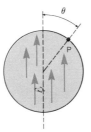

15. In an insulator of dielectric constant $K = 2$ there exists a homogeneous field of magnitude $|\mathbf{E}_0| = 10^4$ V/m. A small flat coinshaped cavity is cut into the dielectric. Calculate magnitude and direction of the electric field in the cavity, if the angle θ (see figure) is (a) $\theta = 90°$; (b) $\theta = 0°$; (c) $\theta = 45°$.

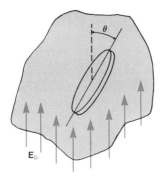

16. What would be the polarization in water if all its molecules were aligned parallel to each other? Could such a polarization be achieved by applying an external electric field? (The dipole moment of the water molecule is $p = 6.2 \times 10^{-30}$ coul-m.)

17. A gas sample is measured to have a dielectric constant $K = 1.00080$. When the gas is liquefied, its density has increased by a factor 1000. What is the dielectric constant of the liquid, if the polarizability is purely electronic?

18. Assuming a typical density of molecules in a liquid of $N \approx 10^{28}$ molecules/m³, what is

the maximum molecular dipole moment for which the Clausius-Mosotti equation could possibly be valid at room temperature?

19. To a sample of dry air (dielectric constant $K = 1.000540$, density $\rho = 1.28$ kg/m³) is added 1% by weight of water vapor (H_2O; molecular weight $M = 18.0$ kg/kg-mole; dipole moment $p = 6.2 \times 10^{-30}$ coul-m). What is the dielectric constant of the mixture? Note that on a hot humid day, the air may contain several percent by weight of water.

20. The dielectric constant of chloroform ($CHCl_3$) is $K_1 = 3.71$ at the temperature $t_1 = 100°C$ and $K_2 = 6.12$ at $t_2 = -40°C$. The density, $\rho = 1.50 \times 10^3$ kg/m³, is practically the same at both temperatures. Determine both the electronic polarizability and the permanent dipole moment of the chloroform molecule.

ELECTRIC CURRENT

1. CHARGES IN MOTION

In the introduction to Chapter 7 we mentioned that charges in a conductor will come to rest in such a way that there is no electric field acting anywhere in the conductor. This was an oversimplified picture which we have to correct. As we said earlier, matter consists of positively charged atomic nuclei and negatively charged electrons. Both types of charge are point charges on the atomic scale; we know today that atomic nuclei have sizes of a few times 10^{-15} m or a few times 10^{-5} Å. On the other hand, nobody has yet succeeded in measuring the size of an electron; all we know is that its size is less than 10^{-16} m $= 10^{-6}$ Å. Since the typical atomic size is a few Ångstroms, we can certainly claim that all charges making up the atom are practically point charges.

We can show that if there exist only point charges acting on each other by their electric fields, then no point charge can be at rest in stable equilibrium unless it is at the same point as a point charge of opposite sign. This theorem has the profound consequence that *none* of the point charges in matter are ever at rest. If any single charge $+Q$ is at rest, it coincides with another charge $-Q'$; the two now form a new point charge $(Q - Q')$ which again can be at rest only at the point of another charge, and so on. Thus, if the electrons and atomic nuclei were in stable equilibrium at rest, each atom would have to collapse to a single point. This is not the case; atoms have finite sizes of the order of 1 Å $= 10^{-10}$ m.

The electrons in a metal either move in the neighborhood of a single atomic nucleus or—if they are free electrons—they move freely through the metal (Fig. 9–1). But even the atomic nuclei are constantly moving, although they are only

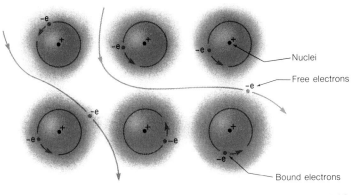

Figure 9–1 Classical picture of a crystal lattice of a conductor. Some electrons are tightly bound, while others can move freely through the crystal.

Nuclei

Free electrons

Bound electrons

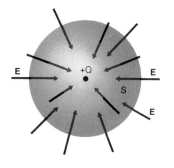

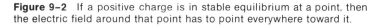

Figure 9-2 If a positive charge is in stable equilibrium at a point, then the electric field around that point has to point everywhere toward it.

oscillating back and forth around their mean positions, never moving far away. Thus, when we think of matter, we have to consider all of its parts to be constantly in motion; our macroscopical measurements will detect only the average behavior of the individual particles.

Even this is not yet the correct picture. We have already mentioned earlier that the electrons—and atomic nuclei—in an atom cannot be described as massive point particles which at any instant of time are at exact positions in space. Quantum mechanics requires that we represent the particles by a wave; the charge can then be considered "smeared out" over the region of the wave. Nevertheless, in quantum mechanics also no particle is ever truly at rest.

■ We will now prove the above-mentioned theorem that no point charge can be at rest in stable equilibrium. Let us assume that the particle which is at rest has the positive charge Q. Stable equilibrium means that if the particle is slightly off its equilibrium position in any direction, there has to be a restoring force pointing toward the equilibrium position. But then the electric field $\mathbf{E}$ in the neighborhood of the charge Q at rest has to point toward Q from all sides. Note that the electric field of Q itself is not to be included in $\mathbf{E}$; no point charge acts on itself.

Let us now surround Q by a small sphere (Fig. 9-2). Since at its surface the electric field points everywhere inward, we have for the integral over the whole sphere

$$\int_{\text{sphere}} \mathbf{E} \cdot d\mathbf{S} < 0$$

(remember, $d\mathbf{S}$ points always outward!). Therefore, because of Gauss's law, a negatively charged particle has to be inside the sphere. Since the sphere can be arbitrarily small, we conclude that a negatively charged particle has to be at *exactly* the same point as the charge $+Q$. Thus, we have proved the statement we asserted. ■

How then are we to interpret the statement, "A charge at rest resides on the conductor always at its surface"? Certainly we do not mean that there are particles—electrons or nuclei—at rest on the surface. What we do mean is that if we average over a few atomic distances—and also average over a sufficiently long time—then the mean charge density will be non-zero at the surface. On the other hand, the mean charge density inside the conductor is always zero in the static case. The atomic nuclei—and most of the electrons—are moving around only inside a single atom. The free electrons carry all the static charge. If the surface charge is negative, then *on the average* there will be more electrons near the surface than in the neutral case; while if the surface charge is positive, there will be on the average fewer electrons near the surface.

2. CONSERVATION AND QUANTIZATION OF CHARGE

In spite of the constant motion of all charged particles, *the total charge of a closed system will never change*. This is a rather trivial statement if we talk only about everyday matter in which the individual particles never disappear. We mean by a closed system a system from which *nothing* leaves and nothing enters, neither energy, nor matter, nor charge. But if no particles disappear and none leave or enter the system, then their total charge also cannot change.

However, this statement becomes less trivial if we consider situations in which the number of particles does not stay constant. This can happen whenever there is enough energy available to create new mass; by the laws of special relativity this happens whenever the energy available satisfies

$$E > mc^2$$

On the other hand, particles can also disappear. An electron, e^-, can be annihilated if it encounters its antiparticle, the positron, e^+. The energy is released as electromagnetic radiation. The reaction

$$e^+ + e^- \rightarrow \text{E.M. radiation}$$

is a typical example of a relativistic reaction in which the number of particles changes. Nevertheless, charge is conserved; the positron carries the charge $+1e$, the electron the charge $-1e$. Thus, the net charge before the reaction occurred was zero, the same as afterward. Today there exists a whole branch of physics — elementary particle physics — which studies such particles and their production and annihilation. In all these processes the net charge is always conserved. Thus, we accept as a fundamental law of nature that this is always true.

Another fundamental fact which emerged from the study of elementary particles is the *quantization of charge:* all known particles are either uncharged or carry the charge

$$Q = \pm Ne = \pm N \times 1.602 \times 10^{-19} \text{ coulomb}$$

where N is an integer $N = 1, 2, 3, \ldots$ This law, like the law of conservation of charge, cannot be derived from any more fundamental law. Since no exception has ever been found, we assume it to be true; but if ever a particle carrying, for example, the charge $+(1/3)e$ should be discovered, we would have to change our assumption that the smallest quantum of charge is equal to the electron charge. Particles of "fractional charge" $(1/3)e$ and $(2/3)e$ have been conjectured to exist by theoretical physicists; but despite an exhaustive search, none of these conjectured particles, called "quarks," [*] has ever been found.

3. CURRENT AND CURRENT DENSITY

There are always electric fields in any atom; however, in the static case which we have discussed so far, the average electric field vanishes. By "average" or "macroscopic" electric field we mean the average taken over many atoms. While each electron is moving, on the average there is also no transport of charge from one place in matter to another. We say there is no macroscopic current. We will from now on understand by "current" or "electric field" the macroscopic quantities averaged over many atoms. We postpone until later the discussion of phenomena on the atomic scale.

We say that there is a *current* in a conductor if there is a net macroscopic

---

[*] The name "quark" was coined by Murray Gell-Mann (1929–), who used a poem by James Joyce (Finnegan's Wake, p. 383). It is believed that Joyce himself did not invent the word, but took it over from the German word "Quark," meaning milk curds and, colloquially, nonsense or empty talk.

transport of charge from one point to another. In a metal the positively charged ions do not move on a macroscopic scale; thus, any current is mediated by the motion of the conduction electrons which are free to move throughout the metal. If there are n_e such electrons per unit volume, their charge density is

$$\rho_e = (-e)n_e = -en_e \qquad (9\text{-}1)$$

In an uncharged conductor – and indeed always *inside* the conductor – this negative charge density is exactly balanced by the positive charge density of the lattice ions

$$\rho_{\text{ion}} = -\rho_e = +en_e \qquad (9\text{-}2)$$

so that the metal as a whole is neutral.

We call *current density* the charge density of the electrons times their mean velocity:

$$\boxed{\mathbf{j} = \rho_e \langle \mathbf{v} \rangle} \qquad (9\text{-}3)$$

Note that since the electrons have negative charge, their charge density is also negative; thus, the current density $\mathbf{j}$ points in the direction opposite to that of the mean velocity $\langle \mathbf{v} \rangle$.

To explain the physical meaning of $\mathbf{j}$, let us disregard for the moment this complication; we will thus now assume that we are talking about positive moving charges. Then $\mathbf{j}$ and $\langle \mathbf{v} \rangle$ have the same direction. Let us further assume that $\mathbf{j} = \rho \langle \mathbf{v} \rangle$ is homogeneous, i.e., everywhere the same. If we imagine (Fig. 9–3) a rectangular area A whose normal forms the angle α with the direction of $\mathbf{j}$, then all the charges in the prism of length $\Delta h = |\langle \mathbf{v} \rangle| \times \Delta t$ will flow through the area in the time Δt. We can show this by imagining that all charges move with the same velocity $\langle \mathbf{v} \rangle$; then any charge traverses the same distance Δh in the time Δt. Thus, any charge inside the prism will have gone through A after the time Δt; but no charge outside will do the same, either because it has not yet reached A or because it misses it. Thus, the charge which has passed through the area A in the time Δt is

$$\Delta Q = \rho \Delta h \, A \cos \alpha = \rho \, |\langle \mathbf{v} \rangle| \, A \, \Delta t \cos \alpha \qquad (9\text{-}4)$$

If we define by $\mathbf{n}$ the unit vector along the normal to A, we have

$$\Delta Q = \rho A \langle \mathbf{v} \rangle \bullet \mathbf{n} \, \Delta t = A(\mathbf{j} \bullet \mathbf{n})\Delta t$$

or

$$I = \frac{\Delta Q}{\Delta t} = (\mathbf{j} \bullet \mathbf{n})A \qquad (9\text{-}5)$$

Thus, the charge flowing through A per unit time is equal to the area A times the normal component of the current density.

The quantity I, the amount of charge flowing through A per unit time, is called the *electric current* through A. Note that the current is defined only if we specify the area A and the direction of the normal $\mathbf{n}$ for which we consider the electric current I to be positive; this is in contrast to the current density, which has a meaning independent of any specific area.

The unit of electric current is one coulomb per one second $= 1$ coul/sec; since it is so frequently used, the name 1 ampere $= 1$ A has been coined for this

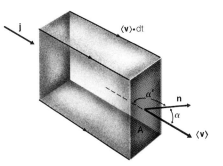

Figure 9-3 Calculating the current through the area segment *A*.

unit. Thus

$$1 \text{ ampere} = 1 \ A = 1 \ \frac{\text{coul}}{\text{sec}} \tag{9-6}$$

The current density, on the other hand, has the unit given by its definition [Eq. (9–3)]:

$$[\mathbf{j}] = \frac{1 \text{ coul}}{\text{m}^3} \frac{1 \text{ m}}{\text{sec}} = \frac{\text{coul}}{\text{sec-m}^2} = \frac{A}{\text{m}^2} \tag{9-7}$$

■ We stress that while we have derived Eq. (9–5) for positive moving charges, it is equally correct for negative moving charges such as electrons. Indeed, let us consider the velocity <v> in Fig. 9–3 to be the velocity of a negative charge density. Then the current will flow in the direction opposite to that indicated in Fig. 9–3; thus, in this case

$$(\mathbf{j} \bullet \mathbf{n}) = |\mathbf{j}| \cos \alpha' = -|\mathbf{j}| \cos \alpha \tag{9-8}$$

since $\alpha' = \pi - \alpha$ and thus $\cos \alpha' = -\cos \alpha$. However, the charge ΔQ flowing through A is also negative, since negative charges move through A. Thus we have

$$-\frac{|\Delta Q|}{\Delta t} = -|\mathbf{j}| \ A \cos \alpha \tag{9-9}$$

which again is equivalent to Eq. (9–5):

$$I = \frac{\Delta Q}{\Delta t} = (\mathbf{j} \bullet \mathbf{n})A$$

We consider the electric current to be the same if positive charge flows from left to right or if negative charge flows from right to left (Fig. 9–4). On a macroscopic scale we cannot

Figure 9-4 The same current density **j** and thus the same current *I* can be caused either by positive charges moving in the same direction or by negative charges in the opposite direction from **j**. Of course, the total current density can have contributions from both positive and negative charges.

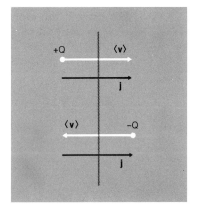

tell the difference; in both cases the total charge on the right will become more positive, either because we take away negative charge or because we add positive charge. ▪

Example 1   A current of $I = 1$ A exists in a copper wire of cross section area $S = 1$ mm². Calculate the mean electron drift velocity in the wire, assuming the current density to be the same everywhere in the wire.

In order to solve this problem we have to know the *free electron density*, i.e., the density of electrons which can move around freely. However, we know that on the average one electron per copper atom is free. As the density of copper is $\rho \approx 9 \times 10^3$ kg/m³ and there are $N_0 = 6.02 \times 10^{26}$ atoms/kg-mole, we have for the electron density

$$n_e = \frac{\rho N_0}{A}$$

where $A = 63.5$ is the atomic weight of copper, implying that the mass of 1 kg-mole $= 6.02 \times 10^{26}$ atoms is 63.5 kg. Substituting into the equation for the electron density, we obtain

$$n_e = 8.5 \times 10^{28} \frac{\text{electrons}}{\text{m}^3}$$

The charge density is the electron density times the charge per electron:

$$\rho_e = n_e(-e) = \left(8.5 \times 10^{28} \frac{\text{electrons}}{\text{m}^3}\right) \times \left(-1.6 \times 10^{-19} \frac{\text{coul}}{\text{electron}}\right) = -1.36 \times 10^{10} \frac{\text{coul}}{\text{m}^3}$$

The current density is equal to the charge density times the electron mean drift velocity:

$$j = |\mathbf{j}| = |\rho_e| \langle v \rangle$$

and the total current I is equal to the current density times the wire cross section (Fig. 9–5):

$$I = jS = |\rho_e|S \langle v \rangle$$

From this, one obtains for the electron velocity

$$\langle v \rangle = \frac{I}{\rho_e S} = \frac{1}{1.36 \times 10^{10} \times 10^{-6}} = 0.73 \times 10^{-4} \frac{\text{m}}{\text{sec}}$$

The simple relation [Eq. (9–5)] between current density and current is true only if the current density is everywhere the same. However, it is rather easy to generalize it to an arbitrary current density distribution on arbitrary (non-flat) surfaces. In Fig. 9–6 we have sketched such a surface. We subdivide it into small elements dS, and then the expression

$$dI = (\mathbf{j} \bullet \mathbf{n}) \, dS$$

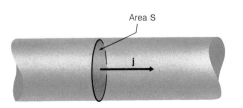

Area S

Figure 9–5 The current *I* in a wire is the product of the current density and the wire cross section area.

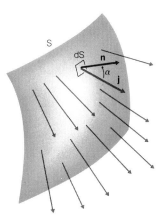

Figure 9-6 An arbitrary surface can be subdivided into small elements to calculate the total current.

is the current flowing through the surface element dS. We can also, as we did earlier in Eq. (7–3b), define the differential vector $d\mathbf{S} = dS\,\mathbf{n}$, and we then have

$$dI = \mathbf{j} \bullet d\mathbf{S} \tag{9–10}$$

The current through the whole surface can then be obtained by summing over all small elements dS:

$$I = \int dI = \int_{\text{surface}} \mathbf{j} \bullet d\mathbf{S} \tag{9–11}$$

Let us consider in particular a closed surface (Fig. 9–7) and define the normal $\mathbf{n}$ to point always outward; then the total outward current, the charge flowing out per unit time, is

$$\left(\frac{dQ}{dt}\right)_{\text{out}} = I = \int_{\substack{\text{closed} \\ \text{surface}}} \mathbf{j} \bullet d\mathbf{S} \tag{9–12a}$$

However, the fundamental law of conservation of charge demands that if the region inside the surface is losing the charge $(dQ/dt)_{\text{out}}$ per unit time, the charge inside has to decrease:

$$\frac{dQ_{\text{(inside)}}}{dt} = -\left(\frac{dQ}{dt}\right)_{\text{out}} \tag{9–12b}$$

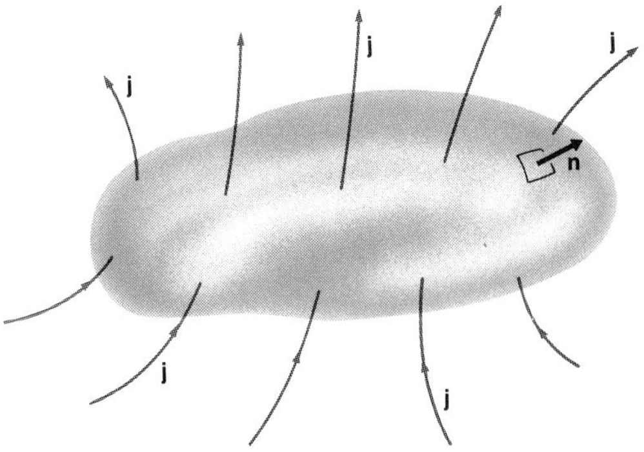

Figure 9-7 If the total charge inside is constant, the total current through a closed surface has to vanish.

From this we conclude that if there is a net current I out of the surface, it has to be equal to the net loss of total charge inside:

$$\frac{d(Q_{\text{inside}})}{dt} = -\int_{\substack{\text{closed} \\ \text{surface}}} \mathbf{j} \cdot d\mathbf{S} \qquad (9\text{–}13)$$

Example 2   The current density on a spherical surface of radius $r = 2$ cm is observed to have the magnitude $|\mathbf{j}| = 10^{-2}$ A/m² and a direction pointing everywhere into the surface (Fig. 9–8). Assuming that there was no charge inside the sphere initially, what would be the total charge inside after 100 seconds? Is this example realistic?

The total current flowing *into* the sphere is

$$I = -\int \mathbf{j} \cdot d\mathbf{S}$$

Since $\mathbf{j}$ points in the direction opposite to the outward normal $\mathbf{n} = d\mathbf{S}/|d\mathbf{S}|$, we have

$$-\mathbf{j} \cdot d\mathbf{S} = +|\mathbf{j}| \, |d\mathbf{S}|$$

and thus the total current inward is

$$I = j \times \text{surface of sphere} = j \times 4\pi r^2$$

Numerically, we obtain

$$I = 10^{-2} \, \frac{A}{m^2} \times 4\pi (0.02)^2 \, m^2 = 5.0 \times 10^{-5} \, A$$

The charge Q inside the sphere will increase at a rate

$$\frac{dQ}{dt} = I$$

and thus we conclude that after the time $t = 100$ seconds the charge inside the sphere is

$$Q = It = 5.0 \times 10^{-3} \, \text{coulomb}$$

In order to see whether the example is realistic, let us imagine the charge Q divided into two parts; let us separate the two charges $Q' = 2.5 \times 10^{-3}$ coulomb by the diameter of the sphere, $d = 4$ cm $= 0.04$ m. The repulsive force between the two charges is then

$$F = \frac{1}{4\pi\epsilon_0} \frac{Q'^2}{d^2} = 3.5 \times 10^7 \, \text{newtons!}$$

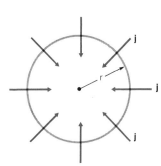

Figure 9–8 Example 2: A converging current density.

Whatever is inside the sphere will explode long before we have been able to accumulate the charge Q inside. Our example is extremely unrealistic.

Example 3   The two plates of an air-filled plane parallel capacitor (area $A = 0.1$ m², separation $d = 1$ mm) are each connected to a wire. Currents $I_1 = 10^{-3}$ A and $I_2 = 2 \times 10^{-3}$ A flow in the two wires in the directions indicated by the arrows in Fig. 9–9. Assuming there was initially no charge on the capacitor plates, how long will it take for the potential difference between the plates to be equal to 100 volts?

This example is a little bit tricky because both plates will have a net positive charge. No current is flowing through the air between the two capacitors; any current inward flows only through the wire. By surrounding each plate by a closed surface S_1 or S_2, we see that after the time t the two plates will carry the charges

$$Q_1 = +I_1 t \qquad (9\text{--}14\text{a})$$

$$Q_2 = +I_2 t \qquad (9\text{--}14\text{b})$$

Thus, we have to calculate the potential difference between two plates when both carry charges of equal sign. Let us consider the two situations shown in Fig. 9–10a and 9–10b. In Fig. 9–10a we have the standard situation of a capacitor in which the two plates are oppositely charged and there is an electric field between the plates. Note that there is no electric field inside either plate, nor is there one outside the plates. In Fig. 9–10b we show the situation in which both plates carry equal charges $+Q'$. The charge resides on the outside; we could fill the space between the plates with metal without changing the situation. There is no electric field between the plates, *or inside the plates*. Finally, in Fig. 9–10c we show the superposition of both charge distributions. The electric field anywhere will be the sum of the electric fields of the two distributions. Thus, there will be no electric field anywhere inside the metal plates. The field between the plates is determined by the charges $\pm Q$, and the field outside is determined by $+Q'$. We have a charge distribution which satisfies the basic requirement that there be no electric field anywhere in the metal.

The potential difference between the plates is given by Q alone. From Eqs. (7–27a) and (7–32) we know that

$$V = \frac{Q}{C} \qquad (9\text{--}15\text{a})$$

where

$$C = \frac{\epsilon_0 A}{d} \qquad (9\text{--}15\text{b})$$

is the capacitance of the system. If we are also to satisfy Eqs. (9–14a) and (9–14b), we have to require

$$Q' - Q = Q_1 = I_1 t$$

$$Q' + Q = Q_2 = I_2 t$$

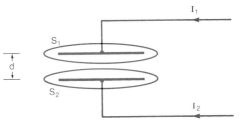

Figure 9-9   Example 3: Currents flowing onto capacitor plates.

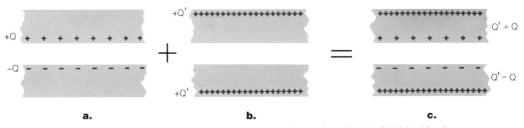

Figure 9–10 Determination of charge distribution on plates: the electric field inside the conductor has to vanish. Since it vanishes both for configuration (a) and (b), it will also vanish for (c).

We now easily obtain

$$Q' = (I_1 + I_2)\frac{t}{2}$$

$$Q = (I_2 - I_1)\frac{t}{2}$$

Thus, the potential will be V if

$$V = \frac{Q}{C} = \frac{(I_2 - I_1)t}{2C} = \frac{(I_2 - I_1)d}{2\epsilon_0 A}t$$

This can be solved for t:

$$t = \frac{2\epsilon_0 AV}{d(I_2 - I_1)} = \frac{2 \times (8.85 \times 10^{-12} \text{ coul/V-m}) \times 0.1 \text{ m}^2 \times 100\text{V}}{0.001 \text{ m} \times [(2 \times 10^{-3} \text{ A}) - (1 \times 10^{-3} \text{ A})]}$$
$$= 1.77 \times 10^{-4} \text{ sec} = 177 \ \mu\text{sec}$$

4. OHM'S LAW

If there is no electric field in a conductor, there is also no electric current; the mean velocity $\langle v \rangle$ of the charge carriers (electrons) vanishes. In many, although by no means all, materials the current density is proportional to the electric field:

$$\boxed{\mathbf{j} = \frac{1}{\rho}\mathbf{E}} \tag{9–16}$$

The quantity ρ is called the *resistivity* of the material; its inverse $1/\rho$ is usually called the *conductivity*. It is a property of the material; in addition, it will vary with the temperature of the conductor.

Eq. (9–16) describes the current density in terms of the electric field at a point in a conductor (Fig. 9–11). It is called *Ohm's law*. Materials that obey Ohm's law are usually called *ohmic conductors*. This relation enables us to calculate the current flowing through a wire of length L which is connected to two terminals — points between which there is a potential difference V. We will

Figure 9–11 Electric field **E**, current density **j**, and mean electron velocity $\langle v \rangle$ in a wire carrying a current.

assume that the wire has the constant cross section A and is made of homogeneous material of resistivity ρ. The total current in the wire is then

$$I = |\mathbf{j}|A \tag{9–17}$$

On the other hand, we know that the potential difference between the two terminals A and B is (Fig. 9–12)

$$V = V(A) - V(B) = -\int_B^A \mathbf{E} \cdot d\mathbf{l} \tag{9–18}$$

along any path. We will choose the path going through the wire, where we know from Eq. (9–16) that

$$\mathbf{E} = \rho\mathbf{j} \tag{9–19}$$

Since the current density $\mathbf{j}$ always points along the wire, so does the electric field. In consequence, Eq. (9–18) can be written as

$$V = -\int \mathbf{E} \cdot d\mathbf{l} = |\mathbf{E}|L \tag{9–20}$$

where L is the length of the wire. From Eqs. (9–17), (9–19), and (9–20), we can eliminate both $|\mathbf{E}|$ and $|\mathbf{j}|$ to obtain

$$V = |\mathbf{E}|L = \rho|\mathbf{j}|L = \rho\frac{I}{A}L$$

or

$$\boxed{V = \frac{\rho L}{A}I = RI} \tag{9–21}$$

We call R the *resistance* of the wire. Note that the resistance depends not only on the material of the wire, but also on its geometric properties; it is proportional to the length of the wire and inversely proportional to its cross section.

The resistance $R = V/I$ of a wire is measured in units of volts/ampere; another name for the same unit is the *ohm* (abbreviated Ω); thus

$$1 \text{ ohm} = 1\ \Omega = 1 \text{ volt/ampere}$$

The resistivity, as can be seen from Eq. (9–16), is measured in units of $\dfrac{\text{V/m}}{\text{A/m}^2}$. We

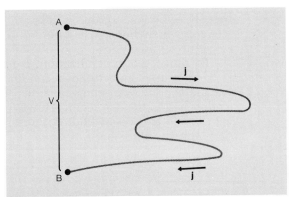

Figure 9–12 To obtain the electrostatic potential difference V between A and B, we can integrate the electric field along the wire, where we know that it is proportional to the current density.

have

$$1\frac{V/m}{A/m^2} = 1\frac{V\text{-}m}{A} = 1\ \Omega\text{-}m$$

Its numerical values for materials at room temperature varies between $\approx 10^{-8}\ \Omega$-m (silver) and $10^{+18}\ \Omega$-m (very good insulators such as amber).[*]

Example 4   An unknown length of insulated copper wire is wound on a coil. The wire has a circular cross section of diameter $d = 1/16$ inch. If one connects it to a 2.0 volt battery, the current through the wire is $I = 0.5$ A. How long is the wire?

The resistance of the wire is

$$R = \frac{V}{I} = \frac{2\ V}{0.5\ A} = 4\ \Omega$$

From Eq. (9–21) we see that the resistance of a wire is

$$R = \frac{\rho L}{A} \tag{9–22}$$

The resistivity of copper is given in Table I (Appendix B) as

$$\rho_{Cu} = 1.75 \times 10^{-8}\ \Omega\text{-}m$$

The length of the wire must be, according to Eq. (9–22),

$$L = \frac{AR}{\rho} = \frac{\pi d^2 R}{4\rho}$$

The diameter d is given as $1/16$ inch = 0.159 cm = 1.59×10^{-3} m. Thus,

$$L = \frac{3.14 \times (1.59 \times 10^{-3}\ m^2)^2 \times 4\ \Omega}{4 \times (1.75 \times 10^{-8}\ \Omega\text{-}m)} = 454\ m$$

We have proved Eq. (9–21), which says that the current between two terminals is proportional to the potential difference between them, only for a straight long wire between them. However, we can prove it for a completely arbitrarily shaped conductor, as long as we know that any moving charge starts out by leaving one terminal and ends by reaching the other. Thus, in our proof we have to exclude such possibilities as the existence of a third terminal. Common terminals in practical circuits are, for example, the terminals of a battery, dry cell, or power supply; in general, we define as terminals any pair of points between which there exists a potential difference for a finite period of time while a current is flowing from one to another.

To prove the generalization, consider the two terminals A and B in Fig. 9–13. They are connected by a conductor of arbitrary shape. We will not assume that the conductor is homogeneous; the resistivity can vary from one place to another within the conductor. However, we will assume that everywhere Ohm's law [Eq. (9–16)] is satisfied. Without loss of generality, we then can assume that the potential difference

$$V = V(A) - V(B)$$

---

[*] See Tables I and IIa (Appendix B) for resistivities of some common substances.

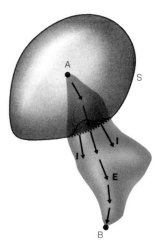

Figure 9–13 The total current through an arbitrarily shaped conductor is proportional to the potential difference between the two terminals.

is positive: $V > 0$. A current will flow from A to B because of Eq. (9–16). One can now surround the terminal A by a closed surface S which does not also enclose terminal B. We can calculate the current flowing from A to B:

$$I = \int_{\substack{\text{closed} \\ \text{surface}}} \mathbf{j} \cdot d\mathbf{S}$$

By Ohm's law [Eq. (9–16)] we then know that

$$I = \int_{\substack{\text{closed} \\ \text{surface}}} \frac{1}{\rho} \mathbf{E} \cdot d\mathbf{S} \tag{9–23a}$$

Note that we cannot take the conductivity $1/\rho$ in front of the integral; it can be different at different points of the surface S.

We also know that the potential difference V is

$$V = V(A) - V(B) = -\int_B^A \mathbf{E} \cdot d\mathbf{l} \tag{9–23b}$$

along some path C leading from B to A.

We now want to show that

$$\boxed{V = RI} \tag{9–24}$$

whatever the potential difference V between the terminals may be. Our argument is really quite simple because, as Eqs. (9–23a) and (9–23b) show, both the potential difference V and the total current I are proportional to the electric field $\mathbf{E}$ on an arbitrary closed surface. Thus, if the electric field everywhere is increased by an arbitrary factor λ, then V is increased by the same factor λ and so is I. In other words, V and I are proportional to each other; but this is exactly what Eq. (9–24) claims. However, it is essential that the electric field everywhere be multiplied by the *same factor;* this is not possible if there is a third terminal C (Fig. 9–14) whose potential difference V' relative to B can be varied independently from the potential difference V between A and B. If we change V without changing V', the currents I_1, I_2, and I_3 will change; but Eq. (9–24) will not be necessarily obeyed by either I_1, I_2, or I_3. Nevertheless, Ohm's law can

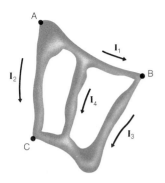

Figure 9-14 If there are more than two terminals, all currents may change if the potential difference between any two terminals changes.

also be used in such a case, as can be seen from the following example:

Example 5   Three wires, whose resistance calculated by Eq. (9–22) are $R_1 = R_2 = R_3 = 10 \ \Omega$, are connected to three terminals A, B, and C (Fig. 9–15a). The potential differences are $V(A) - V(B) = V_{AB} = 5$ V and $V(C) - V(B) = V_{CB} = 1$ V. Calculate the current in each wire.

Since only potential differences are given, we can arbitrarily set $V(B) = 0$. Then $V(A) = V_{AB} = 5$ V and $V(C) = V_{CB} = 1$ V. We do not have two terminals alone; thus, we cannot use Eq. (9–24) directly. However, we can separate the problem into three easier problems by adding an artificial fourth terminal at D, as shown in Fig. 9–15b. We split the three wires, and then we can apply Eq. (9–24). However, D is not really a terminal; no current can flow in or out of it. Since no charge can be stored at D, the law of conservation of charge allows us to say that

$$I_2 = I_1 + I_3 \tag{9-25}$$

if we count the currents positive in the directions indicated by the arrows in Fig. 9–15a. From Eq. (9–24) we have

$$V_{AD} = V(A) - V(D) = R_1 I_1 \tag{9-26a}$$

$$V_{CD} = V(C) - V(D) = R_3 I_3 \tag{9-26b}$$

$$V_{DB} = V(D) - V(B) = R_2 I_2 \tag{9-26c}$$

$V(D)$ is unknown; it has to be such that Eq. (9–25) is satisfied. However, we can calculate I_1, I_2, and I_3 from the three equations (9–26) and substitute into Eq. (9–25). We then have

$$\frac{V(D) - V(B)}{R_2} = \frac{V(A) - V(D)}{R_1} + \frac{V(C) - V(D)}{R_3}$$

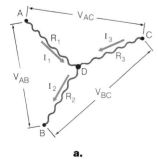

a.

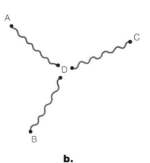

b.

Figure 9-15 Example 5: (a) The real circuit; (b) introducing an artificial fourth terminal D.

or, solving for $V(D)$,

$$V(D) = \frac{\dfrac{V(A)}{R_1} + \dfrac{V(B)}{R_2} + \dfrac{V(C)}{R_3}}{\dfrac{1}{R_1} + \dfrac{1}{R_2} + \dfrac{1}{R_3}} \tag{9-27}$$

In our special case $R_1 = R_2 = R_3 = R$, and thus

$$V(D) = \frac{V(A) + V(B) + V(C)}{3} = \frac{5 \text{ volts} + 0 \text{ volts} + 1 \text{ volt}}{3} = 2 \text{ volts}$$

We can now calculate easily the currents

$$I_1 = \frac{V(A) - V(D)}{R} = 0.3 \text{ A}$$

$$I_2 = \frac{V(D) - V(B)}{R} = 0.2 \text{ A}$$

$$I_3 = \frac{V(C) - V(D)}{R} = -0.1 \text{ A}$$

We see that the current I_3 has the direction opposite to what we assumed when we drew Fig. 9–15; the current is flowing into terminal C and not out of it.

■ From Eq. (9–23a) we can also conclude that even if a steady-state current is flowing in a *homogeneous* conductor, the charge density is zero everywhere inside it. Indeed, consider an arbitrary closed surface S inside a homogeneous conductor, as shown in Fig. 9–16. We assume that neither of the terminals A and B lies inside the surface. Because of charge conservation, the net current into and out of the inside of S has to vanish:

$$\int_S \mathbf{j} \cdot d\mathbf{S} = 0$$

Because the material is homogeneous this implies, because of Ohm's law,

$$\int_S \frac{1}{\rho} \mathbf{E} \cdot d\mathbf{S} = \frac{1}{\rho} \int_S \mathbf{E} \cdot d\mathbf{S} = 0 \tag{9-28}$$

We can take $1/\rho$ in front of the integral; for a homogeneous material the resistivity is the same at any point of the surface S. By Gauss's law we then have

$$\int_S \mathbf{E} \cdot d\mathbf{S} = \frac{Q_{\text{inside}}}{\epsilon_0} = 0 \tag{9-29}$$

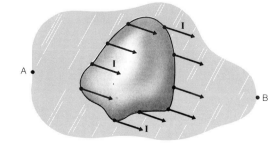

Figure 9–16 Inside a homogeneous conductor there is no net charge density, even in the presence of currents.

Thus, there is no charge inside S. Since the surface S is arbitrary, there can be no charge anywhere inside the conductor. Note that the proof fails if the surface S crosses the surface of the conductor; there can be a surface charge on the surface of a homogeneous conductor. ■

Example 6   Calculate the current in a circular slab of material of resistivity ρ, as depicted in Fig. 9–17. A potential difference V_0 is maintained between the outer rim of radius b and the inner rim of radius a.

We have here the situation of a conductor of arbitrary shape, as we have just discussed. The two "terminals" are the inner and the outer rim. Thus, we want to use Eqs. (9–23a) and (9–23b). We introduce as the closed surface of integration a cylindrical surface of radius r. Because of the symmetry of the problem, the current density will be the same at any point of the surface. The area of the surface where there is a current density $\mathbf{j}$ is

$$A = 2\pi rd$$

The ends will not contribute to the integral

$$I = \int_S \mathbf{j} \bullet d\mathbf{S} = |\mathbf{j}|\, 2\pi rd$$

because there is no current outside the conductor. From Ohm's law we thus conclude that

$$|\mathbf{E}| = \rho\, |\mathbf{j}| = \frac{\rho I}{2\pi rd}$$

and we know that $\mathbf{E}$ will point radially out—parallel to $\mathbf{j}$—if the inner rim is at a higher potential. The value of $|\mathbf{E}|$ depends only on the distance r from the axis. Thus, the potential difference is

$$V = V(\text{inner rim}) - V(\text{outer rim}) = \int_a^b |\mathbf{E}(r)|\, dr$$

or

$$V = \frac{\rho I}{2\pi d} \int_a^b \frac{dr}{r} = \frac{\rho I}{2\pi d} \ln \frac{b}{a}$$

Thus, we have calculated the resistance

$$R = \frac{\rho}{2\pi d} \ln \frac{b}{a}$$

of the slab; given the potential difference V, we can calculate the current $I = V/R$.

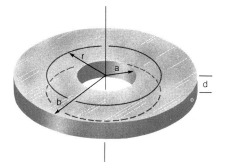

Figure 9–17 Example 6.

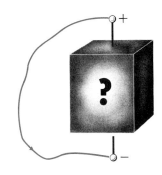

Figure 9–18 The "black box" picture of an electromotive force.

5. ELECTROMOTIVE FORCE

Frequently the physicist or engineer has to use batteries, power supplies, motor-driven generators, and other such devices. These are all devices which have two terminals between which there is a potential difference even in the presence of electric currents. The internal construction of the device in question is irrelevant if we want to use it as a source for electric current. Such devices are called *electromotive sources* or sources of *electromotive force*, abbreviated emf. We have to stress that the name is somewhat misleading, because the emf is *not* a force or an electric field but the integral of electric field. We define as the emf of such a device the quantity

$$\mathscr{E} = \text{emf} = \int \mathbf{E} \cdot d\mathbf{l} \qquad\qquad (9\text{–}30)$$

along a path C (Fig. 9–18) going from the positive terminal to the negative terminal. The emf is measured in units of volts just as is the potential difference. It is important to note that the path in Eq. (9–30) has to be completely outside the device; the inside is to be treated as a "black box" which simply provides us with an emf. We can use a mechanical analogy to the emf and a possible current flowing between the two terminals if they are connected by a conductor of resistance R. The emf is analogous to the pressure difference between the ends of the pipe (Fig. 9–19); the quantity analogous to the electric current would be the liquid flowing through the pipe per unit time. The liquid flow will depend on the diameter and roughness of the pipe and the properties of the liquid; thus, the pipe presents a resistance to the flow of liquid.

One should remember that this is only an analogy and should not be taken too literally; the flow rate in a pipe is in general not proportional to the pressure

Figure 9–19 Mechanical analogy to electrical emf: the flow through a pipe depends on the pressure difference between its ends.

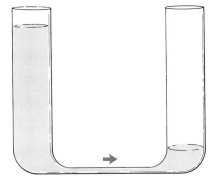

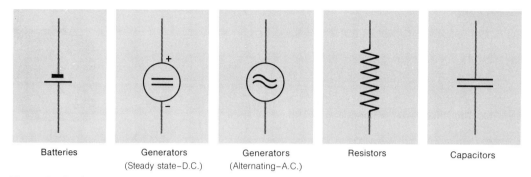

| Batteries | Generators (Steady state–D.C.) | Generators (Alternating–A.C.) | Resistors | Capacitors |

Figure 9-20 Some engineering symbols.

difference between the ends of the pipe, while the electric current is proportional to the emf for all substances obeying Ohm's law. Nevertheless, the analogy helps our intuition if we discuss sources of emf.

There are also sources of emf in which the emf varies with time. The wall plug—or the power station from which the wires lead to the household wall plug—are sources of a periodically oscillating emf:

$$\mathscr{E} = \mathscr{E}_0 \cos \omega t = \mathscr{E}_0 \cos \left(\frac{2\pi}{T} t\right)$$

The period T for household current is 1/60 sec; thus, the frequency $\nu = 1/T = 60$ cycles/second = 60 Hertz.

In engineering design there are certain standard symbols for sources of emf, as well as for resistors and capacitors. They are shown in Fig. 9–20. For an ideal source, the emf is completely independent of the current between the terminals. For real sources this is not the case. In Fig. 9–21 we show how the emf's of a common battery and of a commercial power supply change with the current drawn from the source.

A source of emf is always also a source of electric power. If the charge Q flows from the positive terminal to the negative one, it has traversed the potential difference $\mathscr{E}$ because of Eq. (9–27); thus, the energy

$$W = Q\mathscr{E}$$

has been drawn from the power supply. Therefore, if a steady-state current $I = dQ/dt$ flows from the positive terminal to the negative one, the power de-

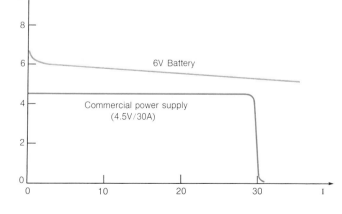

Figure 9-21 Change of emf with current drain for a 6 V battery and a good quality commercial power supply.

livered by the source is

$$P = \frac{dW}{dt} = \frac{dQ}{dt}\mathscr{E}$$

or

$$\boxed{P = \mathscr{E}I} \tag{9–31}$$

Automobile batteries are sometimes rated in terms of "ampere-hours." How much energy is stored in a 12 volt battery advertised at "100 ampere-hours"?

A rating of 100 ampere-hours means — if the advertisement is honest — that, for example, a current of 10 A can be drawn for 10 hours before the battery has to be recharged. The total energy stored is then

$$W = I\mathscr{E}t = 10\,\text{A} \times 12\,\text{V} \times 10\,\text{hours} \times 3600\,\text{sec/hour} = 4.32 \times 10^6\,\text{J}$$

which is quite a respectable amount of energy. Frequently, the unit of kWh (kilo-watt-hour) is also used for electrical energy. It is the energy accumulated by storing the power of 1 kW = 1000 watts for one hour. Thus,

$$1\,\text{kWh} = 1000\,\text{watts} \times 3600\,\text{sec} = 3.6 \times 10^6\,\text{J}$$

The energy stored in the battery is therefore

$$W = 1.2\,\text{kWh}$$

If we want to transform this example into economic terms, let us note that electrical power is sold to households at the cost of ~3 cents/kWh. The cost of recharging the battery would be therefore about 6 cents if we assume a 50% efficiency in recharging. (The rest will be lost in the transformer and rectifier system of the charging apparatus.)

6. RESISTOR CIRCUITS

We have shown in Eq. (9–24) that the potential difference between the two ends of a resistor is proportional to the current through the resistor:

$$V = RI$$

Here we will study the situations in which there are several resistors connected together in circuits. We will, of course, also include at least one source of emf in any circuit, because otherwise there would be no electric currents at all.

Let us first consider the simplest possible circuit, a source of emf whose terminals are connected by a simple resistor R, as shown in Fig. 9–22. The cur-

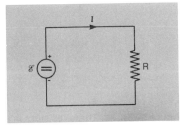

Figure 9–22 The simplest possible resistive circuit: we have $I = \mathscr{E}/R$.

rent I will be given by

$$I = \mathscr{E}/R \qquad (9\text{--}32)$$

We mentioned earlier [Eq. (9–31)] that a source of emf delivers power whenever a current is drawn from it. If the current flows through a resistor, this power is converted into thermal power: the resistor will warm up. We say that the power is dissipated in the resistor. From Eq. (9–31) we know that the power drawn from the emf source is

$$P = \mathscr{E}I$$

Because the current and emf are related by Eq. (9–32), we conclude that the power dissipated in a resistor is proportional to the square of the current flowing through it:

$$\boxed{P = RI^2} \qquad (9\text{--}33)$$

Let us now connect the terminals to a somewhat more complicated arrangement, shown in Fig. 9–23a. Three resistors, R_1, R_2, and R_3, are connected in series between terminals. We want to calculate the current drawn from the emf source. This is easily done if we note that the potential drop* across each resistor is given by Eq. (9–24). The total voltage drop, of course, has to be equal to the emf. Thus we have

$$\mathscr{E} = V_1 + V_2 + V_3 = R_1 I + R_2 I + R_3 I$$

because the same current I flows through each resistor. We can thus write

$$\mathscr{E} = (R_1 + R_2 + R_3)I = RI$$

For resistors in series, the resistances add. We could replace the three resistors

---

* *Potential drop* or *voltage drop* is defined as the potential difference between two points in the circuit, if a current flows from one point to the other. Of course, the current between any two points in a circuit is determined by the configuration of the entire circuit as well as by the externally applied emf.

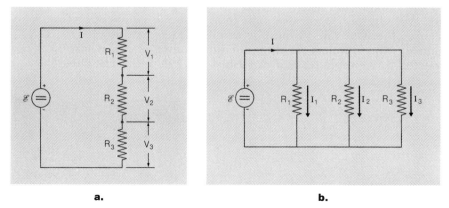

a.   **b.**

Figure 9–23   (a) Three resistors in series. (b) Three resistors in parallel.

by a single resistor

$$R = R_1 + R_2 + R_3$$

(9–34)

In Fig. 9–23b we have a different arrangement: the three resistors R_1, R_2, and R_3 are in parallel. The total current I drawn from the emf source will be distributed among the resistors:

$$I = I_1 + I_2 + I_3$$

The full emf is applied across each resistor; thus, for example, the current I_1 will be equal to $\mathscr{E}/R_1$. We therefore obtain

$$I = I_1 + I_2 + I_3 = \frac{\mathscr{E}}{R_1} + \frac{\mathscr{E}}{R_2} + \frac{\mathscr{E}}{R_3}$$

or

$$I = \left(\frac{1}{R_1} + \frac{1}{R_2} + \frac{1}{R_3} \right) \mathscr{E}$$

Comparing this with Eq. (9–32), we see that the total current I is the same as if we had used a single resistor R whose resistance is given by

$$\frac{1}{R} = \frac{1}{R_1} + \frac{1}{R_2} + \frac{1}{R_3}$$

(9–35)

For resistors in parallel, the inverse resistances add to yield the inverse of the total resistance.

With the two relations (9–34) and (9–35), one is frequently able to reduce complicated arrangements of resistors into a single equivalent resistor.

Example 8   What is the resistance R equivalent to the circuit shown in Fig. 9–24, if the resistors have values $R_1 = R_2 = 1\ \Omega$ and $R_3 = R_4 = 3\ \Omega$?

We replace resistors by equivalent resistors, starting with R_3 and R_4. They are in parallel, so they can be replaced by a single resistor

$$R_{34} = \frac{1}{\dfrac{1}{R_3} + \dfrac{1}{R_4}} = \frac{R_3 R_4}{R_3 + R_4} = \frac{3}{2}\,\Omega$$

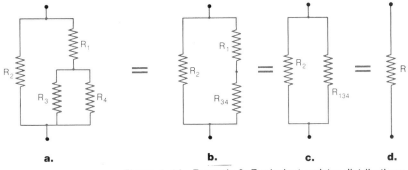

Figure 9–24 Example 8: Equivalent resistor distributions.

The resistance R_{34} is in series with the resistor R_1, forming the resistance R_{134}:

$$R_{134} = R_1 + R_{34} = R_1 + \frac{R_3 R_4}{R_3 + R_4} = \frac{5}{2}\,\Omega$$

Finally, the resistor R_2 is in parallel with R_{134}. Thus, the total resistance R of the circuit is given by

$$\frac{1}{R} = \frac{1}{R_2} + \frac{1}{R_{134}} = \frac{7}{5}\,(\Omega)^{-1}$$

or

$$R = \frac{5}{7}\,\Omega$$

Not all resistor circuits can be reduced this way; the method fails if there are several emf sources or if the connections between the resistors are sufficiently complicated. Thus, neither of the circuits shown in Fig. 9–25 can be reduced by a succession of applications of Eqs. (9–34) and (9–35). However, any such circuit can be considered as consisting of *loops* formed by resistors and emf sources. The loops come together in *nodes* or *junctions* where two or more circuit elements are joined at a single point. One can use two simple rules to analyze such systems. They are known as Kirchhoff's rules.

Kirchhoff's Node Rule. *The algebraic sum of all currents leaving any node is zero.* Currents coming into the node are to be considered negative, and currents leaving the node are positive (Fig. 9–26a). This rule is nothing more than the law of conservation of charge [Eq. (9–13)]. Since no charge can be stored at the node, the total current flowing through a closed surface S around the node has to vanish.

Kirchhoff's Loop Rule. *The algebraic sum of all potential differences around a loop vanishes.* The potential difference is to be considered positive if it has the same direction as the current or if its direction is from the positive terminal of an emf source to the negative one. Thus, V_2 in Fig. 9–26b is positive, while V_4 is negative. The loop rule is a reformulation of Maxwell's first equation that the integral of the electric field over a closed loop vanishes:

$$\int_{\substack{\text{closed} \\ \text{loop}}} \mathbf{E} \cdot d\mathbf{l} = 0$$

Indeed, if we take any closed loop which goes through the four points $A, B, C,$

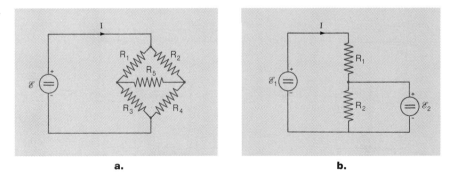

Figure 9–25 Examples of circuits which cannot be replaced by a single equivalent resistance using Eqs. (9–31) and (9–32).

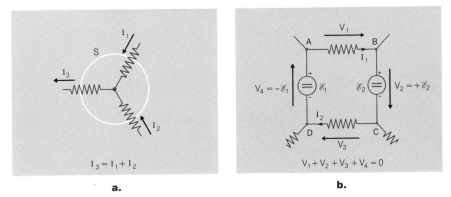

Figure 9-26 Kirchhoff's rules. (a) The node rule: Since no charge can be stored at a node between resistors, there can be no net current flowing through the surface S. (b) The loop rule: When going around a loop, one has to end up at the same electrostatic potential one started out from.

and D we have

$$\int_A^B \mathbf{E} \bullet d\mathbf{l} + \int_B^C \mathbf{E} \bullet d\mathbf{l} + \int_C^D \mathbf{E} \bullet d\mathbf{l} + \int_D^A \mathbf{E} \bullet d\mathbf{l} = 0$$

which is exactly what Kirchhoff's loop rule affirms.

Example 9   Calculate the power drawn from each of the two power supplies, as well as the power dissipated by each resistor, for the arrangement in Fig. 9–27. The three resistors each have the value $R_1 = R_2 = R_3 = R = 2\ \Omega$; $\mathscr{E}_1 = 10$ volts and $\mathscr{E}_2 = 2$ volts.

We define the three currents I_1, I_2, and I_3 to be positive in the direction of the arrows. Kirchhoff's node rule then says that

$$I_2 - I_3 - I_1 = 0 \tag{9-36}$$

Kirchhoff's loop rule applied to the two loops in the direction of the arrows yields two relations:

$$-\mathscr{E}_1 + RI_1 + RI_2 = 0 \tag{9-37a}$$

$$-\mathscr{E}_2 + RI_3 + RI_2 = 0 \tag{9-37b}$$

We now have three equations for the three unknown quantities I_1, I_2, and I_3. Adding the two equations (9–37) and substituting $I_1 + I_3 = I_2$ from Eq. (9–36) yields

$$I_2 = \frac{\mathscr{E}_1 + \mathscr{E}_2}{3R}$$

Figure 9-27 Example 9: A circuit involving two sources of emf.

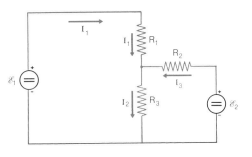

and substituting this into Eqs. (9–37a) and (9–37b) we obtain

$$I_1 = \frac{2\mathscr{E}_1 - \mathscr{E}_2}{3R}$$

$$I_3 = \frac{2\mathscr{E}_2 - \mathscr{E}_1}{3R}$$

With our numerical values we obtain $I_1 = 3$ A, $I_2 = 2$ A, and $I_3 = -1$ A. Thus, the current I_3 flows into the emf source $\mathscr{E}_2$; not only do we not draw power from the source, we actually deliver power to the emf source:

$$P_2 = +\mathscr{E}_2 I_3 = -2 \text{ W}$$

On the other hand, we do draw power from the other emf source:

$$P_1 = \mathscr{E}_1 I_1 = 30 \text{ W}$$

The power dissipated in each resistor is

$$RI_1{}^2 = 18 \text{ W}$$

$$RI_2{}^2 = \ 8 \text{ W}$$

$$RI_3{}^2 = \ 2 \text{ W}$$

Of course, energy conservation requires that

$$P_1 + P_2 = RI_1{}^2 + RI_2{}^2 + RI_3{}^2$$

which is indeed true. The power P_1 drawn from the emf source $\mathscr{E}_1$ is partially dissipated in the resistors; the rest is delivered as $-P_2$ to the emf source $\mathscr{E}_2$.

Example 10   A cube is built out of resistors, as sketched in Fig. 9–28a. Find the resistance between the two opposite points A and B, if all resistors have the same value $R_0 = 10 \ \Omega$.

First we note that, for reasons of symmetry, the three points P_1, P_2, and P_3 are at the same potential, if we connect the points A and B to the terminals of an arbitrary emf source $\mathscr{E}_0$.[°] The same statement is true for the three points R_1, R_2, and

---

[°] Each point is one resistor away from point A, so the potential drops must be equal to each other.

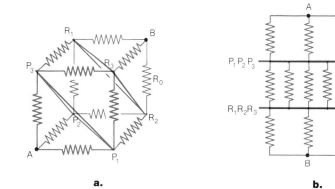

a.   **b.**

Figure 9–28 Example 10: (a) The actual arrangement; (b) an equivalent arrangement.

R_3. We can therefore connect P_1, P_2, and P_3 together by short wires, and do the same for R_1, R_2, and R_3. We can now redraw the circuit as shown in Fig. 9–28b. There are three resistors in parallel between A and the equipotential "plane" $P_1P_2P_3$; there are six resistors in parallel between $P_1P_2P_3$ and $R_1R_2R_3$; and there are again three parallel resistors between $R_1R_2R_3$ and B. Thus, the total resistance is

$$R = \frac{R_0}{3} + \frac{R_0}{6} + \frac{R_0}{3} = \frac{5}{6}R_0 = 8.333 \ \Omega$$

7. RC CIRCUITS

Capacitors are storage devices for charge and electric energy. If the two plates of a capacitor are connected to a resistor by closing a switch (Fig. 9–29), the charges $\pm Q$ stored on the plates are no longer prevented from neutralizing each other. A current will flow in the resistor, until the energy stored in the capacitor is dissipated in the resistor and there is no charge left on the capacitor plates. Let us assume that at the time $t = 0$, when we close the switch, the charge stored on the plates is $\pm Q_0$. Then there is a potential difference $V_0 = Q_0/C$ between the plates.

The question before us is how fast the capacitor will discharge. For this purpose, let us consider the situation at some time t after the switch has been closed. We, denote by $Q(t)$ the charge left on the capacitor plates and by $I(t)$ the current through the resistor R. If we introduce a closed surface S enclosing one plate and crossing the resistor R, then the law of conservation of charge requires that

$$-\frac{dQ}{dt} = \int_S \mathbf{j} \cdot d\mathbf{S} = I(t) \tag{9–38}$$

since the only current anywhere on the surface S is through the resistor. On the other hand, we know that the current $I(t)$ is proportional to the potential difference across the resistor

$$V(t) = R I(t) \tag{9–39a}$$

But $V(t)$ is also the potential difference between the capacitor plates:

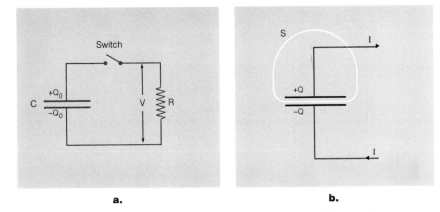

Figure 9–29 Discharging a capacitance through a resistor: (a) the physical arrangement; (b) the total current I through the surface S is equal to the decrease of the charge Q on the plate per unit time.

$$V(t) = \frac{Q(t)}{C} \qquad\qquad (9\text{--}39\text{b})$$

Combining the three equations, we obtain the differential equation

$$-\frac{dQ(t)}{dt} = \frac{Q(t)}{RC} \qquad\qquad (9\text{--}40)$$

It is easy to convince ourselves by substituting into Eq. (9–40) that the solution to Eq. (9–40) is

$$Q(t) = Ae^{-t/RC} \qquad\qquad (9\text{--}41)$$

where A is an arbitrary constant which has to be determined from the initial conditions at $t = 0$. We know that $Q(t = 0) = Q_0$; thus, $A = Q_0$ and we have

$$\boxed{Q(t) = Q_0 e^{-t/RC}} \qquad\qquad (9\text{--}41\text{a})$$

We can then easily calculate $V(t)$ and $I(t)$ using Eqs. (9–39a) and (9–39b):

$$\boxed{\begin{aligned} V(t) &= \frac{Q_0}{C} e^{-t/RC} = V_0 e^{-t/RC} \qquad\qquad &(9\text{--}41\text{b}) \\[2mm] I(t) &= \frac{V(t)}{R} = \frac{Q_0}{RC} e^{-t/RC} \qquad\qquad &(9\text{--}41\text{c}) \end{aligned}}$$

Eq. (9–41c) is plotted in Fig. 9–30.

■   We can show that Eq. (9–41) gives the most general solution in the following way: Multiply Eq. (9–40) by dt and divide by $-Q(t)$ to obtain

$$+\frac{dQ}{Q} = -\frac{dt}{RC}$$

Both sides can be integrated to yield

$$\ln Q = -\frac{t}{RC} + A_0$$

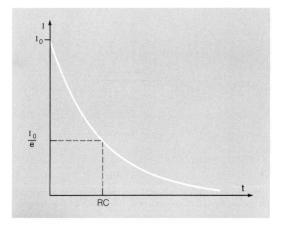

Figure 9–30 The discharging current plotted against time.

where A_0 is an arbitrary constant. Taking the exponential on both sides, we obtain

$$Q = e^{A_0} e^{-t/RC}$$

which is indeed identical to Eq. (9–41) if we write $A = e^{A_0}$. ■

From the three equations (9–41) we can see that the power initially stored in the capacitor is dissipated in the resistor. At any time the dissipated power is

$$P(t) = R\,[I(t)]^2 = \frac{RQ_0^2}{(RC)^2} e^{-2t/RC} \qquad (9\text{–}42)$$

The total energy dissipated is

$$W = \int_{t=0}^{\infty} P(t)\,dt = \frac{Q_0^2}{RC^2} \int_0^{\infty} e^{-2t/RC}\,dt \qquad (9\text{–}43)$$

The integral can be simplified by substituting $u = 2t/RC$. We then have $du = (2/RC)\,dt$ and

$$\int_0^{\infty} e^{-2t/RC}\,dt = \frac{RC}{2} \int_0^{\infty} e^{-u}\,du = \frac{RC}{2}$$

Thus, the total energy dissipated in the capacitor is

$$W = \frac{Q_0^2}{RC^2} \frac{RC}{2} = \frac{Q_0^2}{2C}$$

which indeed is equal to the initially stored energy $Q_0^2/2C = \dfrac{C}{2} V_0^2 = \dfrac{Q_0 V_0}{2}$ in the capacitor.

Other problems involving capacitors can be solved in a similar way. The crucial relations are Eqs. (9–38) and (9–39b), which are always true for a capacitor.

Example 11   An emf source of $\mathscr{E} = 10$ volts is connected through a resistor $R = 1000\ \Omega$ to a capacitor $C = 10\ \mu$F (Fig. 9–31). The switch S is closed at $t = 0$. Calculate the charge Q on the capacitor at any time t.

We define the current $I(t)$ through the resistor to be positive in the direction of the arrow. Then we have to replace Eq. (9–38) by

$$\frac{dQ}{dt} = +I(t)$$

because the current I flows into the capacitor plates. The potential drop across the resistor is

$$RI = \mathscr{E} - V(t)$$

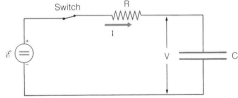

Figure 9–31 Example 11: Charging up of a capacitance from a generator.

where

$$V(t) = \frac{Q(t)}{C}$$

is the voltage across the capacitor plates. From the three equations we obtain

$$R\frac{dQ}{dt} = \mathscr{E} - \frac{Q}{C} \tag{9-44}$$

Obviously, if $dQ/dt = 0$, one has $Q = C\mathscr{E}$; thus, we cannot try to use a solution $Q = Q_0 e^{-\lambda t}$, because simple intuition tells us that after a long time we will have $Q = C\mathscr{E} = $ constant and the capacitor will be fully charged. Instead, we try the solution

$$Q = A + Be^{-t/RC}$$

where A and B are some constants. Substituting into Eq. (9-44), we obtain

$$R\left(-\frac{B}{RC}e^{-t/RC}\right) = \mathscr{E} - \frac{A}{C} - \frac{B}{C}e^{-t/RC} \tag{9-45}$$

and we see that we have a solution (the terms $\mathscr{E} - \dfrac{A}{C}$ disappear) if

$$A = C\mathscr{E}$$

The constant B cannot be determined from Eq. (9-45), since it is satisfied for any B. However, at $t = 0$ there was no charge on the capacitor. Thus,

$$Q(t = 0) = A + Be^0 = A + B = 0$$

and we obtain $B = -A = -C\mathscr{E}$. Thus, we have

$$Q(t) = C\mathscr{E}_0[1 - e^{-t/RC}]$$

In our example,

$$C\mathscr{E}_0 = 10^{-5} \text{ F} \times 10 \text{ V} = 10^{-4} \text{ coul}$$

and

$$RC = 10^3 \ \Omega \times 10^{-5} \text{ F} = 10^{-2} \ \Omega\text{F}$$

and since $1 \ \Omega = 1$ V/A and 1 F $= 1$ coul/V, we have

$$RC = 10^{-2}\frac{\text{V}}{\text{coul/sec}} \times \frac{\text{coul}}{\text{V}} = 10^{-2} \text{ sec} = 10 \text{ msec}$$

Summary of Important Relations

Relation between current density **j**, charge density ρ_e and mean charge velocity $\langle \mathbf{v} \rangle$

$$\mathbf{j} = \rho_e \langle \mathbf{v} \rangle$$

Current through surface S
$$I = \int_S \mathbf{j} \cdot d\mathbf{S}$$

Charge conservation for closed surface S

$$\frac{dQ}{dt}(\text{inside}) = - \int_S \mathbf{j} \cdot d\mathbf{S}$$

Ohm's law $\qquad \mathbf{j} = \frac{1}{\rho}\mathbf{E} \qquad \rho = \text{resistivity}$

Current through a conductor with emf $\mathscr{E}$ across it

$$I = \frac{\mathscr{E}}{R} \qquad R = \text{resistance}$$

For a wire with length L and cross section A $\qquad R = \frac{\rho L}{A}$

For two resistors in series $\qquad R = R_1 + R_2$

in parallel $\qquad \frac{1}{R} = \frac{1}{R_1} + \frac{1}{R_2}$

Capacitor C, initially charged up to potential V_0, discharging through resistor R: the current through the resistor is

$$I(t) = \frac{V_0}{R}e^{-t/RC}$$

Questions

1. A beam of electrons is being accelerated along its direction of motion. Will the charge density in the beam be everywhere the same?

2. Electrons in a conductor move very slowly (see Example 1). Nevertheless, if you turn on a light switch, the light comes on immediately. Explain. Does a similar effect occur for liquid flowing through a tube?

3. In discussing Figs. 9–11 and 9–12, we said that the electric field inside a wire always points along the wire. What about outside the wire? Discuss the charge distribution on the wire surface.

4. We said that there is no net charge density inside a homogeneous ohmic conductor even if there is a current. Is the same true for a non-homogeneous conductor? Explain.

Problems

1. The average current in a typical thunderbolt is 2×10^4 A; its typical duration is 0.1 sec. What is the total charge transmitted to the earth and what is the energy released, if the potential difference between the clouds and ground is 5×10^8 V?

2. A metallic sphere carries the charge $Q = 2 \times 10^{-8}$ coul. After 15 seconds one observes that the charge has decreased by 7%. Assuming that the charge is carried away

equally in all directions, what is the current density 30 cm away from the center of the sphere?

3. An electron beam has the diameter $d = 1$ mm; the electrons in it have the kinetic energy $T = 10$ keV. What is the charge density in the beam, if the total current is 2×10^{-2} A?

4. The two-mile-long linear accelerator at Stanford (SLAC) is capable of accelerating electrons to a kinetic energy of 2×10^{10} eV. The electrons are accelerated in 360 pulses per second, each pulse lasting one microsecond. If 10^{14} electrons are accelerated by each pulse, what is the peak current? What is the average current (averaged over many pulses)? Assuming that the beam diameter is 3 mm, what are the peak and the average current densities? The electrons move at essentially the velocity of light, $c = 3 \times 10^8$ m/sec.

5. A coil of copper wire has a mass of 1.2 kg and a total resistance $R = 0.8\ \Omega$. How long is the wire?

6. Because of the high price of copper, aluminum is sometimes used for wires. What is the ratio of weight between two equally long copper and aluminum wires, if they have the same resistance?

7. Calculate the total resistance between the points A and B in the figure, if all resistors have the same value $R_0 = 1\ \Omega$. Can one remove any of the resistors without affecting the overall resistance?

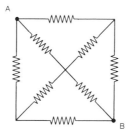

8. For household wiring, one usually uses #10 insulated copper wire (wire diameter $d = 0.10$ inch). What will be the power dissipated in such a wire 50 feet long, if a current of 25 A flows through it?

9. You are given two boxes, A and B, each containing an assortment of resistors, several each of value 10, 15, 18, 20, 22, 27, 33, 39, 47, and 51 Ω (each $\pm 0.2\%$ at room temperature; these are standard values). However, the two boxes contain resistors of different types: in box A all resistances increase with temperature

$$R = R_{\text{room}}[1 + \alpha(T - T_{\text{room}})] \qquad \alpha = 0.2\%/°C$$

while all resistance values in box B decrease with temperature

$$R = R_{\text{room}}[1 - \beta(T - T_{\text{room}})] \qquad \beta = 0.6\%/°C$$

You are asked to assemble from these components a resistance of 100 Ω ($\pm 1\%$), which is completely independent of temperature. (Note: There are several solutions!)

10. Show that if the applied voltage varies sinusoidally with time

$$V = V_0 \sin \omega t$$

then the power averaged over an integer number of cycles dissipated in a resistor is exactly 1/2 of the power dissipated if a D.C. (direct current) voltage V_0 is applied.

11. A hollow hemisphere (inner radius R_1, outer radius R_2) is made of material having a resistivity of 1 Ω-m. The inner and outer spherical surfaces are coated with gold (resistivity 2×10^{-8} Ω-m). Calculate the resistance between the two layers. Choose $R_1 = 1$ cm and $R_2 = 2$ cm.

Gold

°12. A truncated cone (length L, radii R_1 and R_2) is made of material with resistivity ρ. Calculate the resistance between the ends of the cone by subdividing it into an infinite number of very thin disks.

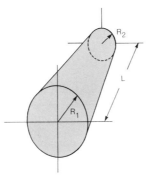

13. A copper wire of diameter $d = 2$ mm is soldered end-to-end to a stainless steel wire of equal diameter. A current $I = 10$ A flows through the wires. Show that there is a surface charge density on the solder joint and calculate its magnitude. (Hint: Use a coin-shaped Gaussian surface straddling the solder joint.)

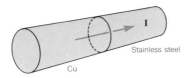

Stainless steel

Cu

14. A charged plane parallel capacitor (area A, separation d) is filled with glass of resistivity $\rho = 10^{13}$ Ω-m and dielectric constant $K = 4.2$. How long will it take until it loses 10% of its charge?

15. A coaxial cable (inner diameter $D_1 = 1$ mm, outer diameter $D_2 = 8$ mm, length $L = 10$ m) is filled with an insulator of resistivity $\rho = 10^{12}$ Ω-m. What is the leakage current in the cable, if the potential difference is 600 V?

°16. An infinitely long series-parallel chain is built up of equal resistors R_0 as shown in the figure. Calculate the total resistance between A and B. Note that the part of the chain to the right of A' and B' is the same as the whole chain because the chain is infinite.

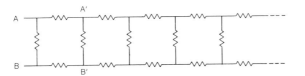

17. An emf source $\mathscr{E} = 15$ V is connected to two resistors in series. R_1 has the value $10\,\Omega$. At what value of R_2 will the power dissipated in R_2 be maximal, and what is that power?

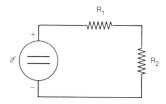

18. Consider the circuit shown in the figure. Using Kirchoff's laws, obtain the equations relating the currents to the voltages. Solve for the current flowing through R_3.

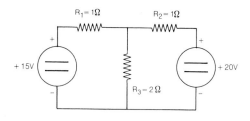

19. During peak load periods, power companies reduce the line voltage by up to 10%. What will then be the output of a water heater which delivers 500 watts at the nominal line voltage?

20. Two capacitors of equal capacitance C are connected in parallel through a resistor and switch. The charge $\pm Q_0$ is deposited on one capacitor, and the switch S is then closed. What is the final distribution of charges? What is the difference between the initial and final stored energy? Where did the energy go?

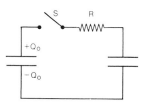

°21. Calculate the time dependence of the current $I(t)$ flowing through the resistor R in Problem 20. Also calculate explicitly the total energy dissipated in the resistor.

22. The emf $\mathscr{E} = \mathscr{E}_0 \cos \omega t$ is applied to an RC circuit, as shown in the figure. Calculate the voltage across the capacitor, assuming that it takes the form $V = V_0 \cos (\omega t - \delta)$. Determine V_0 and δ.

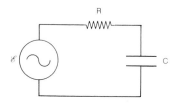

°23. A copper strip of width $a = 5$ cm, thickness $b = 0.5$ mm, and overall length $2L = 2$ meters is folded as shown in the sketch. The two layers are separated by $d = 1$ mm. A current $I = 10$ A is flowing through the strip. Calculate the total charge stored on resulting "capacitor." (Hint: You first have to calculate the voltage $V(x)$ between the two layers at any point x.)

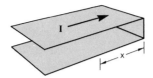

CHAPTER 10

MAGNETIC FIELD

1. INTRODUCTION

It was known to the ancient Greeks that certain materials attract or repel each other, depending on their relative orientation. These materials are called *magnetic*. The tendency of a magnet to orient itself so that one side points roughly to the north pole of the earth has also been known for a long time; since the twelfth century A.D., sailors have been using magnetic compasses to orient themselves on the high seas. However, it was only in 1819 that H. C. Oersted discovered that there is a connection between magnetic forces and electric currents; he noticed that a compass needle will orient itself perpendicularly to a current-carrying wire (Fig. 10–1). One year later, A. M. Ampère discovered that two parallel current-carrying wires also act on each other by a force which is not electrostatic in nature. If the currents in the two wires have the same direction, the wires attract each other; while two wires carrying currents in opposite directions repel each other. If the current in either wire is interrupted, there is no force. More detailed studies in the nineteenth century showed that in the neighborhood of a permanent magnet or a current-carrying wire there exists a new field, called the magnetic field **B**. It has no effect on a point charge at rest; however, a moving charge is acted upon by a force

$$\mathbf{F} = q(\mathbf{v} \times \mathbf{B}) \tag{10–1}$$

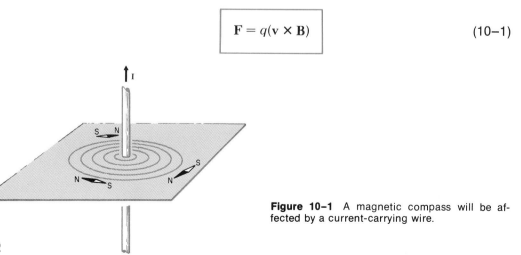

Figure 10–1 A magnetic compass will be affected by a current-carrying wire.

where q is the charge and $\mathbf{v}$ is its velocity. We can use Eq. (10–1) as a definition of the magnetic field $\mathbf{B}$; however, it's much more than a definition. The force is proportional to the *vector product* of the velocity $\mathbf{v}$ and the magnetic field. Thus, we learn from Eq. (10–1) that the force is always perpendicular to both the velocity and the magnetic field; we also see that the force will reverse its direction if the velocity of the charge reverses its direction. In particular, there is no force on a charge at rest in a magnetic field. The force is also proportional to the charge itself; reversing the sign of the charge reverses the direction of the force.

We will assume for the moment that we can measure the magnetic field at any point using Eq. (10–1), and we will study the magnetic fields of various current distributions. For qualitative tests of the magnetic field direction, we can also use a compass needle; it will point in the direction of the magnetic field. We save for later a discussion of the magnetic forces acting on moving charges or currents, as well as the problem of magnetic materials.

2. MAGNETIC FIELD OF A STRAIGHT WIRE

There will be a magnetic field around a thin straight wire carrying a current I. A compass needle shows that the magnetic field points along a circle around the wire. Reversing the current reverses the direction of the magnetic field. If we are able to measure the magnetic field, we find that at any point it has the magnitude

$$|\mathbf{B}| = \frac{\mu_0 I}{2\pi r} \qquad (10\text{–}2a)$$

where r is the distance from the point to the wire and μ_0 is a fundamental constant:

$$\mu_0 = 1.2566 \times 10^{-6} \frac{\text{V-sec}}{\text{A-m}} \qquad (10\text{–}2b)$$

The direction of the magnetic field is such that if the thumb of the *right hand* points in the direction of the current in the wire, then the other fingers encircle the wire in the same direction as the magnetic field does (Fig. 10–2). If we have several straight wires, each will produce a magnetic field; the total magnetic field is obtained as the vector sum of the contributions due to each wire.

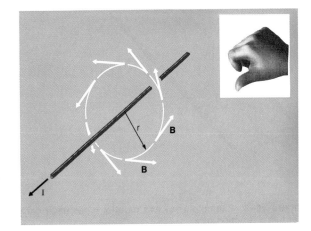

Figure 10–2 Direction of magnetic field near a current-carrying wire; the right-hand rule.

Example 1   Two parallel thin straight wires separated by a distance $2d$ carry the same current I in the same direction. Calculate the magnetic field in the plane formed by the two wires.

The magnetic field of each wire will be pointing either up or down perpendicularly to the plane formed by the two parallel wires. Let us introduce a coordinate system (Fig. 10–3a) with the x-axis in the plane and perpendicular to the wires, the y-axis parallel to the wires, and the z-axis normal to the plane. We choose the origin $x = y = 0$ halfway between the two wires. Then **B** will point in the positive or negative z-direction at any point in the plane.

Between the two wires the magnetic fields of the wires point in opposite directions; thus, we have for $-d < x < +d$

$$B_z = \frac{\mu_0 I}{2\pi}\left(+\frac{1}{d-x} - \frac{1}{d+x}\right) = \frac{\mu_0 I}{2\pi}\frac{2x}{d^2 - x^2} \tag{10–3a}$$

Outside the two wires, both magnetic fields reinforce each other; we have for $x > d$

$$B_z = \frac{\mu_0 I}{2\pi}\left(-\frac{1}{x-d} - \frac{1}{x+d}\right) = \frac{\mu_0 I}{2\pi}\frac{-2x}{x^2 - d^2} \tag{10–3b}$$

and for $x < -d$

$$B_z = \frac{\mu_0 I}{2\pi}\left(\frac{1}{(-x)+d} + \frac{1}{(-x)-d}\right) = \frac{\mu_0 I}{2\pi}\frac{2x}{d^2 - x^2} \tag{10–3c}$$

Comparing the three equations (10–3) we see that everywhere in the plane we have

$$B_z = \frac{\mu_0 I}{\pi}\frac{x}{d^2 - x^2} \tag{10–3d}$$

We have plotted $B_z(x)$ in Fig. 10–3b. Note that we are assuming the wires to be very thin; this is why in Eq. (10–3d) the magnetic field seemingly becomes infinite for $x = \pm d$. In reality we cannot expect Eq. (10–3d) to be correct once we reach the surface of one of the wires.

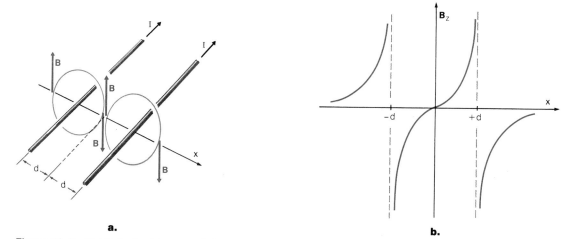

a.   **b.**

Figure 10–3 Example 1: a) two parallel wires each carrying the same current; b) the magnetic field as a function of x.

3. UNITS OF MAGNETIC FIELD

We have said earlier that we measure charges in coulombs, currents in amperes = coul/sec, and that otherwise we will use MKS units. The unit called the volt was defined as

$$1 \text{ volt} \times 1 \text{ coulomb} = 1 \text{ Joule}$$

We can determine the units for the magnetic field from Eq. (10–1). We have

$$[B] = \frac{[F]}{[q][v]} = \frac{\text{N}}{\text{coul-m/sec}}$$

Since 1 N = 1 J/m = 1 V-coul/m, we can also write

$$[B] = \frac{\text{V-sec}}{\text{m}^2} \tag{10–4}$$

Physicists have the tendency to honor the names of their famous colleagues by naming units after them. Thus, the definition

$$1 \text{ volt-second} = 1 \text{ weber} = 1 \text{ Wb} \tag{10–5}$$

Therefore, the magnetic field is measured in units of webers/m². Another frequently used unit is

$$1 \text{ G} = 1 \text{ gauss} = 10^{-4} \text{ weber/m}^2 \tag{10–6}$$

The earth's magnetic field is roughly 0.5 gauss; laboratory magnets typically have maximum fields of 1 to 2 Wb/m², and some go as high as 10 Wb/m². Magnetic fields as high as 200 to 400 Wb/m² can be achieved for very short times (a few microseconds or less) in the laboratory. Much higher fields can exist in the universe. It is believed that at the surface of some very old stars, which have collapsed to a diameter of a few kilometers, the magnetic field can be as high as 10⁴ Wb/m². These are the so-called "neutron stars" and their somewhat larger cousins, the "white dwarfs."

In the units which we use here, the fundamental constant μ_0, called the magnetic permeability of vacuum, has a value which is easy to remember:

$$\mu_0 = 4\pi \times 10^{-7} \frac{\text{V-sec}}{\text{A-m}} \tag{10–7}$$

We had earlier chosen the coulomb as the fundamental electric unit; but we had then not defined its value in terms of any easily available laboratory standard. We have one now; the coulomb is defined so that μ_0 has exactly the numerical value $4\pi \times 10^{-7}$ in Eq. (10–2a).

Example 2   What must be the current through a wire, if it is to produce a magnetic field of 1 Wb/m² at a distance $r = 10$ cm?

From Eq. (10–2) we have

$$I = \frac{2\pi r B}{\mu_0} = \frac{2\pi \times (0.1 \text{ meter}) \times 1 \text{ V-sec/m}^2}{4\pi \times 10^{-7} \text{ V-sec/A-m}} = 5 \times 10^5 \text{ A!}$$

We see that very large currents are required to produce magnetic fields of 1 Wb/m².

4. AMPÈRE'S LAW

We saw in electrostatics the usefulness of the two Maxwell's equations (7–18) and (7–19). They enabled us to solve a wide variety of problems. We now propose to find similar laws for the magnetic field.

Let us first choose a closed loop and calculate the integral

$$\int_{\substack{\text{closed} \\ \text{loop}}} \mathbf{B} \bullet d\mathbf{l}$$

We will use the magnetic field of a straight wire as an example. The simplest path we can choose is a circle of radius r with its center on the wire (Fig. 10–4). Everywhere along the circle the magnetic field has the same magnitude

$$|\mathbf{B}| = \frac{\mu_0 I}{2\pi r}$$

and is parallel to the line element $d\mathbf{l}$ along the loop if we go around the circle in the direction shown in Fig. 10–4. Thus, the integral has the value

$$\int_{\text{circle}} \mathbf{B} \bullet d\mathbf{l} = \frac{\mu_0 I}{2\pi r} \int |d\mathbf{l}| = \frac{\mu_0 I}{2\pi r} \times 2\pi r = \mu_0 I \qquad (10\text{--}8)$$

This is very different from the situation for the electric field; there the integral over any closed loop vanishes.

We have calculated the integral only for a very special closed loop—a circle around the wire. We want a more general result; thus, let us calculate the integral $\int \mathbf{B} \bullet d\mathbf{l}$ along some arbitrary closed loop, say C_1 or C_2 in Fig. 10–5. We show a small segment of the path in Fig. 10–6. In both figures we assume that the

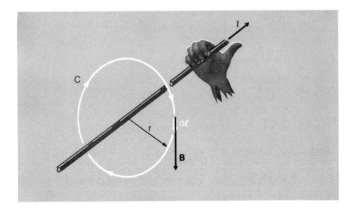

Figure 10–4 Ampère's law: integrating the magnetic field along a circle.

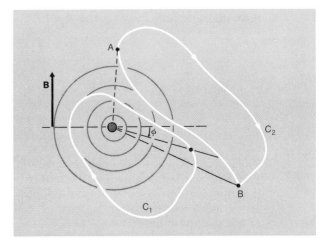

Figure 10-5 Ampère's law: integrating the magnetic field along an arbitrary curve around a single straight wire.

current in the straight wire goes into the paper. From Fig. 10–6 we see that since the magnetic field is always tangential to a circle around the wire, we have

$$\mathbf{B} \cdot d\mathbf{l} = |\mathbf{B}|\, d\ell \cos \alpha = |\mathbf{B}|\, r\, d\phi \qquad (10\text{–}9)$$

where the small angle $d\phi$ is counted positive if it has a clockwise direction around the wire. If we use the known value of $|\mathbf{B}|$ we obtain

$$\mathbf{B} \cdot d\mathbf{l} = \frac{\mu_0 I}{2\pi r}\, r\, d\phi = \frac{\mu_0 I}{2\pi}\, d\phi$$

Thus, the integral $\int \mathbf{B} \cdot d\mathbf{l}$ can be written as

$$\int_{\substack{\text{closed}\\\text{loop}}} \mathbf{B} \cdot d\mathbf{l} = \frac{\mu_0 I}{2\pi} \int_{\substack{\text{closed}\\\text{loop}}} d\phi \qquad (10\text{–}10\text{a})$$

Along the closed loop C_1 in Fig. 10–5 the angle ϕ increases by 2π if we go once around. Thus,

$$\int_{C_1} \mathbf{B} \cdot d\mathbf{l} = \mu_0 I \qquad (10\text{–}10\text{b})$$

We have thus shown Eq. (10–8) or Eq. (10–10) to be true for any closed loop encircling the wire. However, if we choose C_2 as the closed loop, we obtain a different result. As we move from A to B along C_2, the angle ϕ increases; however, as we go back from B to A, the angle again decreases. When we have gone

Figure 10-6 Calculating $\mathbf{B} \cdot d\mathbf{l} = |\mathbf{B}|\ |d\mathbf{l}| \cdot \cos \alpha.$

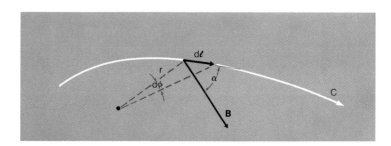

through the whole loop C_2, the angle ϕ has the original value. Thus,

$$\int_{C_2} d\phi = \phi(A) - \phi(A) = 0$$

and because of Eq. (10–10a) we thus have

$$\int_{C_2} \mathbf{B} \cdot d\mathbf{l} = 0 \tag{10–11}$$

The only difference between the two closed loops C_1 and C_2 is that C_1 encircles the wire, while C_2 does not. We can express this also in another way by saying that the current I goes *through* the loop C_1, but not through the loop C_2. We thus arrive at the statement of *Ampère's law:* For any closed loop C the line integral of the magnetic field is

$$\int_{\substack{\text{closed} \\ \text{loop}}} \mathbf{B} \cdot d\mathbf{l} = \mu_0 I_{\text{through loop}} \tag{10–12}$$

where I is the total current flowing through a surface which has the loop C as its border line. The current is to be counted positive if it follows the right hand rule relative to the direction of integration along C, as shown in Fig. 10–7.

We stress that we have *not* derived Ampère's law; we have only shown it to be true for a single straight long wire. By extension, we have also shown it to be true for any current arrangement consisting of individual straight wires. But the generalization even for a single curved wire is a generalization which we *assume* to be true. Ampère's law cannot be derived as we derived Gauss's law in electrostatics, because there exists no magnetic analogy to the point charges out of which we can build up arbitrary charge distributions. One cannot build up an arbitrary current distribution out of linear currents such as exist in a straight wire. Ampère's law is thus a guess which has to be confirmed by experiment. This has been done many times, and as far as we know today, Ampère's law is satisfied for all steady-state currents, i.e., currents which are either time-independent or change only very slowly with time.

■   Ampère's law as stated says that the total current through any surface which has the closed loop as border determines the line integral of the magnetic field around the loop.

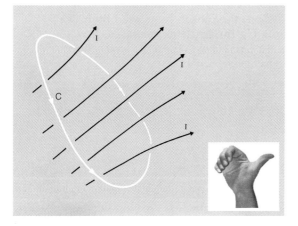

Figure 10–7 The right-hand rule in Ampère's law. The integral $\int \mathbf{B} \cdot d\mathbf{l}$ along C is positive.

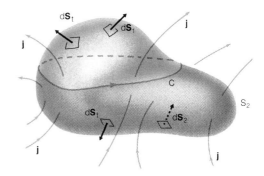

Figure 10-8 The total current through any closed surface has to vanish for Ampère's law to make sense.

However, there exists an infinity of such surfaces; can Ampère's law be true for all of them? To answer this question, consider (Fig. 10–8) a closed loop C and two surfaces S_1 and S_2 which have C as their border. The current through S_1 is given by an integral of the current density:

$$I(S_1) = \int_{S_1} \mathbf{j} \bullet d\mathbf{S}_1$$

and the same equation is true for S_2. Thus, if Ampère's law is to make sense, we need to prove that $I(S_1) = I(S_2)$ or

$$\int_{S_1} \mathbf{j} \bullet d\mathbf{S}_1 = \int_{S_2} \mathbf{j} \bullet d\mathbf{S}_2 \tag{10–13}$$

The direction of $d\mathbf{S}$ is always defined by the right hand rule with respect to the curve C; if our four fingers curve in the direction we go around C, then the outstretched thumb of the right hand points in the direction in which we define the current to be positive. This implies that in Eq. (10–13) $d\mathbf{S}_1$ points outward and $d\mathbf{S}_2$ points into the closed surface S_t formed by both S_1 and S_2. Thus, if we define $d\mathbf{S}_t$ always to point outward, we have $d\mathbf{S}_t = d\mathbf{S}_1$ on S_1 and $d\mathbf{S}_t = -d\mathbf{S}_2$ on S_2. If we now calculate the total integral $\int \mathbf{j} \bullet d\mathbf{S}_t$ over the whole closed surface S_t, we have

$$\int_{S_t} \mathbf{j} \bullet d\mathbf{S}_t = \int_{S_1} \mathbf{j} \bullet d\mathbf{S}_1 + \int_{S_2} \mathbf{j} \bullet (-d\mathbf{S}_2) \tag{10–14}$$

The expression (10–14) will vanish if and only if Eq. (10–13) is satisfied. However, the law of conservation of charge [Eq. (9–13)] demands that

$$\int_{S_t} \mathbf{j} \bullet d\mathbf{S}_t = \frac{-dQ(\text{inside } S_t)}{dt} \tag{10–15}$$

For time-independent currents we demand also time-independent electric fields, and thus also time-independent charges. In consequence, $dQ/dt = 0$ in Eq. (10–15); thus, the expression (10–14) vanishes and Eq. (10–13) is satisfied. In other words, Ampère's law is consistent if there are only steady-state currents; but we can already predict that it will have to be modified to cover the situation of varying currents. We will return to this problem in Chapter 13. ■

Ampère's law [Eq. (10–12)] in itself is not sufficient to determine uniquely the magnetic field of an arbitrary current distribution. We need another law, Maxwell's fourth equation, which states that the integral of **B** over any closed surface vanishes:

$$\int_{\substack{\text{closed} \\ \text{surface}}} \mathbf{B} \cdot d\mathbf{S} = 0 \tag{10-16}$$

Let us once write all four of Maxwell's equations for steady-state electric and magnetic fields together:

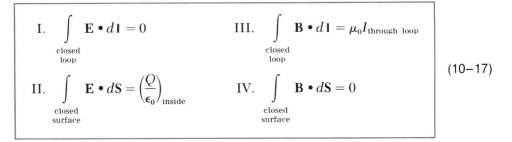

$$
\begin{aligned}
&\text{I.} \quad \int_{\substack{\text{closed} \\ \text{loop}}} \mathbf{E} \cdot d\mathbf{l} = 0
&&\text{III.} \quad \int_{\substack{\text{closed} \\ \text{loop}}} \mathbf{B} \cdot d\mathbf{l} = \mu_0 I_{\text{through loop}} \\[2em]
&\text{II.} \quad \int_{\substack{\text{closed} \\ \text{surface}}} \mathbf{E} \cdot d\mathbf{S} = \left(\frac{Q}{\epsilon_0}\right)_{\text{inside}}
&&\text{IV.} \quad \int_{\substack{\text{closed} \\ \text{surface}}} \mathbf{B} \cdot d\mathbf{S} = 0
\end{aligned}
\tag{10-17}
$$

We see that the equations come in pairs, but that the *source terms* Q/ϵ_0 and $\mu_0 I$ appear in different places for the electric and magnetic fields. The first Maxwell equation derived in Eq. (6-22) expresses the fact that there exists an electrostatic potential; the third is Ampère's law—we conclude that we cannot define a magnetic potential. The second equation is Gauss's law [Eq. (7-5)], stating how the electric field is affected by electric charges. The fourth equation states that there are no magnetic charges.

Actually, the question of whether magnetic charges (also called "magnetic monopoles") exist can be considered still open. Their existence has been proposed to explain certain puzzles in the physics of elementary particles. Many

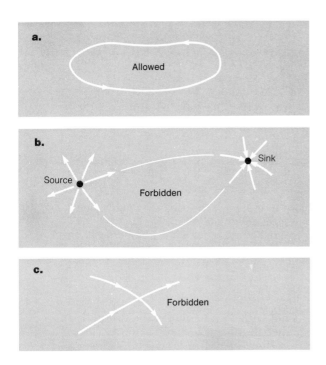

Figure 10-9 Magnetic field lines: allowed and forbidden configurations.

experimenters have tried to detect them. If there were any magnetic monopoles, they would be accelerated in a homogeneous magnetic field in the same way that an electric charge is accelerated in a homogeneous electric field. No experimental search for magnetic monopoles has found them; and that is strong evidence against their existence.

As we will see later on, the second and fourth of Maxwell's equations are complete as we have written them down here. The first and the third equations, both involving integrals over closed loops, will have to be modified for time-dependent electromagnetic fields.

In a region of space in which there are neither electric charges nor currents, the two pairs (I, II) and (III, IV) of Maxwell's equations are identical to each other. This enables us to talk about magnetic flux lines in the same way that we discussed electric flux lines on page 210. However, there is an essential difference between the behavior of magnetic flux lines and that of electric flux lines, as we can see by comparing Fig. 10–9 with Fig. 7–19. *All magnetic flux lines are closed lines.* Thus, the flux line in Fig. 10–9a is allowed – actually, it has the *only* allowed topology. Magnetic flux lines diverging from a point, as in Fig. 10–9b, are forbidden because there are no magnetic charges which could act as sources or sinks to the flux lines. Fig. 10–9c shows a forbidden configuration for either magnetic or electric field lines – they cannot cross in empty space, because then the field at the intersection point would not be defined.

Example 3   Calculate the magnetic field inside and outside a long straight cylindrical wire carrying the current I. The radius of the cylinder is R (Fig. 10–10). Assume that the current density is the same everywhere in the wire.

This problem is a simple generalization of the basic law for the magnetic field of a thin straight wire, Eq. (10–2). We are now dropping the assumption that the wire is thin. Because of the cylindrical symmetry of the problem, the field can have only an azimuthal component B_ϕ. Indeed, if it had a radial component B_r, it would have to be the same everywhere around. But then, choosing a long cylinder as a closed surface one would have

$$\int_{\text{cylinder}} \mathbf{B} \cdot d\mathbf{S} = 2\pi\rho L \, |\mathbf{B}_r| \qquad (10\text{–}18)$$

if r is the cylinder radius and L is its length. Because of the fourth Maxwell's equation (10–17.IV), the left-hand side of Eq. (10–18) has to vanish. This implies

$$\mathbf{B}_r = 0$$

One can obtain the azimuthal component just as easily. Let us choose a circle of radius ρ as a closed loop. Then, if $\rho > R$,

$$\int \mathbf{B}_\phi \cdot d\mathbf{l} = 2\pi\rho \, |\mathbf{B}_\phi| = \mu_0 I \qquad (10\text{–}19)$$

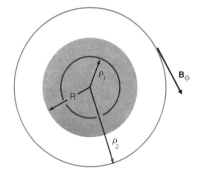

Figure 10–10   Example 3. Calculating the magnetic field of a cylindrical wire.

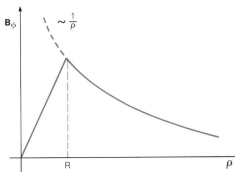

Figure 10–11 Example 3. The magnetic field inside and outside a cylindrical wire.

or

$$B_\phi = \frac{\mu_0 I}{2\pi\rho} \tag{10–20}$$

which is exactly the same as Eq. (10–2). Outside a cylindrical straight wire, the magnetic field depends only on the total current, not on the wire radius. This is analogous to what we found in Chapter 6, where we showed that outside a charged sphere the electric field depends only on the total charge and not on the radius of the sphere.

For a loop of radius $\rho < R$, i.e., inside the wire, Eq. (10–20) is no longer valid. We have to replace the right-hand side $\mu_0 I$ of Eq. (10–19) by the current going *through the circle*. Since the current density is $j = I/\pi R^2$, we have

$$\int \mathbf{B}_\phi \cdot d\mathbf{l} = 2\pi\rho B_\phi = \mu_0 j \pi \rho^2 = \frac{\mu_0 I \rho^2}{R^2}$$

From this we obtain

$$B_\phi = \frac{\mu_0 I}{2\pi} \frac{\rho}{R^2} \tag{10–21}$$

Fig. 10–11 shows the radial dependence of the magnetic field.

A wire is tightly wound on a toroid of major radius R (Fig. 10–12). There are N turns per unit length. Calculate the magnetic field at the center inside the torus, if a current I is flowing through the wire. The minor radius r of the toroid is much smaller than the major radius: $r \ll R$.

Because of the symmetry of the problem, the magnetic field in the torus will be everywhere the same. Thus, if we use as our closed curve a circle of radius R, we have

$$\int_C \mathbf{B} \cdot d\mathbf{l} = 2\pi R \, |\mathbf{B}| \tag{10–22}$$

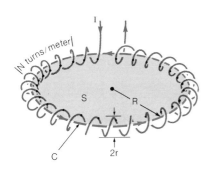

Figure 10–12 Example 4. Magnetic field in a toroid.

A surface through C will cut through each of the wire turns exactly once — on the inside of the toroid. Thus, the total current flowing through the surface will be

$$I_{\text{tot}} = I \times \underbrace{N \times 2\pi R}_{\substack{\text{total number} \\ \text{of turns}}}$$

(10–23)

Because of Ampère's law, the two expressions (10–22) and (10–23) must be proportional to each other. We thus obtain

$$\boxed{|\mathbf{B}| = \mu_0 NI}$$

(10–24)

Note that N is not the total number of turns in the coil, but the number of turns per unit length!

Example 5   A very long straight coil has $N = 30$ turns/cm and carries the current $I = 10$ A. Calculate the magnetic field inside the coil.

We can consider the straight coil as a section of the toroidal coil from the previous example, if the major radius R is very large. The result [Eq. (10–23)] does not depend on the radius; thus, it should also be valid for $R \to \infty$.

We can also obtain the same result by introducing (Fig. 10–13) a rectangular loop. Only the part inside the coil will contribute; the radial segments will cancel (actually there is no radial component of $\mathbf{B}$ because of the fourth Maxwell equation). The return segment can be moved so far away that we are sure there is no magnetic field along it. We thus obtain

$$\int_C \mathbf{B} \cdot d\mathbf{l} = |\mathbf{B}| h$$

where h is the length of the segment of C inside the coil. By Ampère's law, this has to be proportional to the total current through the loop. There are N turns per unit length; thus, a total of Nh turns go through the inside of loop C. Therefore, we have from Ampère's law

$$|\mathbf{B}| h = \mu_0 (Nh) I$$

or

$$|\mathbf{B}| = \mu_0 NI$$

as we obtained in Eq. (10–24). Numerically, we have

$$|\mathbf{B}| = 4\pi \times 10^{-7} \frac{\text{V-sec}}{\text{A-m}} \times \frac{30 \text{ turns}}{\text{cm}} \times \frac{100 \text{ cm}}{\text{m}} \times 10 \text{ A} = 3.77 \times 10^{-2} \frac{\text{Wb}}{\text{m}^2}$$

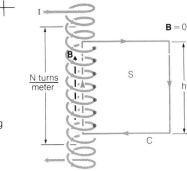

Figure 10–13 Example 5. Calculating magnetic field of a straight long coil.

Note that the closed loop C does not have to go through the axis of the coil; we therefore conclude that inside a long, straight, tightly wound coil the field is everywhere the same, at least as long as we are far away from either end. We have thus found a method to produce a homogeneous magnetic field, albeit of only moderate strength.

5. BIOT-SAVART'S LAW

Ampère's law and "Gauss's law of magnetism," stating that there are no magnetic charges, enable us in principle to determine the magnetic field due to an arbitrary steady-state current distribution. However, actual calculations can be very difficult unless the problem has a high degree of symmetry (as we had in Examples 3 to 5). Fortunately, we can formulate another law, named after two French physicists, Jean Baptiste Biot (1774–1862) and Félix Savart (1791–1841). They started out with the idea that the magnetic field at a point P near a current-carrying wire is a superposition of small fields $d\mathbf{B}$, each due to a small length $d\mathbf{l}$ of the wire. Each element $d\mathbf{l}$ contributes to the total magnetic field the quantity

$$d\mathbf{B} = \frac{\mu_0 I}{4\pi} \frac{d\mathbf{l} \times \mathbf{r}}{r^3} \tag{10–25a}$$

where $\mathbf{r}$ is the vector pointing from the wire element $d\mathbf{l}$ to the point P at which the magnetic field is being calculated (Fig. 10–14). The total magnetic field can then be obtained by integration from Eq. (10–25a):

$$\mathbf{B} = \int d\mathbf{B} = \frac{\mu_0 I}{4\pi} \int_{\substack{\text{wire} \\ \text{path}}} \frac{d\mathbf{l} \times \mathbf{r}}{r^3} \tag{10–25b}$$

We will merely assert here that Eq. (10–25a), called the Biot-Savart law, is equivalent to the two of Maxwell's equations (10–17.III) and (10–17.IV). It has meaning, however, only when we integrate it and obtain Eq. (10–25b). A single element $d\mathbf{l}$ of current cannot exist in a steady-state situation; consequently, only the integral over a closed wire loop can have any physical meaning. We can, of course, in a "thought experiment," continue a wire straight on to infinity;

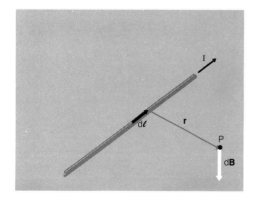

Figure 10–14 Biot-Savart's law.

but all this means is that the return path of the loop is so far away that its detailed shape does not matter. Even the magnetic field in a laboratory electromagnet exists only because of the steady current flowing from the generator
over the transmission line to the magnet and back to the generator. Permanent
magnets — pieces of steel around which there is always a magnetic field — derive
their magnetic field from steady-state currents in their individual atoms.

Example 6   Calculate the magnetic field of a straight, infinitely long wire,
using Biot-Savart's law.

From this calculation we should get back Eq. (10–2); but the exercise is useful
because it helps us to convince ourselves that Biot-Savart's law is indeed equivalent to Maxwell's equations. As we can see from Fig. 10–15, all infinitesimal components $d\mathbf{B}$ at the point P will be parallel and point into the paper, because both
$d\mathbf{l}$ and $\mathbf{r}$ lie in the drawing plane. The magnitude of $d\mathbf{B}$ will be, according to Eq.
(10–24),

$$dB = |d\mathbf{B}| = \frac{\mu_0 I}{4\pi} \frac{|d\mathbf{l}|\, r \sin \alpha}{r^3} \tag{10–26}$$

We can introduce a coordinate x along the wire so that $x = 0$ at Q, the point of the
wire nearest P. We than have $|d\mathbf{l}| = dx$ and

$$\frac{\rho}{r} = \sin(\pi - \alpha) = \sin \alpha \tag{10–27a}$$

$$\frac{x}{r} = \cos(\pi - \alpha) = -\cos \alpha \tag{10–27b}$$

From this we deduce that $x = -\rho \cot \alpha$ and thus

$$|d\mathbf{l}| = dx = -\rho\, d(\cot \alpha) = \frac{\rho\, d\alpha}{\sin^2 \alpha}$$

Substituting this into Eq. (10–26) and using Eq. (10–27a), we obtain

$$dB = \frac{\mu_0 I}{4\pi} \underbrace{\frac{\rho\, d\alpha}{\sin^2 \alpha}}_{dx} \underbrace{\frac{\sin^2 \alpha}{\rho^2}}_{1/r^2} \sin \alpha$$

or

$$dB = \frac{\mu_0 I}{4\pi\rho} \sin \alpha\, d\alpha$$

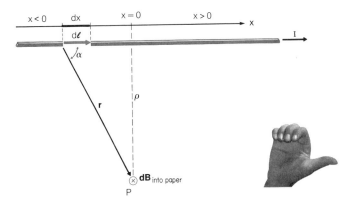

Figure 10–15 Example 6. Calculating
magnetic field of a straight wire using
Biot-Savart's law.

At $x = -\infty$ we have $\alpha = 0$, and at $x = +\infty$ we have $\alpha = 180° = \pi$; thus, to obtain the magnetic field of the infinitely long straight wire we have to calculate

$$|\mathbf{B}| = \frac{\mu_0 I}{4\pi\rho} \int_0^\pi \sin\alpha\, d\alpha = \frac{\mu_0 I}{2\pi\rho}.$$

which is indeed exactly the same result as Eq. (10–2). The direction of $\mathbf{B}$—into the paper in Fig. 10–15—is also the same as we obtained earlier using the right-hand rule.

Calculate the magnetic field of a circular current loop at a point along the axis of the circle (Fig. 10–16).

The differential magnetic field

$$d\mathbf{B} = \frac{\mu_0 I}{4\pi} \frac{d\mathbf{l} \times \mathbf{r}}{r^3}$$

does not point along the axis; however, the components perpendicular to the axis will cancel after we have integrated over the loop. Thus, only the component

$$dB_z = |d\mathbf{B}| \cos\alpha = \frac{\mu_0 I}{4\pi} \frac{d\ell}{(a^2 + z^2)} \cos\alpha$$

will contribute to the total integral. We have used the relation $r^2 = a^2 + z^2$, which can be read off Fig. 10–16. From the same figure we see also that $\cos\alpha = a/\sqrt{a^2 + z^2}$. Furthermore, the contribution of each element $d\ell$ to B_z is equal to all of the others. Thus,

$$|\mathbf{B}| = B_z = \frac{\mu_0 I}{4\pi} \frac{a}{(a^2 + z^2)^{3/2}} \int d\ell$$

and since $\int d\ell = 2\pi a$, we obtain

$$B_z = \frac{\mu_0 I}{2\pi} \frac{\pi a^2}{(a^2 + z^2)^{3/2}} \tag{10–28}$$

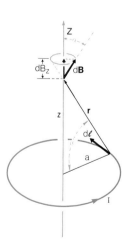

Figure 10–16 Example 7: The magnetic field on axis of a circular current loop.

6. MOTION OF A POINT CHARGE IN MAGNETIC FIELD

We now want to study the effect of the force $\mathbf{F} = q(\mathbf{v} \times \mathbf{B})$ which every charged particle feels in a magnetic field. We will first assume a homogeneous magnetic field and later consider the more general case.

We first note that the force is always perpendicular to the velocity; thus, the scalar product

$$(\mathbf{F} \cdot \mathbf{v}) = 0 \qquad (10\text{-}29)$$

From this we can immediately conclude that the magnetic force [Eq. (10–1)] does no work; in other words, the kinetic energy

$$T = \frac{m\mathbf{v}^2}{2}$$

will always stay the same. Indeed, consider the change of T per unit time

$$\frac{dT}{dt} = \frac{m}{2}\frac{d(\mathbf{v}^2)}{dt} = m\left(\mathbf{v} \cdot \frac{d\mathbf{v}}{dt}\right) \qquad (10\text{-}30)$$

However, by Newton's second law we have

$$m\frac{d\mathbf{v}}{dt} = \mathbf{F} = q(\mathbf{v} \times \mathbf{B}) \qquad (10\text{-}31)$$

and thus from Eq. (10–29) we conclude that indeed $dT/dt = 0$; the kinetic energy does not change. This, of course, does not imply that the velocity $\mathbf{v}$ itself does not change; while $|\mathbf{v}|$ will stay the same, the direction of $\mathbf{v}$ changes with time.

Let us first assume that initially the velocity is perpendicular to the magnetic field. Since from Eq. (10–1) $d\mathbf{v}/dt$ is *always* perpendicular to $\mathbf{B}$, we can then conclude that the whole motion will proceed in a single plane. In Fig. 10–17 we show the motion; the magnetic field is pointing out of the paper. The force has the constant magnitude

$$|\mathbf{F}| = q|\mathbf{v}||\mathbf{B}|$$

and its direction is always perpendicular to the velocity. Such a force results in

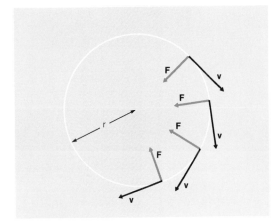

Figure 10–17 A force of constant magnitude, which is always perpendicular to the velocity, leads to circular motion.

uniform circular motion. The acceleration in such a motion points always toward the center of the circle and has the magnitude[*]

$$|\mathbf{a}| = \frac{v^2}{r}$$

where r is the radius of the circle. From Newton's second law we conclude

$$m|\mathbf{a}| = \frac{mv^2}{r} = F = qvB$$

or

$$r = \frac{mv}{qB} \qquad (10\text{--}32)$$

The radius of the circle is proportional to the magnitude of the momentum $\mathbf{p} = m\mathbf{v}$ of the particle. In consequence, if we know the magnetic field and the charge, we can determine the momentum of the particle by measuring the radius of the circular orbit.

An interesting consequence of Eq. (10–32) is that the period of motion T (i.e., the time required for the particle to go once around the circle) is independent of the speed v of the particle (Fig. 10–18). Indeed, the speed is

$$v = \frac{\text{distance}}{\text{time}} = \frac{2\pi r}{T}$$

or, according to Eq. (10–32),
$$T = \frac{2\pi r}{v} = \frac{2\pi m}{qB} \qquad (10\text{--}33\text{a})$$

The period of circular motion in a magnetic field T is called the *Larmor period;* the corresponding angular frequency

$$\omega = \frac{2\pi}{T} = \frac{qB}{m} \qquad (10\text{--}33\text{b})$$

is called the *Larmor frequency.*

---

[*] Compare Volume I, p. 64.

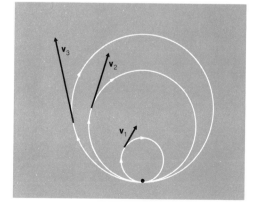

Figure 10–18 Charged particles leaving a point in a magnetic field with varying velocities all return after the same time interval.

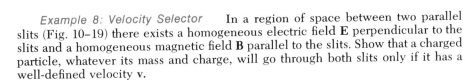

Figure 10-19 Example 8: Velocity selector.

Example 8: Velocity Selector   In a region of space between two parallel slits (Fig. 10–19) there exists a homogeneous electric field **E** perpendicular to the slits and a homogeneous magnetic field **B** parallel to the slits. Show that a charged particle, whatever its mass and charge, will go through both slits only if it has a well-defined velocity **v**.

The particle will go through both slits only if there is no net force on it so that it continues on a straight line. Let us assume $q > 0$. Then there is (Fig. 10–19) a force acting upward due to the electric field

$$\mathbf{F}_1 = q\mathbf{E}$$

and a magnetic force acting downward

$$\mathbf{F}_2 = q(\mathbf{v} \times \mathbf{B})$$

The particle will continue on a straight line if $|\mathbf{F}_1| = |\mathbf{F}_2|$ or if

$$q|\mathbf{E}| = q|\mathbf{v}|\,|\mathbf{B}|$$

or if

$$v = \frac{E}{B} \tag{10–34}$$

Note that as long as the particle is charged, the selected velocity is independent of both the mass and the charge of the particle.

A stream of particles of varying masses, charges, and velocities enters a mass spectrograph. This instrument consists of a velocity selector as in the previous example, which is followed by a region in which there is only the magnetic field **B** (assumed out of the paper in Fig. 10–20). A detector D is situated a distance d from the second slit. Which of the particles will hit the detector?

The velocity selector will pass only particles with a velocity given by Eq.

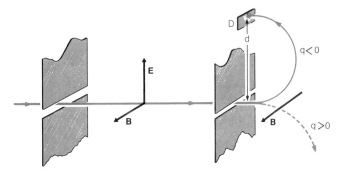

Figure 10-20 Example 9: Mass spectrograph.

(10–34); thus, all particles leaving the second slit will have a momentum

$$p = mv = \frac{mE}{B}$$

If their charge is positive, they will be bent downward in Fig. 10–20, where the magnetic field **B** points out of the paper. They will hit the detector if their radius of curvature as given by Eq. (10–32) is equal to one-half the distance d:

$$d = 2r = \frac{2p}{qB}$$

Thus, only those particles will hit the detector whose velocity is E/B and whose charge/mass ratio is

$$\frac{q}{m} = \frac{2E}{B^2 d} \qquad (10\text{–}35)$$

In practice, an apparatus such as that shown in Fig. 10–20 is used to measure the masses of charged ions or of atomic nuclei. Ions can be produced by an electric discharge in a gas, strong heating, or a beam of electrons which ionize the atoms by colliding with them. The produced ions are accelerated by a longitudinal electric field and are then passed through the apparatus. By varying **E** and **B**, one can measure the ratio of charge to mass. The charge has to be an integer multiple of 1 $e = 1.602 \times 10^{-19}$ coul. Usually one chooses singly charged ions: for instance, Hg$^+$ ions will be produced copiously in a mercury vapor lamp. If one passes them through a mass spectrograph, one finds that the ions do not all have the same mass. There are several *isotopes* of mercury which all have identical chemical properties. The atomic nuclei of different isotopes have the same charge, but not the same mass or molecular weight. Indeed, mercury has seven different stable isotopes.

Let us now relax the condition that the charged particles start out their motion perpendicularly to the magnetic field. We will thus assume that a particle of charge q and mass m starts out at $t = 0$ with an arbitrary velocity $\mathbf{v}_0$; we ask what its motion will be. We still assume for the moment that the magnetic field **B** is homogeneous, i.e., everywhere the same.

We can decompose **v** into a component $\mathbf{v}_\parallel$ parallel to the magnetic field and a component $\mathbf{v}_\perp$ perpendicular to **B** (Fig. 10–22):

$$\mathbf{v}_0 = \mathbf{v}_\parallel + \mathbf{v}_\perp \qquad (10\text{–}36)$$

The force on the particle then is

$$\mathbf{F} = q(\mathbf{v}_\parallel \times \mathbf{B}) + q(\mathbf{v}_\perp \times \mathbf{B}) = q(\mathbf{v}_\perp \times \mathbf{B}) \qquad (10\text{–}37)$$

because the vector product of two parallel quantities vanishes. The force itself is also perpendicular to the magnetic field. We thus have for the accelerations parallel and perpendicular to the magnetic field

$$m\frac{d\mathbf{v}_\parallel}{dt} = 0 \qquad (10\text{–}38a)$$

$$m\frac{d\mathbf{v}_\perp}{dt} = q(\mathbf{v}_\perp \times \mathbf{B}) \qquad (10\text{–}38b)$$

Comparing this to Eq. (10–31), we conclude that the projection $\mathbf{v}_\perp$ of the velocity into a plane perpendicular to **B** obeys the same law. On the other hand, the

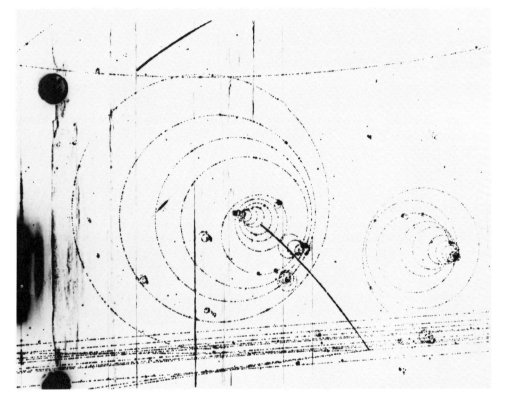

Figure 10–21 If a charged particle moves through a liquid which is at (or barely above) its boiling temperature, bubbles of gas form along the track of the particle. This is used in bubble chambers to investigate reactions between elementary particles. The tracks shown here from a liquid hydrogen bubble chamber are curved because the chamber is inserted into a large magnetic field. From the curvature one can determine the momentum of each particle.

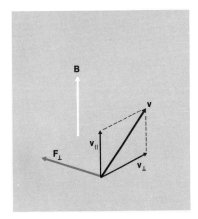

Figure 10–22 If a charged particle starts out with an arbitrary velocity, we decompose the velocity into a component parallel and a component perpendicular to the magnetic field.

velocity parallel to **B** stays constant:

$$m\mathbf{v}_\| = \text{const} = m\mathbf{v}_{0\|} \qquad (10\text{–}39)$$

We thus have the superposition of the motion along a circle and a linear motion perpendicular to the plane of the circle: The particle will move along a spiral as shown in Fig. 10–23.

Let us now consider the completely general case in which the magnetic field is arbitrary. As an example, let us consider the magnetic field of a straight wire. As long as the velocity of the charged particle is small, it will move in a tight circle around the magnetic field lines. Locally — over the size of the orbit — the field is nearly homogeneous; thus, the motion is nearly the same as that in a homogeneous field. However, the particle will follow the magnetic field line as it slowly changes its direction; thus, the particle moves on a spiral orbit around the wire, as shown in Fig. 10–24. We have trapped a charged particle; it will never leave the closed magnetic field line.

The idea that a charged particle will follow magnetic field lines has led in the recent past to various attempts to create a magnetic confinement of very hot gases. Such gases, called plasmas, consist of free electrons and positive ions, (that is, only of charged particles); in principle, one should be able to keep them confined by suitably shaped magnetic fields. It is, of course, impossible to keep such gases confined in conventional containers; their temperature is so high (up to 10^6 °K) that any container wall would also evaporate. This has led to the invention of a *magnetic bottle*, in which the magnetic field is shaped as shown in Fig. 10–25. The magnetic field is higher at the ends — the magnetic field lines come together. But there is then also a radial component B_r of the magnetic

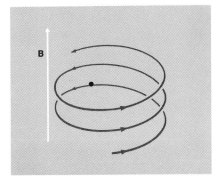

Figure 10–23 In a homogeneous magnetic field the most general path a charged particle can follow is a spiral.

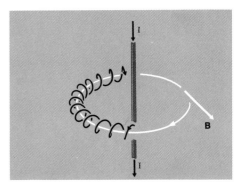

Figure 10-24 Particles can be trapped by magnetic field lines.

field; as the particle spirals toward the end of the bottle, it feels, in addition to the force

$$\mathbf{F}_1 = q(\mathbf{v}_\perp \times \mathbf{B}_z) \qquad (10\text{-}40a)$$

which forces it to move along a circle, also a force

$$\mathbf{F}_2 = q(\mathbf{v} \times \mathbf{B}_r) \qquad (10\text{-}40b)$$

which slows down the advance of the spiral. The particle will, after a while, reverse the direction of the spiral and move away from the "end of the bottle."

Unfortunately, this simple scheme does not work. If one tries to keep many charged particles (i.e., a plasma) inside the bottle, the many moving particles can collide; after the collision they will move on a new spiral offset from the center. Sooner or later any particle will leave the bottle.

These collision losses would be tolerable if one could make the field sufficiently strong and the "bottle" large. However, the spiraling particles themselves constitute currents and thus produce their own magnetic field, which changes the field of the bottle. More sophisticated schemes have been developed in recent years because such hot plasmas are essential to replicate the nuclear fusion processes occurring inside the sun; however, it has not yet been possible to produce hot plasmas of sufficient density for a long enough time to make this feasible.

If the plasma has a low density, even rather weak magnetic fields are sufficient to produce the bottle effect. The earth has a magnetic field strong enough to trap particles which are ejected by the sun as solar wind. The particles congregate mainly around the magnetic equator. These charged particles are mostly protons (hydrogen nuclei) and electrons; the protons congregate closer to the earth, while the much lighter electrons have a maximum intensity at 3 to 4 earth

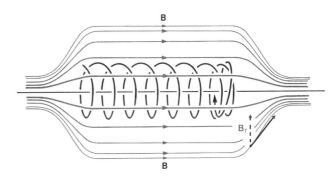

Figure 10-25 A magnetic bottle.

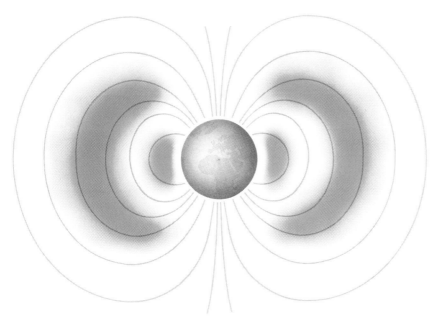

Figure 10–26 Charged particles are trapped by the earth's magnetic field, forming the Van Allen belts.

radii (Fig. 10–26). These "Van Allen belts"[*] are really a terrestrial application of the magnetic bottle principle. The charged particles are spiraling around the magnetic field lines; the progress of the spiral is stopped by the larger magnetic fields near the magnetic poles.

One should not imagine the Van Allen belts as a gas similar to the atmosphere; the particle densities are 10^4 to 10^5 particles/cm³, which is very much less than the density of air (2.5×10^{19} molecules/cm³). Thus, the "plasma" corresponds in particle density to the best vacuum achievable in the laboratory. In spite of the low density, the particles do collide occasionally and there is a constant loss of particles from the belts. On the other hand, they are constantly replenished by the solar wind particles ejected by the sun.

7. MAGNETIC FORCES ON CURRENTS

Let us now study forces as they occur in a wire which carries a current in a magnetic field. Each electron will be acted upon by a force

$$\mathbf{F}_e = (-e)\mathbf{v} \times \mathbf{B} \tag{10–41}$$

If there are n electrons/m³ participating in the current, the current density in the wire is

$$\mathbf{j} = (-e)n\mathbf{v} \tag{10–42}$$

The total force per unit volume will be n times the force on a single electron:

$$\frac{\text{force}}{\text{volume}} = \mathbf{F}_v = (-e)n(\mathbf{v} \times \mathbf{B})$$

---

[*] Predicted by James Van Allen (1914–   ), and confirmed by the first U.S. satellite, Explorer I, in 1958.

or

$$\boxed{\mathbf{F}_v = (\mathbf{j} \times \mathbf{B})} \qquad (10\text{-}43)$$

Would the force $\mathbf{F}_v$ per unit volume be any different if the current had been produced by positive charges instead of the negative electrons? The answer is no, as one can easily see by inspection of Eqs. (10–41) and (10–42). Both the force on each charge and the current density are proportional to the product of the charge and of the velocity $\mathbf{v}$; if we change the sign of both, then neither the current density $\mathbf{j}$ nor the force per unit volume $\mathbf{F}_v$ will change. Thus, currents produced by positive or negative charge carriers produce the same force $\mathbf{F}_v$ per unit volume, as long as the current density is the same for both cases.

If we have a wire carrying a total current I (Fig. 10–27), we can recast Eq. (10–43) into a more convenient form. Assume that the wire has a cross section A. Then the total current in the wire is

$$I = |\mathbf{j}|A \qquad (10\text{-}44)$$

We will now calculate the force $d\mathbf{F}$ on a very short length $d\ell$ of the wire. Let us define the vector $d\mathbf{l}$ as having the magnitude $d\ell$ and the direction of the current density $\mathbf{j}$; thus, $d\mathbf{l}$ is parallel to the wire and points in the direction the current is flowing. Since the short segment of length $d\ell$ of the wire has the volume $d\tau = A\,d\ell$, the force on it will be

$$d\mathbf{F} = \mathbf{F}_v d\tau = A\,d\ell\,(\mathbf{j} \times \mathbf{B}) \qquad (10\text{-}45a)$$

Because of the definition of the vector $d\mathbf{l}$, we have

$$\mathbf{j}\,d\ell = |\mathbf{j}|\,d\mathbf{l}$$

and thus, using Eq. (10–44), we can recast the expression for the force $d\mathbf{F}$ into the form

$$\boxed{d\mathbf{F} = I(d\mathbf{l} \times \mathbf{B})} \qquad (10\text{-}45b)$$

If we want to calculate the total force on a wire of finite length, we have to integrate Eq. (10–45b) over the path of the wire:

$$\mathbf{F} = \int_{\text{wire}} d\mathbf{F} = I \int_{\text{wire}} (d\mathbf{l} \times \mathbf{B}) \qquad (10\text{-}45c)$$

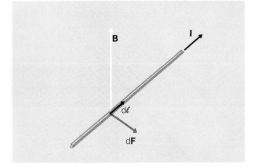

Figure 10–27 Force on a wire element in a magnetic field.

Example 10   Calculate the force between two parallel wires, each carrying the current $I = 10$ amps, if they are a distance $d = 1$ cm apart (Fig. 10–28). The wires are $L = 1$ m long.

This is a simple application of Eq. (10–45). Wire 1 will produce a magnetic field

$$|\mathbf{B}| = \frac{\mu_0 I}{2\pi d} \qquad (10\text{–}46)$$

at the position of wire 2. The magnitude of the force $\mathbf{F}$ acting on wire 2 will then be

$$|\mathbf{F}| = I \left| \int d\mathbf{l} \times \mathbf{B} \right| = IL|\mathbf{B}|$$

since the magnetic field is perpendicular to wire 2. Substituting from Eq. (10–46), we obtain

$$|\mathbf{F}| = \frac{\mu_0 I^2 L}{2\pi d}$$

As shown in Fig. 10–28, the force

$$d\mathbf{F} = I(d\mathbf{l} \times \mathbf{B})$$

will be attractive if the currents are in the same direction. If the two currents go in opposite directions, the force is repulsive.

Of course, the magnetic field of current of wire 2 will also produce a force acting on wire 1. It will be exactly opposite and equal to the force acting on wire 2, as demanded by Newton's third law. Numerically, the magnitude of the force in our example is

$$|\mathbf{F}| = \frac{4\pi \times 10^{-7}\ \text{V-sec/A-m}}{2\pi \times (0.01\ \text{m})} \times (10\ \text{A})^2 \times 1\ \text{m} = 2 \times 10^{-3}\ \text{N}$$

This is a small force—there is no danger that a household extension cord will fly apart because of magnetic forces. However, when many wires are in parallel, e.g., in coils of electromagnets, the forces can become substantial.

The magnetic force on a current in a wire is the net sum of the forces on all individual moving electrons. But we have not yet answered the question of why the electrons continue to move along the wire, although a transverse force $\mathbf{F} = (-e)(\mathbf{v} \times \mathbf{B})$ is acting on them in a magnetic field. The answer is that a transverse *electric* field establishes itself and cancels out the magnetic force. This is known as the *Hall effect*.

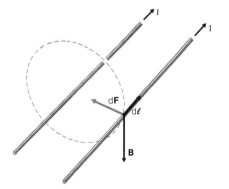

Figure 10–28 Example 10: Two parallel wires.

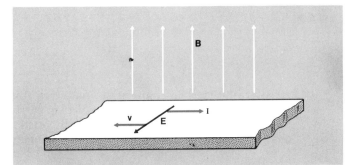

Figure 10-29 Hall effect.

To calculate the effect, let us consider a metallic wire of rectangular cross section, as shown in Fig. 10–29. A current I is flowing along the wire; a magnetic field **B** exists which points parallel to the shorter side d of the rectangle. In equilibrium the current has to flow in the direction of the wire; there can be no perpendicular current. However, there is a perpendicular force on the electrons because of the magnetic field. The force $\mathbf{F}_1 = (-e)(\mathbf{v} \times \mathbf{B})$ is trying to force the electrons to the side. This will lead to a small excess of electrons on one side of the wire; a negative surface charge will build up, while a positive surface charge will exist on the other side. The net effect is to produce a transverse electric field whose force on the electrons is equal to $\mathbf{F}_2 = -\mathbf{F}_1$; then the electrons experience no net sidewise force. Thus, in equilibrium we have, since $\mathbf{F}_2 = -e\mathbf{E}$,

$$\mathbf{E} + (\mathbf{v} \times \mathbf{B}) = 0 \tag{10-47}$$

The electrons are negatively charged; thus, their velocity is opposite to the direction of the current. The electric field will lead to a potential difference

$$V = |\mathbf{E}|a = |\mathbf{v}|\,|\mathbf{B}|a$$

between the two opposite sides of the wire. Since the current density is

$$\mathbf{j} = (-e)n_e\mathbf{v} \tag{10-48}$$

where n_e is the number of electrons per unit volume, we can write for the potential difference

$$V = \frac{|\mathbf{j}|Ba}{(e)n_e}$$

or, since the total current is $I = |\mathbf{j}|ad$, we have

$$V = \frac{1}{(e)n_e}\frac{IB}{d} \tag{10-49}$$

Note that sign of the potential difference depends on the sign of the charges responsible for the current. This is not clearly visible from Eq. (10–49), where the sign of the potential is defined arbitrarily. However, substituting Eq. (10–48) directly into Eq. (10–47), we see that

$$\mathbf{E} = -\frac{(\mathbf{j} \times \mathbf{B})}{(-e)n_e} \tag{10-50}$$

The quantity

$$C_{\text{Hall}} = \frac{1}{(-e)n_e}$$ (10–51)

is called the Hall constant of the material; it is negative if the current is carried by negative charges and positive if it is carried by positive charges. Note that by experimentally measuring the Hall constant one can directly determine the charge carrier density n_e, if all the moving charge carriers have the same polarity.

8.   MAGNETIC DIPOLES

We said early in this chapter (Section 4) that there exist no magnetic charges. However, there does exist the magnetic analogue to the electric dipole. It is an analogue in the sense that the magnetic field of a magnetic dipole—*far away from the dipole*—is the same as the electric field of an electric dipole. The analogy extends even further; just as there is a torque on an electric dipole in an electric field, there is also a torque on a magnetic dipole in a magnetic field. Thus, the magnetic dipole behaves in any respect very similarly to the electric dipole. This will become very important when we discuss magnetic properties of materials from the atomic point of view. To an excellent approximation we are able to treat the atom as a magnetic dipole, just as we approximated the charge distribution of a neutral molecule by a dipole of equal dipole moment. The electrons circulating in an atom constitute a current and thus produce a magnetic field; for many purposes this magnetic effect can be calculated in the dipole approximation.

The simplest magnetic dipole is a flat current loop, such as a rectangular loop through which a current I is flowing (Fig. 10–30). We define the magnetic moment as a vector whose magnitude is the product of the area of the loop and the current flowing around it

$$|\mathbf{m}| = AI$$ (10–52)

and whose direction follows the right-hand rule: if the four fingers of the right hand encircle the loop in the direction in which the current I is flowing, then the thumb points in the direction of $\mathbf{m}$ (Fig. 10–30).

We now have to prove the assertion that such a current loop behaves like a dipole. We will prove this only in rather special cases; the general proof requires a rather detailed study of line integrals and is beyond our scope. Let us first

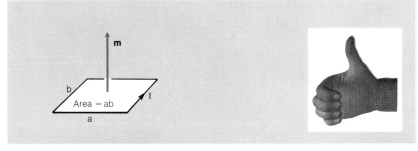

Figure 10–30   Magnetic dipole moment.

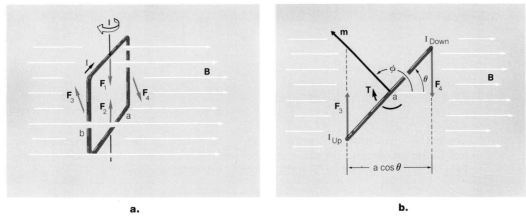

Figure 10–31   a) Forces acting on segments; b) calculating the torque.

assume that we have a rectangular current loop in a magnetic field **B**. For simplicity, we will orient the loop as shown in Fig. 10–31. There will be a force on each side of the loop, whose magnitude can be obtained from Eq. (10–45). The simplicity of our particular arrangement lies in the fact that two sides are perpendicular to the magnetic field, while the forces on the other two sides cancel exactly and produce no torque. Indeed, consider first the top and the bottom sides; the forces on them, $\mathbf{F}_1$ and $\mathbf{F}_2$, are opposite and of equal magnitude:

$$\mathbf{F}_1 = -\mathbf{F}_2$$

Furthermore, they act on the loop along the line going through the center of the loop and point along that line; thus, they also produce no torque.

The forces on the two vertical sides of the current rectangle are also opposite and of equal magnitude

$$\mathbf{F}_3 = -\mathbf{F}_4$$

However, since they do not attack on the same line, they will give rise to a torque. As we can see from Fig. 10–31b, the torque has the magnitude

$$|\mathbf{T}| = |\mathbf{F}_3|d = |\mathbf{F}_3|a \cos \theta$$

The two sides under discussion are perpendicular to the magnetic field; therefore, the force $\mathbf{F}_3$ has the magnitude

$$|\mathbf{F}_3| = IbB$$

and thus we arrive at a value of the torque:

$$|\mathbf{T}| = IabB \cos \theta \qquad\qquad\qquad \text{(10–53)}$$

The dipole moment of the loop has, according to Eq. (10–52), the magnitude

$$|\mathbf{m}| = Iab$$

and the direction shown in Fig. 10–31. Since the angle between **m** and **B** is $\phi = \pi/2 + \theta$, we can write for the torque (using the identity $\cos \theta = \cos (\phi - \pi/2) = \sin \phi$)

$$\boxed{\mathbf{T} = \mathbf{m} \times \mathbf{B}} \qquad\qquad (10\text{--}54)$$

This is the exact analogue to Eq. (6–49) for the torque of an electric dipole in an electric field:

$$\mathbf{T} = \mathbf{p} \times \mathbf{E} \qquad\qquad (10\text{--}55)$$

Since the torque on a magnetic dipole is given by the same equation as the torque on an electric dipole, we conclude that in analogy to Eq. (6–51) the magnetic dipole in a magnetic field has the potential energy

$$\boxed{U = -(\mathbf{m} \cdot \mathbf{B})} \qquad\qquad (10\text{--}56)$$

Example 11   A circular wire loop (radius r, current I) is positioned in a homogeneous magnetic field **B**. The field direction is in the plane of the loop (Fig. 10–32a). Calculate the torque on the loop.

We could use Eq. (10–54) directly. The magnetic moment has the magnitude

$$|\mathbf{m}| = \pi r^2 I$$

and is perpendicular to the magnetic field. Thus, the torque has the magnitude

$$|\mathbf{T}| = \pi r^2 I B \qquad\qquad (10\text{--}57)$$

and rotates the loop in the direction shown in Fig. 10–32a.

We can, however, also easily calculate the torque directly, using Eq. (10–45b). Consider Fig. 10–32b. The small element $d\mathbf{l}$ is acted upon by a force

$$d\mathbf{F} = I(d\mathbf{l} \times \mathbf{B})$$

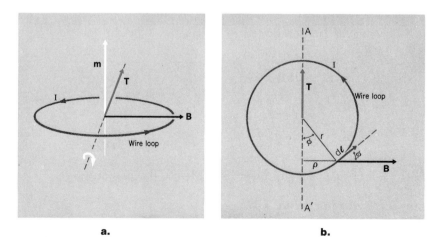

a.   **b.**

Figure 10–32 Torque on a circular loop in magnetic field: a) using Eq. (10–54); b) adding the torques due to wire elements d**l**.

which points into the paper and has the magnitude

$$dF = I \, d\ell \, B \sin \phi$$

It contributes to the torque around the axis AA' the amount

$$dT = \rho \, dF = r \sin \phi \, dF$$

Since $d\ell = r \, d\phi$, we thus have

$$dT = IBr^2 \sin^2 \phi \, d\phi$$

If we integrate from $\phi = 0$ to $\phi = 2\pi$, we obtain the total torque T. Since

$$\int_0^{2\pi} \sin^2 \phi \, d\phi = \int_0^{2\pi} \cos^2 \phi \, d\phi = \pi$$

we have

$$T = IB\pi r^2$$

which is indeed identical to the result, Eq. (10–57), which we obtained from Eq. (10–54). Thus, we have shown from first principles that Eq. (10–54) is also valid for a circular loop.

Let us now turn our attention to the magnetic field produced by a magnetic dipole. We have already calculated a special case in Example 7 of this chapter, where we determined the magnetic field of a circular loop on the axis (Fig. 10–16). We obtained the result [Eq. (10–28)]

$$|\mathbf{B}| = \frac{\mu_0 I}{2\pi} \frac{\pi a^2}{(a^2 + z^2)^{3/2}}$$

If we are far away ($z >> a$), we can simplify this expression by neglecting a^2 in the denominator as being small compared to z^2. Since the magnetic moment of the loop has the magnitude $|\mathbf{m}| = \pi a^2 I$, we can write for the field on the axis of the dipole (Fig. 10–33)

$$\mathbf{B} = \frac{\mu_0}{2\pi} \frac{\mathbf{m}}{r^3} \qquad (10\text{–}58)$$

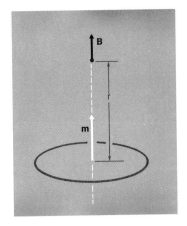

Figure 10–33 The on-axis field of a circular magnetic dipole.

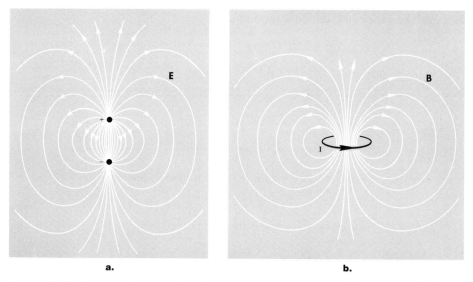

Figure 10–34   Far away, the electric field of an electric dipole (a) and the magnetic field of a magnetic dipole (b) are the same.

if we call the distance r instead of z. Comparing this with Eq. (6–38), we see that we have the identical expression if we replace $1/\epsilon_0$ by μ_0. We will merely assert here without an exact proof that *far away* the analogy is exact: The magnetic far field of a magnetic dipole has the form

$$\mathbf{B} = \frac{\mu_0}{4\pi}\left(3\frac{(\mathbf{m}\cdot\mathbf{r})}{r^5}\mathbf{r} - \frac{1}{r^3}\mathbf{m}\right) \tag{10–59}$$

in exact analogy to Eq. (6–46). We stress that this is true only far away from the dipole; the *near fields* of the electric and the magnetic dipole are quite different. As sketched in Fig. 10–34, "inside" the electric dipole the electric field of the electric dipole points opposite to $\mathbf{p}$—from the positive charge to the negative

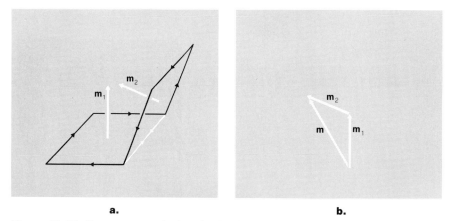

Figure 10–35   If we want to calculate the dipole moment of an arbitrary current loop, we decompose it into individual plane loops.

charge. However, the field of a magnetic dipole points in the same direction as m; the magnetic field lines circle through the dipole.

We have yet to answer the question of how to calculate the dipole moment of an arbitrary current loop. For a loop lying completely in a plane we have the solution; the magnetic moment has the magnitude of the circling current times the enclosed area. More complicated loops can be calculated by decomposing them into "flat" loops. Thus, if we want to calculate the magnetic moment of the loop in Fig. 10–35, we decompose it into two flat loops; the imagined "short" (white in Fig. 10–35) does not contribute to any magnetic effect because it has no net current flowing through it. After calculating the dipole moment of each flat section, one simply adds the two moments vectorially.

Example 12   According to the Bohr model, the hydrogen atom consists of an electron circling a proton at a distance $a_0 = 0.53$ Å. Calculate: (i) the velocity of the electron, (ii) its period of motion, (iii) the average current, and (iv) the magnetic dipole moment due to the electron motion.

The electron will move around the proton under the influence of the attractive electric force of the proton (Fig. 10–36); this force has to be equal to the centripetal acceleration of a uniform motion on a circle:

$$|\mathbf{F}| = \frac{1}{4\pi\epsilon_0} \frac{e^2}{a_0^2} = \frac{mv^2}{a_0}$$

where $+e$ is the charge of the proton (and $-e$ is the charge of the electron). From this we obtain the velocity

$$v = \frac{e}{\sqrt{4\pi\epsilon_0 m a_0}}$$

Numerically, we obtain $v = 2.19 \times 10^6$ m/sec, which is ~0.7% of the velocity of light. The period T is the time required for the electron to traverse the distance $2\pi a_0$. Thus, we have

$$T = \frac{2\pi a_0}{v} = \frac{2\pi a_0^{3/2}}{e} \sqrt{4\pi\epsilon_0 m} = 1.52 \times 10^{-16} \text{ sec}$$

The average current is the charge flowing through a point on the orbit per second; since the charge e traverses it once during each period T, we have

$$I = \frac{e}{T} = \frac{e^2}{2\pi a_0^{3/2} \sqrt{4\pi\epsilon_0 m}} = 1.05 \times 10^{-3} \text{ A}$$

The magnetic dipole moment is the product of the circle's area times the current:

$$|\mathbf{m}| = \pi a_0^2 I = \frac{e^2}{2} \sqrt{\frac{a_0}{4\pi\epsilon_0 m}} = 0.93 \times 10^{-23} \text{ A-m}^2$$

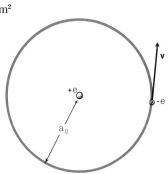

Figure 10–36   Example 12: Classical picture of hydrogen atom.

■   We need to calculate, whenever we encounter the atomic scale, the magnitude of every quantity. We need this to form an idea of what is "big" or "small" on the atomic scale. In normal life we have enough experience to know that 1 meter (the distance from your belt to the floor, if you are of average height) is a reasonable length. We have a feeling for reasonable magnitudes even for electric quantities—we all have learned that household fuses are rated at 15 A and that the household current has a voltage of 110 V; thus, we "know" that, for example, 10^{-4} V is pretty small. On the atomic scale, we lack this intuition. But we have just learned that 10^{-16} sec is a "reasonable" time, just as 1 mA is a "reasonable" current on the atomic scale. The only quantity we had to assume is the hydrogen atom radius—but we could estimate even that from macroscopic quantities such as the density of liquid hydrogen. The masses of the proton and of the electron give us a reasonable scale of mass. We should not be scared by the many powers of ten that we encounter in numerical calculations. They are there because the units of the MKSC system that we are using are ridiculously large on the atomic scale; we are like a crazy astronomer who tells his tailor that he has a waist of 1.21×10^{-16} light years (this is 45 inches—our astronomer is rather stout).

Thus, we have learned that a magnetic moment of 10^{-23} A-m² is "reasonable" on the atomic scale. You should perform this exercise—calculating the magnitude of everything in sight—whenever you are dealing with atoms or other very small (or very large) systems. You will then soon get a feeling of what is "right" and what is "crazy" and obviously wrong.   ■

In the last example we calculated the magnetic dipole moment from the given size of the orbit. We had to assume that the orbit was circular; besides, our calculation was valid only for an orbit under the influence of an electric attractive force. We can do much better than this, because we can show that there is a connection between the orbital angular momentum and the magnetic moment of an electron. In Fig. 10–37 we have sketched the orbit under the influence of an arbitrary *central* force. Since the force is central, the orbital angular momentum is conserved:

$$m(\mathbf{r} \times \mathbf{v}) = \mathbf{L} = \text{constant} \tag{10-60}$$

where m is the mass of the particle executing the orbit, i.e., the electron.

Now consider, on the other hand, the area swept by the electron in the time dt. It is equal to the shaded triangle in Fig. 10–37:

$$dA = \frac{1}{2} r^2 \, d\phi$$

But we also have

$$r \, d\phi = |\mathbf{v}| \, dt \cos \psi = |\mathbf{v}| \, dt \sin \alpha$$

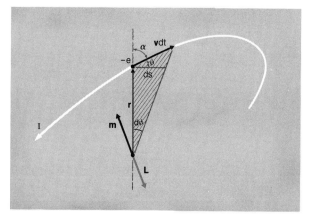

Figure 10–37 Calculating magnetic dipole moment for any orbit of an electron.

because $\alpha = (\pi/2) - \psi$. Thus, we can write

$$dA = \frac{1}{2}|\mathbf{r}|\,|\mathbf{v}|\,\sin\alpha\,dt \qquad (10\text{–}61)$$

Since $|\mathbf{r} \times \mathbf{v}| = |\mathbf{r}|\,|\mathbf{v}|\,\sin\alpha$, we can write, because of Eq. (10–60),

$$dA = \frac{|\mathbf{L}|}{2m}\,dt \qquad (10\text{–}62)$$

The angular momentum is a constant of motion. Therefore, Eq. (10–62) implies that the same area element is swept per unit time anywhere on the orbit. But then the whole area of the orbit must equal

$$A = \int dA = \frac{|\mathbf{L}|}{2m}\,T \qquad (10\text{–}63)$$

where T is the period of the orbit. On the other hand, repeating the argument of Example 12, we have for the average current due to the electron

$$I = \frac{e}{T}$$

and we therefore obtain a very simple relation for the magnitude of the magnetic dipole moment:

$$|\mathbf{m}| = AI = \frac{e}{2m}|\mathbf{L}| \qquad (10\text{–}64)$$

If we consider Fig. 10–37, we can also relate the direction of $\mathbf{m}$ to the direction of $\mathbf{L}$. We note that because the electron has a negative charge, the current I flows in the opposite direction from the velocity $\mathbf{v}$. Using the right-hand rule, we then obtain

$$\boxed{\mathbf{m} = -\frac{e}{2m}\mathbf{L}} \qquad (10\text{–}65)$$

We have obtained a very important result in Eq. (10–65). Whenever the electrons in an atom have a net angular momentum, the atom forms a magnetic dipole. The magnetization of a permanent magnet – as well as the magnetization induced by external currents, such as in an electromagnet – are the effects of many atomic magnetic dipoles aligned in parallel.

As we will study in more detail later on, quantum mechanics teaches us that the component of orbital angular momentum $\mathbf{L}$ along any direction is *quantized;* i.e., it can take only the values

$$\boxed{L_z = n\hbar}$$

where n is an arbitrary integer $n = 0, \pm 1, \pm 2, 3, \ldots$ and

$$\hbar = \frac{h}{2\pi} = 1.05 \times 10^{-34}\,\text{Joule-sec}$$

is called Dirac's constant (the symbol is pronounced "h-bar"). Thus, the smallest magnetic dipole moment an atom can have owing to the orbital motions of its electrons is

$$\mu_B = \frac{e}{2m}\hbar = 0.929 \times 10^{-23} \quad \text{A-m}^2 = 0.929 \times 10^{-23} \quad \frac{\text{J-m}^2}{\text{Wb}}$$

which is about the same as we obtained in Example 12. The quantity μ_B is called the *Bohr magneton*.

However, the orbital angular momentum is not the only one appearing in atomic physics. The electron itself has an intrinsic angular momentum of magnitude

$$|\mathbf{S}| = \frac{1}{2}\hbar$$

This angular momentum is called the *spin*, because the picture of the electron that one gets is one of a spinning top. It is usually denoted by **S** to distinguish it from the orbital angular momentum **L**. Note that the value of **S** is one-half of the minimum non-vanishing orbital angular momentum, or

$$|\mathbf{S}| = \frac{n}{2}\hbar \quad \text{with} \quad n = 1$$

With this angular momentum there is also associated a magnetic moment

$$\mathbf{m}_e = -\frac{e}{m}\mathbf{S} \tag{10-66a}$$

which has the magnitude

$$|\mathbf{m}_e| = 1 \ \mu_B = \frac{e\hbar}{2m} \tag{10-66b}$$

Note that the factor 1/2 is missing in Eq. (10–66a); the ratio of magnetic moment and angular momentum, called the gyromagnetic ratio, is twice as large for the electron spin moment as it is for its orbital motion.

Summary of Important Relations

Ampère's law $\qquad \displaystyle\int_{\substack{\text{closed} \\ \text{loop}}} \mathbf{B} \cdot d\mathbf{l} = \mu_0 I_{\text{through loop}}$

Maxwell's fourth equation $\qquad \displaystyle\int_{\substack{\text{closed} \\ \text{loop}}} \mathbf{B} \cdot d\mathbf{S} = 0$

Magnetic field near straight wire with current I $\qquad B = \dfrac{\mu_0 I}{2\pi r}$

Magnetic field in coil of N turns per meter, current I $\qquad B = \mu_0 N I$

Biot-Savart law $\qquad dB = \dfrac{\mu_0 I}{4\pi} \dfrac{d\mathbf{l} \times \mathbf{r}}{r^3}$

Force on charge q moving with velocity $\mathbf{v}$ $\qquad \mathbf{F} = q(\mathbf{v} \times \mathbf{B})$

Larmor period of mass m with charge q in field $\mathbf{B}$ $\qquad T = \dfrac{2m}{qB}$

Force on wire element $d\mathbf{l}$ carrying current I $\qquad d\mathbf{F} = I(d\mathbf{l} \times \mathbf{B})$

Magnetic dipole moment of current loop with area A $\qquad |\mathbf{m}| = AI$

Torque in magnetic field $\qquad \mathbf{T} = \mathbf{m} \times \mathbf{B}$

Potential energy in magnetic field $\qquad U = -(\mathbf{m} \cdot \mathbf{B})$

Magnetic field of dipole $\qquad \mathbf{B} = \dfrac{\mu_0}{4\pi}\left[3\dfrac{(\mathbf{m} \cdot \mathbf{r})}{r^5}\mathbf{r} - \dfrac{\mathbf{m}}{r^3}\right]$

Magnetic moment of electron
in orbit of angular momentum $\mathbf{L}$ $\qquad \mathbf{m} = -\dfrac{e}{2m}\mathbf{L}$

Magnetic moment of electron $\qquad |\mathbf{m}| = \mu_B = \dfrac{e\hbar}{2m} = 0.929 \times 10^{-23}$ A-m^2

Spin angular momentum of electron $\qquad |\mathbf{S}| = \dfrac{\hbar}{2} = 0.53 \times 10^{-34}$ J-sec

Questions

1. Given the definition (10–7) of μ_0, design an experiment which would enable you to measure the current in a wire in terms of mechanical measurements (length, time, mass, force, etc.). Can you use your setup to define the unit of $1\,\text{A} = 1$ coulomb/second?

2. You observe that in some part of an experimental setup an electron beam deviates from a straight path by a small amount. How would you determine whether a stray electric field or a stray magnetic field is responsible? The electron gun which produces the beam is movable.

3. The magnetic field in a magnet points in the z-direction. It is not exactly homoge-

neous, but changes slightly with x:

$$B_z = B_0(1 + \alpha x) \qquad \alpha x \ll 1 \quad \text{over the size of the magnet}$$

A particle of charge $+q$ starts out with initial velocity $\mathbf{v}_0$ in the x direction. Describe its motion.

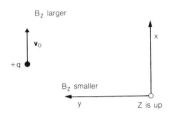

4. Sketch the magnetic field lines in a loosely wound straight long coil carrying the current I. Copy the figure on a sheet of paper before drawing the lines.

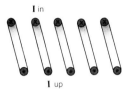

Problems

1. A household wire of diameter $d = 2$ mm carries the current $I = 10$ A. What is the magnetic field on its surface?

2. Two long straight wires are perpendicular to each other; their minimum separation is $a = 10$ cm. A current $I = 25$ A is flowing through each wire. Calculate the magnitude and direction of the magnetic field at a point P halfway along their line of minimum separation.

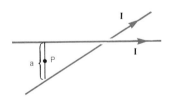

3. You are to design a straight cylindrical coil of inner radius $a = 5$ cm and length $L = 40$ cm which, if driven by a current $I \leqslant 80$ A, produces a magnetic field of 0.02 weber/m². You are to use #10 insulated wire: the wire diameter is 0.1 inch, the outer insulation diameter 0.15 inch.

   (a) How many layers of wire will you have to use, assuming that you wind each layer tight (the wires are touching)?

   (b) What is the resistance of the coil, and what power is dissipated in the coil at a magnetic field of 0.02 weber/m²?

4. A coaxial cable consists of an inner conductor of radius R_1 and an outer conductor

extending from $r = R_2$ to $r = R_3$. The current I flows in opposite directions in the two conductors. Calculate the magnetic field for

    (a) $r < R_1$
    (b) $R_1 < r < R_2$
    (c) $R_2 < r < R_3$
    (d) $r > R_3$

Plot $B(r)$.

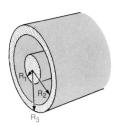

5. A thin long metallic strip of width a carries the current I. What is the magnetic field at the point P which lies in the plane of the sheet a distance a away from one edge? (Hint: Decompose the strip into many parallel wires.)

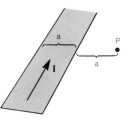

6. Two circular coils of radius $a = 0.5$ m, each with $N = 20$ turns, carry the current $I = 100$ A each. The coils are parallel above each other and are separated by the distance a. Calculate the magnetic field at the point P on their common axis halfway between the two coils. (This arrangement, called *Helmholtz coils*, is used in the laboratory to establish a known approximately uniform magnetic field in the vicinity of P.)

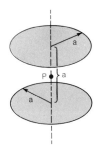

7. A square wire loop of side $a = 10$ cm, with $N = 10$ turns, carries the current $I = 1$ A. Calculate the magnetic field at the center of the loop.

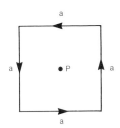

8. Calculate the magnetic field at an arbitrary point P' on the axis through the loop from Problem 7. Show that for $z \gg a$ the result agrees with what one would obtain if one considers the loop to be a dipole.

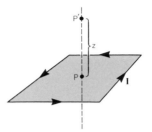

9. A thin disk of radius R carries the charge Q homogeneously distributed over its surface. The disk rotates N times per second around an axis perpendicular to its surface. Calculate the magnetic field at the center of the disk.

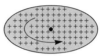

°10. Calculate the magnetic field at an arbitrary point on the axis of a cylindrical coil of finite length L and radius r. The coil has N turns and carries the current I. (Hint: Consider the coil as a superposition of individual circular loops.) What will be the ratio of the on-axis field at the center and at one end of the coil?

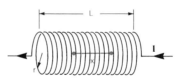

11. An infinitely long coil has one layer of N turns/meter and carries a current I. Its radius is R. Will the wires be under tension or under compression? What is the magnitude of the force on a short stretch $d\ell$ of wire? Be careful to take into account that the mean magnetic field at the wire position is not the same as the field inside the coil. To calculate the mean field, approximate the coil by a cylindrical sheet of finite small thickness d which carries the current I.

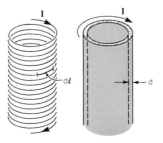

°12. Calculate the magnetic field of a rectangular current loop at a faraway point which lies in the plane of the loop and on one of the two axes of the rectangle.

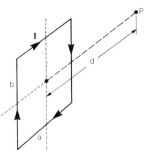

13. The earth's magnetic field is about 0.5 gauss. Calculate the Larmor period and the radius of curvature of an electron of kinetic energy $E_0 = 1$ keV in such a magnetic field. Such electrons are quite abundant in the "aurora," a glow phenomenon seen near the earth's poles.

14. A beam of electron represents a current $I = 1000$ A with a diameter of $d = 5$ mm. The electrons are moving with a velocity $v = 10^6$ m/sec. Calculate the electrostatic (due to the other electrons) and the magnetic (due to the current) force on an electron at the outer perimeter of the beam. What would the electron velocity v have to be for the two forces to cancel?

15. Calculate the Larmor frequency and Larmor period for electrons in a magnetic field of 1.5 webers/m². What is the radius of curvature for electrons having a kinetic energy of 5 eV in this field? Note: As we will discuss in detail in Chapter 15, 5 eV is a typical kinetic energy of the free electrons in a metal.

16. A proton having kinetic energy of 6 MeV (6×10^6 eV) enters a region of magnetic field $B_0 = 1$ weber/m² (pointing into the paper) under an angle $\theta = 45°$, as shown in the sketch. Where and in what direction does it leave the magnetic field? Give x and θ'.

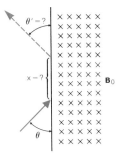

17. A mass separator such as the one shown in Fig. 10–20 has the following working conditions: $E = 10^3$ V/m, $B = 0.1$ weber/m². What should be the location of a detector (distance d) to detect singly charged $_{200}Hg^+$ ions? By how much should one change the magnetic field, keeping the detector in place, if one wants to detect $_{201}Hg^+$ ions? Note: To a good approximation, a neutral atom of atomic weight A has a mass A times that of the proton.

18. One way to determine atomic masses precisely is to measure the time they take to complete a circle in a magnetic field. If the magnetic field is 1000 gauss, and time can be measured to $\pm 1 \times 10^{-10}$ sec, to what accuracy can one determine the mass of

a singly charged ion? If one is able to let the ions go around many times, how many times do they have to circle around if the ion mass is to be determined to ±0.1 electron mass? How can one make this experimentally feasible?

19. In a 0.2 mm thick silicon sample carrying the current $I = 10$ mA in a magnetic field $B_0 = 1$ weber/m^2, one observes a Hall voltage of $V_H = 10$ mV. What is the charge carrier density of the sample?

20. Protons with kinetic energy $T = 50$ MeV (5×10^7 eV) are moving in the x direction and enter a magnet with magnetic field $B_0 = 0.5$ weber/m^2 entending from $x = 0$ to $x = L = 1$ m. Calculate the y-component of the proton momentum at the exit from the magnet and the angle α of the proton direction relative to the x-axis.

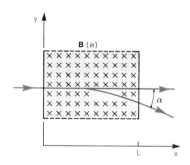

21. A wire loop is constructed of two squares as shown in the figure and carries the current $I = 2$ A. The length a is 10 cm.

    (a) Calculate the dipole moment; give magnitude *and* direction.

    (b) If the loop is in a magnetic field $B_0 = 5000$ gauss pointing up as shown, what is the torque on the circuit?

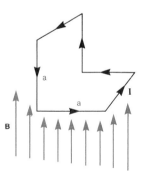

22. A cylindrical coil has a radius $a = 1$ cm and length $L = 20$ cm and a total of $N = 200$ turns. An unknown current I flows through the coil so that the magnetic field inside is $B_0 = 100$ gauss. What is the total dipole moment of the coil?

23. A crude estimate of the magnetic moment of the proton can be obtained by imagining that half of the time the proton separates into an uncharged neutron and a π^+ meson (pion mass $m_\pi = 0.15\ m_p$, where m_p = mass of proton); the pion then moves with angular momentum $|L| = \hbar$ around the neutron on a radius $r \approx 10^{-15}$ m. Estimate the proton magnetic moment on this basis and compare your estimate with the experimental result (see Table III of Appendix B).

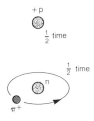

24. *Ballistic Galvanometer.* A circular coil of 100 turns and radius $r = 1$ cm is oriented as shown in a magnetic field $B = 0.5$ weber/m^2. The moment of inertia of the coil is $I = 10$ g-cm^2. With what angular velocity will the coil start rotating, after a current pulse of 100 μA flows through it for 1 msec? The coil has no time to move during the duration of the pulse.

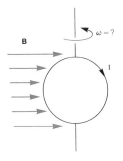

25. What is the magnetic field at the position of the proton in a hydrogen atom, if the electron is moving on a circular orbit of orbital angular momentum $|\mathbf{L}| = \hbar$ and radius $a_0 = 0.53$Å?

CHAPTER 11

MAGNETIC FIELDS IN MATTER

1. PARAMAGNETISM AND DIAMAGNETISM

As we said in the last chapter, atoms or molecules can possess permanent magnetic dipole moments. Because they are produced either by the orbital angular momentum of the electrons or by their intrinsic spin magnetic moment, these atomic moments will be of the order of magnitude of 1 Bohr magneton. Such materials, whose atoms possess permanent magnetic moments, are called *paramagnetic*.

From what we just said, it would seem that all materials should be paramagnetic; after all, each electron has a spin-magnetic moment and most electrons in an atom are likely to have an orbital angular momentum. Thus, it would seem to be a rare accident if the magnetic moments of all electrons in an atom were to cancel each other. But this is exactly what happens. As we will discuss later in quantum mechanics, a fundamental principle discovered by W. Pauli states that if two electrons are very close to each other, they have to have their spin angular momenta antiparallel to each other. From the Pauli principle, whose detailed discussion we postpone until later, one can also derive the conclusion that the electrons in a single atom are also likely to arrange themselves so that the net orbital angular momentum vanishes (Fig. 11–1). Thus, many molecules have no net magnetic moments. Such substances are called *diamagnetic*.

In the presence of an external magnetic field, paramagnetic and diamagnetic substances behave quite differently. In paramagnetic materials, because of the potential energy of a magnetic dipole in a magnetic field

$$U = -(\mathbf{m} \bullet \mathbf{B}) \qquad (11\text{--}1)$$

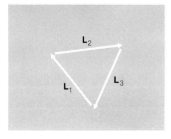

Figure 11–1 In many atoms the individual orbital angular momenta of the electrons add up to zero.

the permanent magnetic dipoles will tend to align themselves parallel to the magnetic field. Random thermal motion will prevent a complete alignment, however. We can also calculate the *mean magnetic moment per unit volume*, **M**, which we call the magnetization. Let us assume that the magnetic field produced by the dipoles themselves is too weak to have any effect. Then we can repeat practically word for word the derivation of Eq. (8–58), substituting everywhere "magnetic" for "electric." The relative probability will be proportional to the Boltzman factor

$$e^{-U/kT} = e^{(\mathbf{m}\cdot\mathbf{B})/kT} \approx 1 + \frac{(\mathbf{m} \bullet \mathbf{B})}{kT}$$

because $(\mathbf{m} \bullet \mathbf{B})/kT \ll 1$. If we calculate the mean angle between the magnetic moment and the magnetic field, we obtain

$$\langle \cos \theta \rangle = \frac{|\mathbf{m}|\,|\mathbf{B}|}{3kT} \tag{11–2}$$

We can now define the magnetization **M** as the magnetic dipole moment per unit volume in the material. If there are in a given volume s such dipoles, each with the magnetic dipole moment $\mathbf{m}_i$ $(i = 1, \ldots s)$, then the magnetization is

$$\boxed{\mathbf{M} = \frac{\sum\limits_{i=1}^{s} \mathbf{m}_i}{\text{volume}}} \tag{11–3}$$

In our case, the dipole moments will have a mean component along the magnetic field of $|\mathbf{m}|\,\langle \cos \theta \rangle$; any perpendicular component will be distributed randomly and thus, on the average, it will vanish. If we define n as the number of dipoles per cubic meter, each having an average dipole moment $|\mathbf{m}|\,\langle \cos \theta \rangle$ along the direction of **B**, then we obtain for the magnetization

$$|\mathbf{M}| = n|\mathbf{m}|\,\langle \cos \theta \rangle = \frac{n|\mathbf{m}|^2}{3kT}\,|\mathbf{B}| \tag{11–4}$$

The quantity

$$\boxed{\chi = \frac{\mu_0 n|\mathbf{m}|^2}{3kT}} \tag{11–5}$$

is called the magnetic susceptibility of a paramagnetic material; we have added the factor μ_0 to make χ a dimensionless quantity. Indeed,

$$[\chi] = \frac{[\mu_0][n][m]^2}{[kT]}$$

and since kT has the dimension of energy and n has the dimension of (number/volume), we have

$$[\chi] = \frac{\text{V-sec}}{\text{A-m}} \frac{\text{m}^{-3}(\text{A-m}^2)^2}{\text{J}} = \frac{\text{V-A-sec}}{\text{J}} = 1$$

because 1 A-sec = 1 coul and 1 coul × 1 V = 1 J.

We stress again that Eq. (11–5) is valid only if the magnetic field due to the magnetic dipoles themselves is negligible. We will investigate this problem in more detail; but first let us briefly discuss diamagnetic substances. Although their atoms have no permanent magnetic dipole moment, they will acquire a weak magnetic moment in the presence of a magnetic field because of the force

$$\mathbf{F} = (-e)\mathbf{v} \times \mathbf{B} \tag{11–6}$$

which acts on any electron in a magnetic field. We can create a crude model of a diamagnetic atom; we imagine that it has a pair of electrons encircling the nucleus in opposite directions. In addition, we imagine that their orbits are in a plane which is perpendicular to the applied magnetic field. In Fig. 11–2 we have drawn the two orbits separately; we imagine the magnetic field **B** to point into the paper. We also show in the figure the direction of the additional force due to the magnetic field and given by Eq. (11–6). In the absence of a magnetic field there will be the attractive force of the nucleus; its magnitude will be of the order of

$$|\mathbf{F}_0| = \frac{e^2}{4\pi\epsilon_0 r^2}$$

(In reality, if the atomic nucleus has a charge Ze and is surrounded by $(Z-1)$ other electrons, the force will lie somewhere between $e^2/4\pi\epsilon_0 r^2$ and $Ze^2/4\pi\epsilon_0 r^2$; it will be the sum of the attractive force of the nucleus and the repulsive force of all the other electrons.) In the absence of a magnetic field this force will just be equal to the centripetal acceleration mv^2/r:

$$\frac{mv^2}{r} = F_0 \tag{11–7}$$

In the presence of a magnetic field, the additional force given by Eq. (11–6) will modify Eq. (11–7):

$$\frac{mv'^2}{r} = F_0 \pm ev'B \tag{11–8a}$$

where the (+) sign is valid for the left orbit° and the (−) for the right orbit in Fig. 11–2. This will change the velocity by a small amount Δv. Since we write

---

° Remember, the electron charge $-e$ is negative!

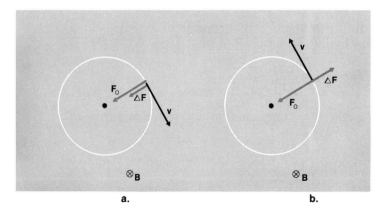

a.   b.

Figure 11–2 Crude model of diamagnetism: two electrons moving along the same orbit in opposite directions. One speeds up and the other slows down in the presence of an external magnetic field.

$v' = v + \Delta v$, we have

$$\frac{m(v + \Delta v)^2}{r} = F_0 \pm e(v + \Delta v)B \qquad (11\text{--}8b)$$

Since we assume

$$|evB| \ll F_0 \qquad (11\text{--}8c)$$

we can neglect Δv on the right side in Eq. (11–8b) and $(\Delta v)^2$ on the left side and obtain

$$\frac{mv^2}{r} + \frac{2mv\,\Delta v}{r} = F_0 \pm evB \qquad (11\text{--}9)$$

from which we obtain with the help of Eq. (11–7)

$$\Delta v = \pm \frac{eBr}{2m} \qquad (11\text{--}10)$$

The electron in Fig. 11–2a will be speeded up and the one in Fig. 11–2b will be slowed down. The magnetic moment has the magnitude

$$|\mathbf{m}| = \pi r^2 I = \pi r^2 \frac{e}{T}$$

and since the period T is equal to $T = 2\pi r/v$, we have

$$|\mathbf{m}| = \frac{\pi r^2 e v}{2\pi r} = \frac{erv}{2}$$

Thus, the magnetic moment in Fig. 11–2a, which points out of the paper, is increased; the one in Fig. 11–2b is reduced. In both cases we obtain a change in magnetic moment which points in a direction opposite to $\mathbf{B}$; thus, the net magnetic moment of both electrons, which was zero for $\mathbf{B} = 0$, now has the magnitude

$$|\Delta\mathbf{m}| = 2\frac{er\,\Delta v}{2} = \frac{e^2 B r^2}{2m}$$

Since $\Delta\mathbf{m}$ is antiparallel to $\mathbf{B}$, we can write

$$\Delta\mathbf{m} = -\frac{e^2 r^2}{2m}\mathbf{B} \qquad (11\text{--}11)$$

Thus, if there are n diamagnetic atoms per unit volume, the magnetization will be

$$\mathbf{M} = -\frac{ne^2 r^2}{2m}\mathbf{B} \qquad (11\text{--}12a)$$

and in analogy to Eq. (11–5) we can write for diamagnetic material

$$\boxed{\chi = -\frac{\mu_0 n e^2 r^2}{2m}} \qquad (11\text{--}12b)$$

TABLE 11–1 MAGNETIC SUSCEPTIBILITIES OF SOME SUBSTANCES AT ROOM TEMPERATURE

MATERIAL	DENSITY (10^3 kg/m³)	10^5 χ
Aluminum	2.70	2.2
Arsenic	5.72	−2.2
Chromium	7.20	32.
Copper	8.9	−0.96
Diamond	3.52	−2.2
Gold	19.3	−0.36
Oxygen (gas at 1 atm)	0.0014	0.21
Manganese	7.20	90.
Mercury	13.5	−3.2
Nitrogen (gas at 1 atm)	0.0012	−0.0005
Osmium	22.5	1.4
Palladium	12.0	80.
Silver	10.5	−2.6
Sulfur	2.0	−1.2

Example 1   Estimate the average magnetic dipole moment per atom for a diamagnetic substance. State all assumptions you make. Estimate also the order of magnitude of the magnetic susceptibility for a diamagnetic solid.

The purpose of this example is to check whether or not the explanations we found for paramagnetic and diamagnetic behavior make sense. We have used only crude models to explain the magnetization in substances; however, we should have obtained in Eqs. (11–5) and (11–12b) estimates which can be tested by experiment. We first discuss diamagnetism. The typical diameter of an atom is $d = 2$ to 3 Å; if we did not know this, we could have obtained it from the typical density of solids of 3 to 10×10^3 kg/m³, from the assumption that atoms touch each other in a solid, and from the known value of Avogadro's number, $N_0 = 6 \times 10^{26}$ atoms/kg-mole (that is, if we have atoms of atomic weight A, then N_0 atoms weigh A kilograms). We will thus assume $r = d/2 \approx 1$ Å $= 10^{-10}$ m in Eq. (11–11). We then have in a magnetic field of, say, 1 Wb/m² [from Eq. (11–11)]:

$$|\Delta \mathbf{m}| = \frac{e^2 r^2 B}{2m} = \frac{(1.6 \times 10^{-19} \text{ coul})^2 \times (10^{-10} \text{ m})^2 \times (1 \text{ Wb/m}^2)}{2 \times (0.9 \times 10^{-30} \text{ kg})}$$

$$\approx 1.4 \times 10^{-28} \text{ A-m}^2$$

This is a very small magnetic moment compared to the "typical value" of one Bohr magneton, 1 $\mu_B = 0.93 \times 10^{-23}$ A-m². For the magnetic susceptibility, we need to know the number of atoms per cubic meter. Since the center-to-center separation of atoms is typically about 2 Å under our earlier assumption about r, we can estimate

$$n \approx \left(\frac{1}{2 \text{ Å}}\right)^3 = \frac{1}{(2 \times 10^{-10} \text{ m})^3} = \frac{10^{30}}{8} \text{ m}^{-3} \approx 10^{29} \text{ m}^{-3}$$

Thus, a typical diamagnetic substance should have a magnetic susceptibility

$$\chi = -\frac{\mu_0 n e^2 r^2}{2m}$$

$$= -\frac{\left(4\pi \times 10^{-7} \dfrac{\text{V-sec}}{\text{A-m}}\right) \times (10^{29} \text{ m}^{-3}) \times (1.6 \times 10^{-19} \text{ coul})^2 \times (10^{-10} \text{ m})^2}{2 \times (0.9 \times 10^{-30} \text{ kg})}$$

$$\approx -1.77 \times 10^{-5}$$

We would consider any value which lies within a factor of 5 of our estimate reasonable. From Table 11–1 we see that indeed most solid substances with negative susceptibilities have values in this range; thus our explanation of the effect which was based on a very crude model is qualitatively correct.

Example 2   Repeat the estimates of Example 1 for a paramagnetic solid; estimate its average magnetic moment per atom and its magnetic susceptibility.

We have derived Eqs. (11–3), (11–4), and (11–5) without specifying the value of $|\mathbf{m}|$; thus, we have to "guess" a reasonable value of the permanent magnetic moment of an atom. We have said that each electron has a magnetic moment

$$\mathbf{m} = -\frac{e}{2m}\mathbf{L} - \frac{e}{m}\mathbf{S}$$

where $\mathbf{L}$ is its orbital momentum (which has a magnitude of some integer times $\hbar$) and $\mathbf{S}$ is the spin (which has the magnitude $\frac{1}{2}\hbar$). Thus, we would expect for each electron a magnetic dipole moment of a few times the Bohr magneton

$$\mu_B = \frac{e\hbar}{2m} = 0.93 \times 10^{-23} \text{ A-m}^2$$

However, if there are several electrons in an atom, their magnetic moments will tend to cancel because of the Pauli principle; thus, we will assume the net magnetic moment of an atom to be one Bohr magneton, but we could be off by a factor of 2 to 3. We then obtain the mean magnetic moment in the direction of the magnetic field from Eq. (11–4):

$$<\mathbf{m}_{||}> = |\mathbf{m}|<\cos\theta> = \frac{|\mathbf{m}|^2}{3kT}|\mathbf{B}|$$

At room temperature, $T = 300°$ K; from Table III of Appendix B we obtain $k = 1.38 \times 10^{-23}$ J/° K. Thus, at room temperature we have

$$kT \approx 4 \times 10^{-21} \text{ J}$$

We now find that, in a magnetic field of 1 Wb/m², the magnetic moment is

$$|\mathbf{m}_{||}| = \frac{(0.93 \times 10^{-23} \text{ A-m}^2)^2}{3 \times (4 \times 10^{-21} \text{ Joule})} \times 1 \text{ V-sec/m}^2 \approx 7 \times 10^{-27} \text{ A-m}^2$$

If we assume the same density of atoms, $n = 10^{29}$ atoms/m³, as in the previous example, we obtain a magnetic susceptibility of

$$\chi = \frac{\mu_0 n |\mathbf{m}|^2}{3kT} = \frac{\left(4\pi \times 10^{-7} \dfrac{\text{V-sec}}{\text{A-m}}\right) \times (10^{29} \text{ m}^{-3})(0.93 \times 10^{-23} \text{ A-m}^2)^2}{3 \times (4 \times 10^{-21} \text{ Joule})}$$

$$= 9 \times 10^{-4} = 90 \times 10^{-5}$$

Thus, we conclude that the typical paramagnetic susceptibility should be about 50 to 150 times larger in absolute value than the typical diamagnetic susceptibility calculated in the previous example. This is indeed roughly correct, as an inspection of Table 11–1 shows. The typical positive susceptibility is of the order 10^{-4} to 10^{-3}; but there are some solid substances which have positive susceptibilities of 10^{-5} to 10^{-4}, or a whole order of magnitude smaller. These values cannot be explained by our simple classical theory. The discrepancy is again due to the Pauli principle; the individual electrons nearly, but not completely, cancel out each other's magnetic moments because they have spins pointing in opposite directions. The effective magnetic dipole moment $|\mathbf{m}|$ which should be inserted in Eq. (11–11) is then only a fraction (~ 0.1) of a Bohr magneton.

2. MAGNETIC FIELD IN MAGNETIZED SUBSTANCES

We have derived Eqs. (11–4) and (11–12a) under the assumption that the magnetic dipoles in the material (i.e., the magnetization) have a negligible contribution to the total magnetic field **B**. We will now show this to be indeed true, and we will also learn how to treat ferromagnetic substances such as iron, in which the magnetization contributes very strongly to the total magnetic field.

For this purpose, let us consider a long cylindrical coil which has N_0 turns per meter and in which a current I is flowing (Fig. 11–3a). The coil is filled by a material of magnetization **M**, and we propose to calculate the magnetic field **B** inside the material.

We can imagine the magnetization **M** as being produced by a large number of dipoles, each having the dipole moment **m**. If the number of dipoles per unit volume is n, then the magnetization is

$$\mathbf{M} = n\mathbf{m} \tag{11–13a}$$

For simplicity we assume each dipole to consist of a circular current loop of radius a and current i. Then we have

$$|\mathbf{m}| = \pi a^2 i \tag{11–13b}$$

We can now calculate the magnetic field using Ampère's law. For this purpose we define a closed loop C, as shown in Fig. 11–3a. The return path can be chosen arbitrarily far away and thus will not contribute. From Ampère's law [Eq. (10–12)] we then have

$$\int_C \mathbf{B} \cdot d\mathbf{l} = \int_A^B \mathbf{B} \cdot d\mathbf{l} = \mu_0(I_{\text{inside}}) \tag{11–14}$$

The current I_{inside} is not only the current I flowing through the coil; it also has a

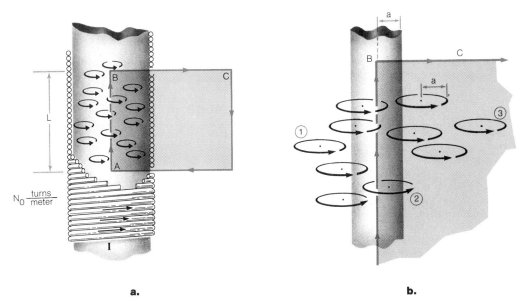

a.   **b.**

Figure 11–3 Calculating the effect of magnetization on the loop integral $\int \mathbf{B} \cdot d\mathbf{l}$. a) The coil filled with magnetized material. b) The individual dipoles, of which some circle the path C.

contribution I_{mag} from the dipoles in the material. If the length of the path AB is L, we can write

$$I_{inside} = I_{free} + I_{mag} = N_0LI + I_{mag} \qquad (11\text{--}15)$$

because N_0L turns of the coil go through the inside of the loop C.

We can determine the magnetization current I_{mag} by determining which of the many circular dipole currents contribute to it. In Fig. 11–3b we have numbered some of the dipoles. Dipole 1 will not contribute because it lies completely outside the loop formed by C. Dipole 3 will not contribute either, although it lies in the plane of the loop and inside the loop. The current i associated with dipole 3 crosses the surface twice — once going in and once going out, so its net contribution to I_{mag} vanishes. The only magnetic dipoles which contribute to I_{mag} are those which, like dipole 2, encircle the path C. Then there is a current I going in through the surface bordered by C, but no compensating current going out. This will be true for all dipoles whose centers are less than the radius a away from the path segment AB of the path C. (Either the rest of C is outside the material, or the path is perpendicular to the individual dipoles $\mathbf{m}$.) Thus, all dipoles in the volume $V = \pi a^2 L$ contribute, each with a current i. Since there are n such dipoles per unit volume, we have

$$I_{mag} = n\pi a^2 Li$$

Because of Eqs. (11–13a) and (11–13b), we can write this as

$$I_{mag} = n|\mathbf{m}|L = |\mathbf{M}|L \qquad (11\text{--}16)$$

From Fig. 11–3a we deduce that, since the magnetization is parallel to the path C from A to B, we can write

$$|\mathbf{M}|L = \int_A^B \mathbf{M} \cdot d\mathbf{l} = \int_C \mathbf{M} \cdot d\mathbf{l} \qquad (11\text{--}17)$$

and we thus arrive, using Eq. (11–15), at the result that

$$\int_C (\mathbf{B} - \mu_0\mathbf{M}) \cdot d\mathbf{l} = \mu_0 N_0 L I_0 \qquad (11\text{--}17a)$$

or that

$$|\mathbf{B}| = \mu_0(N_0 I_0 + |\mathbf{M}|) \qquad (11\text{--}17b)$$

In the previous section we defined the susceptibility χ by Eq. (11–5), which implies

$$\mu_0\mathbf{M} = \chi\mathbf{B}$$

and we warned there that this is true only if the additional magnetic field produced by the magnetization $\mathbf{M}$ is negligible. Now we have in Eq. (11–17b) a confirmation of this for paramagnetic and diamagnetic substances for which $|\chi| \ll 1$. The magnetic field in a coil filled with material of *small* susceptibility χ can be calculated from Eq. (11–17b) as

$$B = \mu_0 NI + \chi B$$

or, since χB is small compared to B, we have

$$B \approx \mu_0 (1 + \chi) NI \approx \mu_0 NI$$

which is nearly the same as if there were no magnetization at all.

However, we now have to discuss how to generalize Eq. (11–18) if the contribution of **M** to the total magnetic field is no longer negligible. This does occur in ferromagnetic substances, such as iron, for which χ can be as large as 10,000. We define the susceptibility using our example of the coil filled with magnetic material; the susceptibility is then defined by the equation

$$\mu_0 \mathbf{M} = \chi \mathbf{B}' \tag{11–18a}$$

where **B'** is the magnetic field in the coil with the magnetic material removed:

$$B' = \mu_0 NI_0 \tag{11–18b}$$

For diamagnetic and paramagnetic substances, the distinction between **B** and **B'** is irrelevant; for ferromagnetic substances it is important.

We now introduce a new field, the *magnetizing field* **H**, which, except for the units it is measured in, is identical to the field **B'**:

$$\mathbf{H} = \frac{\mathbf{B}'}{\mu_0} \tag{11–18c}$$

The proportionality constant $1/\mu_0$ signifies only that we measure **H** in the same units as **M**, i.e., in amperes/m; if **B'** is equal to 1 Wb/m², then **H** is $10^7/4\pi$ A/m $\approx 8 \times 10^5$ A/m.* We can then write

$$\mathbf{M} = \chi \mathbf{H} \tag{11–19a}$$

and the magnetic field **B** is, because of Eq. (11–17a),

$$\mathbf{B} = \mu_0 \mathbf{H} + \mu_0 \mathbf{M} = \mu_0 (1 + \chi) \mathbf{H} \tag{11–19b}$$

We also define the *relative magnetic permeability*

$$\mu_r = 1 + \chi \tag{11–20a}$$

and the *absolute magnetic permeability*

$$\mu = \mu_r \mu_0 \tag{11–20b}$$

so that we can write

$$\mathbf{B} = \mu_r \mu_0 \mathbf{H} = \mu \mathbf{H} \tag{11–21}$$

---

* We use **H** instead of **B'** for a purely historical reason. There is a *physical* difference between **B** and **B'**, but only a difference in the units used between **H** and **B'**.

Figure 11–4 Example 3.

Example 3   A coil of length $L = 30$ cm has a total of $N = 100$ turns (Fig. 11–4). A current of $I = 0.5$ A flows through the coil. Calculate the magnetizing field and the magnetic field inside the coil if it is filled with air or with steel of susceptibility $\chi = 5000$.

The magnetizing field is always the same whether there is material or not, because it is defined as $1/\mu_0$ times the field in the absence of material. Thus,

$$|\mathbf{H}| = \frac{N}{L}I = \frac{100 \text{ turns}}{0.3 \text{ m}} \times 0.5 \text{ A} = 167\frac{\text{A}}{\text{m}}$$

The magnetic field $\mathbf{B}$ will be different in the presence of iron, however; if there is no iron, we have

$$|\mathbf{B}| = \mu_0 |\mathbf{H}| = 4\pi \times 10^{-7}\frac{\text{Wb}}{\text{A-m}} \times 167\frac{\text{A}}{\text{m}} = 2.1 \times 10^{-4}\frac{\text{Wb}}{\text{m}^2}$$

while if there is iron of susceptibility $\chi = 5000$, the magnetic field is

$$\mathbf{B} = \mu_r\mu_0|\mathbf{H}| = (1 + \chi)\mu_0\mathbf{H}$$

or

$$\mathbf{B} = 5001 \times 4\pi \times 10^{-7}\frac{\text{Wb}}{\text{A-m}} \times 167\frac{\text{A}}{\text{m}} = 1.05\frac{\text{Wb}}{\text{m}^2}$$

We see that the presence of the iron increases the magnetic field by a factor

$$\mu_r = 1 + \chi \approx \chi$$

because when $\chi \gg 1$ the magnetization contributes nearly all of the magnetic field in the coil. Nevertheless, the magnetization exists only because of the small magnetizing field $\mathbf{H}$.

Let us note that Eq. (11–19b) shows that the magnetization $\mathbf{M}$ produces a magnetic field $\mathbf{B}$ which points in the same direction as $\mathbf{M}$. This is in contrast to what happens in dielectrics; there, the electric polarization $\mathbf{P}$ reduces the electric field $\mathbf{E}$ in the material. We will return to this in Section 4, where we will discuss Maxwell's equations in matter.

3. FERROMAGNETIC MATERIALS

We now turn our attention to the properties of ferromagnetic substances. The detailed theory of ferromagnetic substances is very complex, and by no means have all questions been answered. We can here present only qualitative arguments about why ferromagnetism occurs in nature and sketch some properties of ferromagnets.

The atomic angular momenta in ferromagnetic substances are coupled to each other; these atomic angular momenta are the spin angular momenta of some of the atom's electrons. A complex interaction between neighboring atoms causes the spins of each atom to align themselves preferentially so that they are parallel to each other. We stress that this alignment is *not* due to the magnetic interaction between the electrons' magnetic moments; this effect would be far too weak to produce any alignment at room temperature.

The net effect is that when some of the atomic magnetic moments are pointing in some direction, the others will tend to do the same. Thus, instead of Eq. (11–19a) we have to write

$$\mathbf{M} = \chi_0(\mathbf{H} + \alpha\mathbf{M}) \tag{11–22}$$

where the presence of the term $\alpha\mathbf{M}$ reflects the mutual interactions between the electrons. If the temperature is sufficiently high, the material is still paramagnetic; the intrinsic magnetic susceptibility χ_0 is still given by Eq. (11–5):

$$\chi_0 = \frac{\mu_0 N |\mathbf{m}|^2}{3kT} = \frac{C}{T} \tag{11–23}$$

However, the effective magnetic susceptibility, as obtained by any macroscopic measurement, is by definition the proportionality constant between $\mathbf{M}$ and $\mathbf{H}$:

$$\mathbf{M} = \chi\mathbf{H}$$

From Eq. (11–22) we obtain, using the value of Eq. (11–23) for χ_0 and solving for $\mathbf{M}$,

$$\mathbf{M} = \frac{\chi_0}{1 - \alpha\,\chi_0}\mathbf{H} = \frac{C/T}{1 - \alpha(C/T)}\mathbf{H}$$

which we can also write as

$$\mathbf{M} = \frac{C}{T - \alpha C}\mathbf{H} \tag{11–24}$$

We have thus calculated the effective magnetic susceptibility

$$\boxed{\chi = \frac{C}{T - \alpha C} = \frac{C}{T - T_c}} \tag{11–25}$$

Here we have introduced the new quantity $T_c = \alpha C$, which is called the Curie temperature; the constant C is called the Curie constant. Both are named after Pierre Curie, who is today (maybe wrongly) mainly remembered as the husband of Marie Curie, with whom he shared the 1903 Nobel prize for the discovery of radioactivity and of the radioactive elements radium and polonium.[*] Pierre Curie, together with Pierre Weiss (another French physicist), also founded the

_____

[*] The third winner in the same year was Henri Becquerel, who was the first person who actually observed natural radioactivity.

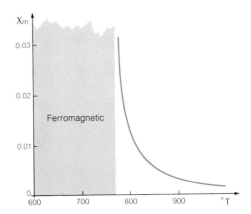

Figure 11-5 Magnetic susceptibility of iron as a function of temperature.

modern theory of magnetism. Eq. (11–25) is thus also known as the Curie-Weiss law. It tells us that for temperatures $T > T_c$ the material is paramagnetic, with a magnetic susceptibility which depends strongly on temperature. This is shown for iron in Fig. 11–5. At the temperature $T = T_c$, the susceptibility becomes infinitely large; for $T < T_c$ the material is spontaneously magnetized (Table 11–2). The influence of neighboring atoms is then sufficiently large to overcome the thermal disordered motion, and over some region all spins will be aligned in parallel, as sketched in Fig. 11–6. Thus, ferromagnetic substances have regions, called domains, which show maximum magnetization; the sizes of individual domains vary between 10^{-3} and 10^{-1} cm.

TABLE 11-2. CURIE TEMPERATURES FOR VARIOUS FERROMAGNETIC ELEMENTS

	T_c	
	°K	°C
Fe	1043	770
Co	1404	1131
Ni	631	358
Gd	293	20

In the absence of any external magnetic fields, the magnetization in the various domains will point in different directions. The size of the domains will be strongly dependent on the past history of the material: how fast it cooled from the melt, whether it was forged or not, and so forth. In iron, whose crystals have cubic symmetry (Fig. 11–8), the magnetization in each domain is in general

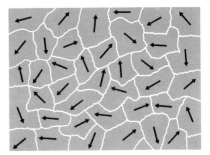

Figure 11-6 Ferromagnetic materials consist of irregular domains; each is magnetized in a different direction.

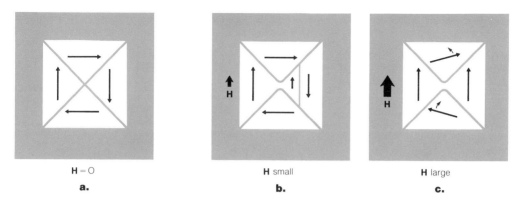

H = O	**H** small	**H** large
a.	**b.**	**c.**

Figure 11–7 Upon applying a magnetizing field **H**, the domains magnetized along **H** grow at the expense of those whose magnetization is opposite to **H**; at larger magnetizing fields domains start to rotate, aligning their magnetization along **H**.

parallel to one of the cube's edges as shown in Fig. 11–7a. If a small magnetizing field **H** is applied, the domains parallel to **H** will grow at the expense of the domains which are antiparallel to **H**. The process is not fully reversible except for very small fields; if the magnetizing field is turned off, the overall mean magnetization does not revert to zero. At large magnetizing fields, even those domains which were originally oriented perpendicularly to the magnetizing field will start changing their direction.

Macroscopically, the magnetization **M** as a function of **H** will behave as sketched in Fig. 11–9. If we start with unmagnetized material and apply a slowly increasing magnetizing field **H**, the magnetization will follow the "virgin" curve 1. The permeability μ_r is defined as the slope of the curve in the initial linear region; this is the only region in which the process is reversible, i.e., where **M** will revert to zero if we switch **H** off. When we increase the magnetizing field, after a while the magnetization will reach its *saturation value* M_s; all elementary magnetic dipoles are now aligned parallel to **H**, and increasing **H** further will no longer increase **M**. If we now start reducing **H**, the magnetization will follow curve 2. Even after the magnetizing field is zero, the magnetization is still finite and has the value M_r, called the *remanence magnetization*. We have to apply the *coercive field* H_c in the opposite direction before the magnetization vanishes. On making **H** more negative than $-H_c$, the material reaches saturation in the opposite direction. This phenomenon is known as *hysteresis*.

A large number of alloys have been developed which show a wide range of the important parameters M_s, M_r, and H_c. The purer the iron, the lower in general will be the magnetic remanence M_r and the coercive field H_c. On the other hand, special alloys which are used for permanent magnets have a mag-

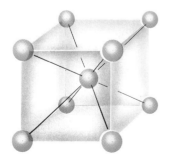

Figure 11–8 Crystal structure of iron: the body-centered cubic lattice.

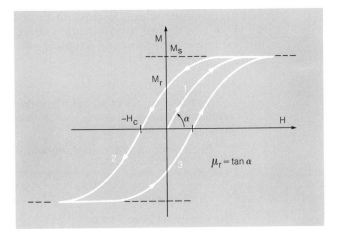

Figure 11-9 Definition of saturation magnetization M_s, remanent magnetization M_r, and coercive magnetizing field H_c. The relative magnetic permeability μ_r is the initial slope of the "virgin" curve 1.

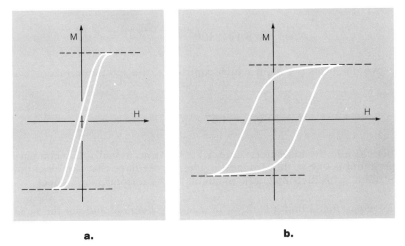

Figure 11-10 Magnetization curves for (a) cold rolled steel, and (b) a permanent magnet alloy.

TABLE 11-3 MAGNETIZATION CURVE PARAMETERS FOR PURE IRON AND A "HARD" MAGNETIC MATERIAL

	PURE IRON (99.99+ %)	ALNICO VI (48% Fe, 8% Al, 16% Ni, 24% Co, 3% Cu, 1% Ti)
μ_r	18,000	$\sim 10°$
M_s (A/m)	1.7×10^6	6×10^5
M_r (A/m)	10^2–10^3	5×10^5
H_c (A/m)	4.0	6×10^4

° Depends strongly on previous heat treatment.

netic remanence which is close to their saturation magetization. In Fig. 11–10 we show two typical magnetization curves, while Table 11–3 gives the parameters for rather pure iron and a typical alloy out of which permanent magnets are made.

Example 4   Let us calculate the saturation magnetization M_s in soft iron, assuming that two electrons per atom contribute with their spins to the magnetic moment.

Each electron will contribute with its spin magnetic moment of

$$1 \, \mu_B = \frac{e\hbar}{2m} = 0.929 \times 10^{-23} \text{ A-m}^2$$

The density of pure iron is

$$\rho = 7.86 \text{ g/cm}^3 = 7.86 \times 10^3 \, \frac{\text{kg}}{\text{m}^3}$$

The atomic weight of iron is $A = 55.85$; thus, one kg-mole has a mass of 55.85 kg. Since there are $L = 6.02 \times 10^{26}$ atoms/kg-mole, we have

$$N = \frac{\rho L}{A} = 8.47 \times 10^{28} \, \frac{\text{atoms}}{\text{m}^3}$$

As there are two electrons per atom which contribute, the saturation magnetization should be

$$M_s = N(2\mu_B) = 1.57 \times 10^6 \, \frac{\text{A}}{\text{m}}$$

which is rather close to the measured value, 1.7×10^6 A/m. Actually, this comparison with experiment is one of the strongest evidences we have that two electrons per atom contribute with their magnetic moments to the ferromagnetic effect.

■   In our example, we assumed that it is the spin magnetic moment and not the magnetic moment due to orbital motion which is responsible for ferromagnetism in metal. That this is so was shown in 1915 by Einstein and de Haas by demonstrating an effect which is named after them. Let us discuss their experiment and its interpretation. Somewhat simplified, their experimental setup consists of an iron cylinder suspended on a thin fiber (Fig. 11–11). A coil is placed around the cylinder. Upon applying a current to the coil, the iron cylinder becomes magnetized. However, the cylinder also begins to *rotate;* the magnetization gives it an angular momentum **L**. The ratio between the total magnetic dipole moment of the rod

$$\mathbf{m} = \int_{\substack{\text{rod} \\ \text{volume}}} \mathbf{M} \, d\tau$$

and the angular momentum **L** was found to be[*]

$$\frac{|\mathbf{m}|}{|\mathbf{L}|} = 1.65 \times 10^{11} \, \frac{\text{coul}}{\text{kg}} \tag{11–26}$$

_____

[*] That these are the right units can be seen if we insert the units for magnetic dipole moment and angular momentum:

$$\frac{\text{A-m}^2}{\text{kg-m}^2\text{-sec}} = \frac{\text{A-sec}}{\text{kg}} = \frac{\text{coul}}{\text{kg}}$$

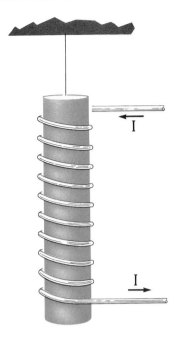

Figure 11-11   The Einstein-de Haas experiment.

The ratio $|\mathbf{m}|/|\mathbf{L}|$ was found to be independent of the size of the rod and of the magnetization imparted to it.

We can understand the effect if we remember that the electrons in the cylinder atoms have an orbital angular momentum around the cylinder axis of

$$L_z = n\hbar$$

where $n = 0, \pm 1, \pm 2, \pm 3, \ldots$ With this orbital angular momentum there is associated a magnetic dipole moment

$$m_z(\text{orbit}) = n\frac{e\hbar}{2m} = \frac{e}{2m}L_z \tag{11-27}$$

The electrons also have spin angular momentum, whose value around the fiber axis is

$$S_z = \pm\frac{1}{2}\hbar$$

but with this, there is associated a magnetic moment whose z-component is

$$m_z(\text{spin}) = \pm\frac{e\hbar}{2m} = \frac{e}{m}S_z \tag{11-28}$$

Originally, there were many magnetic domains, each oriented in a random direction. After magnetization not only are the magnetic moments aligned, but so are the angular momenta. Thus, all the parallel atomic angular momenta now give a nonvanishing total angular momentum in the direction of the magnetization. However, the overall angular momentum of the rod cannot have changed, and thus it still has to vanish. Therefore, the rod must begin to rotate, so that its macroscopic angular momentum of rotation compensates for the atomic angular momenta which have suddenly aligned themselves parallel to each other.[*] We see that if the orbital motion is responsible for the magnetiza-

---

[*] The effect is in reality very small and quite difficult to measure. See Problem 11-6.

tion, we should have from Eq. (11–27)

$$\frac{|\mathbf{m}|}{|\mathbf{L}|} = \frac{e}{2m} \tag{11–29a}$$

while if the spin is responsible, we have to use Eq. (11–28):

$$\frac{|\mathbf{m}|}{|\mathbf{L}|} = \frac{e}{m} \tag{11–29b}$$

Numerically, we have $e/m = 1.76 \times 10^{11}$ coul/kg. We can now rewrite the experimental result (11–26) as

$$\frac{|\mathbf{m}|}{|\mathbf{L}|} = g \frac{e}{2m} \tag{11–30}$$

where g should be equal to 1 if the orbital motion is responsible, and 2 if the electron spins are responsible for the ferromagnetic magnetization. From Eq. (11–26) we find

$$g = \frac{1.65 \times 10^{11}}{\frac{1}{2} \times 1.76 \times 10^{11}} = 1.88$$

Thus, g is nearly equal to 2, although a little smaller; this tells us that the electron spins give the dominant contribution to ferromagnetism, but that there is also a minor effect due to the orbital motion of the electrons. ■

Another side effect of the property of ferromagnetism is *magnetostriction*. If we carefully measure the length of a rod, we will find that upon magnetization the rod has become a few parts in 10^6 longer. On the other hand, if we stretch an originally unmagnetized rod, we will find that the rod has become magnetized. We can qualitatively understand the effect quite easily. The mutual interaction of the individual atomic spins will not only make them tend to be oriented parallel to each other, but also will slightly deform the crystal lattice. Thus, even if there is no macroscopic deformation, the iron crystals are not exactly cubic; the length of the "cube" in the direction of magnetization is slightly longer. When the spins rotate into the direction of the external magnetizing field (compare Fig. 11–7), the lattice deformation will be affected by the rotation. The net macroscopic effect is the slight change of the rod length upon magnetization.

4. MAXWELL'S EQUATIONS IN MATTER

We now want to summarize what we have learned about electromagnetic fields in matter. We still consider only the situation in which both the electric and the magnetic fields are either time-independent or at least change very slowly with time. When either the electric or the magnetic field changes rapidly, new phenomena occur; these will be discussed in the next chapters.

We have discussed earlier, in Chapter 8, the problem of electric fields in dielectric insulators. We said there that in a homogeneous large insulator in the presence of electric charges (e.g., on some conductors enclosed in the dielectric) the electric field is everywhere a factor $1/K$ smaller than if that dielectric was replaced by air (or vacuum). How then are we to modify Maxwell's equations, which in vacuum take on the form given in Eq. (10–17)?

The first of Maxwell's equations is unchanged because it is equivalent to the statement that there exists an electrostatic potential. Thus, we will still write

I.
$$\oint_{\substack{\text{closed} \\ \text{loop}}} \mathbf{E} \cdot d\mathbf{l} = 0 \qquad\qquad (11\text{--}31)$$

even inside a dielectric. The second Maxwell's equation has to be changed, however. Instead of Eq. (10–17.II), we have to write inside a homogeneous dielectric

$$\oint_{\substack{\text{closed} \\ \text{surface}}} \mathbf{E} \cdot d\mathbf{S} = \frac{Q_{\text{inside}}}{K\epsilon_0} \qquad\qquad (11\text{--}32)$$

because then the electric field will be everywhere $1/K$ times the value it would have in vacuum.[*]

Eq. (11–32) is valid only if K is everywhere the same. More generally, we should write it in the form

$$\oint_{\substack{\text{closed} \\ \text{surface}}} K\epsilon_0 \mathbf{E} \cdot d\mathbf{S} = Q_{\text{inside}} \qquad\qquad (11\text{--}33)$$

because now K can vary on the closed surface over which we are integrating. Since we have

$$K\epsilon_0 \mathbf{E} = \epsilon_0 \mathbf{E} + (K - 1)\epsilon_0 \mathbf{E}$$

and we can use Eq. (8–21) to obtain

$$K\epsilon_0 \mathbf{E} = \epsilon_0 \mathbf{E} + \mathbf{P}$$

we can then write Maxwell's second equation in the form

II.
$$\oint (\epsilon_0 \mathbf{E} + \mathbf{P}) \cdot d\mathbf{S} = Q_{\text{inside}} \qquad\qquad (11\text{--}34)$$

Let us now turn our attention to the third and fourth Maxwell's equations, which involve the magnetic field. The third equation in free space reads

$$\oint_{\substack{\text{closed} \\ \text{loop}}} \mathbf{B} \cdot d\mathbf{l} = \mu_0 I_{\text{through loop}} \text{ (in free space)} \qquad\qquad (11\text{--}35)$$

In the presence of a magnetization, we have to include the magnetic field of the dipoles. We can repeat word for word the derivation of Eq. (11–17a), only now choosing an arbitrary current distribution instead of the currents in a long

---

[*] Note that Q_{inside} is only the *free* charge residing on conductors.

straight coil. We then obtain in exact analogy

III.

$$\int_{\substack{\text{closed} \\ \text{loop}}} (\mathbf{B} - \mu_0 \mathbf{M}) \cdot d\mathbf{l} = \mu_0 I_{\text{through loop}}$$

(11–36)

Finally, the fourth of Maxwell's equations is the same inside materials and in free space:

IV.

$$\int_{\substack{\text{closed} \\ \text{surface}}} \mathbf{B} \cdot d\mathbf{S} = 0$$

(11–37)

because it expresses the fact that there are no point sources of magnetic field (magnetic charges). This is true whether we are inside materials or not.

The four Maxwell equations which we wrote down are not complete; we have the relation between **P** and **E**

$$\mathbf{P} = (K - 1)\epsilon_0 \mathbf{E}$$

(11–38a)

and between **M** and **H**

$$\mathbf{M} = (\mu_r - 1)\mathbf{H} = (\mu_r - 1)\left(\frac{\mathbf{B}}{\mu_0} - \mathbf{M}\right)$$

(11–38b)

which we can consider as definitions of the dielectric constant K and the relative magnetic permeability μ_r. With the help of these quantities, we can recast Eq. (11–36) into a different form. From Eq. (11–38b) we obtain

$$(\mu_r - 1)\frac{\mathbf{B}}{\mu_0} = \mu_r \mathbf{M}$$

and thus

$$\mathbf{B} - \mu_0 \mathbf{M} = \mathbf{B} - \mu_0 \frac{\mu_r - 1}{\mu_r \mu_0} \mathbf{B} = \frac{\mathbf{B}}{\mu_r}$$

We can therefore write Eq. (11–36) in the form

III'.

$$\int_{\substack{\text{closed} \\ \text{loop}}} \frac{\mathbf{B}}{\mu_r \mu_0} \cdot d\mathbf{l} = I_{\text{through loop}}$$

(11–39a)

and thus obtain an equation analogous to Eq. (11–33):

II'.

$$\int_{\substack{\text{closed} \\ \text{surface}}} K\epsilon_0(\mathbf{E} \cdot d\mathbf{S}) = Q_{\text{inside}}$$

(11–39b)

In ferromagnetic substances we have $\mu_r > 1$, while in all dielectrics $K > 1$; thus, we see from Eqs. (11–39a,b) that the presence of ferromagnetic materials increases the magnetic field, while the presence of a dielectric reduces the electric field. Indeed, it is $\mathbf{B}/\mu_r$ and $K\mathbf{E}$ which stay the same.

In Chapter 8 we also discussed how to calculate electric fields if one or several media of different dielectric constants are involved. From the first Maxwell's equation we derived Eq. (8–26), which says that the tangential component of the electric field is the same on both sides of an interface between two media:

$$(E_t)_{\text{medium 1}} = (E_t)_{\text{medium 2}}$$

In exactly the same way, we can deduce from Eq. (11–39a) that on the interface between two media the tangential component of $\mathbf{B}/\mu_r$ is the same on both sides. Indeed, consider Fig. 11–12. We choose a closed loop going from A to B in medium 1 and back to A inside medium 2. Because of Eq. (11–39a), we must have

$$\int_{\text{loop}} \frac{\mathbf{B}}{\mu_r} \cdot d\mathbf{l} = 0$$

because we assume that there is no real electric current on the interface itself. But this implies, if we denote by $\mathbf{d}$ the vector from A to B, that

$$\left[\left(\frac{\mathbf{B}}{\mu_r} \right)_1 - \left(\frac{\mathbf{B}}{\mu_r} \right)_2 \right] \cdot \mathbf{d} = 0$$

and since $\mathbf{d}$ lies in the interface itself, we have for the tangential components

$$\boxed{\left(\frac{\mathbf{B}_t}{\mu_r} \right)_1 = \left(\frac{\mathbf{B}_t}{\mu_r} \right)_2} \qquad (11\text{–}40)$$

In Chapter 8 we also showed, using Gauss's law, that the normal components of $(K\mathbf{E})$ are the same on both sides of an interface. We can rederive this result very simply using Eq. (11–39b). Consider Fig. 11–13, where we have drawn a coin-like closed surface whose two sides lie on different sides of the interface. If there are no free charges on the interface, then Eq. (11–39b) says that

$$\int_{\text{coin}} K\mathbf{E} \cdot d\mathbf{S} = 0$$

But the coin is very flat, and thus all the contributions to the integral come from the two faces:

$$\int K\mathbf{E} \cdot d\mathbf{S} = (KE_n)_1 \pi r^2 - (KE_n)_2 \pi r^2 = 0$$

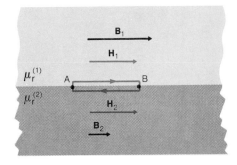

Figure 11–12 The tangential component of $\mathbf{B}/\mu = \mathbf{H}$ is the same on both sides of an interface between two different materials.

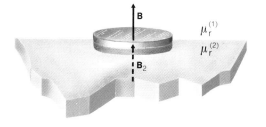

Figure 11-13   The normal component of **B** is the same on both sides of an interface.

from which we immediately conclude that

$$(\mathbf{KE}_n)_1 = (\mathbf{KE}_n)_2 \tag{11-41}$$

We can derive a similar result for the magnetic field. Using the fourth Maxwell's equation (11–37),

$$\int \mathbf{B} \bullet d\mathbf{S} = 0$$

we immediately obtain the result that

$$\boxed{(\mathbf{B}_n)_1 = (\mathbf{B}_n)_2} \tag{11-42}$$

In Table 11-4 we have summarized our results.

Example 5   Two good conductors carry each the charge $\pm Q$; they are completely enclosed by material of dielectric constant K and resistivity ρ (Fig. 11–14). Show that the current between the two is independent of the shape and separation of the two conductors.

This rather surprising statement is easily proved. We enclose either of the two charges by a closed surface S. By the second Maxwell's equation we have, if we choose the surface shown in Fig. 11–14,

$$\int_S \mathbf{E} \bullet d\mathbf{S} = \frac{Q}{K\epsilon_0}$$

Now let us calculate the total current I leaving the same conductor. At any point on the surface S there exists a current density $\mathbf{j}$. The total current I leaving the conductor inside S is

$$I = \int_S \mathbf{j} \bullet d\mathbf{S}$$

However, in material of resistivity ρ one has everywhere

$$\mathbf{j} = \frac{1}{\rho}\mathbf{E}$$

Since the resistivity is everywhere the same on S, we have

$$I = \int_S \frac{1}{\rho}\mathbf{E} \bullet d\mathbf{S} = \frac{1}{\rho}\int \mathbf{E} \bullet d\mathbf{S} = \frac{Q}{\rho K\epsilon_0}$$

As we had asserted, the current depends only on Q, ρ, and K; it is independent of the shape or separation of the two conductors.

TABLE 11–4 COMPARISON OF ELECTRIC AND MAGNETIC FIELDS INSIDE MATTER

	ELECTRIC	MAGNETIC
Free space closed loop integral	$\displaystyle\int_{\substack{\text{closed}\\\text{loop}}} \mathbf{E} \cdot d\mathbf{l} = 0$	$\displaystyle\int_{\substack{\text{closed}\\\text{loop}}} \mathbf{B} \cdot d\mathbf{l} = \mu_0\,(I_{\text{through loop}})$
Free space closed surface integral	$\displaystyle\int_{\substack{\text{closed}\\\text{surface}}} \mathbf{E} \cdot d\mathbf{S} = \dfrac{Q_{\text{inside}}}{\epsilon_0}$	$\displaystyle\int_{\substack{\text{closed}\\\text{surface}}} \mathbf{B} \cdot d\mathbf{S} = 0$
Sources	charge Q	current I
Dipole moment/unit volume	polarization $\mathbf{P}$	magnetization $\mathbf{M}$
Model used for calculating effect of dipole moments	"as if" surface charge density σ_d	many circular current loops throughout material
Maxwell's equations in matter	$\displaystyle\int_{\substack{\text{closed}\\\text{loop}}} \mathbf{E} \cdot d\mathbf{l} = 0$ $\displaystyle\int \left(\mathbf{E} + \dfrac{\mathbf{P}}{\epsilon_0}\right) \cdot d\mathbf{S} = \dfrac{Q_{\text{free}}}{\epsilon_0}\,(\text{inside})$	$\displaystyle\int_{\substack{\text{closed}\\\text{loop}}} (\mathbf{B} - \mu_0\mathbf{M}) \cdot d\mathbf{l} = \mu_0\,(I_{\text{free}})_{\text{(through loop)}}$ $\displaystyle\int \mathbf{B} \cdot d\mathbf{S} = 0$
Material properties	dielectric susceptibility χ_e dielectric constant $K = 1 + \chi_e$	magnetic susceptibility χ_m relative magnetic permeability $\mu_r = 1 + \chi_m$
Relations between dipole moment/unit volume and field	$\mathbf{P} = \chi_e \epsilon_0 \mathbf{E} = (K\text{-}1)\epsilon_0 \mathbf{E}$	$\mathbf{M} = \chi_m \mathbf{H} = \chi_m \left(\dfrac{\mathbf{B}}{\mu_0} - \mathbf{M}\right)$
If sources are equal, presence of dipoles	reduces $\mathbf{E}$ by factor $1/K$	increases $\mathbf{B}$ by factor μ_r
At interface between two different media one has continuity of	E_t = tangential component of $\mathbf{E}$ $(KE)_n$ = normal component of $K\mathbf{E}$	$\dfrac{B_t}{\mu_r}$ = tangential component of $\dfrac{\mathbf{B}}{\mu_r}$ B_n = normal component of $\mathbf{B}$

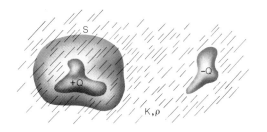

Figure 11-14 Example 5: Two charged metallic objects separated by a poorly conducting dielectric.

Example 6: Electromagnet   In Fig. 11–15 we show a typical electromagnet. The two coils each have $N = 100$ turns and the current $I = 10$ A flows through each coil. The gap height h is 2 cm. Estimate the magnetic field in the gap. The overall magnet size is $L = 30$ cm and the steel has a relative magnetic permeability $\mu_r > 3000$.

This seems a formidable problem, but is really a simple application of Maxwell's equations. From Eq. (11–39a) we know that if we choose the path C in Fig. 11–15, we have

$$\int_C \frac{\mathbf{B} \cdot d\mathbf{l}}{\mu_r \mu_0} = I_{\text{through}} \tag{11-43}$$

and since there are $2N$ turns in both coils and each carries a current I, we have

$$I_{\text{through}} = 2NI$$

Now consider Fig. 11–16, where we show in detail the bottom edge of the gap. We know from Eq. (11–42) that the normal component of $\mathbf{B}$ is the same inside the iron as it is in the air of the gap. We conclude that $\mathbf{B}$ in the iron will be roughly the same as in the air of the gap. We can now divide the integral in Eq. (11–43) into two parts:

$$\int_C \frac{\mathbf{B} \cdot d\mathbf{l}}{\mu_r \mu_0} = \frac{1}{\mu_0} \int_{\text{gap}} \mathbf{B} \cdot d\mathbf{l} + \frac{1}{\mu_r \mu_0} \int_{\text{iron}} \mathbf{B} \cdot d\mathbf{l} \tag{11-44}$$

The first integral is equal to

$$\frac{1}{\mu_0} \int_{\text{gap}} \mathbf{B} \cdot d\mathbf{l} \approx \frac{1}{\mu_0} |\mathbf{B}| h \tag{11-45a}$$

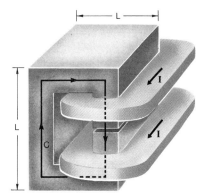

Figure 11-15 Example 6: Electromagnet.

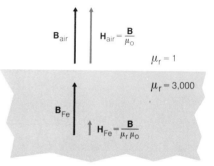

Figure 11–16 Calculating the magnetic field inside the gap of an electromagnet.

where h is the gap height. The second integral is much smaller, as we can see by the following argument: The total path in the iron is less than $4L$; thus, the second integral in Eq. (11–44) is roughly equal to

$$\frac{1}{\mu_r \mu_0} \int_{\text{iron}} \mathbf{B} \cdot d\mathbf{l} \approx \frac{|\mathbf{B}|}{\mu_r \mu_0} 4L = \left(\frac{|\mathbf{B}|}{\mu_0} h\right) \frac{4L}{\mu_r h} \tag{11–45b}$$

and thus is smaller than the expression in Eq. (11–45a) by a factor

$$\frac{4L}{\mu_r h} < \frac{4 \times 0.3 \text{ m}}{3000 \times 0.02 \text{ m}} = 0.02 = 2\%$$

Thus, to 2% accuracy we can write, instead of Eq. (11–43),

$$\int_{\text{gap}} \frac{\mathbf{B} \cdot d\mathbf{l}}{\mu_r \mu_0} \approx \frac{1}{\mu_0} |\mathbf{B}| h = I_{\text{through}} = 2NI$$

or

$$|\mathbf{B}| = \frac{2\mu_0 NI}{h}$$

Numerically, we obtain

$$|\mathbf{B}| = 2 \times 4\pi \times 10^{-7} \frac{\text{Wb}}{\text{A-m}} \times \frac{100 \text{ turns} \times 10 \text{ A}}{0.02 \text{ m}} = 0.126 \frac{\text{Wb}}{\text{m}^2}$$

Of course, ours was not an exact calculation; nevertheless, the estimate, good to 2% in our case, gives us a very useful method for calculating magnetic fields in electromagnets. Note that the exact value of the magnetic permeability of the iron does not matter; as long as it is sufficiently large, the contribution to the loop integral from inside the iron is small.

5. FERROELECTRICS

There exists an electric effect in some materials which is analogous and quite similar to ferromagnetism. The so-called ferroelectric materials show an extremely large dielectric constant. Examples of ferroelectric materials are barium titanate ($BaTiO_3$) and potassium dihydrogen phosphate (KH_2PO_4). If we plot the polarization $\mathbf{P}$ in a sample of such material, we obtain a curve such as that shown in Fig. 11–17. Just as we did for ferromagnetics, we can define the saturation polarization P_s, the remanent polarization P_r which persists after the electric field is switched off, and the coercive field E_c which is needed to reduce the polarization to zero.

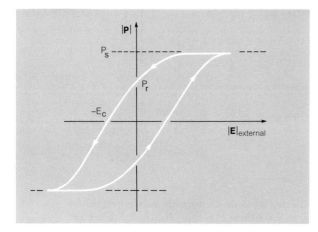

Figure 11-17 For ferroelectric materials one can define the saturation polarization P_s, the remanent polarization P_r, and the coercive electric field $-E_c$.

Even on the microscopic level, ferroelectrics are somewhat similar to ferromagnetic substances. We can understand the phenomenon if we analyze more closely the Clausius-Mosotti equation (8–45):

$$\frac{K-1}{K+2} = \frac{N\alpha}{3\epsilon_0} \tag{11-46}$$

We can solve it for K to obtain

$$K = \frac{1 + (2N\alpha/3\epsilon_0)}{1 - (N\alpha/3\epsilon_0)} \tag{11-47}$$

we see that if the polarizability α is sufficiently large so that

$$\frac{N\alpha}{3\epsilon_0} \simeq 1 \tag{11-48}$$

then the dielectric constant K becomes infinite. If α is larger than the limit value given by Eq. (11–48), spontaneously polarized domains develop in the crystal. As it happens in ferromagnetic substances, in the presence of an external electric field the domains oriented parallel to the field will grow at the expense of those whose polarization vectors point in other directions. The net macroscopic effect is a very large dielectric constant ($K \approx 3000$ for $BaTiO_3$) for small applied electric fields. However, if the field is increased, the polarization saturates once all domains are oriented parallel to the electric field. The net polarization, once acquired, will partially persist even when the external electric field is removed.

There exists even an electric analogue to the magnetostrictive effect; upon applying pressure on a single crystal of a ferroelectric substance, the material acquires a net polarization. Similarly, the crystal will deform slightly if it is polarized (e.g., by being inserted between the plates of a charged capacitor). This phenomenon is known under the name of *piezoelectric effect*.

While all ferroelectrics are also piezoelectrics, there are substances which show the piezoelectric effect although they have dielectric constants in the "normal" range $K = 2$ to 10. Thus, quartz (SiO_2) is piezoelectric, while its dielectric constant at room temperature is only $K = 4$. Indeed, quartz is the oldest known piezoelectric substance—mainly because single crystals can be found in nature. The piezoelectric effect is used in "quartz clocks" in which a crystal is made to vibrate at its natural frequency (standing sound waves) by applying an oscillating electric field. Positive feedback from the crystal to the electric oscillator keeps the oscillator frequency exactly at resonance. The result is a harmonic oscillation at a frequency which depends only on the crystal size, its intrinsic properties, and the ambient temperature. The frequency can be kept constant to one part in 10^8 over long periods of time; therefore, such crystal oscillators are used as time standards.

Summary of Important Relations

$\mathbf{M}$ = magnetization = magnetic dipole moment per unit volume

Paramagnetic materials $\mathbf{M} = \dfrac{N|\mathbf{m}|^2}{3kT}\mathbf{B}$

    susceptibility $\chi = \dfrac{\mu_0 N|\mathbf{m}|^2}{3kT}$

Diamagnetic materials $\mathbf{M} \approx -\dfrac{Ne^2 r^2}{2m}\mathbf{B}$ $\chi \approx -\dfrac{\mu_0 Ne^2 r^2}{2m}$

In general: $\mathbf{M} = \chi\mathbf{H}$ where $\mathbf{H}$ = magnetizing field

$\mathbf{B} = (1 + \chi)\mu_0\mathbf{H} = \mu_r\mu_0\mathbf{H} = \mu_0(\mathbf{H} + \mathbf{M})$

Ferromagnetic materials:

$$\chi = \frac{C}{T - T_C} \qquad \begin{array}{l} C = \text{Curie constant} \\ T_C = \text{Curie temperature} \end{array}$$

See also Table 11–4 for a summary of Maxwell's equations in matter.

Questions

1. In a newspaper interview, a man proposes a new theory that the earth's magnetic field is due to the permanent magnetization of the molten iron core of the earth. Is the man a crackpot?

2. Permanent magnets and electromagnets frequently have conical pole pieces. Explain why this increases the magnetic field in the gap.

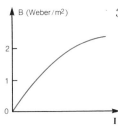

3. If one plots the magnetic field in an electromagnet against the current through the coil, one typically obtains a graph like the one shown in the figure. Does this contradict Example 6? Explain.

4. If a box made of ferromagnetic material is in the magnetic field of the earth, the magnetic field inside the box is much smaller. This is called magnetic shielding. Explain the effect.

5. Show that the magnetic susceptibilities for oxygen and nitrogen given in Table 11–1 are reasonable. Use arguments similar to those given in Examples 1 and 2.

Problems

1. A cylindrical tantalum rod ($\chi = +1.8 \times 10^{-4}$) of 1 cm diameter and 5 cm length is inserted in a magnetic field of 1.5 webers/m². What is the total magnetic dipole moment of the rod?

2. You are asked to mix powdered silver and aluminum to produce a powder of exactly vanishing magnetic susceptibility. What ratios by weight would you use for the mixture?

3. In going from Eq. (11–8) to Eq. (11–9), we had to assume that Δv is small (compared to what?). Show that this is indeed true for any reasonable magnetic field.

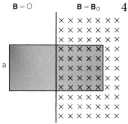

4. A sheet of tantalum ($\chi = +1.8 \times 10^{-4}$, width $a = 10$ cm, thickness $d = 1$ mm) is partially in and partially out of a magnet of magnetic field $B_0 = 1.2$ webers/m². Show that there is a force on the sheet pulling it into the magnet and calculate its magnitude. (Hint: Calculate the change in potential energy $-(\mathbf{m} \cdot \mathbf{B})$ due to the induced magnetization in the sheet if the rod is moved in by a small distance dx.)

5. Give an estimate of the Curie constant C for iron, assuming that two electrons per atom participate with their spin magnetic moments of one Bohr magneton. What would you predict to be the susceptibility of iron at a temperature of 800° C? Compare with Fig. 11–5.

°6. *Einstein-de Haas Effect.* An iron cylinder of radius $R = 0.5$ cm and length $L = 10$ cm is suddenly magnetized to a magnetization $M = 10^6$ A/m.
 (a) What will be the change in total macroscopic angular momentum of the cylinder?
 (b) If the cylinder was at rest before the magnetization, what is its angular velocity when magnetized? The moment of inertia of a cylinder of mass m and radius R around its axis is $I = mR^2/2$.

°7. You are asked to produce a rigid block of dimensions $10 \times 20 \times 10$ cm out of electrically insulating material with relative magnetic permeability $\mu_r = 3.5$. How would you solve this problem? Be as specific as you can.

8. A compass needle (iron) has a mass $m = 5$ g and a magnetization $M = 2 \times 10^5$ A/m. Calculate the torque on the needle if it points east-west. Assume a horizontal component of the earth's magnetic field of 0.3 gauss.

9. An electromagnet has a gap size $d = 1.0$ cm. It is driven by two coils of 100 turns each. What current should one use to obtain a magnetic field of 8000 gauss in the gap?

10. A cold-rolled steel ($\mu_r = 3500$) bar is surrounded by a coil of much larger diameter with $N = 100$ turns/meter. A current $I = 1$ A flows through the coil. What is

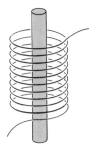

    (a) the magnetic field in the iron,

    (b) the magnetization of the iron,

    (c) the magnetic field inside the coil but outside the iron bar?

11. The magnetization M in terms of the magnetizing field H can be approximated by the relation

$$M = a \tan^{-1}(bH \pm c)$$

where the $\pm$ sign is valid for the two legs of the magnetization curve. Calculate the saturation magnetization M_s, the remanent magnetization M_r, and the coercive field H_c in terms of the parameters a, b, and c.

12. A permanent cylindrical bar magnet (diameter $r = 1$ cm, length $L = 5$ cm) has a magnetization $M = 5 \times 10^5$ A/m along its axis. Estimate at what distance away from the bar (along its axis) the magnetic field has the magnitude of one gauss. (Use the dipole approximation.)

13. Show that at the surface of a ferromagnetic material ($\mu_r \gg 1$) the magnetic field is always nearly perpendicular to the surface.

14. A block of iron ($\mu_r = 5000$) is magnetized by an external coil so that the magnetic field in the iron is $B_0 = 0.3$ weber/m². A small cavity in the shape of a flat disk is cut inside the iron. Calculate the magnetic field inside the cavity

    (a) If the face of the cavity is exactly parallel to $\mathbf{B}_0$ ($\theta = 0$);

    (b) if the angle θ between the face of the disk and $\mathbf{B}_0$ is $\theta = 1°$.

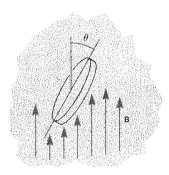

15. A toroid of minor radius $r = 1$ cm and major radius $R = 20$ cm is made of iron of relative permeability $\mu_r = 5000$. A coil with 100 turns is wound around the toroid. What current should flow through the coil to produce an average magnetization of $M = 2 \times 10^5$ A/m in the toroid?

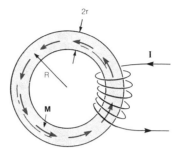

16. A permanent magnet is shaped like a toroid and has a small gap. The magnetization of the magnet is $M = 5 \times 10^5$ A/m. What is the magnetic field in the gap?

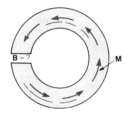

CHAPTER 12

FARADAY'S LAW

1. INDUCED CURRENTS

Up to now we have considered only charges which were either at rest or moving with constant velocities, thus producing steady-state currents. In this and the next chapter we will drop this restriction. We will learn that many new phenomena occur when rapidly changing currents — or rapidly changing electromagnetic fields — are involved.

We can perform a few very simple experiments to find out what new types of phenomena we have to discuss. Let us use a coil of wire with some 10 to 100 turns; we connect its ends to some current-measuring device, such as a galvanometer. We now take a permanent magnet — an ordinary bar magnet is adequate — and move one of its ends past the coil. In Figs. 12–1 and 12–2 we have

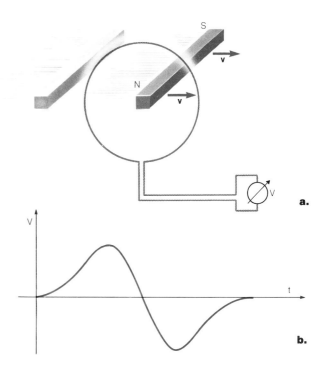

Figure 12–1 An electromotive force will appear on a closed loop or a coil whenever a permanent bar magnet is moved past it.

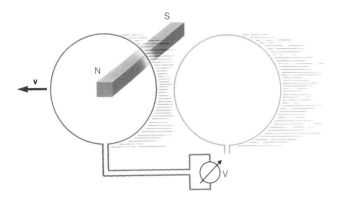

Figure 12-2 The same emf will appear if the magnet is held steady and the coil is moved instead.

drawn the coil as consisting of a single loop. As the magnet moves past the loop, we will see the galvanometer needle first swing one way; when the magnet has passed the center of the loop and is again moving outward, the galvanometer needle will now swing in the other direction. We conclude that the magnetic field of the bar magnet produced a current in the coil.

We can repeat the experiment in a slightly different fashion: we hold the bar magnet steady and move the coil past the magnet. The galvanometer needle will indicate the same current as in the previous experiment: first in one direction, later in the opposite direction.

Let us now perform a third experiment: we install our coil with the galvanometer opposite another coil (Fig. 12–3) which is connected to a battery B over the switch S. When we close the switch, the galvanometer needle will briefly swing in one direction. When we then open the switch again, the galvanometer will again briefly swing out, but in the opposite direction. If we keep the switch either open or closed for a long period of time, no indication of any current will be seen on the galvanometer. We can also modify this setup by pushing an iron bar through both coils. Again the galvanometer will swing from its quiescent position at the instant we either close or open the switch. However, the swing is much larger with the iron bar than without it.

We conclude that the changing magnetic field in the neighborhood of the coil produces an electromotive force, which in turn produces the current through the coil. We could convince ourselves that it is the emf and not directly the current which is produced by the time-dependent magnetic field; for this

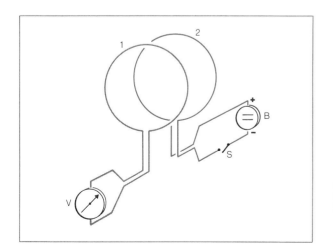

Figure 12-3 An electromotive force appears for an instant in coil 1 whenever the switch S connecting coil 2 to a power supply is opened or closed.

purpose we would vary the resistance R of the circuit and find that the current is indirectly proportional to R; thus,

$$RI = \mathscr{E}$$

is the same whatever the resistance of the coil, if only the magnetic field always changes in the same way.

By further detailed study we can convince ourselves that the emf in a simple loop of wire is *proportional to the rate of change of magnetic flux* through the loop. We define the magnetic flux in the following way (Fig. 12–4): If we traverse the loop in the direction given by the arrow, then for a surface which has the loop as a boundary we can define a unique direction for the normal to the surface using the right-hand rule: curve the right hand so that the four fingers point in the direction the loop is being traversed; then the normal points in the direction of the thumb. With this definition of the normal, we can define the flux Φ through the surface

$$\Phi = \int_S \mathbf{B} \cdot d\mathbf{S} = \int_S (\mathbf{B} \cdot \mathbf{n}) dS \qquad (12\text{–}1)$$

Faraday's law states that the electromotive force calculated around the loop in the direction of the arrow is equal to the decrease in flux through the loop:

$$\boxed{\mathscr{E} = -\frac{d\Phi}{dt} = -\frac{d}{dt} \int_S \mathbf{B} \cdot d\mathbf{S}} \qquad (12\text{–}2a)$$

It forms a new law, which modifies Maxwell's equations as we have learned to know them so far.

If the coil in which we measure the emf has N turns, the emf will also be N times larger:

$$\mathscr{E} = -N\frac{d\Phi}{dt} = -N\frac{d}{dt} \int_S \mathbf{B} \cdot d\mathbf{S} \qquad (12\text{–}2b)$$

Figure 12–4 Definition of magnetic flux through a loop. Once the direction of the loop is defined arbitrarily, the direction of the normal $d\mathbf{S}/|d\mathbf{S}|$ is uniquely defined by the right-hand rule.

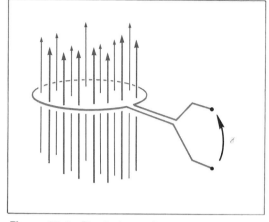

Figure 12-5 Physical meaning of the emf produced by a changing magnetic flux; if the loop is open, the emf is the integral $\int \mathbf{E} \cdot d\ell$ between the terminals.

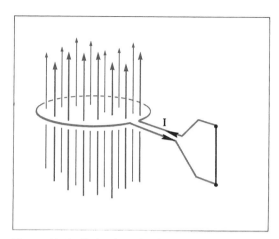

Figure 12-6 If the loop is closed, a current $I = \mathscr{E}/R$ will flow through the loop resistance R.

We should refresh our memory as to the meaning of an electromotive force. It was defined as the integral of the electric field

$$\mathscr{E} = \int \mathbf{E} \cdot d\mathbf{l} \qquad (12\text{-}3)$$

between two terminals. Thus, if the wire loop is interrupted somewhere, $\mathscr{E}$ will be the potential difference between the two ends (Fig. 12-5). If, however, the wire ends touch each other (Fig. 12-6), a current will flow and the internal resistance will determine the electric field distribution. Faraday's law simply states that whatever the actual physical arrangement, it is as if the emf was distributed along the wire in such a form that over any closed loop

$$\Sigma \, \mathscr{E}_i = \mathscr{E} = -\frac{d\Phi}{dt} \qquad (12\text{-}4)$$

We will postpone for the next chapter a detailed discussion of the origin of the emf.

We also stress here that Faraday's law prohibits us from defining an electrostatic potential V if there is a time-dependent magnetic field. In Chapter 6 we defined the electrostatic potential after we had convinced ourselves that the electric field of stationary charges is conservative, i.e., that for any closed loop

$$\int_{\substack{\text{closed} \\ \text{loop}}} \mathbf{E} \cdot d\mathbf{l} = 0$$

However, Faraday's law states that this is no longer true if there are time-dependent magnetic fields; thus, in this case we have to replace the concept of an electrostatic potential by the concept of electromotive force. Both have the same dimension and are measured in units of volts.°

° We repeat again that the emf is *not* a force; it is the integral of an electric field over a closed loop.

Example 1   A coil of $N = 10$ turns is placed inside an electromagnet (Fig. 12-7). The magnetic field increases linearly with time and reaches the value $B_0 = 1$ Wb/m² at the time $T = 10$ sec; it remains constant thereafter. The coil has an area $A = 100$ cm² and is oriented perpendicularly to the magnetic field. (a) What is the total induced electromotive force in the coil? (b) The coil has a resistance $R = 0.5\ \Omega$. If the ends of the coil are shorted, what will be the current in the coil? (c) What is the total energy dissipated in the wire during the time it takes to turn the magnet on?

a. First we calculate the emf. The flux through the coil is

$$\Phi = |\mathbf{B}|A \tag{12-5a}$$

where $\mathbf{B}$ is increasing linearly with time:

$$\mathbf{B} = \mathbf{B}_0 \frac{t}{T} \tag{12-5b}$$

Thus, the total electromotive force is

$$\mathscr{E} = -N\frac{d\Phi}{dt} = -NA\frac{d|\mathbf{B}|}{dt} \tag{12-5c}$$

Substituting from Eq. (12-5b), we obtain[°]

$$\mathscr{E} = NA\frac{B_0}{T} = 10 \times 10^{-2}\ \text{m}^2 \times \frac{1\ \text{Wb/m}^2}{10\ \text{sec}} = 0.01\ \text{V} \tag{12-6}$$

b. If the coil is shorted, the voltage drop due to the current I has to be equal to the total electromotive force:

$$\mathscr{E} = RI$$

Solving for I, we obtain

$$I = \frac{|\mathscr{E}|}{R} = 0.02\ \text{A} = 20\ \text{mA} \tag{12-7}$$

c. The dissipated energy can be calculated as follows. The current will flow for the time T; thus, the total charge flowing through any point of the coil will be

$$Q = IT = \frac{\mathscr{E}T}{R} = 0.2\ \text{coul}$$

[°] We remind the reader that 1 Wb = 1 V-sec. Since we are not asked for the direction (sign) of $\mathscr{E}$, we can use the absolute value, i.e., ignore the (−) sign in Eq. (12-5c).

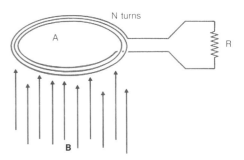

Figure 12-7 Example 1.

We can think of this charge Q having traversed the "potential difference" (actually emf) of $\mathscr{E}$. The total dissipated energy is then

$$W = \mathscr{E}Q = 0.01\,\mathrm{V} \times 0.2\,\mathrm{coul} = 0.002\,\mathrm{Joule}$$

We can calculate instead the dissipated power, i.e., the energy dissipated per unit time at any instant:

$$P = \mathscr{E}I = RI^2 \qquad (12\text{–}8)$$

and obtain the total energy as the time integral of the dissipated power

$$W = \int P\,dt = RI^2T$$

which gives us again

$$W = \mathscr{E}IT = \mathscr{E}Q$$

We have not yet discussed the importance of the minus sign in Eq. (12–2). It gives us the sign of the electromotive force

$$\mathscr{E} = \int_{\text{loop}} \mathbf{E} \cdot d\mathbf{l}$$

The direction of the integral over a loop is defined by the right-hand convention, as sketched in Fig. 12–4. Then $\mathscr{E}$ will be positive if the magnetic flux through the loop decreases, and negative if the flux increases. We stress that the emf is considered to be positive in the direction that a current would flow if the circuit were closed; but the emf is there whether or not a current actually flows through the wire loop.

There exists a rule which is easy to remember and helps us to keep the

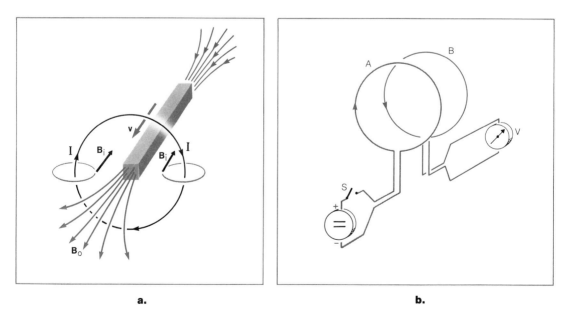

a.   b.

Figure 12–8 *Lenz's law* states that the magnetic field of the induced current opposes the change in flux through the loop. Thus (a) as the bar magnet is pushed toward the loop, the induced magnetic field $\mathbf{B}_i$ opposes the field $\mathbf{B}_0$; (b) upon closing or opening of the switch S, the current in coil B has the direction that will minimize the change of magnetic flux through B.

direction of current flow straight in our minds. It is called *Lenz's law*. It states that *the induced current always produces a magnetic flux through the loop which opposes the externally applied change*. To illustrate what we mean by this law, we show in Fig. 12–8a a bar magnet moving toward the coil and the direction of the field $\mathbf{B}_0$ of the bar magnet at the coil. Since the externally applied field increases, the induced current is oriented so that the magnetic field $\mathbf{B}_i$ that it produces opposes $\mathbf{B}_0$ *inside the coil*. Thus, the current comes out of the page at the top and goes into the page at the bottom. The same situation occurs if the external magnetic field is produced by the current in a wire loop A which is situated close to the test loop B (Fig. 12–8b). If we increase the current in A, the induced current in B will be opposite to the current in A; if we decrease the current in A, the induced current in B will point in the direction the current in A had originally. In both cases the induced current is oriented so as to preserve as much as possible the *status quo ante*; i.e., the total current through both loops and with it the total magnetic flux through the two loops.

> *Example 2*   A rectangular coil is moved out of a region where there is a magnetic field $\mathbf{B}_0$. What will be the direction of the flow of current? The magnetic field in Fig. 12–9 points into the paper.
>
> The flux through the coil is decreasing; thus, the induced current in the coil should produce a magnetic field which points *inside the coil* in the same direction as $\mathbf{B}_0$ (into the paper). Using the right-hand rule for determining the magnetic field of a straight wire, we easily see that the current flows in the direction of the arrow.

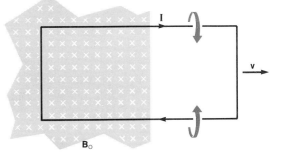

Figure 12–9 Example 2.

> *Example 3*   A wire loop, shown in Fig. 12–10, is connected to a parallel plate capacitor. The magnetic field $\mathbf{B}$ in the region shown is increasing in magnitude; its direction is into the paper. Which of the two capacitor plates is positively charged?
>
> The magnetic field is increasing; thus, if the capacitor is replaced by a short (dotted line in Fig. 12–10), the induced current would produce a magnetic field which, inside the loop, points up (out of the paper). Thus, the direction of the emf

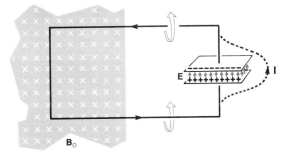

Figure 12–10 Example 3.

is counterclockwise around the loop. Since no current can flow in the actual circuits shown (except briefly at the start to charge up the capacitor plates), there can be no electric field in the wire; the whole emf

$$\mathscr{E} = \int_{\text{loop}} \mathbf{E} \cdot d\mathbf{l}$$

is concentrated in the capacitor. The electric field has to point in the direction shown (up); thus, the bottom plate will be positively charged.

2. ELECTROMAGNETIC GENERATORS AND MOTORS

Faraday's law gives us the means to convert mechanical energy into electrical energy. The simplest electromagnetic generator consists of a coil which is being cranked around an axis perpendicular to an external magnetic field (Fig. 12–11). We have redrawn the coil in Fig. 12–12 to show the angle ϕ between the plane of the coil and the magnetic field $\mathbf{B}_0$. If the area of the coil is A and the coil has N turns, the magnetic flux at any instant through the area of the coil is

$$\Phi = B_0 A \cos \theta = B_0 A \sin \phi \qquad (12\text{–}9)$$

If the coil is cranked at a constant rate, the angle ϕ will increase linearly with time:

$$\phi = \omega t + \phi_0$$

Since the point at which we start measuring is arbitrary, we can set $\phi_0 = 0$. The period of revolution T is the time for the angle ϕ to increase by 2π. Thus, we have $2\pi = \omega T$ or

$$T = \frac{2\pi}{\omega} \qquad (12\text{–}10)$$

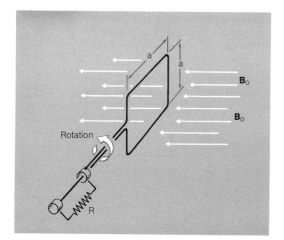

Figure 12–11 Electromagnetic generator: a coil is rotating in a magnetic field. A current is produced in the resistance R.

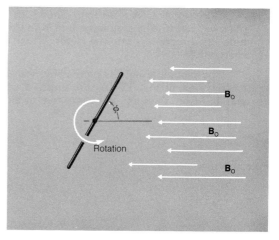

Figure 12–12 Calculation of flux through the coil of the electromagnetic generator.

Since the flux through the coil changes with time, there will be an induced emf

$$\mathscr{E} = -N\frac{d\Phi}{dt} = -N\frac{d}{dt}(B_0 A \sin \omega t)$$

and performing the differentiation we obtain

$$\mathscr{E} = -NAB_0\omega \cos \omega t \qquad (12\text{--}11a)$$

Note that the emf will vary periodically with time. The peak value is

$$\mathscr{E}_{max} = NAB_0\omega \qquad (12\text{--}11b)$$

The peak value is indicated in Fig. 12–13, where we show the time dependence of the emf.

As long as the coil is open, there is no current flowing and thus no electric energy has been produced. To study the energy production, let us imagine that we connect the coil to a resistor of resistance R (e.g., a light bulb). Then the current

$$I = \frac{\mathscr{E}}{R} = -\frac{NAB_0\omega}{R}\cos \omega t \qquad (12\text{--}12)$$

will flow through the resistor. The dissipated power at any instant is

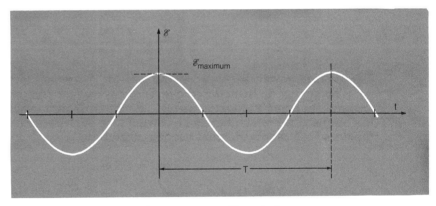

Figure 12–13   The emf of the generator oscillates in time.

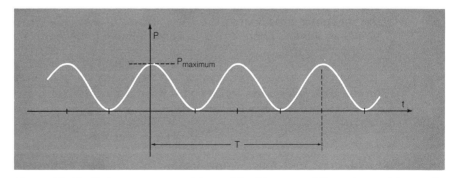

Figure 12–14   If the generator emf is used to drive current through a resistance, the power delivered by the generator also varies with time.

$$P = RI^2 = \frac{\mathscr{E}^2}{R} = \frac{N^2 A^2 B_0{}^2 \omega^2}{R} \cos^2 \omega t \qquad (12\text{--}13)$$

The dissipated power is again not constant; it varies between the peak power

$$P_{\max} = \frac{N^2 A^2 B_0{}^2 \omega^2}{R} \qquad (12\text{--}14)$$

and zero, as sketched in Fig. 12–14. The average power can be obtained as the energy deposited in the resistor during a full period T divided by the period itself:

$$<P> = \frac{1}{T} \int_0^T P \, dt \qquad (12\text{--}15)$$

Since we have

$$\int \cos^2 (\omega t) \, dt = \frac{1}{2\omega} (\sin \omega t \cos \omega t + \omega t)$$

we obtain for the average power

$$<P> = \frac{P_{\max}}{T} \frac{1}{2\omega} \left[\sin \omega t \cos \omega t + \omega t \right]_{t=0}^{t=T}$$

and since $\sin (0) = \sin (2\pi) = 0$, we have because of Eq. (12–10):

$$<P> = \frac{P_{\max}}{2\pi/\omega} \frac{1}{2\omega} \left[\omega\left(\frac{2\pi}{\omega}\right) \right] = \frac{P_{\max}}{2} \qquad (12\text{--}16)$$

The average power dissipated in the resistor is one-half of the peak power.

The energy dissipated in the resistor has to come from somewhere; we have to supply energy by cranking the coil, thus supplying a torque. We can calculate the torque if we remember that a current loop such as our coil forms a magnetic dipole whose dipole moment has the magnitude

$$|\mathbf{m}| = (NI)A \qquad (12\text{--}17a)$$

The direction of the dipole moment can be seen from Fig. 12–15. In the position shown, the angle ϕ is less than $\pi/2$ $(= 90°)$ and increasing. The flux through the coil is increasing; thus, by Lenz's law the emf and thus the current have to produce a magnetic flux in opposite directions: the current is flowing in the direction of the arrow. Referring back to the definition of the magnetic dipole (Chapter 10, p. 328), we note that the dipole moment vector points in the same direction as the magnetic field inside the loop. Thus, in Fig. 12–15b the dipole moment points as shown. There exists in consequence a mechanical torque

$$\boldsymbol{\tau} = \mathbf{m} \times \mathbf{B}_0 \qquad (12\text{--}17b)$$

on the coil, which has to be counteracted by the external crank. We therefore have to supply externally the torque

$$\boldsymbol{\tau}' = -\boldsymbol{\tau} = -\mathbf{m} \times \mathbf{B}_0$$

to keep the coil moving (neglecting any friction in the mechanical parts). Its magnitude is

$$|\boldsymbol{\tau}| = |\mathbf{m}| B_0 \sin \alpha = |\mathbf{m}| B_0 \cos \phi \qquad (12\text{--}17c)$$

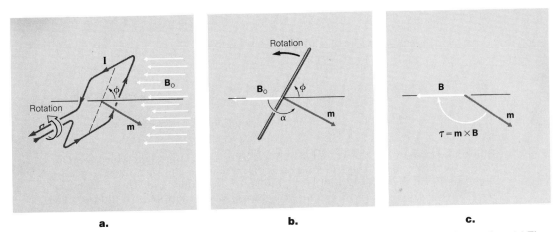

Figure 12–15 The current in the generator produces a torque which opposes the rotation. (a) The direction of the dipole moment vector of the coil; (b) the angles α and ϕ; (c) determining the direction of the torque.

and using Eq. (12–17a) for the magnetic dipole moment and Eq. (12–12) for the current I, we obtain

$$|\tau| = NIAB_0 \cos \omega t = \frac{N^2 A^2 B_0^2}{R} \omega \cos^2 \omega t \qquad (12\text{–}18)$$

The mechanical energy expended by rotating the coil by a small angle $d\phi$ is

$$dW = |\tau| \, d\phi$$

and thus the mechanical power, the energy expended per unit time, is

$$P_m = \frac{dW}{dt} = |\tau| \frac{d\phi}{dt} = |\tau| \omega$$

Using Eq. (12–18), we obtain

$$P_m = \frac{N^2 A^2 B_0^2 \omega^2}{R} \cos^2 \omega t \qquad (12\text{–}19)$$

which is exactly the same as we obtained in Eq. (12–13) for the electric power. This has to be so if the law of energy conservation is to hold: the electric power extracted from the generator has to exactly equal the mechanical power driving the generator at any instant.

Example 4   A generator consists of a coil with $N = 20$ turns and has a circular area of radius $R = 20$ cm (≈ 8 inches). How fast does it have to turn to deliver a peak emf of 160 volts if the magnetic field is 0.2 Wb/m²?

The peak emf is given by Eq. (12–11b)

$$\mathscr{E}_{\max} = NAB_0 \omega$$

from which we obtain for the period

$$T = \frac{2\pi}{\omega} = \frac{2\pi NAB_0}{\mathscr{E}_{\max}}$$

Numerically, we obtain

$$T = \frac{6.28 \times 20 \times 3.14 \times (0.2 \text{ m})^2 \times 0.2 \text{ Wb/m}^2}{160 \text{ V}} = 1.97 \times 10^{-2} \text{ sec}$$

Thus, the generator will run at 50.7 cycles/sec or

$$50.7 \times 60 \frac{\text{sec}}{\text{min}} = 3040 \text{ r.p.m. (rotations per minute)}$$

Household current has a peak emf of $115 \sqrt{2}$ volts $= 162$ volts at 60 cycles/sec. Our values are rather close to this; it seems clear that all parameters are quite reasonable. A coil of 40 cm diameter is a bit large, but we could reduce the size of the coil if we are willing to increase the number of turns, N. We see that it would be quite easy to build a generator for household use.

Example 5   We have shown up to now only that the electric power of a generator is equal to the supplied mechanical power if the current I at any instant proportional to the emf $\mathscr{E}$. Show that mechanical power equals electric power for any current $I(t)$.

We could ask ourselves whether we can find an arrangement such that $I(t)$ is not proportional to $\mathscr{E}(t)$. A simple arrangement of this type is shown in Fig. 12–16: the generator charges a capacitor C through a resistance R. The resistor will reverse its current each half period; but the current now will depend not only on the emf but also on the charge already stored on the capacitor. We conclude that the current $I(t)$ can have a time dependence different from that of the emf given by Eq. (12–11a)

$$\mathscr{E}(t) = -NAB_0\omega \cos \omega t$$

However, the electric power supplied by the generator at any instant is always equal to

$$P_{\text{el}} = \mathscr{E}(t)I(t) \tag{12–20}$$

Now let us go back to the derivation of Eq. (12–19), the expression for the mechanical power (torque × angular velocity) which we had to supply to keep the generator running. The torque we have to *supply* is still given by

$$\boldsymbol{\tau} = -\mathbf{m} \times \mathbf{B}_0$$

and the magnetic moment is still given by Eq. (12–17a)

$$|\mathbf{m}| = NAI(t)$$

The magnitude of the torque thus is still given by Eq. (12–17c)

$$|\boldsymbol{\tau}| = |\mathbf{m}|B_0 \cos \phi$$

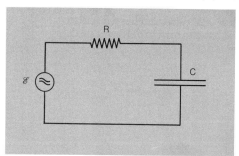

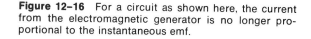

Figure 12–16 For a circuit as shown here, the current from the electromagnetic generator is no longer proportional to the instantaneous emf.

from which we conclude that

$$|\boldsymbol{\tau}| = (NAB_0 \cos \omega t)I(t)$$

The mechanical power thus is

$$P_{\text{mech}} = |\boldsymbol{\tau}|\omega = (NAB_0\omega \cos \omega t)I(t)$$

or, because of Eq. (12–11a),

$$P_{\text{mech}} = \pm\mathscr{E}(t)I(t) \qquad\qquad (12\text{–}21)$$

We have to put a $\pm$ sign in Eq. (12–21) because we have calculated only the absolute value of the mechanical power; we have not yet checked whether we have to supply the power or whether we can obtain mechanical power from the generator — in which case we would have invented a "perpetual motion machine"! Let us thus look again carefully at Fig. 12–15. As long as the current is in the same direction as the emf $\mathscr{E}$, the torque produced by the magnetic moment of the coil

$$\boldsymbol{\tau} = + (\mathbf{m} \times \mathbf{B})$$

opposes the direction of the rotation. Thus, we have to supply a torque *driving the coil*, as long as $\mathscr{E}(t)$ and $I(t)$ have the same sign. But this means that if the electric power obtained *from the generator*

$$P_{\text{el}} = \mathscr{E}(t)I(t)$$

is positive, then the mechanical power *supplied to the generator* is also positive

$$P_{\text{mech (supplied)}} = P_{\text{el (obtained)}} \qquad\qquad (12\text{–}22)$$

Thus we have no perpetuum mobile; energy is conserved.

The last example points out a very important fact: Suppose that by connecting the terminals of the generator to an external source of emf we force a current through the generator such that

$$\mathscr{E}(t)I(t) < 0$$

We are now supplying electric energy to the "generator"; the energy conservation law (Eq. 12–22) now states that we can obtain mechanical energy. In other words, the generator has become an electric motor. The torque $\mathbf{m} \times \mathbf{B}$ produced by the current drives the coil — the coil will accelerate if there is no external torque. Note that, in general, the driving torque of such a simple motor will be time-dependent; we would have to rely on the inertia of the rotor (coil + shaft) to keep the angular velocity nearly constant if the machine (e.g., a lathe) driven by the motor requires a constant speed. Because such considerations make the discussion a bit tricky, we will instead consider a simpler arrangement, called the linear motor.

Example 6: Linear Motor   A rod of length $L = 10$ cm lies atop two highly conducting rails (Fig. 12–17); there is a potential difference $V_0 = 15$ V between the rails. The resistance of the rod is $R = 0.1\ \Omega$. The rod is connected via a rope over a pulley to a mass $m = 1.2$ kg. Calculate the steady-state velocity of the rod along the rail, if the magnetic field is $B_0 = 1.0$ Wb/m² in the direction shown.

Our example is a linear (one-dimensional) analog of a rotating coil driven by an external voltage. However, the voltage in our case is constant. When a current I is

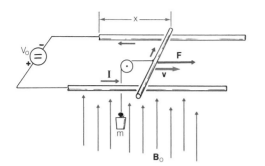

Figure 12–17 Example 6: The linear motor.

flowing through a wire there will be a force (compare p. 325)

$$|\mathbf{F}| = (IL)B_0 \qquad\qquad (12\text{–}23a)$$

acting toward the right. (The reader should convince himself that in Fig. 12–17 the resulting force is indeed toward the right.) In equilibrium we will have

$$|\mathbf{F}| = |\mathbf{F}_0| = mg \qquad\qquad (12\text{–}23b)$$

However, the moment that the rod starts moving, the total area $A = xL$ encircled by the current will change. Thus, the magnetic flux through the loop will also change:

$$\frac{d\Phi}{dt} = B_0\,\frac{d}{dt}\,(xL) = B_0 L |\mathbf{v}|$$

This will produce an induced emf which will act against the applied voltage V_0:

$$\mathscr{E} = -\frac{d\Phi}{dt} = -B_0 L v$$

The net emf around the loop determines the current I flowing through it:

$$RI = V_0 + \mathscr{E} = V_0 - B_0 L v \qquad\qquad (12\text{–}24)$$

We can now substitute Eq. (12–24) into Eq. (12–23b) to obtain

$$mg = F_0 = LB_0 I = \frac{LB_0}{R}(V_0 - LB_0 v)$$

from which we obtain for the velocity

$$v = \frac{1}{LB_0}\left(V_0 - \frac{Rmg}{LB_0}\right) \qquad\qquad (12\text{–}25)$$

With the numerical values of our example we obtain

$$v = \frac{1}{0.1 \text{ m} \times (1 \text{ Wb/m}^2)}\left(15 \text{ V} - \frac{0.1\ \Omega \times 1.2 \text{ kg} \times (9.8 \text{ m/sec}^2)}{0.1 \text{ m} \times (1.0 \text{ Wb/m}^2)}\right) = 32 \text{ m/sec}$$

This is a very high velocity. Let us note, however, that if we had chosen a resistance only slightly higher, namely $R = 0.128\ \Omega$, we would have obtained

$$v = 0$$

Thus, we see that the steady-state velocity depends critically on the resistance R; its value limits the maximum current and thus the maximum pulling force if the rod is held fixed.

Example 7   What fraction of the electric power supplied by the battery is converted into mechanical power in the previous example?

The total power supplied by the battery is

$$P_1 = V_0 I$$

and from the two Eqs. (12–23) we obtain

$$P_1 = V_0 mg/LB_0 \qquad (12-26)$$

This has to be equal to the sum of mechanical power $Fv = mgv$ and the power RI^2 dissipated in the resistor. From Eqs. (12–23) and (12–25) we obtain

$$P_2 = mgv + RI^2 = \frac{mg}{B_0 L}\left(V_0 - \frac{Rmg}{B_0 L}\right) + R\left(\frac{mg}{B_0 L}\right)^2$$

and we see that this is indeed equal to P_1 given by Eq. (12–26). However, only the fraction $Fv = mgv$ has been usefully converted into mechanical power; thus, the efficiency of the linear motor is

$$\text{Eff} = \frac{mgv}{P_1}$$

From Eqs. (12–26) and (12–25) we obtain for the efficiency

$$\text{Eff} = \frac{LB_0}{V_0} v = 1 - \frac{mgR}{B_0 L V_0}$$

which yields a value

$$\text{Eff} = 1 - \frac{1.2 \text{ kg} \times (9.8 \text{ m/sec}^2) \times 0.1 \text{ }\Omega}{(1 \text{ Wb/m}^2) \times 0.1 \text{ m} \times 15 \text{ V}} = 1 - 0.784 = 22\%$$

Thus, only 1/5 of the electric energy supplied by the battery is converted into mechanical energy; 4/5 is dissipated as heat in the resistor. Our "motor" is rather inefficient.

3. EDDY CURRENTS

Whenever we move a conductor in an inhomogeneous magnetic field, the induced emf will cause currents to flow in the conductor. We can demonstrate this effect if we rotate a copper disk which is partially between the pole pieces of a permanent magnet (Fig. 12–18). These induced currents—called eddy currents—can be so intense as to heat the copper disk until it is red hot. The same effect can be noticed also if we rotate a conductor in a completely homogeneous magnetic field. We can think of the conductor as a continuous distribution of wire loops (Fig. 12–19); this helps us to visualize the process.

The currents produced in the conductor heat the conductor; this dissipated energy has to come from the mechanical energy of the moving object. Thus, there will be a strong mechanical resistance to the motion of the disk in Fig. 12–18. This effect is sometimes used to produce "magnetic brakes"; by turning on a small electromagnet, one can rapidly stop a rotating engine. On the other hand, the effect can be injurious if we are trying to build an engine (e.g., an electric fan) which is supposed to work inside a large magnetic field. In such a case one would either use insulating material or build the fan in such a way as to minimize these eddy currents.

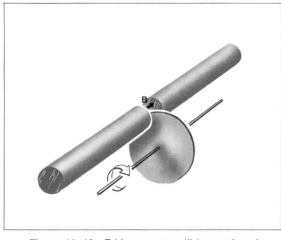

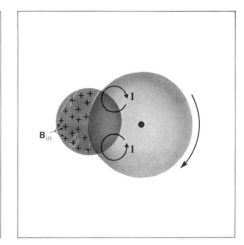

Figure 12-18 Eddy currents will be produced in conducting materials whenever they are moving in or out of a magnetic field.

Figure 12-19 Orientation of eddy currents.

4. INDUCTANCE AND SELF-INDUCTANCE

If two coils are near each other, the magnetic field produced by a current in one coil will produce a magnetic flux through the other coil. If this magnetic flux changes with time, an emf will be produced in the second coil. Let us discuss this, using as a specific example two long coils on a common axis (Fig. 12–20) which have radii R_1 and R_2, equal length ℓ, and a total number of N_1 and N_2 turns, respectively. Both coils are wound in the same direction. Let us assume that the current I_1 flows through the first coil. It will produce inside the first coil a magnetic field of magnitude [according to Eq. (10–24)]

$$|\mathbf{B}_1| = \mu_0 \frac{N_1}{\ell} I_1 \tag{12-27}$$

since there are N_1/ℓ turns per unit length in the coil. If the coils are long ($\ell \gg R_1, R_2$), there will be nearly no magnetic field between the two coils (we assume $R_2 > R_1$). Thus, the flux through any single turn of the second coil is

$$\Phi_1 = \pi R_1^2 B_1 = \mu_0 \pi R_1^2 \frac{N_1}{\ell} I_1$$

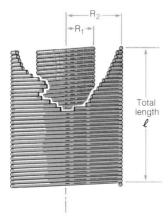

Figure 12-20 Mutual inductance between two coils with a common axis.

If the current I_1 changes with time, an emf will be induced in the second coil:

$$\mathscr{E}_2 = -N_2 \frac{d\Phi_1}{dt} = -\mu_0 \pi R_1{}^2 \frac{N_1 N_2}{\ell} \frac{dI_1}{dt} \qquad (12\text{--}28)$$

If we define the *mutual inductance* between the two coils as

$$\boxed{L_{12} = \mu_0 \pi R_1{}^2 \frac{N_1 N_2}{\ell}} \qquad (12\text{--}29)$$

we can write Eq. (12–28) in the simple form

$$\mathscr{E}_2 = -L_{12} \frac{dI_1}{dt} \qquad (12\text{--}30)$$

Let us now modify our experiment by driving the current I_2 through the second coil instead of the first. The magnetic field inside both coils is now

$$|\mathbf{B}_2| = \mu_0 \frac{N_2}{\ell} I_2 \qquad (12\text{--}31)$$

The flux through a single turn of the first coil is then

$$\Phi_2 = \pi R_1{}^2 B_2 = \mu_0 \pi R_1{}^2 \frac{N_2}{\ell} I_2$$

The first coil has a smaller radius ($R_1 < R_2$); thus, only the flux through the area $\pi R_1{}^2$ will produce an emf if the current I_2 is time-dependent. The emf will then be, according to Faraday's law

$$\mathscr{E}_1 = -N_1 \frac{d\Phi_2}{dt} = -\mu_0 \pi R_1{}^2 \frac{N_1 N_2}{\ell} \frac{dI_2}{dt}$$

which we again can write in the form

$$\mathscr{E}_1 = -L_{12} \frac{dI_2}{dt} \qquad (12\text{--}32)$$

Note that exactly the same mutual inductance L_{12} occurs in both Eqs. (12–30) and (12–32); thus, if a given rate of change dI_1/dt in the first coil produces a certain emf in the second coil, the same rate of change of current in the second coil will produce the same emf in the first coil. The two coils affect each other in a completely symmetrical way. This is quite general and does not depend on our rather special geometry. If we have an arbitrary arrangement of two coils (Fig. 12–21), one can define the mutual inductance between the two coils by the emf produced in one coil by a current change in the other

$$\boxed{\mathscr{E}_2 = -L_{12} \frac{dI_1}{dt}} \qquad (12\text{--}33a)$$

The same mutual inductance will give the emf in coil 1 if the current in coil 2

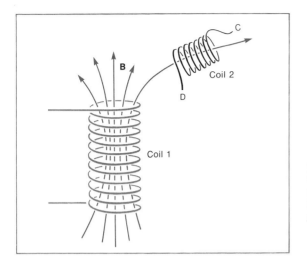

Figure 12-21 A mutual inductance will exist between any two coils which are sufficiently close to each other, so that the magnetic field produced by one coil produces a magnetic flux through the other coil.

changes

$$\mathscr{E}_1 = -L_{12}\frac{dI_2}{dt} \qquad (12\text{--}33b)$$

We should add a warning here concerning the (−) sign in Eqs. (12–33). It is usual to define the mutual inductance always as a positive quantity: $L_{12} > 0$. The relative sign between $\mathscr{E}_2$ and dI_1/dt, on the other hand, is quite arbitrary. It depends on the direction in which we arbitrarily define I_1 to be positive, as well as on the direction in which we define $\mathscr{E}_2$ to be positive; obviously, if we interchange the two leads C and D in Fig. 12–21, the sign of $\mathscr{E}_2$ will change. Thus, we really should write

$$\mathscr{E}_2 = \pm L_{12}\frac{dI_1}{dt}$$

and we have to determine which sign is valid in any specific case after we have defined the directions of I_1 and $\mathscr{E}_2$. However, once we have chosen these two directions so that the (−) sign in Eq. (12–33a) is correct, then the sign in Eq. (12–33b) will also be correct, as long as $\mathscr{E}_1$ and I_1 are considered to be positive in the same direction and the same is true of the pair $\mathscr{E}_2$ and I_2.

Inductance is not only a phenomenon between two different coils. Let us again study this problem on a simple example, a tightly wound coil (Fig. 12–22)

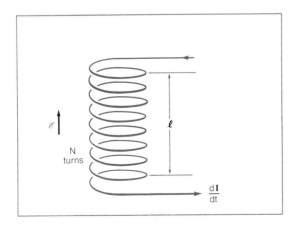

Figure 12-22 A single coil will experience a self-induced emf if the current through the coil changes.

of radius R, length ℓ, and a total of N turns. If a current I flows through the coil, the magnetic field

$$B = \mu_0 \frac{N}{\ell} I$$

in the coil yields a flux Φ through any single turn of the coil:

$$\Phi = B\pi R^2 = \mu_0 \pi R^2 \frac{N}{\ell} I \qquad (12\text{--}34)$$

If the current I changes, there will be an induced emf in the same coil.

$$\mathcal{E} = -N\frac{d\Phi}{dt} = -\mu_0 \pi R^2 \frac{N^2}{\ell}\frac{dI}{dt} \qquad (12\text{--}35)$$

This we can write

$$\boxed{\mathcal{E} = -L\frac{dI}{dt}} \qquad (12\text{--}36)$$

if we define the *self-inductance* of our special coil

$$\boxed{L = \mu_0 \pi R^2 \frac{N^2}{\ell}} \qquad (12\text{--}37)$$

The $(-)$ sign in Eq. (12–36) is unambiguous; we always define $\mathcal{E}$ and I as positive in the same direction, so the $(-)$ sign tells us that the self-induced emf in the coil always opposes the change of the current.

Both the mutual inductance and the self-inductance are measured in the same units. We have

$$[L] = \frac{[\mathcal{E}]}{\left[\dfrac{dI}{dt}\right]} = \frac{\text{V}}{\text{A/sec}}$$

and we define the unit of inductance as

$$\boxed{1 \text{ henry} = 1 \text{ H} = 1\frac{\text{V-sec}}{\text{A}}}$$

Example 8   A coil of length $\ell = 8$ inches and diameter $d = 1$ inch has 400 turns. Calculate its self-inductance (a) if the coil is empty and (b) if an iron cylinder of $\mu_r = 3000$ has been inserted into the coil. The cylinder just fits snugly into the coil.

a. Let us first calculate the self-inductance of the empty coil. From Eq. (12–37) we have

$$L = \mu_0 \pi R^2 \frac{N^2}{\ell}$$

We are given the distances in inches, so we should first convert into metric units. Since 1 inch = 2.54 cm = 0.0254 m, we have ℓ = 8 inches = 0.203 m, and for the radius we obtain $R = d/2 = 0.5$ inch = 0.0127 m. Thus, we have

$$L_{air} = 4\pi \times 10^{-7} \left(\frac{V\text{-sec}}{A\text{-m}}\right) \times 3.14 \times (0.0127 \text{ m})^2 \times \frac{(400)^2}{0.203 \text{ m}} = 5.02 \times 10^{-4} \text{ H}$$

or about 0.5 millihenry = 0.5 mH.

b. We now have to calculate the self-inductance if the coil is filled with iron. Thus, let us review the derivation of Eq. (12–37) to see where and how the iron has an effect. The magnetic field in the coil will be larger by the factor $\mu_r = 3000$. This will produce a flux Φ and thus also an emf which are larger by the same factor. In consequence, the self-inductance will be increased by the same factor μ_r:

$$L_{iron} = \mu_r \mu_0 \pi R^2 \frac{N^2}{\ell} = \mu_r L_{air} \qquad (12\text{–}38)$$

Thus, the self-inductance of the coil filled with iron will be

$$L_{iron} = 3000 \times 0.5 \times 10^{-3} \text{ H} = 1.5 \text{ H}$$

We still have to get a feeling as to what is a large inductance and what is a small one. Thus, let us calculate the induced emf if the current changes by a "reasonable" amount of 1 A in 0.1 sec. We then have for the air-core coil

$$\mathscr{E} = -L_{air}\frac{dI}{dt} = -5 \times 10^{-3} \text{ V} = -5 \text{ mV}$$

while for the coil filled with iron we have

$$\mathscr{E} = -L_{iron}\frac{dI}{dt} = -15 \text{ V}$$

For comparison, let us assume that the wire of the coil is made of copper with resistivity $\rho = 1.7 \times 10^{-8}$ Ω-m. The copper wire will reasonably have a radius r such that $\ell = (2r)N$; this would mean a tightly wound one-layer coil. The length of the wire, on the other hand, has to be equal to $s = 2\pi RN$. Thus, we have for the resistance

$$\text{Resistance} = \frac{\rho s}{\pi r^2} = \frac{\rho 2\pi RN}{\pi(\ell/2N)^2}$$

or

$$\text{Resistance} = \frac{8\rho RN^3}{\ell^2} = 2.7 \text{ Ω}$$

Thus, at a current of 1 A the emf needed to overcome the resistance would be 2.7 V; the self-inductive effect of the air coil is negligible, but for the coil with the iron core the effect is very large.

5.   ENERGY DENSITY OF A MAGNETIC FIELD

Whenever we turn on the current in a coil with a nonzero self-inductance, we have to overcome the self-induced emf; that is, we have to supply energy. This energy is not dissipated; it is stored in the coil and can be taken out later when the current is reduced.

If we change the current I in a coil by a small amount dI in a time interval dt, we have an emf

$$\mathscr{E} = -L\frac{dI}{dt}$$

If $dI/dt > 0$, then $\mathscr{E}$ will oppose the current I. Thus, in the time interval dt we have to supply the energy

$$dW_{\text{supplied}} = |\mathscr{E}| I \, dt$$

or

$$dW = L\frac{dI}{dt} I \, dt = LI \, dI$$

Let us now start with a current $I = 0$ and then increase the current to the value $I = I_0$; the total energy stored in the coil is then

$$W_{\text{stored}} = \int dW = \int_{I=0}^{I_0} LI \, dI$$

or

$$W_{\text{stored}} = \frac{LI_0^2}{2} \tag{12-39}$$

We can recover this energy by decreasing the current I from its full value I_0 to zero; the emf now will be in the same direction as the current, and the coil supplies the power $\mathscr{E}I$ to the outside. This energy is stored in a coil while the current persists; we have a situation analogous to electrostatics, in which a distribution of charges at rest also contains the stored energy given by Eq. (8–5)

$$W_{\text{el}} = \int \frac{\epsilon_0 \mathbf{E}^2}{2} d\tau \tag{12-40}$$

In the same way, we can say that moving charges (currents) contain stored magnetic energy. We can even derive an expression for the magnetic energy density using our example of a straight long cylindrical coil of radius R and length ℓ and having N turns. Using Eq. (12–37), we find for the stored magnetic energy

$$W = L\frac{I_0^2}{2} = \mu_0 \frac{\pi R^2 N^2}{\ell} \frac{I_0^2}{2} \tag{12-41}$$

On the other hand, we know that the magnetic field has the value $B = \mu_0 \, N/\ell \, I_0$ inside the coil and is negligibly small outside it. In analogy to Eq. (12–40), let us calculate the volume integral $\int \mathbf{B}^2 \, d\tau$ over all space. It is equal to

$$\int \mathbf{B}^2 d\tau = B^2 \times (\text{volume of coil})$$

or

$$\int \mathbf{B}^2 d\tau = \mu_0^2 \frac{N^2}{\ell^2} I_0^2 \times \pi R^2 \ell = \mu_0^2 \pi R^2 \frac{N^2}{\ell} I_0^2$$

Comparing this with Eq. (12–41), we see that

$$W_{\mathrm{magn}} = \int \frac{\mathbf{B}^2}{2\mu_0} d\tau \qquad (12\text{–}42)$$

Note that we have "derived" Eq. (12–42), which yields the magnetic energy density

$$\rho_m = \frac{\mathbf{B}^2}{2\mu_0} \qquad (12\text{–}43)$$

only for the special example of a tightly-wound long coil. We merely assert that it is true in general, independently of the particular configuration of the currents.

We can even use our example to derive (actually, to guess at!) the magnetic energy in ferromagnetic materials. The magnetic field will be larger by a factor μ_r; the self-inductance and thus the stored energy will increase by the same factor μ_r, according to Eq. (12–38). Thus, we have, since the volume of the coil is still $\pi R^2 \ell$,

$$W = \mu_r \mu_0 \pi R^2 \frac{N^2}{\ell} \frac{I_0^2}{2} = \rho_m \pi R^2 \ell$$

or

$$\rho_m = \mu_r \mu_0 \frac{N^2}{\ell^2} \frac{I_0^2}{2} = \frac{\mathbf{B}^2}{2\mu_r \mu_0}$$

as the general expression for the magnetic energy density inside materials.

Example 9   A large electromagnet has a gap height $d = 8$ inches, is $\ell = 40$ inches long, and has a gap width $a = 20$ inches (Fig. 12–23). At a current of $I = 1500$ A, the magnetic field in the gap is 1 Wb/m². Estimate the stored magnetic energy and the self-inductance of the magnet.

We are told to estimate; this should warn us that an exact calculation would be prohibitively complicated. Indeed, if we had to calculate the energy density inside the magnet iron, we would be in serious trouble—particularly since we are not even told the exact shape of the steel yoke. However, we have learned earlier

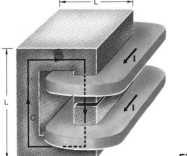

Figure 12–23 Example 9.

(Chapter 11, Example 6) that the magnetic field **B** inside the iron is about the same as in the gap. But this implies that the magnetic energy density is a factor $\mu_r \approx 2000$ to 3000 smaller than in the gap. As a rough estimate, the volume of the iron yoke may be as much as 10 or 20 times as large as the volume of the gap; but even then the total magnetic energy in the iron is only a few percent of the energy stored in the gap. Thus, we have a good approximation for the total magnetic energy stored:

$$W = \frac{\mathbf{B}^2}{2\mu_0} a\ell d$$

where $a\ell d$ is the volume of the gap. Since we have $d = 8$ inches ≈ 0.2 m, $\ell = 40$ inches ≈ 1 m, and $a = 20$ inches ≈ 0.5 m, we have a total volume of 0.1 m³. Thus, the total energy stored is

$$W = \frac{(1 \text{ Wb/m}^2)^2 \times 0.1 \text{ m}^3}{2 \times (4\pi \times 10^{-7} \text{ Wb/A-m})} \approx 4.0 \times 10^4 \text{ J}$$

This is a large energy; it is equal to the kinetic energy of a 220 pound ($= 100$ kg) mass dropped from the height of 40 meters—or from about the 12th floor of a building. We still have to estimate the self-inductance of the magnet; but this is now easy, since we know that

$$W = \frac{L}{2} I^2$$

or

$$L = \frac{2W}{I^2} = \frac{8 \times 10^4 \text{ J}}{(1500 \text{ A})^2} \approx 3.6 \times 10^{-2} \text{ H}$$

This is by no means a very large inductance; although the magnetic energy is large, it was produced by a very large current I.

6. ELECTRIC CIRCUITS WITH INDUCTANCES

We have said that whenever the current in a coil changes, an induced emf

$$\mathscr{E} = -L\frac{dI}{dt}$$

opposes the change in current. We call an *inductor* any circuit element which has a nonnegligible self-inductance. In principle, any piece of wire, such as a wire loop, forms an inductor because there is always a magnetic field associated with any current. However, in such simple situations the self-inductance is usually negligibly small. The self-inductance becomes large usually only when the magnetic field is increased by the presence of ferromagnetic materials. An electromagnet is one example of an inductor; but small inductors are frequently also used in electronic circuits for temporary storage of energy.

A real inductor—since it consists of a wire (wound on an iron core)—always has also a finite resistance. Let us consider such an inductor and ask how fast the current will build up when we apply an external emf; we could, for instance, close a switch that connects the inductor to a battery of emf V_0, as shown in Fig. 12–24a. If there was no self-inductance, we would obtain immediately upon closing the switch the current

$$I = \frac{V_0}{R}$$

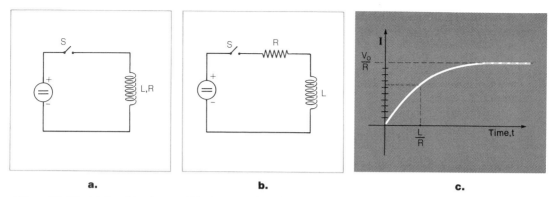

Figure 12–24   (a) A coil having a self-inductance L and resistance R is equivalent to (b) a separate resistance R and a resistance-less coil of self-inductance L. (c) The time dependence of current through the coil upon closing switch S.

However, the self-inductance prevents the current from reaching its final value instantaneously; this would imply an infinite dI/dt and thus an infinite emf opposing the emf of the applied battery. Thus, not all of the applied emf $= V_0$ is available immediately to drive the current I through the resistance R; only the fraction $V_0 - L\dfrac{dI}{dt}$ is available for this purpose:

$$V = V_0 - L\frac{dI}{dt} = RI \qquad (12\text{–}44a)$$

We can rewrite this in the form

$$V_0 = L\frac{dI}{dt} + RI \qquad (12\text{–}44b)$$

and we note that exactly the same relation would hold if we had the arrangement in Fig. 12–24b, where a resistance R is in series with a resistanceless inductor of self-inductance L. This is because the voltage drop across the resistor is RI; and the emf which has to be applied across the inductor is equal to $\mathscr{E} = L\,dI/dt$. This point is important, because we now know that we are also solving the problem of an inductor and an external resistance R_{ext} in series. If we define the total resistance

$$R = R_{\text{ext}} + R_{\text{inductor}}$$

then Eq. (12–44b) will be unchanged; all we have to do is to insert the correct total resistance R.

Before we can solve Eq. (12–44b), we have to specify the initial conditions: we close the switch at the time $t = 0$. Since before we closed the switch there was no current, and since I has to change smoothly, we have $I(t = 0) = 0$. On the other hand, after a long time we expect that the current will be equal to $I(t = +\infty) = V_0/R$. Since V_0/R is a constant, we can write instead of Eq. (12–44b)

$$\frac{d[I - (V_0/R)]}{dt} = -\frac{R}{L}[I - (V_0/R)] \qquad (12\text{–}45)$$

Now, since $d(e^{-\alpha t})/dt = -\alpha e^{-\alpha t}$, we see that the solution to Eq. (12–45) is

$$I - \frac{V_0}{R} = Ce^{-Rt/L} \tag{12–46}$$

where C is an as yet unknown constant which we can determine from the initial condition $I(t=0) = 0$. We then have

$$0 = I(t=0) = \frac{V_0}{R} + Ce^{-0}$$

which yields $C = -V_0/R$. By substitution into Eq. (12–46), we obtain

$$I(t) = \frac{V_0}{R}(1 - e^{-Rt/L}) \tag{12–47}$$

We have plotted this curve in Fig. 12–24c.

Example 10   Assume that the coil in Example 8 has a resistance of $R = 2.7\ \Omega$; this was our estimate derived there. A switch connecting it to a $V = 5$ volt battery is closed at $t = 0$. What will be the steady-state current through the coil, and how long will it take for the current to reach 99% of its final value?

The steady-state current is given as

$$I = \frac{V}{R} = 1.85\ \text{A}$$

The time T required to reach 99% of this value, according to Eq. (12–47), is given by

$$1 - e^{-RT/L} = 0.99$$

or

$$T = -\frac{L}{R}\ln{(0.01)} = +\frac{L}{R}\ln{(100)}$$

Since $\ln{(100)} \cong 4.6$, we obtain for the empty coil from Example 8

$$T = 4.6 \times \frac{5.02 \times 10^{-4}\ \text{H}}{2.7\ \Omega} = 8.6 \times 10^{-4}\ \text{sec} = 0.86\ \text{msec}$$

This is a very short time, measurable with an oscilloscope, but negligible on the "human" scale of one second.

On the other hand, for the coil filled with iron we have to wait

$$T = 4.6 \times \frac{1.5\ \text{H}}{2.7\ \Omega} = 2.56\ \text{sec}$$

before the current reaches 99% of its final value.

A somewhat different situation arises if we connect a charged capacitor to an inductor. We will treat here only the somewhat idealized situation in which the resistance of the inductor is negligibly small. Let us first discuss qualitatively what happens if we charge up the capacitor C (Fig. 12–25) and then connect it to an inductor L by closing the switch S. Since there is a potential difference V between the plates, a current will start to flow through the inductor. Because of the self-inductance L, a finite time is required for the current to reach

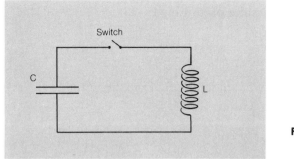

Figure 12-25   An LC circuit.

its maximum value. This will occur when the capacitor is completely discharged. One should note that since there is no resistance in the circuit, none of the initial energy

$$U_e = \frac{Q_0^2}{2C} \tag{12-48a}$$

stored in the capacitor is dissipated (converted into heat). The LC system does not lose any energy. The energy stored in the electric field inside the capacitor is converted into magnetic field energy inside the inductor.

Once all energy has been transferred to the inductor, the process does not stop; the energy

$$U_m = \frac{L}{2} I_0^2 \tag{12-48b}$$

is stored—but only as long as there is a current in the inductor. This current will now start charging up the capacitor again, but with opposite polarity. Negative charge accumulates on the plate which was initially positively charged. This continues until all the magnetic energy is converted back into electric energy stored in the capacitor. We now have the same situation as at the beginning, except that the capacitor is charged with opposite polarity. Thus, current will start flowing in the opposite direction and the electric energy will be reconverted into magnetic energy. Finally, the initial situation is exactly restored—the process repeats itself indefinitely. We have an oscillation of energy between electric energy in the capacitor and magnetic energy in the inductor. In Fig. 12-26a, we show a series of "snapshots" of the current I and stored charge Q.

There is a close analogy between the oscillation of energy in an LC circuit and the oscillation between potential and kinetic energy of a mass m attached to a spring. If we stretch the spring and then release it, we have the initial potential energy

$$U = \frac{k}{2} x^2 \tag{12-49a}$$

Because the spring exerts a force on the mass, the mass m will accelerate and the spring will slacken. When the spring has its unstretched length, all the potential energy has been converted into kinetic energy

$$T = \frac{m}{2} v^2$$

However, the mass will continue moving and compress the spring, decelerating

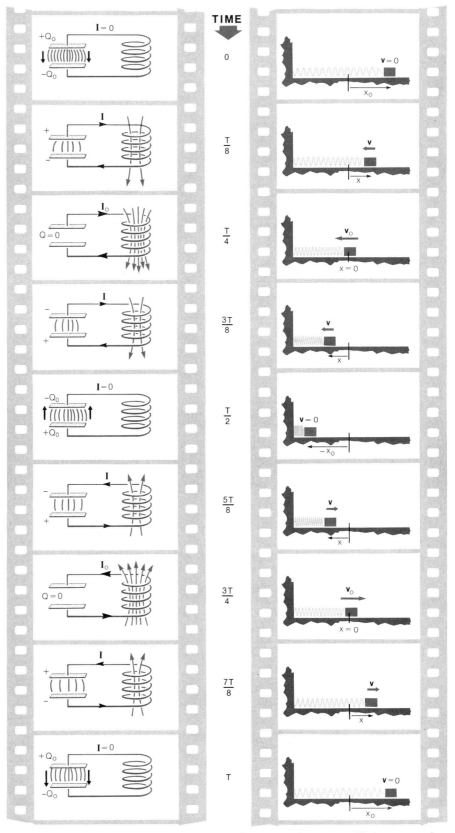

TIME

0

$\frac{T}{8}$

$\frac{T}{4}$

$\frac{3T}{8}$

$\frac{T}{2}$

$\frac{5T}{8}$

$\frac{3T}{4}$

$\frac{7T}{8}$

T

Figure 12-26 ''Film strips'' of an oscillating LC circuit and a mass oscillating on a spring.

TABLE 12–1 ANALOGOUS QUANTITIES FOR MASS ON SPRING AND LC OSCILLATOR

SPRING	LC OSCILLATOR
spring constant k	$\dfrac{1}{\text{capacitance}}\quad\dfrac{1}{C}$
mass m	inductance L
position x	charge Q
velocity v	current I
potential energy $\dfrac{k}{2}x^2$	electric energy $\dfrac{Q^2}{2C}$
kinetic energy $\dfrac{m}{2}v^2$	magnetic energy $\dfrac{L}{2}I^2$
period $T = 2\pi\sqrt{\dfrac{m}{k}}$	period $T = 2\pi\sqrt{LC}$

in the process (Fig. 12–26b). This will continue until the mass stops and the spring is fully compressed—the kinetic energy has been reconverted into potential energy. The mass then starts moving in the opposite direction, and oscillates forever (neglecting friction).

We can even extend our analogy further by identifying the individual variables for the two problems. We have $v = dx/dt$ and $I = dQ/dt$; the two pairs of time-dependent quantities (x,v) and (Q,I) have the same relation to each other. Extending the analogy even further, we can "identify" the electric with the potential energy, the magnetic with the kinetic energy. Thus, $1/C$ is the electric analogon to the spring constant k, and L is the electric analogon to the mass m. In mechanics (Vol. I, p. 367) we learned that the period of oscillation for a spring is

$$T = 2\pi\sqrt{\frac{m}{k}}$$

and we thus conclude that the LC oscillator has a period of

$$T = 2\pi\sqrt{\frac{L}{1/C}} = 2\pi\sqrt{LC} \qquad (12\text{–}50)$$

In Table 12–1 we list the analogous quantities for the two situations.

Example 11   An LC circuit has $L = 0.1$ H and $C = 100\ \mu$F. Initially there was a potential difference $V_0 = 100$ volts across the capacitor plates and no current I. Calculate the period of oscillation, and the maximum current I_0 through the inductor.

We can calculate the period directly from Eq. (12–50):

$$T = 2\pi\sqrt{LC} = 2\pi\sqrt{(0.1\text{ H}) \times 10^{-4}\text{ F}} = 1.99 \times 10^{-2}\text{ sec} = 19.9\text{ msec}$$

We can calculate the peak current by equating the peak magnetic energy to the peak (initial) electric energy:

$$\frac{L}{2}I_0^2 = \frac{Q_0^2}{2C} = \frac{CV_0^2}{2}$$

and this yields

$$I_0 = \sqrt{\frac{C}{L}}V_0 = \sqrt{\frac{10^{-4}\ \mathrm{F}}{10^{-1}\ \mathrm{H}}} \times 100\ \mathrm{V} = 3.16\ \mathrm{A}$$

Let us now discuss the LC oscillator more mathematically (Fig. 12–27). There is a potential difference

$$V = \frac{Q}{C}$$

across the capacitor plates; this acts as an electromotive force driving the current through the inductor L. It has to be equal to the self-induced emf

$$|\mathscr{E}| = L\frac{dI}{dt}$$

which opposes the increase in current. We thus have

$$\frac{Q}{C} = L\frac{dI}{dt} \tag{12–51}$$

However, the current as defined by the arrow in Fig. 12–27 is equal to the *decrease* of charge on the capacitor plates per unit time:

$$I = -\frac{dQ}{dt} \tag{12–52}$$

Combining the two equations and rearranging slightly leads us to the differential equation

$$\frac{d^2Q}{dt^2} + \frac{1}{LC}Q = 0 \tag{12–53}$$

Note that this is again the exact analogue to the differential equation for the mass on a spring. Since the force is $F = -kx$ and since also $F = m(d^2x/dt^2)$, we have there

$$\frac{d^2x}{dt^2} + \frac{k}{m}x = 0$$

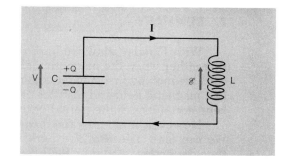

Figure 12–27 Calculating the current through the coil: the voltage V across the capacitor has to equal the emf $\mathscr{E}$ in the coil.

We thus look for a solution which is similar to the solution found for the oscillating spring; we write

$$Q = Q_0 \cos (\omega t - \delta) \tag{12-54}$$

We calculate the first and second derivatives:

$$\frac{dQ}{dt} = -Q_0\omega \sin (\omega t - \delta) \quad \text{and} \quad \frac{d^2Q}{dt^2} = -Q_0\omega^2 \cos (\omega t - \delta)$$

and substitute d^2Q/dt^2 as well as Q into Eq. (12–53):

$$-Q_0\omega^2 \cos (\omega t - \delta) + \frac{1}{LC} Q_0 \cos (\omega t - \delta) = 0$$

This equation will be obeyed for any Q_0 and δ as long as

$$\boxed{\omega = \frac{1}{\sqrt{LC}}} \tag{12-55}$$

This is identical to Eq. (12–50). Indeed, we have

$$T = \frac{2\pi}{\omega} = 2\pi \sqrt{LC}$$

since the period is defined as the time difference T for which $\omega T = 2\pi$, i.e., for which the argument of the cosine increases by one full period 2π.

Example 12   What is the time dependence of the current through the inductor of Fig. 12–27, if the charge on the capacitor is given by Eq. (12–54)?

From Eq. (12–52) we have

$$I = -\frac{dQ}{dt} = +Q_0\omega \sin (\omega t - \delta)$$

The peak current

$$I_0 = \omega Q_0 = \sqrt{\frac{1}{LC}} Q_0$$

will occur when $\omega t - \delta = \pi/2$; exactly at the same time, $Q = 0$. We are thus confirming what we stated earlier: When the current is maximal, there is no charge on the capacitor; all the electric energy in the capacitor has been converted into mechanical energy in the inductor.

7. SUMMARY

With Faraday's law we have discovered a new range of phenomena occurring either when the magnetic field is time-dependent or when objects such as coils are moving in a magnetic field. We have learned that there is energy stored in a magnetic field and that this energy is free in the sense that it can be stored, kept stored for a finite period of time, and then again removed.

However, up to now the theory we have presented is highly unsatisfactory. We can (and, because of experimental evidence, must) accept the fact that a time-dependent magnetic field generates an electric field. This is really the only way that we can interpret the phenomenon of self-inductance. The induced emf must be produced in a coil by a real electric field. However, let us return to the electromagnetic generator discussed in Section 2. We had there a constant homogeneous magnetic field in which a coil is rotating. With the coil removed — or not turning — there is no electric field. How are we then to interpret the fact that, when the coil starts turning, an emf appears and can produce a measurable current? Are we to understand that rotating the coil produces an electric field? Actually, if we succeeded in measuring the electric field near the coil we would find a highly surprising result: if the ends of the coil are open so that no current can flow, there exists an electric field near the surface. However, when we short the coil so that the emf can produce a current, the electric field decreases; in the limit where the resistance of the coil becomes zero, the electric field vanishes completely.

We will discuss these problems in the next chapter, where we will find that Faraday's law is closely related to the fact that a moving charge in a magnetic field is acted upon by the Lorenz force $q(\mathbf{v} \times \mathbf{B})$. We will also learn that while the force acting on a charge is independent of who observes the charge, the separation into electric force $q\mathbf{E}$ and magnetic force $q(\mathbf{v} \times \mathbf{B})$ depends on the observer's *relative* velocity.

Summary of Important Relations

Faraday's law
$$\mathscr{E} = -\frac{d\phi}{dt} = -\frac{d}{dt}\int \mathbf{B} \cdot d\mathbf{S}$$

In rotating coil of area A with N turns $\mathscr{E} = -NAB\,\omega\cos\omega t$

Coil of self-inductance L $\mathscr{E} = -L\dfrac{dI}{dt}$

For straight coil of length ℓ with N turns and area A $L = \mu_0 A \dfrac{N^2}{\ell}$

Mutual inductance $\mathscr{E}_2 = -L_{12}\dfrac{dI_1}{dt}$

but also $\mathscr{E}_1 = -L_{12}\dfrac{dI_2}{dt}$

Energy stored in inductance $W = \dfrac{LI^2}{2}$

Energy density of magnetic field
$$\rho_m = \frac{\mathbf{B}^2}{2\mu_0}\quad\text{in vacuum}$$

$$\rho_m = \frac{\mathbf{B}^2}{2\mu_r\mu_0}\quad\text{in material with }\mu_r \neq 1$$

LC oscillator: angular frequency $\omega = \sqrt{LC}$

period $T = \dfrac{2\pi}{\sqrt{LC}}$

See also Table 12–1.

Questions

1. A coil of resistance R is connected to a battery of emf $\mathscr{E}$. A long iron rod is moved with constant velocity through the coil. Sketch the time dependence of the current through the rod from before the rod approaches until it is again far away.

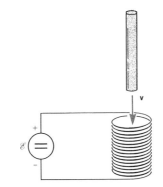

2. A flexible copper wire loop shaped in the form of an ellipse rests on the pole piece of an electromagnet. How will the wire loop deform if

    (a) the magnet was initially off and is then quickly turned on;

    (b) the magnet was initially on and is suddenly turned off?

3. A coil is filled with nonmagnetic conducting material, such as copper. Will the self-inductance of the coil be increased or decreased by the presence of the conducting material? Why?

4. An electromagnet is powered by several coils. If you want to be able to turn the magnet on and off as fast as possible, would you connect the coils to the power supply in parallel or in series? Why?

5. In reality, the current oscillations in an LC circuit do not last forever, but decrease exponentially with time. Explain why.

Problems

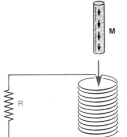

1. A cylindrical permanent magnet of radius $r_1 = 0.5$ cm and magnetization $|\mathbf{M}| = 0.8 \times 10^6$ A/m is moved rapidly into a coil of radius $r_2 = 3$ cm and $N = 10$ turns. The coil is shorted by a resistor $R = 10\ \Omega$. Calculate the current pulse $Q = \int I\, dt$.

2. A coil of $N = 150$ turns and mean area $A = 1$ cm^2 is lying flat on the pole piece of an electromagnet. If it is removed rapidly from the magnet, a total charge of 5×10^{-4} coul flows through a galvanometer connected to the coil. What is the magnetic field in the magnet, if the total circuit resistance (coil and galvanometer) is $15\ \Omega$?

3. A rectangular coil with 10 turns (length $a = 10$ cm, width $b = 2$ cm) is a distance b away from a straight wire carrying an alternating current $I = I_0 \cos \omega t$ with $I_0 = 100$ A and frequency (not angular frequency!) $\nu = 60$ Hz. Calculate the induced emf in the coil.

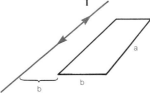

4. A rectangular coil of dimensions 3 cm × 10 cm is in an electromagnet at an angle $\theta = 45°$ relative to the pole pieces. At a magnet current of 20 A, the magnetic field between the pole pieces is 10^4 gauss. Calculate the emf in the coil at the time $t = 2$ sec if the magnet current is

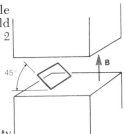

$$I = I_0 \left(1 - e^{-t/t_0}\right)$$

with $I_0 = 15$ A and $t_0 = 4$ seconds.

5. A rectangular coil of dimensions $a = 5$ cm and $b = 30$ cm is moving with velocity $v = 20$ m/sec into a magnetic field $B_0 = 1$ weber/m². The total coil resistance is $R = 0.2 \,\Omega$. Calculate the braking force on the coil.

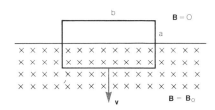

6. Calculate the emf produced in a circular coil (radius $r = 2$ cm, $N = 50$ turns) if it rotates 30 times a second in a magnetic field $B = 0.8$ weber/m² and if the magnetic field forms the angle $\theta = 30°$ with the axis of rotation.

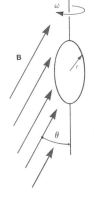

7. Given a length L of wire, construct a circular coil so as to maximize the current in the wire when placed in a time-dependent magnetic field **B**. How many turns will the coil have?

8. Two square coils, each with $N = 10$ turns and side $a = 10$ cm, are mounted perpendicular to each other at a common axle and are rotating with angular frequency $\omega = 80$ radians/sec. The two coils are connected in series to each other, and there is a magnetic field of magnitude $B_0 = 5000$ gauss perpendicular to the axis of rotation. Calculate the peak emf in the assembly.

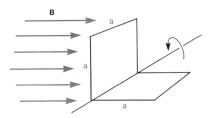

9. *Linear Generator.* In a magnetic field B (out of the paper) we have an arrangement similar to Fig. 12–17. The rod of length L is moved with constant velocity **v.** The resistance of the circuit is R. Calculate the current I and the force F required to keep the rod moving at constant velocity **v.** Show explicitly that energy is conserved.

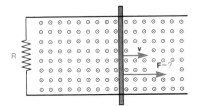

10. A circular coil (radius $r = 1$ cm, $N = 100$ turns) is rotating 30 times a second in a magnetic field $B_0 = 0.6$ weber/m² (perpendicular to the axis of rotation). The coil itself has a resistance $R_1 = 2\ \Omega$ and is connected to an external resistance $R_2 = 10\ \Omega$. Calculate the average power dissipated in R_2.

°11. A circular coil (radius $r = 3$ cm) has $N = 100$ turns. A steady current $I = 1$ m A is flowing through the wire. The coil is rotating 10 times a second in a magnetic field $B = 1.0$ weber/m². Calculate

    (a) the torque on the coil at any instant of time;

    (b) the average torque;

    (c) the voltage necessary to produce the constant current I, if the coil resistance is $R = 1\ \Omega$.

12. A circular coil (radius r, N turns) is connected to a capacitor C. The coil is rotating with constant angular velocity ω in a magnetic field B_0 perpendicular to the axis of rotation. The resistance of the coil is negligibly small. Calculate the instantaneous torque on the coil as well as the average torque.

°13. Repeat the preceeding problem, but this time do not neglect the resistance R of the coil and the leads to the wire. (Hint: Use the result of Problem 9–22.)

14. (a) Two inductances L_1 and L_2 are connected in series; their separation is large. Show that the total inductance is $L = L_1 + L_2$.
    (b) Two inductances L_1 and L_2 have the mutual inductance L_{12}. They are connected in series. Calculate the total inductance. Can you change the total inductance by interchanging leads?

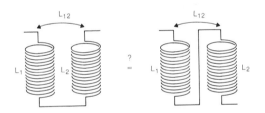

15. An inductor of unknown self-inductance has the resistance $R = 0.5\ \Omega$. When connected by a switch to a battery of emf $= 5$ V, one measures a current after one second of 4 A. Calculate the self-inductance.

16. A choke coil has a laminated iron core ($\mu_r = 5000$) of cross sectional area 1×1 inch. Its overall size is 4×4 inches, as shown in the figure. The coil itself has $N = 400$ turns. Estimate the self-inductance of the coil.

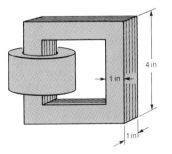

17. An experimental magnet has a gap of 25 cm; the pole piece area is 1 m². At a current $I = 1500$ A the field is $B = 1.0$ weber/m². A safety shunt of 10 Ω is put across the terminals. At time $t = 0$ the switch S is opened.

    (a) What will be the peak voltage across the resistor?

(b) What will be the total energy dissipated in the resistor?

(c) How long will it take for the magnetic field to decrease to 1% of its initial value?

Neglect the internal resistance of the magnet.

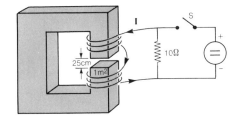

° 18. Calculate the self-inductance per unit length of a coaxial cable of inner radius R_1 and outer radius R_2. Assume the outer shell to be very thin. The inside conductor, however, is a solid piece of metal. Does this affect your result? (Hint: Calculate the magnetic energy.)

19. Two inductors each have the self-inductance L and a mutual inductance L_{12}. One coil is connected to an emf source $\mathcal{E} = \mathcal{E}_0 \cos \omega t$; the other is shorted through a resistor R. Calculate the currents I_1 and I_2 in the two coils. What happens if $L_{12} = L$?

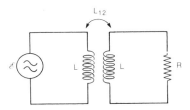

20. An LC circuit has an inductance $L = 10^{-2}$ H and a capacitance $C = 10 \ \mu$F. The total energy stored in the circuit is $E = 1$ Joule. Calculate the maximum current in the inductor and the maximum potential difference across the capacitor plates. What is the time interval between the occurrence of maximum current and maximum potential?

21. Given a capacitance $C = 1 \ \mu$F, what inductance would you choose to obtain oscillations of period $T = 100 \ \mu$sec $= 10^{-4}$ sec?

22. You have a set of 10 capacitors of $C = 10 \ \mu$F each, each rated at 150 volts max. You want to build an LC circuit operating at a frequency $\nu = 10^5$ Hz. What inductance would you use if

(a) all capacitors are in parallel?

(b) all capacitors are in series?

In which case could you store more energy in the circuit?

23. An emf $\mathcal{E} = \mathcal{E}_0 \cos \omega t$ with $\mathcal{E}_0 = 15$ V and $\omega = 2\pi \times 60$ radians/sec is connected to an inductor of self-inductance $L = 0.1$ H and resistance $R = 1 \ \Omega$. Calculate the current I through the coil, assuming I has the form $I = I_0 \cos (\omega t - \phi)$ with I_0 and ϕ to be determined.

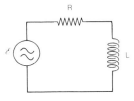

CHAPTER 13

MOVING COORDINATE SYSTEMS

1. INERTIAL REFERENCE FRAMES

We have learned in mechanics that the basic laws of physics are the same in all inertial reference frames. By *reference frame* or *observer*° we mean any observational setup, whether manned or not, which can measure distances, directions, and time intervals. By *inertial reference frame* we mean a reference frame in which Newton's first law is valid: an isolated object not subject to any outside influence (forces) moves with constant velocity. If there is one such inertial reference frame (and Newton's first law asserts that this is so), then all reference frames that move with constant velocity and do not rotate with respect to such an inertial frame are also inertial reference frames.

Let us briefly discuss the arguments: If an observer on a moving train drops a ball (Fig. 13–1), he will see the object falling down. Another observer on the ground will see the same ball starting out with a horizontal velocity and then following a parabolic path. However, both observers will measure the same

---

° We usually talk about an "observer" if some action is to be taken, such as dropping a ball. The name "reference frame" is used if we want to be more abstract. However, there is really no significant difference; the observer could consist of a movie camera and some arrangement to release the ball and start the camera at the same time.

Figure 13–1 Although the path an object follows seems different to observers $\mathscr{A}$ and $\mathscr{B}$, both measure the same acceleration.

acceleration of the ball:

$$\mathbf{a} = \mathbf{g} \tag{13-1}$$

Thus, both will conclude that the same force acts on the ball:

$$\mathbf{F} = m\mathbf{a} = m\mathbf{g} \tag{13-2}$$

From this one concludes that two inertial observers have to observe the same forces acting on any object. There are seeming exceptions, including velocity-dependent forces such as the drag on an object moving through a liquid:

$$\mathbf{F} = -\lambda\mathbf{v} \tag{13-3}$$

However, $\mathbf{v}$ in Eq. (13-3) is not the absolute velocity of the object, but its relative velocity in the liquid

$$\mathbf{v} = \mathbf{v}_{\text{object}} - \mathbf{v}_{\text{liquid}}$$

This difference will be the same to any observer, whatever velocity he himself has with respect to the liquid. If you stand in a convertible going at 30 mph, you feel the same wind velocity and the same force as if you were standing on the ground in a 30 mph wind.

However, one should not conclude from what we have just said that two inertial observers will measure the same value for any physical quantity. As an example, the kinetic energy

$$T = \frac{m}{2}\mathbf{v}^2$$

of a mass m will be measured to have different values by two observers moving with constant velocity relative to each other.

Within the realm of Newtonian mechanics, distances, time intervals, and forces are the same if measured in any inertial reference frame. Let us consider two such inertial frames, as shown in Fig. 13-2. Observer $\mathscr{A}$ has a clock measuring the time t, and he measures positions of a mass m by noting the vector $\mathbf{r}$. Observer $\mathscr{B}$ also has a clock measuring the time t', and he measures the position of the same mass by noting the vector $\mathbf{r}'$. As we see from Fig. 13-2, the two are related:

$$\mathbf{r}' = \mathbf{r} + \mathbf{r}_0$$

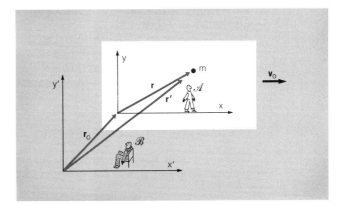

Figure 13-2 Two observers moving with constant velocity relative to each other.

and if $\mathscr{A}$ is moving with velocity $\mathbf{v}_0$ relative to $\mathscr{B}$, we can write $\mathbf{r}_0 = \mathbf{R}_0 + \mathbf{v}_0 t$, obtaining

$$\mathbf{r}' = \mathbf{R}_0 + \mathbf{r} + \mathbf{v}_0 t \tag{13–4}$$

where $\mathbf{R}_0$ is the position at which $\mathscr{B}$ sees $\mathscr{A}$ at the time $t = 0$. Each observer can choose a coordinate origin in his own system such that the two coincide at the time $t = 0$. Then for $t = 0$, we have $\mathbf{r}' = \mathbf{r}$ and thus $\mathbf{R}_0 = 0$. Note that we implicitly assumed that both observers measure the same time t; this is one of the fundamental assumptions in Newtonian dynamics. We can then write

$$\boxed{\begin{aligned} \mathbf{r}' &= \mathbf{r} + \mathbf{v}_0 t \\ t' &= t \end{aligned}} \qquad\begin{aligned}&\text{(13–5a)}\\&\text{(13–5b)}\end{aligned}$$

We call the transformation (13–5) between $\mathbf{r}'$ and $\mathbf{r}$ a *Galilean transformation*. An object at rest to observer $\mathscr{A}$ (i.e., an object for which $\mathbf{r} = $ constant) moves with velocity $\mathbf{v}_0$ to observer $\mathscr{B}$: $\mathbf{r}' = $ constant $+ \mathbf{v}_0 t$.

Both Eqs. (13–5) are true only if the speed $|\mathbf{v}_0|$ is much smaller than the velocity of light. As $|\mathbf{v}_0|$ approaches c, one has to replace the nonrelativistic Galilean transformation by the relativistic Lorentz transformation. If the velocity $\mathbf{v}_0$ is in the direction of the x-axis, the relativistic transformation law is[°]

$$x' = \frac{x + v_0 t}{\sqrt{1 - (v_0^2/c^2)}} \tag{13–6a}$$

$$y' = y \tag{13–6b}$$

$$z' = z \tag{13–6c}$$

$$t' = \frac{t + \dfrac{v_0 x}{c^2}}{\sqrt{1 - (v_0^2/c^2)}} \tag{13–6d}$$

It is easy to see that for $v_0 << c$ this indeed reduces to the Galilean transformation (13–5).

In this chapter we will study how electric and magnetic fields change when seen by two observers moving relative to each other with constant velocity. We will limit ourselves to velocities which are small compared to the velocity of light. This will enable us to use the simpler Galilean transformation. However, we stress that the theory of electromagnetism is a completely relativistic theory. Light, after all, is an electromagnetic phenomenon, and the experimental fact that its velocity is the same as seen by all inertial observers is the basis for the special theory of relativity.

Actually, because we will be using the slightly wrong Galilean transformation, our results will be slightly inconsistent; whenever we obtain expressions such as $1 \pm (v^2/c^2)$, we will have to set them equal to 1. The Galilean transformation is correct only to the extent that $v^2/c^2 << 1$ and can be therefore neglected.

2. TRANSFORMATION OF ELECTRIC AND MAGNETIC FIELDS

When we said that moving charges produce a magnetic field, we did not worry about whether it was the charges or the observer who was moving. Thus,

---

[°] Compare Vol. I, Ch. 9, in particular pp. 241–245.

where one observer measures only an electric field, another will also observe a magnetic field.

To be more specific, let us consider (Fig. 13–3) observer $\mathscr{A}$ sitting on a very long (infinitely long) charged rod; the rod has the charge $+\lambda$ per unit length, and it is moving with constant velocity $\mathbf{v}_0$ along its axis relative to another observer $\mathscr{B}$ sitting on a chair.

Observer $\mathscr{A}$ will say: "The charged rod is at rest. At a point P there will thus be no magnetic field. However, since the rod is charged, according to Eq. (7–17) there will be an electric field of magnitude

$$|\mathbf{E}| = \frac{\lambda}{2\pi\epsilon_0 r} \tag{13–7}$$

and pointing away from the rod."

Observer $\mathscr{B}$, while he watches observer $\mathscr{A}$ flashing by, will disagree and say: "There is indeed an electric field given by Eq. (13–7), and it does point radially away from the rod. However, the charged rod is moving; there is a current

$$I = \lambda\,|\mathbf{v}|$$

flowing along the rod. Since the moving rod constitutes the same current as a long straight wire carrying the current I, one can use Eq. (10–2) to determine the magnetic field; it will have the magnitude

$$|\mathbf{B}| = \frac{\mu_0 I}{2\pi r} = \frac{\mu_0 \lambda v}{2\pi r} \tag{13–8}$$

and point azimuthally around the wire."

Both observers have stated the facts correctly. We have to conclude that we cannot determine the electric and magnetic fields independently of a particular reference frame. We can also see from Eqs. (13–7) and (13–8) that the two fields have magnitudes that are proportional to each other. The magnitude of both fields is proportional to the linear charge density λ, and both fields fall off like $1/r$. Indeed, by inspection of Fig. 13–3 we see that $\mathbf{v}_0$, $\mathbf{E}$, and $\mathbf{B}$ form a right-handed system. Thus, we can interpret both observations of $\mathscr{A}$ and $\mathscr{B}$ if we state: If observer $\mathscr{A}$ sees the electric field $\mathbf{E}$ and no magnetic field $\mathbf{B}$, and if observer $\mathscr{A}$ moves with velocity $\mathbf{v}$ relative to observer $\mathscr{B}$, then $\mathscr{B}$ will see the electric field

$$\mathbf{E}' = \mathbf{E} \tag{13–9a}$$

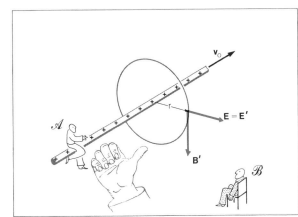

Figure 13–3 Observer $\mathscr{A}$ sitting on a moving charged rod measures only an electric field; observer $\mathscr{B}$ can detect, in addition, a magnetic field.

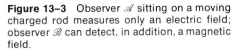

and the magnetic field

$$\mathbf{B}' = \epsilon_0\mu_0(\mathbf{v} \times \mathbf{E}) \qquad (13\text{–}9b)$$

Let us look at the numerical value of the coefficients in front of Eq. (13–9b). We have

$$\mu_0 = 4\pi \times 10^{-7}\frac{\text{V-sec}}{\text{A-m}}$$

$$\epsilon_0 = 8.854 \times 10^{-12}\frac{\text{A-sec}}{\text{V-m}}$$

The product of the two equals

$$\epsilon_0\mu_0 = 1.1126 \times 10^{-17}\frac{\text{sec}^2}{\text{m}^2}$$

which is numerically identical to the inverse of the square of the velocity of light

$$\frac{1}{c^2} = \frac{1}{(2.998 \times 10^8 \text{ m/sec})^2} = 1.1126 \times 10^{-17}\frac{\text{sec}^2}{\text{m}^2}$$

This is no accident. As we will see later, the velocity of light in vacuum is

$$c = \frac{1}{\sqrt{\epsilon_0\mu_0}} \qquad (13\text{–}10)$$

Thus, we can write instead of Eq. (13–9b)

$$c\mathbf{B}' = \frac{\mathbf{v}}{c} \times \mathbf{E} \qquad (13\text{–}11)$$

Since we assumed that all velocities involved are small compared to the velocity of light, we also have to limit $\mathbf{v}$ so that $|\mathbf{v}| \ll c$; thus,

$$c|\mathbf{B}'| \ll |\mathbf{E}'|$$

We conclude that the magnetic field is itself a relativistic effect. Indeed, consider a point charge q moving with velocity $\mathbf{w}$ in the frame of observer $\mathscr{B}$. The force on the charge will be as seen by observer $\mathscr{B}$:

$$\mathbf{F} = q\mathbf{E}' + q(\mathbf{w} \times \mathbf{B}')$$

and since $\mathbf{B}'$ is given by Eq. (13–11), we have

$$\mathbf{F} = q\mathbf{E}' + q\left[\frac{\mathbf{w}}{c} \times \left(\frac{\mathbf{v}}{c} \times \mathbf{E}'\right)\right] = \mathbf{F}_e + \mathbf{F}_m \qquad (13\text{–}12)$$

Since the second term in Eq. (13–12), expressing the magnetic force $\mathbf{F}_m$, is smaller than the electric force $\mathbf{F}_e = q\mathbf{E}'$ by a factor

$$\frac{\mathbf{F}_m}{\mathbf{F}_e} \approx \frac{|\mathbf{w}|}{c}\frac{|\mathbf{v}|}{c} \ll 1$$

we have to neglect it as being a factor (velocity/c)^2 smaller, by our agreement at the beginning of this section.

Why then do we encounter magnetic fields in everyday life? Velocities of electrons or ions in solids or liquids are typically small compared to the velocity of light c. Indeed, if the electric field of the conduction electrons were not screened by the oppositely charged ions in the crystal lattice, their electric effects would be so much larger as to make magnetic effects completely negligible.

Example 1   A copper wire of radius $r = 1$ mm carries the current $I = 10$ A. The wire is charged so that the electric field at the surface has the magnitude $|\mathbf{E}| = 10^6$ V/m (the breakdown field in air). Calculate the fraction of conduction electrons needed to form the surface charge, and compare the electric and magnetic forces on the surface electrons.

We have earlier (p. 274) calculated the density of free electrons in copper: $n_e = 8.5 \times 10^{28}$ electrons/m^3. Thus, in a length $L = 1$ m of the wire there are

$$N_1 = n_e \pi r^2 L = 2.67 \times 10^{23} \text{ electrons}$$

On the other hand, the surface charge per unit length is given from Eq. (13–7)

$$\lambda = \epsilon_0 2\pi r |\mathbf{E}| = 5.56 \times 10^{-8} \text{ coul/m}$$

so that the number of electrons participating in forming the surface charge is

$$N_2 = \frac{\lambda L}{e} = 3.47 \times 10^{11} \text{ electrons}$$

Thus, only the fraction

$$\frac{N_2}{N_1} = 1.3 \times 10^{-12}$$

of the conduction electrons participate in the surface charge. We now compare the electric and magnetic forces on a surface electron. The electric force is

$$F_e = \frac{1}{2} e |\mathbf{E}| = 0.80 \times 10^{-13} \text{ newton}$$

(We discussed the reason for the factor 1/2 at the end of Chapter 7.) In order to calculate the magnetic force on the electron, we have to know its velocity; we can obtain its value from the current density $j = I/\pi r^2$, since it is also equal to the electron charge density times their mean velocity:

$$j = \frac{I}{\pi r^2} = e n_e v$$

We thus obtain

$$v = \frac{I}{e n_e \pi r^2} = 2.34 \times 10^{-4} \text{ m/sec}$$

The magnetic field at the surface has the magnitude

$$|\mathbf{B}| = \frac{\mu_0}{2\pi r} I = 2 \times 10^{-3} \text{ Wb/m}^3$$

and because it is perpendicular to the current and thus to the electron velocity, the magnetic force on a surface electron has the magnitude

$$\mathbf{F}_m = e|\mathbf{v} \times \mathbf{B}| = ev|\mathbf{B}| = 0.75 \times 10^{-25} \text{ newton}$$

The magnetic force on a surface electron is smaller by a factor 10^{12} than the electric force. Magnetic forces are really quite weak.

As we have seen, the electric field seen by one observer gives rise to an electric and a magnetic field as seen by another moving observer. The inverse is also true; if one observer observes only a magnetic field and no electric field, another observer moving with respect to the first will observe an electric field as well. Let us again consider the two observers $\mathscr{A}$ and $\mathscr{B}$; but this time observer $\mathscr{A}$ carries with him a permanent magnet of magnetic field $\mathbf{B}$ (out of the paper in Fig. 13–4). While he is passing by with velocity $\mathbf{v}_0$, observer $\mathscr{B}$ holds a charge q into the magnetic field. To observer $\mathscr{A}$ this charge is moving with velocity $\mathbf{w} = -\mathbf{v}_0$, and he will thus find it to be subject to a Lorentz force

$$\mathbf{F} = q(-\mathbf{v}_0) \times \mathbf{B} \tag{13–13}$$

However, to observer $\mathscr{B}$ the charge q is at rest; the magnetic field can exert no force because it acts only on moving charges. Nevertheless, he must observe the same force acting on the charge, because $\mathscr{A}$ and $\mathscr{B}$ are moving with constant velocity with respect to each other and thus have to observe the same acceleration and the same force causing the acceleration. Observer $\mathscr{B}$ thus will conclude that apart from the magnetic field $\mathbf{B}$ there exists also an electric field $\mathbf{E}'$

$$\mathbf{E}' = -\mathbf{v}_0 \times \mathbf{B} \tag{13–14a}$$

The magnetic field will be the same to both observers

$$\mathbf{B}' = \mathbf{B} \tag{13–14b}$$

because they can measure, for example, the same torque

$$\mathbf{T} = \mathbf{m} \times \mathbf{B}$$

acting on a magnetic dipole, whether it is at rest or not.

What is the situation if observer $\mathscr{A}$ observes both a magnetic and an electric field, as in Fig. 13–5? If he measures an electric field $\mathbf{E}$ and a magnetic field $\mathbf{B}$, what will be the electric field $\mathbf{E}'$ and magnetic field $\mathbf{B}'$ measured by observer $\mathscr{B}$? We can generalize the two previous derivations and conclude that observer $\mathscr{B}$

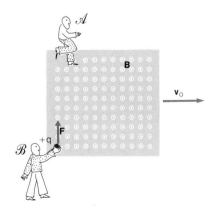

Figure 13–4 Both observers $\mathscr{A}$ and $\mathscr{B}$ observe the same force $\mathbf{F}$ on a charge q which $\mathscr{B}$ is holding; however, while $\mathscr{A}$ considers it to be due to the Lorentz force $q(-\mathbf{v}_0) \times \mathbf{B}$, observer $\mathscr{B}$ considers it to be due to an electric field.

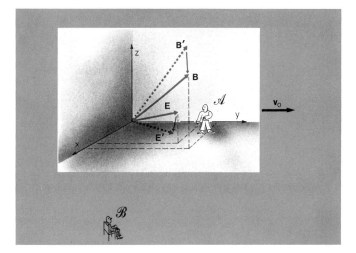

Figure 13–5 If observer $\mathscr{A}$ measures the electric field **E** and magnetic field **B**, observer $\mathscr{B}$ will measure the different fields **E**′ and **B**′.

will measure the fields

$$\mathbf{E}' = \mathbf{E} - \mathbf{v}_0 \times \mathbf{B} \tag{13–15a}$$

$$\mathbf{B}' = \mathbf{B} + \frac{\mathbf{v}_0}{c^2} \times \mathbf{E} \tag{13–15b}$$

However, we repeat that the transformation equations (13–15a) and (13–15b) are valid only if both the velocity $\mathbf{v}_0$, as well as the velocities of any charges as seen by either observer $\mathscr{A}$ or $\mathscr{B}$, are small compared to the velocity of light. They are only the nonrelativistic approximations to the exact relativistic expressions.

A charge q is moving with velocity **w** as seen by observer $\mathscr{A}$. Show that both observers $\mathscr{A}$ and $\mathscr{B}$ will observe the same force acting on the charge, if terms of order (velocity/c)^2 are neglected.

Observer $\mathscr{A}$ will observe the force

$$\mathbf{F}_{\mathscr{A}} = q\mathbf{E} + q(\mathbf{w} \times \mathbf{B})$$

To observer $\mathscr{B}$ the charge will be moving with a velocity

$$\mathbf{w}' = \mathbf{w} + \mathbf{v}_0$$

and he will thus measure the force

$$\mathbf{F}_{\mathscr{B}} = q\mathbf{E}' + q(\mathbf{w}' \times \mathbf{B}')$$

We have to show that the two forces are equal. We have from Eqs. (13–15)

$$q\mathbf{E}' = q\mathbf{E} - q\mathbf{v}_0 \times \mathbf{B}$$

and

$$q(\mathbf{w}' \times \mathbf{B}') = q(\mathbf{w} + \mathbf{v}_0) \times \left(\mathbf{B} + \frac{\mathbf{v}_0}{c^2} \times \mathbf{E} \right)$$

which, multiplied out, yields the result

$$q\mathbf{v}_0 \times \mathbf{B} + q\mathbf{w} \times \mathbf{B} + q\frac{\mathbf{w}}{c} \times \left(\frac{\mathbf{v}_0}{c} \times \mathbf{E} \right) + q\frac{\mathbf{v}_0}{c} \times \left(\frac{\mathbf{v}_0}{c} \times \mathbf{E} \right)$$

The last two terms are of the order of

$$q \times |\mathbf{E}| \times (\text{velocity}/c)^2$$

and should be neglected. We thus obtain

$$\mathbf{F}_{\mathscr{B}} = [q\mathbf{E} - q(\mathbf{v}_0 \times \mathbf{B})] + [q(\mathbf{v}_0 \times \mathbf{B}) + q(\mathbf{w} \times \mathbf{B})]$$

which is indeed equal to $\mathbf{F}_{\mathscr{A}}$ because the two terms $\pm q(\mathbf{v}_0 \times \mathbf{B})$ cancel.

Example 3   A molecule of dipole moment **p** travels with velocity **v** perpendicularly to a homogeneous magnetic field $\mathbf{B}_0$. The vector **p** is parallel to the velocity **v** (Fig. 13–6). Calculate the torque on the molecule.

We argue as follows: To an observer sitting at rest on the molecule, the laboratory magnet producing the magnetic field is moving with velocity

$$\mathbf{v}_0 = -\mathbf{v}$$

Thus, he will observe, in addition to the magnetic field **B**, the electric field

$$\mathbf{E}' = -(\mathbf{v}_0 \times \mathbf{B}) = +\mathbf{v} \times \mathbf{B}$$

The torque will be then equal to

$$\mathbf{T} = \mathbf{p} \times \mathbf{E}' = \mathbf{p} \times (\mathbf{v} \times \mathbf{B}) \tag{13–16}$$

Since $\mathbf{v} \perp \mathbf{B}$, the electric field has the magnitude vB and is perpendicular to **v** and thus to **p**. In consequence, the outer vector product in Eq. (13–16) is also easy to calculate: The torque has the magnitude

$$|\mathbf{T}| = pvB$$

and its direction is as shown in Fig. 13–6.

3. FARADAY'S LAW AND LORENTZ FORCE

In Chapter 12 we introduced Faraday's law by studying the emf produced in wire loops or coils which have a time-dependent magnetic flux passing through them. For a single-turn loop we concluded that the emf had the value given by Eq. (12–2a)

$$\mathscr{E} = -\frac{d\Phi}{dt} = -\frac{d}{dt} \int_S \mathbf{B} \bullet d\mathbf{S} \tag{13–17}$$

where the magnetic flux Φ is equal to the integral $\int \mathbf{B} \bullet d\mathbf{S}$ through a surface having the loop as its border (Fig. 13–7). Somewhat inaccurately, we then said

Figure 13–6 Example 3: Electric dipole moving in a magnetic field.

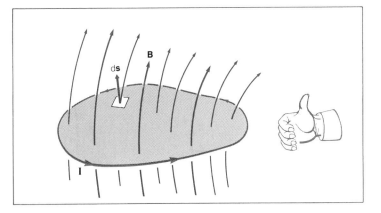

Figure 13-7 Definition of magnetic flux through a loop.

that the emf could be interpreted as the integral of the electric field around the loop

$$\mathscr{E} = \int_{\text{loop}} \mathbf{E} \cdot d\mathbf{l} \tag{13-18}$$

This interpretation of the emf is quite reasonable if the wire loop is stationary and the magnetic field itself is changing. However, we know that Faraday's law also holds if the magnetic field is constant and instead the wire loop itself is moving or rotating. In this case the interpretation (13–18) must be wrong. After all, there is no electric field inside a permanent magnet as long as there is no wire loop (or if there is a wire loop at rest) inside it. If the wire itself carries no net charge, why then should motion or rotation of the loop produce an electric field?

Let us use a specific example to discuss this problem in more detail: A rectangular wire loop of sides a and b is moving with velocity $\mathbf{v}$; at the instant we are looking at it, it is partially in and partially out of the magnetic field $\mathbf{B}$ (Fig. 13-8). Using Faraday's law, we would argue as follows: The magnetic flux through the loop is

$$\Phi = axB$$

where x is the distance between the magnet edge and the loop end. The flux Φ is increasing with time because $dx/dt = v$; thus, there will be an emf of magnitude

$$|\mathscr{E}| = \frac{d\Phi}{dt} = avB \tag{13-19}$$

around the loop.

Faraday's law gives us the global view; we can easily calculate the emf around the whole loop. But we cannot deduce the details from Faraday's law. Thus, let us study the problem from the viewpoint of observer $\mathscr{A}$ in the laboratory and from the viewpoint of an observer $\mathscr{B}$ moving with the wire loop.

Observer $\mathscr{A}$ sees a magnetic field which does not depend on time; thus, he will not be able to detect any electric field: any charge at rest in his frame will be subject to no force at all. However, the loop is moving to observer $\mathscr{A}$; thus, to him any charge q on the loop is subject to a force

$$\mathbf{F}_L = q(\mathbf{v} \times \mathbf{B})$$

at any point where the loop is inside the magnetic field. In the region outside

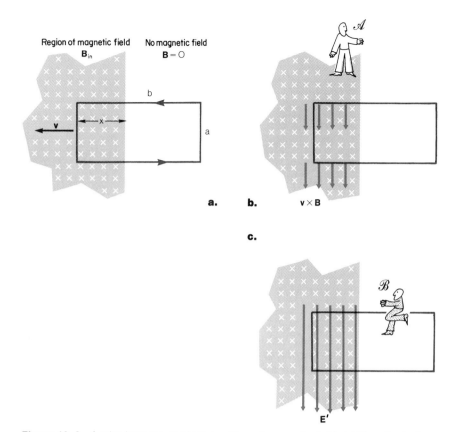

Figure 13-8   A wire loop moving into a region of magnetic field. (a) The physical arrangement. (b) From the viewpoint of $\mathscr{A}$, who is at rest with respect to the magnet, the emf is due to the Lorentz force. (c) From the viewpoint of $\mathscr{B}$, who is at rest with respect to the loop, the emf is due to an electric field **E'** inside the magnet.

$\mathscr{A}$'s magnet there is, of course, no force on any charge moving with the loop. Nevertheless, observer $\mathscr{A}$ will conclude that there is an emf because the Lorentz force is just as effective a force to produce a current in a wire as an electric field would be. The emf will have the value

$$\mathscr{E} = \frac{1}{q} \int\limits_{\text{loop}} \mathbf{F}_L \bullet d\mathbf{l} \qquad (13\text{--}20)$$

Since on both sides of length b the force is perpendicular to the element $d\ell$ of the loop, they do not contribute to the integral. The only nonvanishing contribution is from the side of length a inside the magnetic field, along which

$$\int \mathbf{F}_L \bullet d\mathbf{l} = |\mathbf{F}_L| a$$

Since $|\mathbf{F}_L| = q|\mathbf{v} \times \mathbf{B}| = qvB$, we have

$$\mathscr{E} = \frac{1}{q} \left| qvBa \right| = vBa$$

which is exactly the same result as Eq. (13–19) obtained by applying Faraday's law. Thus to observer $\mathscr{A}$, while Faraday's law is correct, it has nothing to do with any electric field. It is a necessary consequence of the Lorentz force acting on any moving charge in a magnetic field.

Observer $\mathscr{B}$ will measure an electric field; since observer $\mathscr{A}$ is moving with velocity $(-v)$ relative to $\mathscr{B}$, $\mathscr{B}$ will observe the electric field

$$\mathbf{E}' = -(-\mathbf{v}) \times \mathbf{B} = \mathbf{v} \times \mathbf{B} \qquad (13\text{-}21)$$

at any point inside the magnet. He will also see a time-dependent magnetic field, since $\mathscr{A}$'s static magnetic field to $\mathscr{B}$ is moving with velocity $-\mathbf{v}$. On the other hand, the wire loop to $\mathscr{B}$ is at rest; thus, the Lorentz force is unable to produce any emf. The emf to $\mathscr{B}$ is really an integral of the electric field around the loop

$$\mathscr{E} = \int_{\text{loop}} \mathbf{E}' \cdot d\mathbf{l} \qquad (13\text{-}22)$$

Again—as to observer $\mathscr{A}$—there is no contribution to $\mathscr{B}$'s integral from the two sides of length b or from the side a outside the magnet. Thus, he obtains

$$\mathscr{E} = |\mathbf{E}'|a = vBa$$

This is again the same as Eq. (13-19). Observer $\mathscr{B}$, however, will interpret Faraday's law in this special case in a very different way from observer $\mathscr{A}$: "If there is a time-dependent magnetic field, it produces an electric field. It is this electric field which is responsible for the emf given by Faraday's law." Thus, while both $\mathscr{A}$ and $\mathscr{B}$ observe the same emf, they will find "different" physical causes for it.

Let us now consider a situation which will occur a short while later: the rectangular loop has completely entered the magnet and is now moving with velocity $\mathbf{v}$ in a homogeneous magnetic field. What will be the emf? Faraday's law says that the magnetic flux through the loop (Fig. 13–9)

$$\Phi = Bab$$

is constant. Thus, there will be no emf. The same conclusion will also be reached by our two observers, $\mathscr{A}$ sitting inside the magnet and $\mathscr{B}$ riding with the loop, if they investigate the forces on the conduction electrons in the ring and then perform an integration over the loop. Let us consider in detail the viewpoint of observer $\mathscr{B}$ only; he will, because he is moving with the loop, observe everywhere a homogeneous electric field

$$\mathbf{E}' = \mathbf{v} \times \mathbf{B}$$

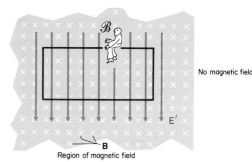

No magnetic field

E'

B

Region of magnetic field

Figure 13–9 Observer $\mathscr{B}$'s viewpoint once the loop is fully inside the magnet; everywhere around the loop there is the same electric field **E'**. Thus, the integral $\int \mathbf{E} \cdot d\mathbf{l}$ around the loop vanishes.

Since the field $\mathbf{E}'$ is everywhere the same, the loop integral around the rectangular loop will vanish:

$$\mathscr{E} = \int \mathbf{E}' \cdot d\mathbf{l} = 0 \qquad (13\text{–}23)$$

Thus, again the final result, the determination of the emf, is the same for all observers.

Example 4   A metallic ring of radius $R = 10$ cm and an axle through its center are connected through a voltmeter V. A rod connected to the axle forms a sliding contact to the ring. The axle and the rod connected to it are rotating at $N = 3$ turns per second. The whole apparatus is inside a magnetic field $|\mathbf{B}_0| = 0.6$ weber/m² whose direction is parallel to the axle (Fig. 13–10a). Calculate the voltage registered by the voltmeter (a) using Faraday's law, and (b) by studying in detail the Lorentz force on the charges.

In Fig. 13–10a we show the apparatus; we assume that the magnetic field points into the paper and that the rod rotates counterclockwise. It might seem surprising that we can use Faraday's law; however, if we consider the shaded area in Fig. 13–10a, we see that it forms a closed loop which is increasing with time. In the time dt the area changes (Fig. 13–10b) by the triangle of size

$$dA = \left(\frac{1}{2}R\right) R\, d\phi$$

and we thus have

$$\frac{dA}{dt} = \frac{1}{2}R^2\frac{d\phi}{dt} = \frac{1}{2}R^2\omega$$

Since $\omega = 2\pi N$, we have for the emf

$$|\mathscr{E}| = \frac{d\Phi}{dt} = \frac{1}{2}R^2\omega B_0 = \pi R^2 B_0 N \qquad (13\text{–}24)$$

The direction of $\mathscr{E}$ can be seen most easily from Lenz's law: the area and thus the flux are increasing, so the induced current will have the direction such that its magnetic field points out of the paper inside the loop. The emf will thus have the direction shown by the arrows in Fig. 13–10a.

We can, instead of using Faraday's law, also consider the Lorentz force on a positive charge q riding on the rod (Fig. 13–11). A charge a distance r away from the axle along the rod moves with the velocity

$$v = r\omega$$

Thus, it will be subject to the Lorentz force

$$\mathbf{F} = q(\mathbf{v} \times \mathbf{B}_0)$$

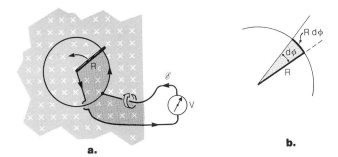

Figure 13–10 Example 4: (a) The physical arrangement and the "closed loop" changing with time; (b) calculating the change in area per unit time.

a.

b.

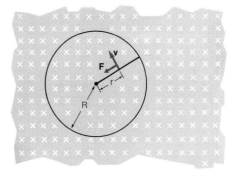

Figure 13–11 Example 4: Using the Lorentz force to calculate the emf.

which has the magnitude

$$|\mathbf{F}| = qvB_0 = q\omega B_0 r$$

and which points toward the axle ($\mathbf{B}$ is into the paper!); this is equivalent to an electric field

$$|\mathbf{E}| = \frac{|\mathbf{F}|}{q} = \omega B_0 r$$

which also points toward the center of the circle. The integral

$$\int \mathbf{E} \bullet d\mathbf{l} = \frac{1}{q} \int \mathbf{F} \bullet d\mathbf{l}$$

around the loop from the rod to the axle, the wire to the voltmeter, and back to the ring will have a contribution only from the rod itself. But there $\mathbf{E}$ is parallel to $d\mathbf{l}$ and $|d\mathbf{l}| = dr$. Thus, we have

$$\mathcal{E} = \int \mathbf{E} \bullet d\mathbf{l} = \int_0^R \omega B_0 r dr = \frac{\omega B_0 R^2}{2} \qquad (13\text{–}25)$$

which is identical to Eq. (13–24) derived from Faraday's law. The direction is also the same; the current flows from rim to axle. Numerically, we obtain either Eq. (13–24) or Eq. (13–25)

$$\mathcal{E} = \pi \,(3\text{ turns/sec})(0.6\text{ Wb/m}^2)(0.1\text{ m})^2 = 0.0565\text{ V} = 56.5\text{ mV}$$

■ *Example 5*   A rectangular wire loop of sides a and b is moving with velocity $\mathbf{v}$ in the x-direction in a magnetic field which is perpendicular to the loop plane and which increases linearly with x (Fig. 13–12):

$$|\mathbf{B}| = B_z = \alpha x$$

Show explicitly that, from the point of view of an observer riding on the loop, Faraday's law is satisfied, i.e., that

$$\int_{\text{loop}} \mathbf{E} \bullet d\mathbf{l} = -\int_{\substack{\text{area} \\ \text{inside}}} \frac{\partial \mathbf{B}}{\partial t} \bullet d\mathbf{S} = -\frac{d}{dt} \int_{\substack{\text{area} \\ \text{inside}}} \mathbf{B} \bullet d\mathbf{S} \qquad (13\text{–}26)$$

To the observer $\mathscr{B}$ riding on the loop, the magnetic field at any point with coordinate x' is increasing with time. Since

$$x = x' + vt$$

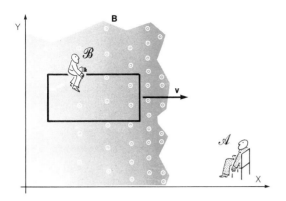

Figure 13–12 Example 5: A wire loop moving in an inhomogeneous magnetic field.

$\mathscr{B}$ will observe the magnetic field in the z-direction

$$B'_z = \alpha \,(x' + vt) = \alpha x' + \alpha v t$$

However, in addition he will observe the electric field given by Eq. (13–15a)

$$\mathbf{E}' = -(-\mathbf{v}) \times \mathbf{B} = \mathbf{v} \times \mathbf{B}$$

because to him the laboratory has a velocity $-\mathbf{v}$. Since $\mathbf{B}$ is pointing "up" (out of paper), the electric field (Fig. 13–13) will point as shown in the negative y-direction. We thus have

$$E'_y = -v\alpha x = -v\alpha x' - \alpha v^2 t$$

Let us now assume that observer $\mathscr{B}$ takes a "snapshot" of the electric field by simultaneously measuring it everywhere. He then sits down and performs the integral $\int \mathbf{E} \cdot d\boldsymbol{\ell}$ in a clockwise direction around the loop. At the position x'_2 in Fig. 13–13, he sees an electric field of magnitude

$$|E_y(x'_2)| = \alpha v x'_2 + \alpha v^2 t$$

which points along his direction of integration. At x'_1 he finds a smaller electric field opposite to his direction of integration:

$$|E_y(x'_1)| = \alpha v x'_1 + \alpha v^2 t$$

Since along the two segments of length b the electric field is perpendicular to the path

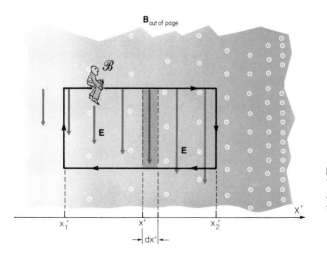

Figure 13–13 Observer $\mathscr{B}$'s viewpoint: There exists an electric field $\mathbf{E} = \mathbf{v} \times \mathbf{B}$ in addition to the time-dependent magnetic field $\mathbf{B}$.

along which he integrates, they do not contribute to his integral. He then finds

$$\int \mathbf{E} \cdot d\mathbf{l} = |E_y(x_2')|a - |E_y(x_1')|a$$

and substituting he finds

$$\int \mathbf{E} \cdot d\mathbf{l} = \alpha va\,(x_2' - x_1') = \alpha abv \qquad (13\text{--}27)$$

To perform the integral $\int \mathbf{B} \cdot d\mathbf{S}$ he has to take into account the right-hand rule which tells him that he should consider $\mathbf{B} \cdot d\mathbf{S}$ positive if $\mathbf{B}$ points into the paper. However, $\mathbf{B}$ points out of the paper, and thus he has

$$\int \mathbf{B'} \cdot d\mathbf{S} = -\int |\mathbf{B'}(x')| \, |d\mathbf{S}| \qquad (13\text{--}28)$$

He can subdivide the area into strips (Fig. 13–13) of height a and width dx'; inside each strip the field is everywhere the same:

$$|\mathbf{B'}(x')| = \alpha x' + \alpha vt \qquad (13\text{--}29)$$

Thus, he has

$$\int \mathbf{B'} \cdot d\mathbf{S'} = -\int_{x'=x_1'}^{x_2'} (\alpha x' + \alpha vt)a\,dx'$$

or

$$\int \mathbf{B'} \cdot d\mathbf{S} = -\frac{\alpha}{2}a(x_2'^2 - x_1'^2) - \alpha avt(x_2' - x_1')$$

Performing the time derivative, he obtains

$$\frac{d}{dt}\int \mathbf{B} \cdot d\mathbf{S} = -\alpha av(x_2' - x_1') = -\alpha abv$$

thus confirming Eq. (13–26). However, he can also take the time derivative directly in Eq. (13–29):

$$\frac{\partial B'}{\partial t} = \alpha v = \text{constant}$$

and then he obtains

$$\int \frac{\partial \mathbf{B}}{\partial t} \cdot d\mathbf{S} = -\int (\alpha v)dS = -\alpha v(ab) \qquad (13\text{--}30)$$

The (−) sign in Eq. (13–30) is the same as the (−) sign in Eq. (13–28); it stems from the correct application of the right-hand rule.   ■

Let us summarize what we have learned in this section. An observer who measures a time-dependent magnetic field will also always detect an electric field. If he chooses a closed loop at rest to himself, he will find the relation

$$\int_{\text{loop}} \mathbf{E} \cdot d\mathbf{l} = -\int_{\text{inside}} \frac{\partial \mathbf{B}}{\partial t} \cdot d\mathbf{S} \qquad (13\text{--}31)$$

where the integral on the right-hand side is to be taken over any surface which has the loop as a boundary. The direction of $d\mathbf{S}$ is chosen with help of the right-hand rule (Fig. 13–14). It is irrelevant whether the loop is a physical loop made

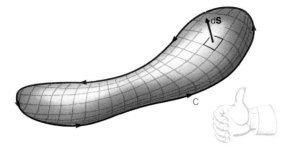

Figure 13–14 In general, a surface having a loop C as its border does not have to be flat; it is not even uniquely defined.

of a wire or a loop chosen by drawing an imagined closed curve in space; the electric field is defined at any point and the integral can always be performed.

If an observer sees a moving *physical wire loop* in a time-independent but nonhomogeneous (position dependent) magnetic field, he will observe an emf

$$\mathscr{E} = -\frac{d}{dt} \int\limits_{\substack{\text{loop} \\ \text{area}}} \mathbf{B} \cdot d\mathbf{S} \tag{13–32}$$

This emf will be completely due to the Lorentz force $\mathbf{F} = q\,(\mathbf{v} \times \mathbf{B})$ acting on the charges in the wire. There is no electric field in the absence of a net nonzero charge density — any possible electric field is due not to the magnetic field but to some charge distribution, and does not contribute to the emf in Eq. (13–32). It is essential that a physical loop be moving; there is no emf in the absence of motion.

If an observer observes a moving physical loop in a time-dependent magnetic field, he will observe an emf

$$\mathscr{E} = -\frac{d}{dt} \int\limits_{\substack{\text{loop} \\ \text{area}}} \mathbf{B} \cdot d\mathbf{S}$$

which is due partially to an electric field

$$\int \mathbf{E} \cdot d\mathbf{l} = -\int \frac{\partial \mathbf{B}}{\partial t} \cdot d\mathbf{S}$$

and partially to the Lorenz force acting on the charges of the moving loop.

We still have not answered one question which is required to make expressions such as Eq. (13–31) meaningful. For any given closed loop there exists an infinity of surfaces which have the loop as their border. When writing down the expression for the flux

$$\Phi = \int\limits_{\substack{\text{loop} \\ \text{area}}} \mathbf{B} \cdot d\mathbf{S} \tag{13–33}$$

we do not specify which of these surfaces we should use. Thus, Eq. (13–33) and Faraday's law make sense only if the resulting integral is the same whatever surface we choose to perform the actual integration.

Let us consider two such surfaces, S_1 and S_2, as shown in Fig. 13–15. They have both the same border, the loop C. Together they form a single closed surface S. Given the sense of direction along the loop C indicated in Fig. 13–15, the right-hand rule dictates that $d\mathbf{S}_1$ is pointing *out* of the closed surface, while $d\mathbf{S}_2$ is pointing *inward*. Thus, if we define $d\mathbf{S}$ for the overall closed surface as

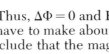

Figure 13-15 Two surfaces S_1 and S_2, having the same loop C as their border, form a closed surface. The direction of $d\mathbf{S}_1$ is out of the closed surface, while $d\mathbf{S}_2$ points inward.

always pointing along the normal outward, then

$$\text{on } S_1 \qquad d\mathbf{S}_1 = d\mathbf{S}$$
$$\text{on } S_2 \qquad d\mathbf{S}_2 = -d\mathbf{S}$$

We now prove that

$$\Phi = \int_{S_1} \mathbf{B} \bullet d\mathbf{S}_1 = \int_{S_2} \mathbf{B} \bullet d\mathbf{S}_2 \qquad (13\text{–}34\text{a})$$

If we write

$$\Delta\Phi = \int_{S_1} \mathbf{B} \bullet d\mathbf{S}_1 - \int_{S_2} \mathbf{B} \bullet d\mathbf{S}_2$$

then since $d\mathbf{S}_1 = d\mathbf{S}$ and $d\mathbf{S}_2 = -d\mathbf{S}$ we have

$$\Delta\Phi = \int_{S_1} \mathbf{B} \bullet d\mathbf{S} + \int_{S_2} \mathbf{B} \bullet d\mathbf{S} = \int_{S} \mathbf{B} \bullet d\mathbf{S} \qquad (13\text{–}34\text{b})$$

where the integral is to go over the complete closed surface S. But according to the fourth Maxwell's equation

$$\int_{S} \mathbf{B} \bullet d\mathbf{S} = 0$$

for any closed surface S. Thus, $\Delta\Phi = 0$ and Eq. (13–34a) is indeed correct. Since the only assumption we have to make about S_1 and S_2 is that they both have the loop C as border, we conclude that the magnetic flux through C

$$\Phi = \int_{\substack{\text{inside} \\ C}} \mathbf{B} \bullet d\mathbf{S}$$

has the same value whatever surface we choose for the actual integration. This is exactly what we tacitly assumed when discussing Faraday's law.

4. DISPLACEMENT CURRENT

We have just concluded that Faraday's law is a necessary consequence of the fact that if one observer measures only a magnetic field, another observer moving with constant velocity with respect to the first will also measure an electric field. But as we have shown earlier in this chapter, the reverse statement is also correct. If one observer measures only an electric field, another observer moving with respect to the first will also measure a magnetic field. Thus, we can conclude that not only is a time-dependent magnetic field always associated with an electric field, but there will always also be a magnetic field in the presence of a time-dependent electric field.

We have earlier (Chapter 10) discussed Ampère's law, which can be written as

$$\int_{\substack{\text{closed} \\ \text{loop}}} \mathbf{B} \cdot d\mathbf{l} = \mu_0 I_{\substack{\text{through} \\ \text{loop}}} = \mu_0 \int_{\substack{\text{through} \\ \text{loop}}} \mathbf{j} \cdot d\mathbf{S} \qquad (13\text{–}35)$$

We will now show that this law cannot be complete; the integral of the magnetic field over a closed loop can be non-vanishing even if no current flows through the loop.

Let us again consider two observers moving with respect to each other. Observer $\mathscr{A}$ is at rest relative to a charged plane parallel capacitor; the capacitor and $\mathscr{A}$ are moving with velocity $\mathbf{v}$ relative to observer $\mathscr{B}$, who is sitting on a chair in the laboratory as shown in Fig. 13–16a. In Fig. 13–16b we show the top view as seen by observer $\mathscr{B}$. Besides the electric field $\mathbf{E}$ (into the paper), observer $\mathscr{B}$ will also see everywhere inside the capacitor the magnetic field

$$\mathbf{B}' = \frac{1}{c^2}(\mathbf{v} \times \mathbf{E})$$

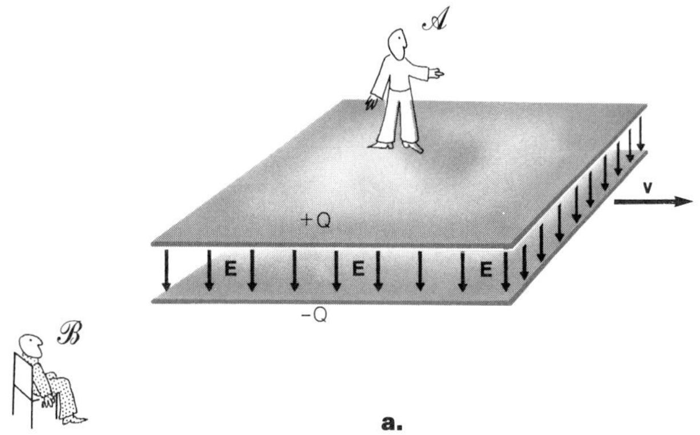

a.

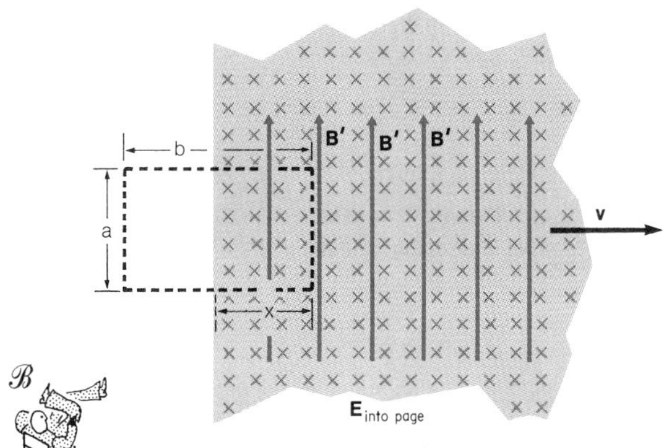

b.

Figure 13–16 A moving electric field. (a) The two observers $\mathscr{A}$ and $\mathscr{B}$; (b) top view from $\mathscr{B}$'s viewpoint: The integral $\int \mathbf{B} \cdot d\mathbf{l}$ around the imagined loop will not vanish, although there is no current flowing through the loop.

Let us now imagine observer $\mathscr{B}$ defining a rectangular loop of sides a and b. After taking a snapshot of the magnetic field, he will perform the integral $\int \mathbf{B} \cdot d\mathbf{l}$ over the loop. Since the loop is partially in the capacitor and partially outside, he will find a nonzero value. But there is only vacuum in the plane of the loop — no current is flowing through the loop. Specifically, he will find that because only the side of length a inside the capacitor contributes to the integral

$$\int_{\text{loop}} \mathbf{B} \cdot d\mathbf{l} = |\mathbf{B}'|a = \frac{1}{c^2} \, avE \qquad (13\text{–}36)$$

Since there is no current through the loop, observer $\mathscr{B}$ concludes that he has to modify Ampère's law [Eq. (13–35)]. Prompted by earlier success with Faraday's law, he calculates the total electric flux through the area of the loop. Because he integrated counterclockwise around the loop, he decides that because of the right-hand rule, he should consider $\mathbf{E} \cdot d\mathbf{S} > 0$ if the electric field points up — out of the paper in Fig. 13–16b. However, it is pointing down, and thus he finds

$$\int_{\substack{\text{loop} \\ \text{area}}} \mathbf{E} \cdot d\mathbf{S} = - |\mathbf{E}|ax$$

In $\mathscr{B}$'s setup, x is decreasing and thus

$$-\frac{dx}{dt} = v$$

Consequently, he finds

$$\frac{d}{dt} \int_{\substack{\text{loop} \\ \text{area}}} \mathbf{E} \cdot d\mathbf{S} = + Eav \qquad (13\text{–}37)$$

He then concludes that in the absence of a current through the loop

$$\int_{\text{loop}} \mathbf{B} \cdot d\mathbf{l} = \frac{1}{c^2} \frac{d}{dt} \int_{\substack{\text{area} \\ \text{of loop}}} \mathbf{E} \cdot d\mathbf{S} \qquad (13\text{–}38)$$

He now wants to combine Eq. (13–38) and the original Ampère's law (13–35) into a single law; because of Eq. (13–10)

$$\frac{1}{c^2} = \epsilon_0 \mu_0$$

he can write as a generalization of Ampère's law

$$\int_{\substack{\text{closed} \\ \text{loop}}} \mathbf{B} \cdot d\mathbf{l} = \mu_0 \int_{\substack{\text{area} \\ \text{inside loop}}} \left(\mathbf{j} + \epsilon_0 \frac{\partial \mathbf{E}}{\partial t} \right) \cdot d\mathbf{S} \qquad (13\text{–}39)$$

The quantity $\epsilon_0 \, \partial \mathbf{E}/\partial t$ is called the *displacement current density*. It is measured in the same units of A/m² as the real current density $\mathbf{j}$.

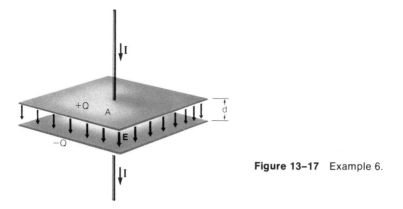

Figure 13–17   Example 6.

Example 6   A plane parallel capacitor (plate spacing d, area A) is charged at a constant rate $dQ/dt = I$ (Fig. 13–17). Calculate the displacement current density and the total displacement current between the capacitor plates.

If the charge $\pm Q$ resides on the capacitor plates, the potential difference is

$$V = \frac{Q}{c} = \frac{Qd}{\epsilon_0 A}$$

The electric field is then

$$|\mathbf{E}| = \frac{V}{d} = \frac{Q}{\epsilon_0 A}$$

and if Q changes with time we have

$$\left|\frac{\partial \mathbf{E}}{\partial t}\right| = \frac{1}{\epsilon_0 A}\frac{dQ}{dt} = \frac{I}{\epsilon_0 A}$$

Thus, the displacement current density is

$$\epsilon_0\left|\frac{\partial \mathbf{E}}{\partial t}\right| = \frac{I}{A}$$

Since $\mathbf{E}$ does not change direction but only magnitude, the displacement current will be everywhere pointing vertically from one capacitor plate to the other. The total displacement current through a plane between the two plates is then

$$I_{\text{disp}} = \epsilon_0 \int \frac{\partial \mathbf{E}}{\partial t}\cdot d\mathbf{S} = \epsilon_0\left|\frac{\partial \mathbf{E}}{\partial t}\right| A$$

or

$$I_{\text{disp}} = I$$

The total displacement current between the capacitor plates is exactly the same as the current charging the capacitor.

■   We have derived the existence of a displacement current using the idea that observers moving with constant velocity with respect to each other should observe the same physical laws. This was not at all obvious in 1860 when Maxwell first proposed the modification [Eq. (13–25)] of Ampère's law. At the time it was believed that electric and magnetic fields were manifestations of mechanical deformation of an ether, which was thought of as a tenuous gas permeating all space. But if one accepts the existence of such an ether, there is no reason why an observer moving with respect to it should observe the same laws as an observer who is at rest to the ether. If one is moving through air, one encounters a resistance which is absent if one is at rest.

Maxwell used a different argument, which we will briefly discuss. His starting point was a consideration analogous to our derivation of Eq. (13–34). Ampère's law in its simple form

$$\underset{\substack{\text{closed}\\\text{loop}}}{\int} \mathbf{B} \cdot d\mathbf{l} = \mu_0 \underset{\substack{\text{surface}\\\text{through loop}}}{\int} \mathbf{j} \cdot d\mathbf{S} \qquad (13\text{–}40)$$

makes sense only if the right-hand side is the same whatever surface we take (as long as it has the closed loop as its border). Using exactly the same argument which led us to Eq. (13–34b), we conclude that this will be true if and only if

$$\underset{\substack{\text{closed}\\\text{surface}}}{\int} \mathbf{j} \cdot d\mathbf{S} = 0 \qquad (13\text{–}41)$$

for any closed surface. Eq. (13–41) is correct, however, only for steady state (time-independent) currents, because it is closely connected to the conservation of charge. The integral in Eq. (13–41) is the total current flowing out of the closed surface. Thus, it has to be compensated by a net decrease of charge inside the same closed surface (Fig. 13–18). Therefore, by conservation of charge we always have for a closed surface S

$$-\frac{dQ_{\text{inside } S}}{dt} = \underset{S}{\int} \mathbf{j} \cdot d\mathbf{S} \qquad (13\text{–}42)$$

This is equivalent to Eq. (13–41) only if the charge distribution does not change. We conclude that we have to modify Eq. (13–40) by writing

$$\underset{\substack{\text{closed}\\\text{loop } C}}{\int} \mathbf{B} \cdot d\mathbf{l} = \mu_0 \underset{\substack{\text{surface}\\\text{inside}}}{\int} (\mathbf{j} + \mathbf{X}) \cdot d\mathbf{S} \qquad (13\text{–}43\text{a})$$

where the unknown quantity $\mathbf{X}$ has to be such that for any closed surface

$$\underset{\substack{\text{closed}\\\text{surface}}}{\int} (\mathbf{j} + \mathbf{X}) \cdot d\mathbf{S} = 0 \qquad (13\text{–}43\text{b})$$

We can find $\mathbf{X}$ with the help of Eq. (13–42). From Gauss's law we know that for any closed surface

$$\underset{\substack{\text{closed}\\\text{surface } S}}{\int} \mathbf{E} \cdot d\mathbf{S} = \frac{Q_{\text{inside}}}{\epsilon_0}$$

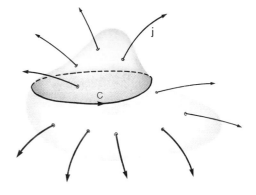

Figure 13–18 The integral $\int \mathbf{j} \cdot d\mathbf{S}$ over a closed surface does not have to vanish, if the total charge inside is changing.

and taking the time derivative on both sides, we find that

$$\epsilon_0 \int_S \frac{\partial \mathbf{E}}{\partial t} \cdot d\mathbf{S} = \frac{dQ_{\text{inside}}}{dt} \tag{13-44}$$

Comparing this with Eq. (13–42), we see that

$$\int \left(\mathbf{j} + \epsilon_0 \frac{\partial \mathbf{E}}{\partial t} \right) \cdot d\mathbf{S} = 0 \tag{13-45}$$

for any closed surface. We have therefore found a quantity $\mathbf{X}$ which satisfies Eq. (13–43b).

We have not shown that the *only* possible choice for $\mathbf{X}$ is $\mathbf{X} = \epsilon_0 \dfrac{\partial \mathbf{E}}{\partial t}$. Indeed, one can find an infinity of other expressions $\mathbf{X}$ which also satisfy Eq. (13–43b). Maxwell chose — correctly, as we know today—the simplest expression which he knew to be consistent with the then available experimental evidence. Substituting

$$\mathbf{X} = \epsilon_0 \frac{\partial \mathbf{E}}{\partial t} \tag{13-46}$$

into Eq. (13–43a), he then obtained Eq. (13–39), which we have derived from the principle of Galilean invariance. ■

5. SUMMARY OF MAXWELL'S EQUATION

Let us now summarize the basic equations which govern all electromagnetic phenomena. There are only five of them, four of which involve integrals of the electric and magnetic fields over closed loops or closed surfaces:

I. $\displaystyle \int_{\substack{\text{closed} \\ \text{loop}}} \mathbf{E} \cdot d\mathbf{l} = - \int_{\substack{\text{surface with} \\ \text{loop as border}}} \mathbf{B} \cdot d\mathbf{S}$

II. $\displaystyle \int_{\substack{\text{closed} \\ \text{surface}}} \epsilon_0 \mathbf{E} \cdot d\mathbf{S} = Q_{\text{inside}}$

III. $\displaystyle \int_{\substack{\text{closed} \\ \text{loop}}} \mathbf{B} \cdot d\mathbf{l} = \mu_0 \int_{\substack{\text{surface with} \\ \text{loop as border}}} \left[\mathbf{j} + \epsilon_0 \frac{\partial \mathbf{E}}{\partial t} \right] \cdot d\mathbf{S}$

IV. $\displaystyle \int_{\substack{\text{closed} \\ \text{surface}}} \mathbf{B} \cdot d\mathbf{S} = 0$

To this should be added the equation giving the force on a charge q moving with velocity $\mathbf{v}$:

V. $\quad \mathbf{F} = q[\mathbf{E} + \mathbf{v} \times \mathbf{B}]$

The first four equations are Maxwell's equations in their full general form. The second one is Gauss's law; the first, Faraday's law of induction. The third equation is Ampère's law generalized to include the displacement current. The fourth equation, which we could call "Gauss's law of magnetism," expresses the fact that there are no magnetic charges. The fifth equation, which gives the electromagnetic force on charged particles, provides the connection to mechanics. If one considers how many of the phenomena we encounter in everyday life (e.g., the greening of a tree, the melting of ice upon heating) are electromagnetic in origin, it is truly amazing to contemplate the consequences of these five simple relations.

If there are neither charges nor currents nearby, the four equations take a particularly simple and symmetrical form:

$$\text{I}'. \int_{\text{loop}} \mathbf{E} \cdot d\mathbf{l} = - \int_{\text{inside}} \frac{\partial \mathbf{B}}{\partial t} \cdot d\mathbf{S} \qquad \text{III}'. \int_{\text{loop}} \mathbf{B} \cdot d\mathbf{l} = \epsilon_0 \mu_0 \int_{\text{inside}} \frac{\partial \mathbf{E}}{\partial t} \cdot d\mathbf{S}$$

$$\text{II}'. \int \mathbf{E} \cdot d\mathbf{S} = 0 \qquad\qquad\qquad \text{IV}'. \int \mathbf{B} \cdot d\mathbf{S} = 0$$

From these equations we will derive the existence of electromagnetic waves. A time-dependent magnetic field produces an electric field according to Eq. I', which in general will also vary with time. But because of Eq. III', the time-dependent electric field produces a magnetic field of its own a little further on, and the process is repeated in time and space. A time-dependent electromagnetic field thus can propagate over large distances in the form of a wave.

However, many electromagnetic waves (e.g., visible light) are produced by individual atoms; thus, one also has to understand atomic physics in order to discuss them, and this cannot be done without a new type of mechanics—quantum mechanics, which replaces the classical Newtonian mechanics on the atomic level. We will learn in quantum mechanics that in many ways an electron or any other particle behaves like a wave. But we will also learn that an electromagnetic wave is composed of particles called *photons*. Thus, the distinction between particles and waves will become quite blurred. It is something of a historical accident that when physicists learned to understand nature, they discovered first the particle nature of some particles (e.g., electrons), while for others (photons) the wave nature was the first to be investigated. We will first discuss the quantum mechanics of electrons and atomic nuclei; later on we will discuss photons.

Summary of Important Relations

If observer $\mathscr{A}$ moves with velocity $\mathbf{v}_0$ relative to observer $\mathscr{B}$ and $\mathscr{A}$ observes the electromagnetic fields $\mathbf{E}$ and $\mathbf{B}$, then $\mathscr{B}$ observes

$$\mathbf{E}' = \mathbf{E} - \mathbf{v}_0 \times \mathbf{B}$$

$$\mathbf{B}' = \mathbf{B} + \frac{\mathbf{v}_0}{c^2} \times \mathbf{E}$$

Displacement current $\qquad\qquad \mathbf{j}_d = \epsilon_0 \frac{\partial \mathbf{E}}{\partial t}$

Complete set of Maxwell's equations in vacuum:

$$\text{I}. \int_{\substack{\text{closed} \\ \text{loop}}} \mathbf{E} \cdot d\mathbf{l} = - \int_{\text{surface}} \mathbf{B} \cdot d\mathbf{S}$$

II. $\displaystyle\int_{\substack{\text{closed}\\\text{surface}}} \epsilon_0 \mathbf{E} \cdot d\mathbf{S} = Q_{\text{inside}}$

III. $\displaystyle\int_{\substack{\text{closed}\\\text{loop}}} \mathbf{B} \cdot d\mathbf{l} = \mu_0 \int_{\text{surface}} \left[\mathbf{j} + \epsilon_0 \frac{\partial \mathbf{E}}{\partial t} \right] \cdot d\mathbf{S}$

IV. $\displaystyle\int_{\substack{\text{closed}\\\text{surface}}} \mathbf{B} \cdot d\mathbf{S} = 0$

In materials, replace everywhere ϵ_0 by $K\epsilon_0$ and μ_0 by $\mu_r\mu_0$.

Questions

1. The kinetic energy $T = mv^2/2$ is different for two observers moving relative to each other. How then can the law of energy conservation be true for both observers? Discuss the elastic collision between two (unequal mass) particles and the motion of a particle under a constant force $\mathbf{F}$. What happens if $\mathbf{F}$ depends on the position of the particle for one observer? Will the total energy then be conserved for the other observer?

2. Observer $\mathscr{A}$ observes a current I in an overall neutral stationary metallic wire. To $\mathscr{A}$ the current is due only to the electrons in the wire. Observer $\mathscr{B}$ is moving with velocity $\mathbf{v}$ relative to $\mathscr{A}$. Discuss what current $\mathscr{B}$ measures and what produces it.

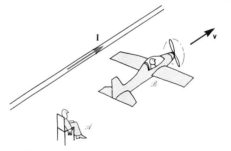

3. We have neglected the displacement current up to now, for instance, when discussing the energy stored in an LC circuit. Was this justified? Give arguments about how rapidly a process would have to occur for the displacement current to become important. Do not forget—practical electric fields are limited to a few times 10^6 V/m.

4. Would you expect magnetic effects between electrons to be important in atoms? Why?

Problems

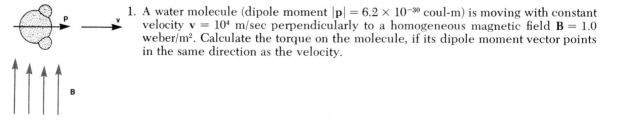

1. A water molecule (dipole moment $|\mathbf{p}| = 6.2 \times 10^{-30}$ coul-m) is moving with constant velocity $\mathbf{v} = 10^4$ m/sec perpendicularly to a homogeneous magnetic field $\mathbf{B} = 1.0$ weber/m². Calculate the torque on the molecule, if its dipole moment vector points in the same direction as the velocity.

2. A water molecule (atomic weight $A = 18.0$ kg/mole, dipole moment $p = 6.2 \times 10^{-30}$ coul-m) is moving with velocity $\mathbf{v} = 5 \times 10^4$ m/sec in the x-direction in an inhomogeneous magnetic field which points in the z-direction:

$$B_z = \alpha x \quad \text{with} \quad \alpha = 10.0 \text{ weber/m}^3$$

Calculate the acceleration of the molecule, if its dipole moment also points in the positive x-direction.

3. An electron (spin $|\mathbf{S}| = \hbar/2$, magnetic moment $\mathbf{m} = -e/m_e \mathbf{S}$) is moving with a velocity $\mathbf{v} = 10^6$ m/sec in the x-direction in a homogeneous field of magnitude $|\mathbf{E}| = 10^6$ V/m that points in the y-direction. At the time $t = 0$, the spin points in the direction of motion. Calculate the acceleration of the electron and the angular velocity Ω with which its spin will precess. Ω is defined as

$$\frac{1}{|\mathbf{S}|} \left| \frac{d\mathbf{S}}{dt} \right|$$

4. An electron starts with velocity $\mathbf{v}$ perpendicular to a magnetic field $\mathbf{B}$. Its initial spin $\mathbf{S}$ is parallel to $\mathbf{v}$. Show that as the electron moves on a circle, its spin vector stays always parallel to its velocity vector.

5. A nonconducting plate made of ferrite ($\mu = 4000$) is moving with velocity $\mathbf{v} = 10^3$ m/sec in an electric field $\mathbf{E}_0 = 10^5$ V/m. Calculate the magnetization in the plate. What will be the magnetization's direction?

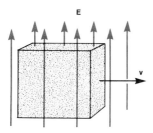

6. A wire is bent as shown in the figure, and is rotating with period T in a magnetic field B_0. Calculate the induced emf

    (a) using Faraday's law.

    (b) with the help of the Lorentz force.

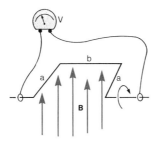

7. A metallic plate is moving with velocity $v = 100$ m/sec perpendicularly to a homogeneous magnetic field of magnitude $B = 30,000$ gauss. The plate is oriented so that

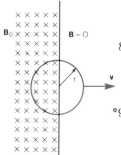

both **v** and **B** are parallel to the plate (see figure following Problem 10.) Calculate the surface charge density on the plate surfaces.

8. A circular wire loop (radius r) is moving with velocity **v** out of a magnet; at the time $t = 0$ the circle's center is exactly at the magnet's edge. Calculate the integral $\int \mathbf{E} \cdot d\mathbf{l}$ of the electric field as seen by an observer at rest to the loop and check that Faraday's law is obeyed.

*9. A thin circular insulating slab (dielectric constant $K = 2.5$) of radius R is rotating around its axis with period $T = 1/30$ sec in a magnetic field **B** ($B = 0.8$ weber/m²) perpendicular to the rotation axis. Calculate the polarization **P** in the slab at an arbitrary point specified by its polar coordinates (ρ, ϕ). Choose $R = 20$ cm.

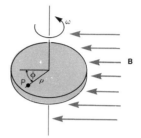

10. An insulating slab of dielectric constant K is moving with velocity **v** in a magnetic field **B**, as shown in the figure. Calculate the electric field in the slab

   (a) as seen by an observer sitting on the slab.

   (b) as seen by an observer at rest to the magnet.

   For numerical purposes, choose $K = 3$, $|\mathbf{B}| = 1.0$ weber/m², and $v = 10^2$ m/sec.

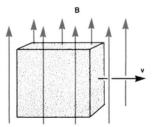

11. Calculate the magnetic field that an electron sees in its rest frame in a circular orbit of 0.53 Å (hydrogen radius) around a proton. From this, calculate the potential energy difference between the situations in which the electron spin is (a) parallel, and (b) antiparallel to the magnetic field.

12. Calculate the magnetic field a distance **r** away from a point charge q moving with velocity **v**. Show that this is identical to what one would obtain from Biot-Savart's law by replacing $I\,d\mathbf{l}$ by $q\mathbf{v}$. Why is this reasonable?

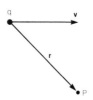

13. An electron is moving with velocity $v = 5 \times 10^6$ m/sec past an atom at rest; the minimum distance is $d = 10$ Å. Calculate the magnetic field $\mathbf{B}(t)$ seen by the atom. What is its maximum value? What is the time interval Δt during which $|\mathbf{B}(t)|$ is more than 1/2 of its maximum value? Does the direction of $\mathbf{B}$ change as the electron moves by?

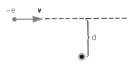

14. An electric field of magnitude $|\mathbf{E}| = 10^5$ V/m and a magnetic field $\mathbf{B} = 0.1$ weber/m² are perpendicular to each other. Does there exist a moving reference frame (velocity v) for which

(a) the electric field vanishes?   (b) the magnetic field vanishes?

If the answer is yes, find the velocity vector v.

15. Between two plates of a plane parallel capacitor separated by a distance d, the electric field has the magnitude $E_y = 10^6$ V/m. The capacitor is in a magnetic field $\mathbf{B}$ which is parallel to the plates, of magnitude $B = 0.1$ weber/m². Electrons with negligible velocity leave the negatively charged capacitor plate. How large can the plate separation d be if they are to reach the other plate? (Hint: Consider the problem in a moving coordinate system in which $\mathbf{E}$ vanishes.)

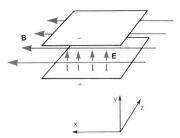

16. Calculate the path of the electron released from the negative plate in Problem 15. What is the maximum and the mean velocity of the electron in the z-direction? Assume a plate separation $d = 2$ cm.

17. A circular coil of radius r is rotating with angular velocity $\omega = d\phi/dt$ around an axis perpendicular to a magnetic field $\mathbf{B}$. Calculate the electric field in the coil's rest frame at any point on the coil

(a) at the instant when $\phi = 0$ (i.e., $\mathbf{B}$ lies in the plane of the coil)

(b) for an arbitrary ϕ

and show that the integral $\int \mathbf{E} \cdot d\mathbf{l}$ around the coil is the same as the emf obtained from Faraday's law.

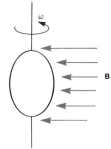

18. A plane parallel capacitor with cylindrical plates of radius $r = 10$ cm and plate separation $d = 3$ mm is being charged with a current of 1 A. Calculate the magnetic field between the plates.

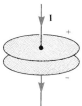

19. A charged sphere is discharging with a spherically symmetric radial current density. Show that there is no magnetic field associated with this current.

CHAPTER 14

QUANTUM MECHANICS

1. UNCERTAINTY RELATION

Before we can properly discuss atomic phenomena, we have to introduce a new type of mechanics, quantum mechanics. The classical description as given by Newtonian mechanics fails for very small—and very light—objects such as electrons. The classical theory fails because some of its basic assumptions are no longer even approximately true.

Let us first review classical mechanics and try to decide which of its "obvious" assumptions really are not necessarily true. Let us talk about a small object, a mass point, which we will call a particle. By this we mean an object which has a mass m, but which is sufficiently small that its shape or mass distribution does not matter (the Earth is certainly a particle on the galactic scale!). In Newtonian mechanics we tacitly assume that we can at any time determine *exactly where* the particle is without disturbing it. For macroscopic objects this is certainly true; all we have to do is to look at them. But for microscopic objects this assumption is not necessarily true. Whenever we try to "see" an object, we have to scatter something, such as light, from it; or we have to let it interact with a fluorescent screen.

These interactions also happen to macroscopic objects, but they can be kept "arbitrarily small." The reflection of a light ray by even a very small ball bearing will have a negligible effect on its motion. However, for microscopic objects such as electrons, the interaction by which we "see" it will affect its future motion in a nonnegligible and unpredictable way. The act of "seeing" will change the particle's direction of motion, or the magnitude of its velocity, or both, by an unknown amount. Even if we knew its momentum exactly before now, we no longer know it so well—we cannot predict where it will be a small time Δt later.

One can claim that this does not prevent us from exactly predicting the path of an electron or any other small particle. We simply have to assume that it is possible to calculate exactly how we are affecting the particle by looking at it. Although the interaction of "seeing" will influence the behavior of the particle, we can correct for it. But the crucial point is that this is no longer obvious; we have to make an additional assumption which may or may not be true. Only by experiment can we either prove or disprove this axiom of classical mechanics.

Modern quantum mechanics started with the realization by W. Heisenberg that the tacit Newtonian assumption could be replaced by a different axiom. Heisenberg's new axiom, called the *uncertainty relation*, states that whenever we try to measure simultaneously the position x and the momentum (in the x-direction) p_x of a particle, there is a lower limit to the accuracy with which we can measure these two quantities; if we measure x to an accuracy of $\pm \Delta x$ and p_x to $\pm \Delta p_x$, then always

$$\Delta x \cdot \Delta p_x \geq \hbar \tag{14-1}$$

where $\hbar$ (pronounced "aitch-bar") is a fundamental constant of nature, called Dirac's constant:

$$\hbar = (1.05421 \pm 0.00003) \times 10^{-34} \text{ Joule-sec}$$

We have written in Eq. (14–1) the sign $\geq$ meaning "larger than or roughly equal to" because without going into details of statistical analysis the concept of "accuracy" is not precisely defined.[*] We here define the error Δx by stating that if we repeat the same experiment many times, most of the times the value found will lie between $x - \Delta x$ and $x + \Delta x$. Since "most of the time" is conveniently vague, we should not be surprised if we find that the product of Δx and Δp_x sometimes is $0.9\ \hbar$; Eq. (14–1) states, however, that we will rarely find a value much less than $\hbar$, say $0.1\ \hbar$.

Eq. (14–1) replaces the earlier-stated tacit Newtonian assumption that, given enough care, both x and p_x can be determined to arbitrary accuracy. At first glance, there does not seem to exist a large difference between the two assumptions; after all, we never in reality measure the position of a particle to infinite accuracy because of limitations of our measuring equipment. However, one tacit assumption of Newtonian mechanics is that for every problem involving a particle there exists a solution in the form of a vector $\mathbf{r}(t)$ which is a function of time (Fig. 14–1a); by finding the explicit form of $\mathbf{r}(t)$, we "solve the problem,"

---

[*] If we define by Δx and Δp_x as the standard deviations of x and p_x as measured in a many times repeated experiment, then the $\geq$ sign can be replaced by an exact inequality, $\Delta x \cdot \Delta p_x \geq \hbar/2$.

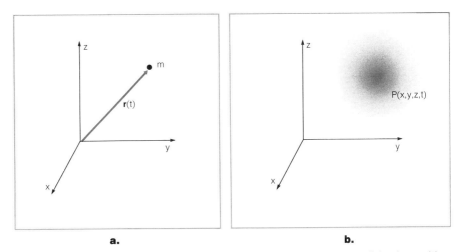

Figure 14–1 (a) Classically we describe a particle at an instant t by specifying its position vector **r**. (b) In quantum mechanics we have to give instead at any instant the probability $P(x,y,z)$ of finding it at the position (x,y,z).

i.e., we now can predict where the particle will be at any time in the future. But in order to calculate $\mathbf{r}(t)$ one has to know the "initial conditions," the position $\mathbf{r}_0$ and velocity $\mathbf{v}_0$ (or momentum $\mathbf{p}_0 = m\mathbf{v}_0$) at a time $t = t_0$. This is in direct contradiction to Eq. (14–1), which says that we can never determine both the initial position and the initial momentum; but not knowing the initial conditions, we cannot calculate $\mathbf{r}(t)$ for later times! As we see, we have to modify more than just some details of Newtonian mechanics; *the basic way of describing nature has to be modified*. We cannot describe the particle by a single vector giving its position; instead, we have to describe it by a probability function $P(\mathbf{r})$ (Fig. 14–1b), which gives the probability that we will find the particle at the point $\mathbf{r}$. This probability will itself change with time if the particle is moving.

Example 1   If the position of an electron is determined to within 1 Å $= 10^{-10}$ m, what is the uncertainty of its velocity?

From Eq. (14–1) we have, since $p = mv$

$$\Delta v = \frac{\hbar}{m\Delta x} = \frac{1.05 \times 10^{-34} \text{ Joule-sec}}{0.91 \times 10^{-30} \text{ kg} \times 10^{-10} \text{ m}} = 1.14 \times 10^6 \text{ m/sec} \qquad (14\text{–}2)$$

This is a large velocity, 0.3% of the velocity of light. Since 1 Å is a typical size of an atom, the uncertainty relation implies that electrons have to have a large velocity inside an atom.

Eq. (14–2) describes the uncertainty in velocity in any direction: x, y, or z. The electron therefore has to have a minimum kinetic energy of roughly

$$T \approx \frac{m}{2}(\Delta v_x{}^2 + \Delta v_y{}^2 + \Delta v_z{}^2) = \frac{3}{2}m \; (\Delta v)^2$$

which numerically is

$$\frac{3}{2} \times 0.91 \times 10^{-30} \text{ kg} \times 1.30 \times 10^{12} \text{ m}^2/\text{sec}^2 = 17.7 \times 10^{-19} \text{ Joule} = 11.1 \text{ eV}$$

This energy has to be compared with the typical atomic binding energy (energy required to remove an electron from an atom) of 10 to 20 eV. We see that the uncertainty principle has a large effect on the atomic scale.

Example 2   Show that for macroscopic objects the uncertainty principle has only a negligible effect. As an example, choose a steel ball of diameter $d = 1$ mm.

Since the density of steel is $\rho = 7.9 \times 10^3$ kg/m³, the mass of the ball bearing will be

$$m = \frac{4\pi}{3}\left(\frac{d}{2}\right)^3 \rho = 4.14 \times 10^{-6} \text{ kg} = 4.14 \text{ mg}$$

This is 10^{24} times larger than the electron mass; thus, a given uncertainty in position will cause a much smaller uncertainty in velocity. Let us assume that the ball bearing is at rest and that we know its position to within 1 micron $= 10^{-3}$ mm $= 10^{-6}$ m; this is about the best we can do with mechanical measuring equipment. Then the uncertainty in its momentum will be

$$\Delta p = \frac{\hbar}{\Delta x} = 1.05 \times 10^{-28} \text{ newton-sec}$$

which corresponds to a velocity uncertainty

$$\Delta v = \frac{\Delta p}{m} = 2.5 \times 10^{-23} \text{ m/sec}$$

This is hardly anything to worry about; if the ball bearing moved with velocity Δv, it would take more than one billion years before the ball bearing had moved another micron—given the unlikely circumstance that our laboratory is still around after this time has elapsed.

2. DE BROGLIE WAVES

In 1924, L. de Broglie proposed that a small particle could be represented by a wave. A wave has the required properties which we discussed in the previous section. Instead of thinking of the particle as localized at a certain point, we think of it as spread out over a region of space as a wave packet.° Specifically, de Broglie proposed that a particle of mass m moving with the momentum $\mathbf{p} = m\mathbf{v}$ is represented by a plane wave moving in the same direction as the particle and having the wavelength

$$\lambda = \frac{2\pi\hbar}{mv} = \frac{h}{mv} = \frac{h}{p} \tag{14-3}$$

Note that we can rewrite Eq. (14–3) using the concept of the wave vector $\mathbf{k}$, which has the magnitude $k = 2\pi/\lambda$ and points in the direction the wave is moving. Since

$$k = \frac{2\pi}{\lambda} = \frac{mv}{\hbar}$$

We can write

$$\boxed{\mathbf{k} = \frac{m\mathbf{v}}{\hbar} = \frac{\mathbf{p}}{\hbar}} \tag{14-4}$$

■ The quantity

$$h = 2\pi\hbar = 6.62 \times 10^{-34} \text{ Joule-sec}$$

is called Planck's constant. Obviously, it is not necessary to define two fundamental constants—h and $\hbar$—which are nearly identical. The quantity h was introduced by M. Planck in 1902 during a discussion of the electromagnetic waves radiated by hot objects; later it was found that $\hbar = h/2\pi$ is the more convenient quantity to use. Both h and $\hbar$ are used in modern literature. ■

We now have to show that such a description of a particle satisfies the uncertainty principle [Eq. (14–1)]. Let us thus first qualitatively discuss a wave packet, shown in Fig. 14–2. Since we assume that the particle is being described by the packet, we have to say that the uncertainty in its position is

$$\pm \Delta x \approx \pm\frac{L}{2} \tag{14-5}$$

where L is the overall length of the packet. On the other hand, the wavelength of such a packet is also not perfectly defined. The wavelength is a property of a harmonic wave which stretches from $x = -\infty$ to $x = +\infty$. A wave packet can be approximated by a harmonic wave; as wavelength we can choose

$$\lambda = \frac{L}{N} \tag{14-6}$$

° We have said earlier (p. 22) that any harmonic wave is really a wave packet, because it has to have a finite extent.

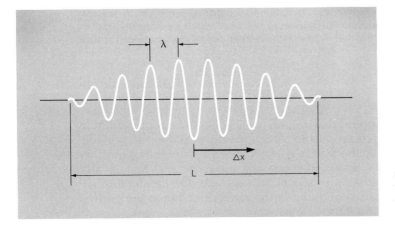

Figure 14-2 A wave packet of finite length L does not have a perfectly well-defined wavelength λ.

where N is the number of maxima of the wave. However, because the field in a wave packet falls off at the ends, N is not perfectly defined, and neither is L; we can estimate this effect by saying that N is uncertain by $\Delta N = \pm 1/2$. The momentum of the particle will also be uncertain, because it is related to the wavelength

$$p = \frac{2\pi \hbar}{\lambda} = 2\pi \hbar \frac{N}{L}$$

Since N is uncertain by $\pm 1/2$, p is uncertain by

$$\Delta p \approx 2\pi \hbar \frac{\Delta N}{L} = \pm \frac{\pi \hbar}{L}$$

But $\Delta x \approx L/2$; thus, we have

$$\Delta p \cdot \Delta x \approx \frac{\pi \hbar}{L} \cdot \frac{L}{2} = \frac{\pi}{2}\hbar \approx \hbar \tag{14-7}$$

As we see, the uncertainty principle is a natural consequence of the wave description of a particle.

We can also deduce the uncertainty principle in a somewhat more precise, although more abstract, way. A harmonic wave has the form

$$\psi_H = A \cos (kx - \omega t)$$

and if we choose an instant in time, say $t = 0$, we can try to build a wave packet out of a sum of harmonic waves

$$\psi = A_1 \cos (k_1 x) + A_2 \cos (k_2 x) + A_3 \cos (k_3 x) + \cdots = \sum_{n=1}^{N} A_n \cos (k_n x) \tag{14-8a}$$

over a range of values k_n around a central value k_0.

We have already discussed (p. 36) the best phenomenon that occurs if we superimpose two harmonic waves of nearly the same wave number. We showed there that if we superimpose two waves of equal amplitude A and with wave numbers $k_0 - \Delta k$ and $k_0 + \Delta k$ to form

$$\psi = A \cos [(k_0 - \Delta k)\, x] + A \cos [(k_0 + \Delta k)\, x] \tag{14-8b}$$

with $\Delta k << k_0$, we obtain

$$\psi = 2A \cos (k_0 x) \cos (\Delta k x)$$

This now has a gross structure of width $\Delta x \sim 1/\Delta k$, as shown in Fig. 14–3a. We can also use more terms to construct

$$\psi = \sum_{n=1}^{N} A_n \cos [(k_0 + \Delta k_n) x] \qquad (14\text{–}9)$$

with all of the $|\Delta k_n| \lesssim |\Delta k|$. In Figs. 14–3b to d we have chosen $N = 4, 8$, and 14 terms; Δk_n in each case was chosen to run through the sequence shown as vertical lines on the left. The lengths of the individual lines indicate the relative magnitudes of the amplitudes A_n. On the right we show the functions $\psi(x)$ ob-

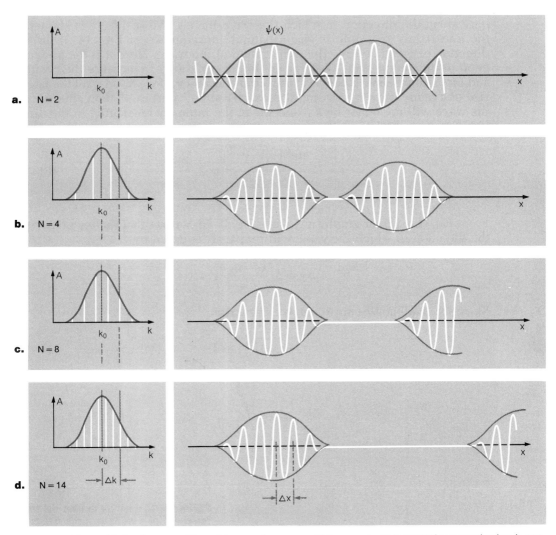

Figure 14–3 Superposition of harmonic waves within a range of k around a central value k_0 produces a wave packet. We can demonstrate this by plotting the result of the superposition at the time $t = 0$. At left we plot the amplitudes (vertical lines) and k values for a superposition of 2, 4, 8, and 14 harmonic waves. At right we plot the resulting wave.

tained by performing the sum on a computer; one can see from the plots that by adding more terms one can move the "side packets" further and further away. In the limit $N \to \infty$, we obtain an isolated wave packet. However, we always have to have a superposition of waves with a range of wave numbers

$$\pm \Delta k \approx \pm 1/\Delta x$$

to produce a wave packet of width Δx. Thus, we can write for a wave the uncertainty relation

$$\Delta k \cdot \Delta x \gtrsim 1 \qquad (14\text{--}10)$$

Since according to the de Broglie relation a purely harmonic wave of wave number k describes a particle of momentum $p = \hbar k$, we conclude that Eq. (14–10) is equivalent to Heisenberg's uncertainty relation.

However, Eq. (14–10) describes the uncertainty in position and momentum only along the direction in which the wave travels. We still have to show that the uncertainty principle is also a natural consequence of Eq. (14–3) for any direction perpendicular to the propagation of the wave. Here it is the phenomenon of diffraction (Fig. 14–4) which limits our ability to measure the position and momentum simultaneously to arbitrary accuracy. If we try to limit the lateral size of a plane wave by passing it through a slit of aperture d, then after the slit the wave will no longer be a plane wave; the intensity emanating in the direction θ is given by Eq. (3–32a):

$$I = I_0 \frac{\sin^2 \left(\dfrac{kd \sin \theta}{2} \right)}{\left(\dfrac{kd \sin \theta}{2} \right)^2} = I_0 \frac{\sin^2 \gamma}{\gamma^2} \qquad (14\text{--}11)$$

From Eq. (14–11), or graphically from Fig. 3–16, we see that the uncertainty in the direction in which the wave will propagate is given by

$$\Delta \gamma = \Delta \left(\frac{kd \sin \theta}{2} \right) \approx 1$$

Since for small angles $\sin \theta \approx \theta$, we can write

$$(k \Delta \theta) \left(\frac{d}{2} \right) \approx 1 \qquad (14\text{--}12)$$

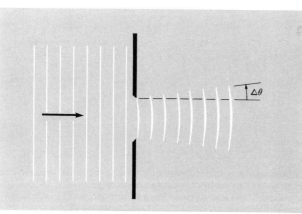

Figure 14–4 If we try to limit the transverse size of a wave by a slit, diffraction introduces an uncertainty $\Delta \theta$ in the direction of propagation of the wave.

According to the de Broglie relation, Eq. (14–3), the perpendicular component of momentum is

$$p_y = p \, \sin \theta = \hbar k \, \sin \theta \approx \hbar k \theta$$

for small angles for which $\theta \approx \sin \theta$. Thus, the uncertainty in p_y is

$$\Delta p_y = \hbar k \Delta \theta \tag{14–13}$$

and since the beam has been limited by the slit to a width $\pm \Delta y = \pm d/2$, we have from Eq. (14–12)

$$\Delta p_y \cdot \Delta y \approx \hbar$$

which again confirms that de Broglie's assumption satisfies the uncertainty principle (14–1).

Example 3   A beam of electrons of velocity $v = 10^6$ m/sec is to have an uncertainty in direction of less than one part in 10^5. How narrow can such a beam be made, using a slit such as that shown in Fig. 14–5, if the uncertainty principle is the only limitation?

Since the electron has a mass of 0.91×10^{-30} kg, the momentum will be

$$p = mv = 0.91 \times 10^{-24} \text{ kg-m/sec}$$

The direction should be accurate to one part in 10^5; thus, we require that the transverse uncertainty be limited by

$$p_\perp < 10^{-5} \, p = 0.91 \times 10^{-29} \text{ kg-m/sec}$$

According to the uncertainty relation, this means that the beam halfwidth has to be at least

$$\Delta y \geq \frac{\hbar}{p_\perp} = \frac{1.05 \times 10^{-35} \text{ J-sec}}{0.91 \times 10^{-29} \text{ kg-m/sec}} = 1.15 \times 10^{-6} \text{ m} \approx 1 \text{ micron}$$

For practical macroscopic electron beams, the uncertainty principle does not provide a limitation; the limit to the accuracy with which one can make parallel electron beams is given by the electrostatic repulsion between the individual electrons in the beam.

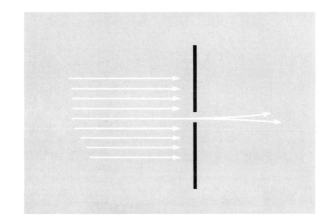

Figure 14–5 Example 3: Uncertainty principle for an electron beam.

3. WAVE-PARTICLE DUALITY

Let us now discuss both the experimental evidence and the implications of the two basic relations, Heisenberg's uncertainty principle [Eq. (14−1)] and the de Broglie relation [Eq. (14−3)]. We have earlier (Chapter 3) shown that one can demonstrate the wave nature of electrons by their diffraction by crystal lattices; in this situation they behave in the same way as x-rays, which are electromagnetic waves.

The historical development of quantum mechanics started out quite differently. In 1905, Albert Einstein proposed that electromagnetic radiation consists of particles called photons. The wave nature of electromagnetic radiation had been known long before, but physicists were unable to explain the *photoelectric effect:* If one shines light on a clean metallic surface, one observes electrons leaving the surface (Fig. 14−6). This in itself is not surprising, because the electromagnetic fields in the wave will act on the electrons. However, it was also found that the kinetic energy T of the emitted electrons is independent of the light intensity, and depends in a simple way on the frequency ν of the incident light:

$$T = \left(\frac{mv^2}{2}\right)_{\text{electron}} = h\nu - W \tag{14−14}$$

where W, called the work function, is different for each metal but is always independent of the nature of the incident light. The quantity $h = 2\pi \hbar$ is exactly the same Planck's constant appearing in Eq. (14−3).

One cannot explain Eq. (14−14) using only the knowledge that electromagnetic fields form the wave. Classically, one would expect the kinetic energy of the emitted electrons to increase with the strength of the field, and thus with the intensity of the electromagnetic radiation. Experimentally, one finds that if one increases the intensity of the impinging light, more electrons leave the metal; but the kinetic energy of each electron is unchanged. As given by Eq.

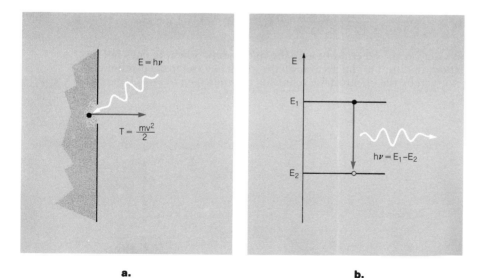

a.   **b.**

Figure 14−6 The particle nature of the photon implies that if a photon is absorbed or emitted, the energy $h\nu$ is transmitted. (a) In the photoeffect, the energy $h\nu$ of the absorbed photon is transmitted to the electron; (b) electrons can "jump" from one energy state to another of lower energy, emitting a photon in the process.

(14–14), the kinetic energy depends only on the *frequency* and not on the intensity of the incident light.

Einstein's solution to the puzzle posed by Eq. (14–14) was to say that light consists of quanta of energy – the photons; each photon has the energy

$$E = h\nu = \hbar\,\omega \qquad\qquad (14\text{–}15)$$

where ν is the frequency and ω is the angular frequency of the radiation. The work function W can be interpreted as the energy required to remove one electron from the metal. The process can then be imagined as the absorption of a *single* photon by *one* of the conduction electrons inside the metal. The electron thus suddenly acquires the additional energy $h\nu$. A fraction W of this energy is required to separate the electron from the metal; the rest is left over as kinetic energy of the electron.

About ten years later, Niels Bohr used the same basic idea to explain another phenomenon which puzzled physicists at the time. It had been known for some time that atoms emit light only at very specific frequencies. This is inconsistent with the classical picture of an atom consisting of a heavy nucleus with the electrons moving in orbits around the nucleus. According to the classical picture, these electrons should radiate electromagnetic energy away, slowly reducing the size of their orbit. Such a classical atom would radiate light of all possible frequencies – not just at a discrete set of frequencies which are specific to each atom.

Bohr's explanation was that electrons can exist only in orbits of very specific energies, in contrast to the classical picture in which any orbit (and thus also any energy) of an electron is allowed. According to Bohr, the atom can suddenly shift its "state" from the energy E_1 to the energy E_2, emitting in the process a single photon of energy

$$h\nu = E_1 - E_2 \qquad\qquad (14\text{–}16)$$

Of course, such a photon will then have the well-defined frequency ν; it will also have the well-defined wavelength $\lambda = c/\nu$.

Bohr further postulated the *duality* of all matter. His idea was that waves and particles are but two aspects of the same natural phenomenon; any particle will show diffraction phenomena which are a specific property of waves, but also every wave can be considered as consisting of particles, of which each carries the energy $h\nu$. Thus, the energy carried by any wave – even by a water wave or an acoustic wave – is always an integer multiple of $h\nu = \hbar\omega$.

■ Bohr's proposal seems at first glance rather strange – even ridiculous. How can anything be a wave and a particle at the same time? Bohr's answer to such a question is that physics is not discussing how nature *is*, but how nature *appears*. There is nothing self-contradictory in stating that a particle "appears" to be or "can be described" by a wave. Nor is there anything contradictory in stating that a sound wave "appears" to consist of particles. Nevertheless, Bohr's proposal started a fundamental change in the view that physicists have of their discipline. Until then it was believed that physics describes nature as it *really is*. Nowadays most physicists believe that the question of reality is beyond the scope of physics. Because the investigation of nature can be done only by experiment, the best we can achieve is knowledge about how nature *appears* – i.e., the ability to predict the outcome of any experiment. ■

Bohr's duality assumption implies that the probability of finding the particle at a certain position has to be calculated from the wave description. If the wave describing the particle has the functional form $\psi(x,y,z,t)$, then the probability of finding the particle at the point $x = x_0$, $y = y_0$, $z = z_0$ is proportional to the square of ψ:

$$P(x_0, y_0, z_0, t) \propto |\psi(x_0, y_0, z_0, t)|^2 \qquad (14\text{–}17)$$

Thus, once we have determined the "wave function" ψ, we can predict the likelihood of finding the particle at any place; by comparing $|\psi|^2$ at different times, we can also determine the likelihood that we measure a certain velocity.

We choose Eq. (14–17) in analogy to Eq. (1–34), which describes the energy density in a harmonic wave. Since each particle carries the energy $\hbar\omega = h\nu$, the energy density at a point should be proportional to the probability of finding the particle at the same spot. However, Eq. (14–17) is an assumption; it has to be – and many times has been – confirmed by experiment.

The duality postulate tells us that the frequency of matter waves should also obey Eq. (14–15). Because the energy of a free particle is equal to its kinetic energy

$$T = \frac{mv^2}{2} = \frac{p^2}{2m}$$

(remember, $\mathbf{p} = m\mathbf{v}$!), the simplest approach would be to describe a free particle of momentum $\mathbf{p}$ in the x-direction by the wave function

$$\psi = A \cos(kx - \omega t) \qquad (14\text{–}18)$$

with

$$k = p/\hbar \qquad (14\text{–}19a)$$

and

$$\omega = \frac{E}{\hbar} = \frac{p^2}{2m\hbar} \qquad (14\text{–}19b)$$

The expression in Eq. (14–18) is not very satisfactory. The probability distribution $P(x_0, t) = A^2 \cos^2(kx - \omega t)$, shown in Fig. 14–7, would be a rapidly oscillating function of position; thus, near a point where the particle is likely to be, there is another point where the probability of finding the particle vanishes. However, it is important to note that the wave function itself is not a physical quantity, but only a means of determining the probability distribution of the particle. Thus, only $|\psi|^2$ but not ψ itself is directly measurable by experiment.

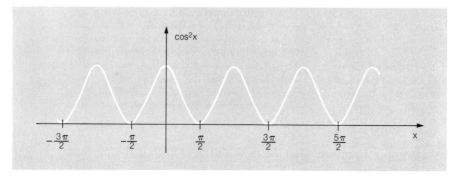

Figure 14–7 The function $\cos^2 x$.

Once we realize this, there is no reason why we cannot assume that ψ can be a complex function. Therefore, instead of Eq. (14–18) we write for a free particle of sharply defined momentum p

$$\psi = e^{i(kx-\omega t)} = \cos(kx - \omega t) + i\sin(kx - \omega t) \qquad (14\text{–}20)$$

Then we have for the square of the absolute value

$$|\psi|^2 = |e^{i(kx-\omega t)}|^2 = \cos^2(kx - \omega t) + \sin^2(kx - \omega t) = 1 \qquad (14\text{–}21)$$

everywhere. This is exactly what we want for the description of a particle which has a perfectly defined momentum p. Because of the uncertainty principle, its position should then be completely unknown. Thus, it should be equally probable to find the particle at any x — but this is exactly what Eq. (14–21) states to be true.

The reader may have gained the impression that we wrote down Eqs. (14–17), (14–19), and (14–20) without any proof that they are correct. This impression is quite valid; these equations are all assumptions, and all we can show is that they are consistent and reasonable. The proof of their correctness can come only from experimental evidence.

So far, we have shown that the wave description of particles leads naturally to the uncertainty relation. But we also need to connect our new theory to classical mechanics — we are talking about particles, and a billiard ball is a particle; how can a billiard ball behave like a wave? More precisely, we have to obtain from our theory the laws of classical Newtonian mechanics, if we "do not look too closely." In this section we are talking only about free particles not subject to any forces; thus, the only relevant law is Newton's first law, which states that free particles move with constant velocity. Since we claimed that Eq. (14–20) describes a particle with momentum p (in the x-direction), it should have the velocity

$$v = p/m = \hbar k/m \qquad (14\text{–}22a)$$

If we have a wave such as $\psi = A \exp i(kx - \omega t)$, we can define the phase velocity

$$v_{\text{ph}} = \frac{\omega}{k} \qquad (14\text{–}22b)$$

and then write the argument in Eq. (14–20) as

$$kx - \omega t = k(x - v_{\text{ph}}t)$$

We see that at a point moving with the velocity v_{ph}, the argument of the complex exponential in Eq. (14–20) stays constant; therefore, ψ is also constant along such a moving point. However, we *cannot* identify the phase velocity given by Eq. (14–22b) with the velocity of the particle associated with the wave. The expression given in Eq. (14–20) describes a particle of absolutely sharp momentum. Because of the uncertainty relation [Eq. (14–1)], this implies that we have no knowledge at all about its position. The velocity of the particle is given by

$$v = \frac{(\text{position at time } t_2) - (\text{position at time } t_1)}{t_2 - t_1}$$

and if we can give no meaning to the particle's position, its velocity is also completely undefined. If we want to describe a particle whose position is reasonably well defined, we have to use wave packets as they were shown in Fig. 14–3.

We then need a superposition of periodic terms to describe the particle, as we did in Eqs. (14–8) and (14–9).

The velocity of the particle – in the sense of classical velocity – then will be the velocity with which the whole packet moves. From Eqs. (14–19) we can obtain a relation between ω and $k = 2\pi/\lambda$:

$$\omega = \frac{\hbar k^2}{2m} = \omega(k) \qquad (14\text{–}23)$$

Thus, the angular frequency $\omega(k)$ depends in a nontrivial way on the wave number k. In order to calculate the velocity with which a wave packet moves, let us write down the simplest "wave packet" in analogy to Eq. (14–8b):

$$\psi = \exp\{i[(k_0 - \Delta k)x - \omega(k_0 - \Delta k)t]\} + \exp\{i[(k_0 + \Delta k)x - \omega(k_0 + \Delta k)t]\} \quad (14\text{–}24)$$

where we have omitted the overall amplitude A. If Δk is sufficiently small, we can write

$$\omega(k_0 \pm \Delta k) = \omega(k_0) \pm \frac{d\omega}{dk}\Delta k$$

and the phases inside the brackets in Eq. (14–24) can be rewritten, if we define $\omega_0 = \omega(k_0)$:

$$(k_0 \pm \Delta k)x - \left(\omega_0 \pm \frac{d\omega}{dk}\Delta k\right)t = \left(k_0 x - \omega_0 t\right) \pm \Delta k\left(x - \frac{d\omega}{dk}t\right) \quad (14\text{–}25)$$

Remembering the relations

$$e^{i(\alpha+\beta)} = e^{i\alpha}e^{i\beta} \quad \text{and} \quad \cos\alpha = \frac{e^{i\alpha} + e^{-i\alpha}}{2}$$

we can then write for ψ in Eq. (14–21)

$$\psi = 2e^{i(k_0 x - \omega_0 t)} \cos\left\{\Delta k\left(x - \frac{d\omega}{dk}\right)t\right\} \qquad (14\text{–}26)$$

If we start adding more terms in order to obtain wave packets as shown in Figs. 14–3b to d, we still obtain for every term a phase such as that in Eq. (14–25); instead of Eq. (14–26), we will end up with

$$\psi = e^{i(k_0 x - \omega_0 t)} f\left(x - \frac{d\omega}{dk}t\right)$$

where $f(x - \frac{d\omega}{dk}t)$ describes the motion of the wave packet as a whole (Fig. 14–8). The wave packet as a whole moves with the velocity

$$\boxed{v_g = \frac{d\omega}{dk}} \qquad (14\text{–}27)$$

which is called the *group velocity*. From Eq. (14–23) we obtain for a free particle

$$v_g = \frac{d\omega}{dk} = \frac{\hbar}{2m}\frac{d(k^2)}{dk} = \frac{\hbar k}{m}$$

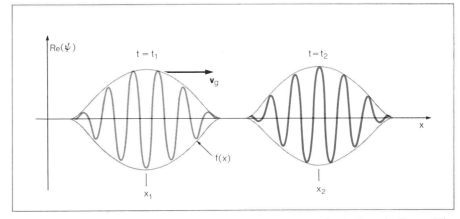

Figure 14–8 The velocity of a particle described by a wave packet is the velocity $\mathbf{v}_g$ of the whole packet.

which, because of $p = \hbar k$, is equivalent to

$$mv_g = p \tag{14–28}$$

Thus, the wave packet as a whole indeed moves with the classical velocity given by Newtonian dynamics. Quite generally, when we consider only the average position and velocity of a particle, they will obey the laws of classical mechanics. Classical mechanics works for any macroscopic object, because its de Broglie wavelength is much smaller than the size of the object; then the whole wave packet which describes the motion of the "particle" can be small compared to the particle size, and the uncertainty principle has a negligible effect.

Example 4   An electron (mass 0.9×10^{-30} kg, size $< 10^{-16}$ m), a hydrogen atom (mass 1.7×10^{-27} kg, size 1 Å), and a protein molecule (molecular weight $A = 5 \times 10^5$ kg/mole, size 200 Å) all have the kinetic energy of 1 eV. Can they be described by classical Newtonian mechanics?

An object of mass m and velocity v has the momentum $p = mv$ and the kinetic energy

$$T = \frac{mv^2}{2} = \frac{p^2}{2m}$$

Thus, if its kinetic energy is given, we can calculate the momentum, and using Eq. (14–3), also the de Broglie wavelength

$$\lambda = \frac{h}{p} = \frac{h}{\sqrt{2mT}}$$

The approximation of classical mechanics will be good if the wavelength is small compared to the size of the object. Since 1 eV $= 1.6 \times 10^{-19}$ J, we have for the electron

$$\lambda_e = \frac{6.6 \times 10^{-34} \text{ J-sec}}{\sqrt{2 \times (0.91 \times 10^{-30} \text{ kg}) \times (1.6 \times 10^{-19} \text{ J})}} = 1.2 \times 10^{-9} \text{ m}$$

This is much larger than the electron size; classical mechanics would give a completely wrong picture. When calculating de Broglie wavelengths for the other two objects, we note that λ is indirectly proportional to $\sqrt{m}$. Thus, for the hydrogen atom we have

$$\lambda_H = \sqrt{\frac{m_e}{m_p}}\, \lambda_e = 2.3 \times 10^{-2}\, \lambda_e = 0.28 \times 10^{-10}\ \text{m} = 0.28\ \text{Å}$$

This is smaller than the hydrogen atom, but only by a small factor; classical mechanics will give qualitatively correct results, but we should not trust any quantitative calculations.

The protein molecule has a molecular weight of 5×10^5 kg/mole; thus, its mass is larger by the factor 5×10^5 than the hydrogen atom mass:

$$m_{pr} = 5 \times 10^5 \times (1.6 \times 10^{-27}\ \text{kg}) = 8 \times 10^{-22}\ \text{kg}$$

Its de Broglie wavelength is thus a factor $\sqrt{5 \times 10^5}$ smaller than that of a hydrogen atom of equal kinetic energy:

$$\lambda_{pr} = \frac{0.28\ \text{Å}}{\sqrt{5 \times 10^5}} = 4 \times 10^{-4}\ \text{Å}$$

This is a factor 2×10^{-6} smaller than the size of the molecule; the overall motion of the molecule will be quite well described by classical mechanics.

4. THE BOHR MODEL OF THE ATOM

As we mentioned earlier, in 1913 Niels Bohr for the first time succeeded in calculating the frequencies of light emitted by atoms. While the *line spectra*— the sets of emitted sharp frequencies of light—are extremely complicated for most atoms, the line spectrum of hydrogen (Fig. 14–9) is rather simple. This is understandable: the hydrogen atom consists of a single electron in an "orbit" around a proton (hydrogen nucleus). In contrast, in heavier atoms there are many electrons, each being electrostatically attracted by the central heavy nucleus; but they also repel each other because they all have the same negative charge. Thus, it is not surprising that the complex many-body problem presented by the heavier atoms is not easily solved. But if quantum mechanics is to be declared successful, it should certainly be able to solve the problem of the hydrogen atom: there is only one electron around the much heavier proton, which we can consider to be at rest.

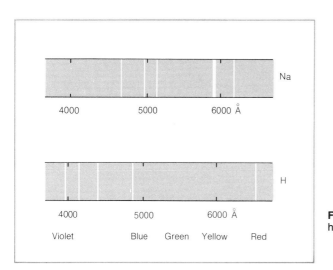

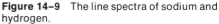

Figure 14–9   The line spectra of sodium and hydrogen.

In 1884, Johann Balmer had already found a very simple relation between the frequencies of all hydrogen lines in the visible part of the spectrum. He found that all of these lines obey the relation

$$\frac{\nu}{c} = \frac{1}{\lambda} = R\left(\frac{1}{4} - \frac{1}{m^2}\right) \qquad \text{(Balmer series)} \qquad (14\text{–}29\text{a})$$

where m is an integer: $m = 3, 4, 5, \ldots$; and R, called the Rydberg constant, can be measured to high accuracy

$$R = 109,677.58 \text{ cm}^{-1} \qquad (14\text{–}30)$$

Balmer could find no explanation for his formula; however, the simple fact that it is satisfied to such a precision convinced other physicists that any exact theory of the atom had to yield Eq. (14–29a) as one of its results. Later on, more lines were found in the hydrogen spectrum whose frequencies obeyed the relation

$$\frac{\nu}{c} = \frac{1}{\lambda} = R\left(\frac{1}{9} - \frac{1}{m^2}\right) \qquad m = 4, 5, 6 \ldots \qquad \text{(Paschen series)} \quad (14\text{–}29\text{b})$$

where R is the same Rydberg constant. Today we know that all lines expressed by the relation

$$\frac{\nu}{c} = \frac{1}{\lambda} = R\left(\frac{1}{n^2} - \frac{1}{m^2}\right) \qquad (14\text{–}31)$$

where n and m are positive integers, exist in the hydrogen spectrum. Of course, since the frequency ν is always positive, we must have $m > n$.

Using Bohr's assumption [Eq. (14–16)], we can explain the hydrogen spectrum given by Eq. (14–31) if we can prove that the hydrogen atom can exist only at a discrete set of energies (Fig. 14–10)

$$E_n = -hcR\,\frac{1}{n^2} \qquad (14\text{–}32)$$

because then the transition of the atom from one such *energy level* to another will be associated with the emission of a photon having an energy of

$$h\nu = E_m - E_n = -hcR\,\frac{1}{m^2} - \left(-hcR\,\frac{1}{n^2}\right)$$

This, as one can easily see, is equivalent to Eq. (14–31). Thus, Eq. (14–31) is explained if we can derive the fact that the allowed energies are as sketched in Fig. 14–10, where each energy is represented as a horizontal line along the vertical energy scale.

In the derivation of Eq. (14–32) we will use a method different from Bohr's. De Broglie's relation [Eq. (14–3)] was not known when Bohr published his fundamental paper; we will use it because it makes the derivation simpler.

We want to find stationary orbits—orbits which persist for a long time. Actually, we should not talk about orbits, but about *states of motion*, because we have to use the wave description of the electron; as we have shown, on the atomic scale the uncertainty principle plays an important role. In particular we talk about *stationary states* when the electron persists in the same state for a long time. Each stationary state has a well-defined total energy. However, several different stationary states could in principle have the same energy; thus, at

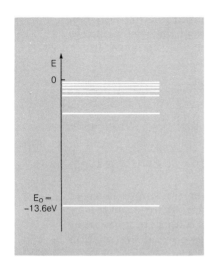

Figure 14-10   The energy levels of the hydrogen atom.

one energy level there may be one or more than one stationary state.

We can, in analogy to what we learned in Chapter 2, assume that the only wave pattern which persists for an arbitrarily long time is a harmonic wave. However, a harmonic wave extends from infinity to infinity; thus, it cannot describe an atom of finite size—*unless it is a standing wave.* Therefore, we are led quite naturally to search for standing waves.

Our model has many features of classical mechanics. We assume that the particle is moving on an orbit. For simplicity, we assume the orbit to be circular; this is not necessary but makes the calculation easier. However, since the electron also is to be described by a standing wave, the orbit cannot have an arbitrary radius. As we can see from Fig. 14–11, it is necessary that the circumference be equal to an integer number times the wavelength λ:

$$2\pi r = n\lambda \qquad n = 1, 2, 3, \ldots \qquad (14\text{–}33a)$$

The wavelength λ is given by the de Broglie relation

$$\lambda = \frac{h}{p} = \frac{h}{mv} \qquad (14\text{–}33b)$$

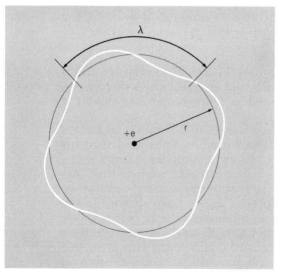

Figure 14-11   The Bohr model: standing waves along a circular orbit.

The two equations (14–33) yield us a relation between the orbit radius r and the velocity v with which the electron moves along it. Another relation is given by Newton's second law, which tells us that the attractive electrostatic force on the electron has to be responsible for the centripetal acceleration mv^2/r on the circular orbit. Since the electron has the charge $(-e)$ and the proton has the charge $(+e)$, we have

$$\frac{e^2}{4\pi\epsilon_0}\frac{1}{r^2} = \frac{mv^2}{r} \tag{14–34a}$$

from which we obtain

$$r = \frac{e^2}{4\pi\epsilon_0}\frac{1}{mv^2} = \frac{e^2}{4\pi\epsilon_0}\frac{m}{p^2} \tag{14–34b}$$

From Eq. (14–34b) and the two equations (14–33) we can calculate both the radius and the momentum (or velocity) of the electron. From Eq. (14–33a) and (14–33b) we obtain

$$r = \frac{n\lambda}{2\pi} = n\frac{h}{2\pi}\frac{1}{p} = \frac{n\hbar}{p} \tag{14–33c}$$

and using Eq. (14–34b) we can eliminate r to obtain

$$mv = p = \frac{e^2}{4\pi\epsilon_0}\frac{m}{\hbar}\frac{1}{n} \tag{14–35a}$$

Substituting Eq. (14–35a) into Eq. (14–33c) then yields

$$r = \frac{\hbar^2}{m}\frac{4\pi\epsilon_0}{e^2}n^2 \tag{14–35b}$$

The total energy will be the sum of the kinetic energy $T = \dfrac{mv^2}{2}$ and the potential energy $-e^2/4\pi\epsilon_0 r$:

$$E_{\text{tot}} = \frac{mv^2}{2} - \frac{e^2}{4\pi\epsilon_0 r} = \frac{m}{2\hbar^2}\left(\frac{e^2}{4\pi\epsilon_0}\right)^2\frac{1}{n^2} - \frac{m}{\hbar^2}\left(\frac{e^2}{4\pi\epsilon_0}\right)^2\frac{1}{n^2}$$

We see that the potential energy is twice as large in absolute value as the kinetic energy. The total energy is thus

$$E_{\text{tot}} = -\frac{m}{2\hbar^2}\left(\frac{e^2}{4\pi\epsilon_0}\right)^2\frac{1}{n^2} = -E_0\frac{1}{n^2} \tag{14–36}$$

By comparing this with Eq. (14–32), we see that not only have we correctly predicted the general form of the hydrogen spectrum [Eq. (14–31)], but we also have a theoretical value for the Rydberg constant:

$$R = \frac{E_0}{hc} = \frac{m}{4\pi c\hbar^3}\left(\frac{e^2}{4\pi\epsilon_0}\right)^2 \tag{14–37}$$

The additional factor 2π appears because $h = 2\pi\hbar$. If one substitutes the best known values for all quantities into Eq. (14–37), one obtains $R = 109,727.2\ \text{cm}^{-1}$, which is some 0.05% larger than the experimental value [Eq. (14–30)]. This small discrepancy can be removed if we take into account the fact that the central proton is not infinitely heavy and therefore is not completely at rest. Instead of the electron mass m, one should use the reduced mass of the electron-proton system

$$\mu = \frac{mM}{m + M} = \frac{m}{1 + m/M}$$

where M is the proton mass. If one uses μ in Eq. (14–37) instead of the electron mass m, one obtains agreement between theory and experiment to better than one part in 10^4. Thus, the Bohr model is able to predict the energies in the hydrogen atom to very high accuracy.

The total energy of the electron in the hydrogen atom is always negative. This is necessary because the stationary states whose energy we have calculated in Eq. (14–36) are *bound states.* The electron is bound to the proton; without supplying energy from the outside, it is impossible for it to move away from the proton to infinity, where the potential energy would be

$$U(r = \infty) = \left.\frac{-e^2}{4\pi\epsilon_0 r}\right|_{r\ =\ \infty} = 0$$

Example 5   According to the Bohr model, what is the radius a_0 and the velocity v_0 of the electron in the hydrogen atom in its ground state (lowest energy state)?

The state of lowest energy is the state with the most negative energy; this occurs when $n = 1$ in Eq. (14–36). Substituting $n = 1$ into Eq. (14–35b), we obtain

$$a_0 = \frac{\hbar^2}{m}\frac{4\pi\epsilon_0}{e^2} = 0.53\ \text{Å}$$

The distance a_0 is called the Bohr radius; it is the radius of the hydrogen atom in its ground state.

The velocity we can obtain from Eq. (14–35a)

$$v_0 = \frac{e^2}{4\pi\epsilon_0}\frac{1}{\hbar} = 2.19 \times 10^6\ \text{m/sec}$$

This velocity is $\sim 0.6\%$ of the velocity of light. According to the theory of relativity, the mass of the electron increases at the velocity v by a factor

$$\frac{1}{\sqrt{1 - (v^2/c^2)}} \approx 1 + \frac{1}{2}\frac{v^2}{c^2}$$

We will therefore expect that relativistic effects will affect the hydrogen spectrum at the level of

$$\frac{1}{2}(0.6\%)^2 \approx 2 \times 10^{-5}$$

These effects, while very small, can actually be measured.

Example 6   Calculate the orbital angular momentum in the nth atomic state of hydrogen.

Since we are assuming circular orbits, the angular momentum is

$$|\mathbf{L}| = m \, |\mathbf{r} \times \mathbf{v}| = mrv$$

From the two equations (14–35), we find

$$|\mathbf{L}| = rp = n\hbar \qquad n = 1, 2, 3, \ldots$$

We see that the angular momentum itself is quantized — it can take on only discrete values, namely integer multiples of $\hbar$. The nth orbit, according to the Bohr model, has the angular momentum $n\hbar$.

Bohr's model of the hydrogen atom, by its success in deriving Eq. (14–31), started the birth of modern atomic physics. Bohr himself can be justly considered the father of modern physics; during the 1920's he started the "Kopenhagen school" of physics after the Danish government founded the Niels Bohr Institute. The brightest young physicists of the time — Heisenberg, Schrödinger, Pauli, to name just a few — all visited frequently, and during their discussions much of what we today call "modern physics" was developed.

Nevertheless, the success of the Bohr model is somewhat of an accident. While it correctly takes into account the wave nature of the electron along the orbit, it completely ignores the fact that the uncertainty principle is also true in the radial direction. When deriving Eq. (14–36) we assumed circular orbits of perfectly well-defined radius; but if the radius is given to an accuracy Δr, then there must be also an uncertainty in the radial component of the momentum (Fig. 14–12) so that

$$\Delta r \cdot \Delta p_r \gtrsim \hbar \qquad (14\text{–}38)$$

We assumed exactly circular orbits, for which there is no radial component of velocity at all — thus, we assume $p_r = 0$ and $\Delta p_r = 0$. But we also assumed a sharp radius and thus $\Delta r = 0$. This violates Eq. (14–38), and we can show that this error has a large effect. For this purpose, let us consider the ground state of the hydrogen atom. Since the radius for the ground state is, according to Example 5, $a_0 = 0.53$ Å, we certainly need $\Delta r < a_0$ if the concept of an orbit is to make sense at all. But then

$$\Delta(mv_r) = \Delta p_r \gtrsim \frac{\hbar}{\Delta r} > \frac{\hbar}{a_0}$$

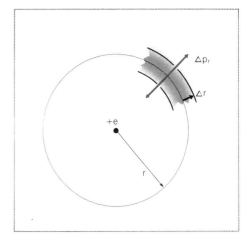

Figure 14–12 The Bohr model neglects the fact that the radius r of the orbit cannot be sharp; if the orbit is defined to $\pm\Delta r$, then the radial momentum is defined only to $\pm\Delta p_r \approx \pm h/\Delta r$.

However, this implies that the kinetic energy $T = mv^2/2$ has a heretofore neglected contribution from the radial motion:

$$\Delta T = \frac{p_r^2}{2m} \gtrsim \frac{(\Delta p_r)^2}{2m} \qquad (14\text{–}39\text{a})$$

since the radial component of the momentum will be at least of the same order as its uncertainty. Substituting for Δp_r, we obtain

$$\Delta T \gtrsim \frac{\hbar^2}{2ma_0^2} \qquad (14\text{–}39\text{b})$$

and since

$$a_0 = \frac{\hbar^2}{m} \frac{4\pi\epsilon_0}{e^2} \qquad (14\text{–}40)$$

we obtain

$$\Delta T \gtrsim \frac{m}{2\hbar^2} \left(\frac{e^2}{4\pi\epsilon_0} \right)^2$$

Comparing this with Eq. (14–36), we see that the error we made is at least of the same order as the total energy of the electron in its ground state:

$$\Delta T \gtrsim E_0$$

Obviously, we therefore cannot believe that we have calculated the ground state energy correctly. The Bohr model can be expected to work reasonably well for large values of n, for which the radius is large enough so that we can have $\Delta r << r$ and $\Delta T << E_0$ simultaneously. But for the ground state ($n = 1$) the exact agreement of the prediction [Eq. (14–36)] with experiment is just a lucky coincidence. Bohr realized this himself, and this is why he spent the next ten years trying to understand the problem; the final answer was given in 1926 by Erwin Schrödinger.

Schrödinger succeeded in calculating exactly and correctly all energy levels in the hydrogen atom. He found that while the energy levels are given correctly by Eq. (14–36), the angular momentum of each level is different from the result predicted by the Bohr model. The *principal quantum number* n determines the energy of each level according to Eq. (14–36); a second integer quantum number, the *angular momentum quantum number* ℓ, determines the total orbital angular momentum of each state:

$$|\mathbf{L}| = \hbar \sqrt{\ell(\ell + 1)} \qquad (14\text{–}41\text{a})$$

In the hydrogen atom ℓ can take on any of the values

$$\ell = 0, 1, \ldots, (n - 1) \qquad (14\text{–}41\text{b})$$

Thus, at the lowest energy ($n = 1$) the orbital angular momentum vanishes; only $\ell = 0$ is allowed.

There exists a third quantum number, the *azimuthal quantum number* m, for each state. Given a value for ℓ, m can take on any integer value between $-\ell$ and $+\ell$:

$$m = -\ell, (-\ell + 1), (-\ell + 2), \ldots, (\ell - 1), \ell \qquad (14\text{–}42\text{a})$$

Its physical interpretation is given by the fact that even if the absolute value of the orbital angular momentum $|\mathbf{L}|$ is given by Eq. (14–41a), its direction can be still further specified. However, according to Schrödinger, we cannot completely specify the direction of the orbital angular momentum; the uncertainty principle makes this impossible. We can only specify the projection of $\mathbf{L}$ along one coordinate axis, e.g., the z-axis; this projection is always equal to an integer multiple of $\hbar$:

$$L_z = m\hbar \qquad (m \text{ integer}) \qquad\qquad (14\text{–}42b)$$

Of course, no component of a vector can be larger than the absolute value of the same vector; thus, we have

$$|L_z| \leqslant |\mathbf{L}|$$

from which one obtains, because of Eq. (14–41a):

$$|m| \leqslant \sqrt{\ell(\ell + 1)}$$

But since m is always an integer, this is equivalent to the range of values given by Eq. (14–42a). Thus, there are in general many states at each energy, which is determined according to Eq. (14–36) purely by the principal quantum number n; for each n there exists a range of allowed values for both the orbital quantum number ℓ and the azimuthal quantum number m.

Example 7   How many states of equal energy are there at the principal quantum number $n = 3$?

For $n = 3$ we can have $\ell = 0, 1$, or 2. However, if $\ell = 1$ there are three states with $L_z = +1\hbar$, 0, and $-1\hbar$. For $\ell = 2$ there are five states with $L_z = +2\hbar, +1\hbar, 0, -1\hbar, -2\hbar$. Thus, the total number of states is $5 + 3 + 1 = 9$.

■ 5. THE SCHRÖDINGER EQUATION

Schrödinger gave a prescription to calculate the wave function $\psi(\mathbf{r},t)$ for any problem. We immediately stress that his equation cannot be derived – it forms a basis of a new theory. Its correctness can be shown only by comparing theoretical predictions with experimental results. As we know today – 40 years later – the Schrödinger equation describes correctly all nonrelativistic quantum mechanical phenomena.

Schrödinger's prescription to calculate the wave function in an arbitrary potential is as follows:

(a) Write down the classical expression for the total energy in terms of the momentum and the position of the particle:

$$E = \frac{\mathbf{p}^2}{2m} + V(\mathbf{r}) = \frac{p_x^2 + p_y^2 + p_z^2}{2m} + V(x,y,z) \qquad\qquad (14\text{–}43)$$

where the terms $p_x = mv_x$ and so forth are the components of the total momentum $\mathbf{p}$.

(b) Replace the momenta in Eq. (14–43) by partial derivatives:

$$p_x \text{ by } \frac{\hbar}{i}\frac{\partial}{\partial x}; \quad p_y \text{ by } \frac{\hbar}{i}\frac{\partial}{\partial y}; \quad p_z \text{ by } \frac{\hbar}{i}\frac{\partial}{\partial z}$$

Products, such as the expression p_x^2, are to be replaced by double derivatives

$$p_x^2 \rightarrow \left(\frac{\hbar}{i}\frac{\partial}{\partial x}\right)^2 = -\hbar^2 \frac{\partial^2}{\partial x^2}$$

(c) Replace E by $i\hbar \dfrac{\partial}{\partial t}$

(d) Now consider Eq. (14–43) to be a partial differential equation for the wave function ψ. This equation is called the *Schrödinger equation:*

$$i\hbar \frac{\partial \psi}{\partial t} = -\frac{\hbar^2}{2m}\left(\frac{\partial^2 \psi}{\partial x^2} + \frac{\partial^2 \psi}{\partial y^2} + \frac{\partial^2 \psi}{\partial z^2}\right) + V(x,y,z)\,\psi \tag{14–44}$$

(e) In general, the solution to Eq. (14–44) will not describe a state of sharply defined energy. A stationary state which will have a well-defined energy E will be described by a wave function

$$\psi(x,y,z,t) = \phi(x,y,z)e^{-iEt/\hbar} \tag{14–45}$$

where ϕ is independent of time.

Substituting Eq. (14–45) into the Schrödinger equation, we obtain an equation for ϕ alone, since the overall factor $e^{-iEt/\hbar}$ cancels:

$$E\phi = -\frac{\hbar^2}{2m}\left(\frac{\partial^2 \phi}{\partial x^2} + \frac{\partial^2 \phi}{\partial y^2} + \frac{\partial^2 \phi}{\partial z^2}\right) + V(x,y,z)\phi \tag{14–46}$$

Example 8   Find the stationary states of a free particle of kinetic energy E in one dimension.

We should, of course, obtain Eq. (14–20). Because we are searching for stationary states of energy E, we know that $\psi(x,t)$ will have the form

$$\psi(x,t) = \phi(x)e^{-iEt/\hbar}$$

From Eq. (14–44) we obtain an equation for ϕ

$$E\phi = -\frac{\hbar^2}{2m}\frac{d^2\phi}{dx^2} \tag{14–47a}$$

which we can rewrite as

$$\frac{d^2\phi}{dx^2} + k^2\phi = 0 \tag{14–47b}$$

with

$$k = \frac{\sqrt{2mE}}{\hbar} \tag{14–47c}$$

Eq. (14–47b) is familiar; it is the same as Eq. (12–53) with x instead of t as the independent variable. Thus, we know that $\phi = \cos kx$ and $\phi = \sin kx$ are solutions. The most general solution will have the form

$$\phi = A\,\cos kx + B\,\sin kx$$

with arbitrary *complex* constants A and B. If we set $B = iA$, we obtain for $\psi(x,t)$

$$\psi(x,t) = A(\cos kx + i\,\sin kx)\,e^{-iEt/\hbar} = Ae^{i[kx-(E/\hbar)t]}$$

which is identical to Eq. (14–20) because of Eq. (14–19b). We are describing a particle of momentum $p = \hbar k$ and energy E. Because of Eq. (14–47c) we have

$$p = \sqrt{2mE} \quad \text{or} \quad E = \frac{p^2}{2m}$$

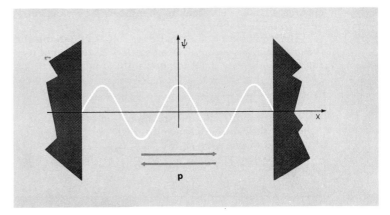

Figure 14-13 Standing wave representing a particle reflected between two rigid walls.

which is exactly the correct relation between momentum and energy. If, on the other hand, we set $B = -iA$, we obtain for ψ

$$\psi(x,t) = A(\cos kx - i \sin kx)\, e^{-iEt/\hbar} = Ae^{i[-kx-(E/\hbar)t]}$$

which describes a particle of negative momentum $-p$ and the same energy $E = p^2/2m$. This is natural; whether a particle moves from left to right or from right to left, it has the same energy as long as the magnitude of its momentum is the same. Of course, we can also choose different combinations of A and B. We can choose, for instance, $B = 0$ and we then have for ψ

$$\psi = A(\cos kx)e^{-iEt/\hbar} = A\frac{e^{i(kx-Et/\hbar)} + e^{i(-kx-Et/\hbar)}}{2}$$

This describes a particle forming a standing wave; there is no classical analogy (Fig. 14–13). The momentum of the particle is not well-defined; it can be either $+\hbar k$ or $-\hbar k$. The closest classical analogy is that of a particle bouncing back and forth between two rigid walls; it will then also have alternately positive and negative momentum.

■ Finding all possible solutions to the Schrödinger equation for a given potential would be equivalent to solving Newton's equations for the same potential. Obviously the problem, even for simple potentials, is vastly more complex in quantum mechanics than in classical mechanics. Luckily the following theorem is true, which enables us to use, with care, Newtonian mechanics even for atomic phenomena.

Theorem 1: If a particle has "reasonably" well-defined position and momentum, these obey the classical equations of motion

$$\frac{d\mathbf{r}}{dt} = \frac{\mathbf{p}}{m}$$

$$\frac{d\mathbf{p}}{dt} = -\text{grad } V = \mathbf{F}$$

By "reasonably well-defined" we mean that $\Delta p/p \ll 1$ and that $|\Delta\mathbf{r}|$ is sufficiently small so that the force $\mathbf{F}$ is essentially the same everywhere in the interval $|\Delta\mathbf{r}|$. This theorem enables us to use classical mechanics for free electrons in vacuum, such as the electron beam in a T.V. tube.

The complexity of partial differential equations such as the Schrödinger equation prevents us from giving here a detailed discussion of its solutions. Instead, we formulate and briefly discuss some important theorems without giving their proofs.

Theorem 2: Bound stationary states (i.e., states for which the probability of finding the particle far away vanishes, or $|\phi|^2 \to 0$ at infinity) exist only for a discrete set of energies $E_1, E_2, \ldots, E_n, \ldots$ The number of such energy levels can be finite or infinite; however, all E_n will have values less than the value $V(x,y,z)$ at infinity.

A bound state corresponds to a bound orbit in classical mechanics. If the force is attractive, there exist orbits (Fig. 14–14) for which the particle always stays nearby. Even classically, such orbits must have energies such that

$$E = \frac{mv^2}{2} + V = T + V$$

is less than the potential $V(x,y,z)$ very far away. Only then does the law of energy conservation prevent the particle from ever moving far away from the center.

The same statement is true in quantum mechanics; if the energy (Fig. 14–15) is less than the value of the potential very far away, then energy conservation makes it impossible for the particle to be at infinity. On the other hand, if the energy is large enough so that the particle can have a positive kinetic energy very far away, then sooner or later it will find itself there; and once it is far away, it will never come back.

Theorem 3: If the potential has spherical symmetry, i.e., if

$$V(x,y,z) = V(\sqrt{x^2 + y^2 + z^2}) = V(r)$$

depends only on the distance from some origin, then the bound states will also have a sharply defined and *quantized* total angular momentum. The possible values for the magnitude of the angular momentum are

$$|\mathbf{L}| = \sqrt{\ell(\ell + 1)}\, \hbar \qquad (14\text{–}48a)$$

where ℓ is a nonnegative integer: $\ell = 0, 1, 2, 3, \ldots$ If the total angular momentum is $\sqrt{\ell(\ell + 1)}\, \hbar$, there will be $2\ell + 1$ bound states at exactly the same energy, each having a well-defined component of angular momentum along some axis (e.g., the z-axis):

$$L_z = \ell\, \hbar,\, (\ell - 1)\, \hbar,\, \ldots,\, -\ell\, \hbar \qquad (14\text{–}48b)$$

This theorem corresponds to the classical statement that for a central (spherically symmetric) potential the total angular momentum is a constant. If the force always points to a center (Fig. 14–16), then the torque of the force $\mathbf{T} = \mathbf{r} \times \mathbf{F}$ vanishes, and therefore the angular momentum $\mathbf{L} = m(\mathbf{r} \times \mathbf{v})$ around the center never changes. However, in quantum mechanics the angular momentum, like the energy, is quantized; i.e., it can take on only certain values given by Eq. (14–48a). Furthermore, the direction of the angular mo-

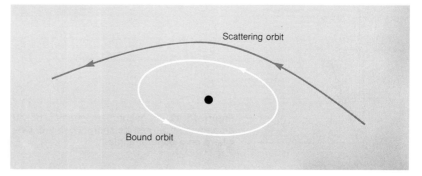

Figure 14–14 Classically an electron in a bound orbit never moves very far away from the nucleus; in a scattering orbit the particle passes by.

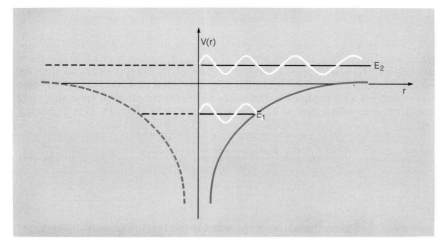

Figure 14–15 In quantum mechanics a bound orbit has an energy E_1 which is less than the potential energy far away. The wave function decreases rapidly for $r > r_0$. A scattering state has an energy E_2 which is larger than the potential energy far away. The wave function continues to be large, even very far away.

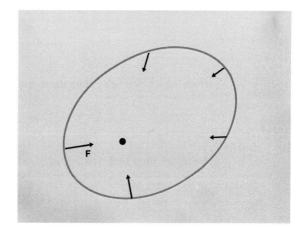

Figure 14–16 In a central potential the force always points to the center; thus, there is never a torque $r \times F$ around the center.

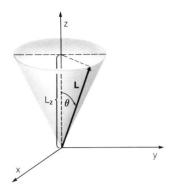

Figure 14–17 In quantum mechanics, only the overall length $|\mathbf{L}|$ and the z-component L_z can be sharply defined. One can imagine $\mathbf{L}$ lying somewhere on a cone of angle θ.

mentum is uncertain; there exists an uncertainty in the angular momentum

$$\mathbf{L} = \mathbf{r} \times (m\mathbf{v}) = \mathbf{r} \times \mathbf{p}$$

because one can never measure both $\mathbf{r}$ and $\mathbf{p}$ simultaneously without violating the uncertainty principle. Thus, while the magnitude of $\mathbf{L}$ can be perfectly defined if it takes on one of the values given by Eq. (14–48a), only one of its components can then be perfectly defined; in addition, according to Eq. (14–18b), that one component can take on only values which are a multiple of $\hbar$. The other two components then are completely unknown; the angular momentum lies anywhere along a cone (Fig. 14–17) of opening angle θ, where

$$\cos \theta = \frac{L_z}{|\mathbf{L}|} = \frac{m}{\sqrt{\ell(\ell+1)}} \qquad (14\text{–}49)$$

Example 9 Given the value of $|\mathbf{L}|$, what is the maximum possible accuracy to which the direction of $\mathbf{L}$ can be defined?

From Eq. (14–48a) we know that $|\mathbf{L}|$ can take on only certain values. Given the value of ℓ in Eq. (14–48a), the maximum component of $\mathbf{L}$ in any direction is, according to Eq. (14–48b),

$$L_{z\,\text{max}} = \ell\,\hbar$$

Thus, the minimum angle obeys the relation

$$(\cos\theta) = \frac{\ell}{\sqrt{\ell(\ell+1)}} = \sqrt{\frac{\ell}{\ell+1}}$$

from which we can calculate the sine

$$(\sin\theta) = \sqrt{1 - \cos^2\theta} = \frac{1}{\sqrt{\ell+1}}$$

Next, we note that to a rather good approximation for $\ell > 1$

$$\ell(\ell+1) \approx \left(\ell + \frac{1}{2}\right)^2 = \ell^2 + \ell + \frac{1}{4}$$

Thus, $|\mathbf{L}| \approx \left(\ell + \frac{1}{2}\right)\hbar$ and we have

$$(\sin \theta)_{\min} \approx \sqrt{\dfrac{\hbar}{|\mathbf{L}| + \dfrac{\hbar}{2}}}$$

Let us now discuss the hydrogen atom. To find all the levels of the hydrogen atom, one should find all values of $E < 0$ for which Eq. (14–46) has a solution if we set

$$V = -\frac{e^2}{4\pi\epsilon_0 r}$$

One can show that all stationary states have energies

$$E = -\frac{m}{2\hbar^2}\left(\frac{e^2}{4\pi\epsilon_0}\right)^2\frac{1}{n^2} \tag{14-50}$$

where $n = 1, 2, 3, \ldots$ is a positive integer which is called *the principal quantum number*. For each principal quantum number there are, in general, several solutions ϕ for the Schrödinger equation, and thus several stationary states of different orbital angular momenta. Using the angular momentum quantum number ℓ from Eq. (14–48a), one can show that the possible values of ℓ are

$$\begin{aligned}\text{for } n &= 1 & \ell &= 0 \\ n &= 2 & \ell &= 0 \text{ or } 1 \\ n &= 3 & \ell &= 0 \text{ or } 1 \text{ or } 2\end{aligned}$$

.
.
.

As mentioned before (Eq. 14–41b), the ground state ($n = 1$) has orbital angular momentum $|\mathbf{L}| = 0$, while the Bohr model yields an orbital angular momentum $|\mathbf{L}| = \hbar$ for

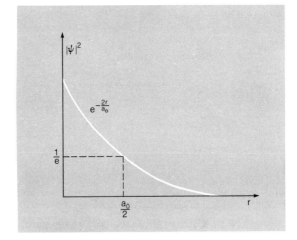

Figure 14–18 The probability distribution in the hydrogen ground state.

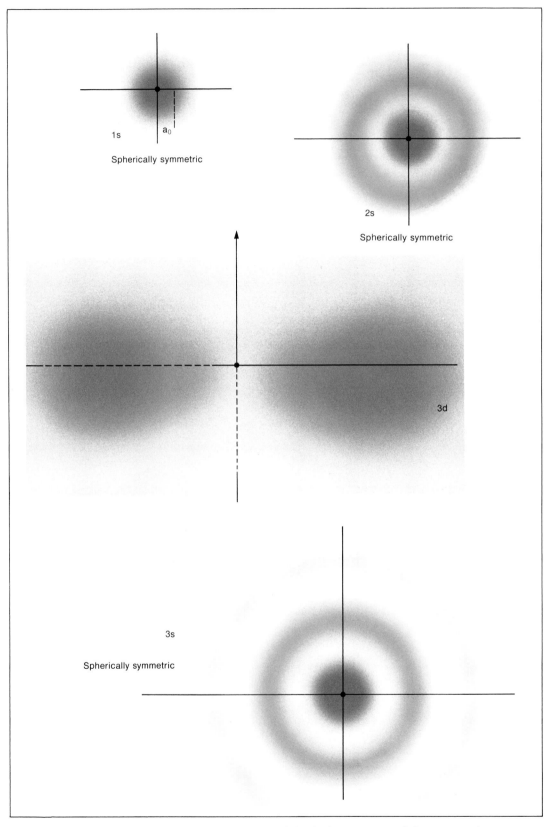

Figure 14-19 Electron distributions in some states of the hydrogen atom, all to same scale. Imagine each rotated around the vertical axis; the 3*d* state has a toroidal distribution.

the same state. The solutions ϕ can all be written in the form

$$\phi(x,y,z) = \text{(Polynominal in } x,y,z) \times \text{(Polynomial in } r) \times e^{-\lambda r} \qquad (14\text{--}51)$$

where $r = \sqrt{x^2 + y^2 + x^2}$ and λ is a quantity which depends on the energy of the state. The probability distributions $|\phi(x,y,z)|^2$ for several of the states of small principal quantum number n are shown in Figs. 14–18 and 14–19.

Example 10   Show that $\phi = e^{-\lambda r}$ (where $r = \sqrt{x^2 + y^2 + z^2}$) is a solution to Eq. (14–46) for the hydrogen atom, and determine both λ and the energy of the state.

We have to find values of λ and E such that

$$-\frac{\hbar^2}{2m}\left(\frac{\partial^2\phi}{\partial x^2} + \frac{\partial^2\phi}{\partial y^2} + \frac{\partial^2\phi}{\partial z^2}\right) - \frac{e^2}{4\pi\epsilon_0 r}\phi = E\phi \qquad (14\text{--}52)$$

Thus, we first calculate the second derivatives. We have for the first derivative

$$\frac{\partial\phi}{\partial x} = \frac{\partial}{\partial x}e^{-\lambda\sqrt{x^2+y^2+z^2}} = -\frac{\lambda x}{\sqrt{x^2 + y^2 + z^2}}e^{-\lambda\sqrt{x^2+y^2+z^2}}$$

To obtain the second derivative, we note that we have a product of three terms: $-\lambda x$, the square root in the denominator, and the exponential. The rule for differentiating products then yields

$$\frac{\partial^2\phi}{\partial x^2} = \frac{-\lambda}{\sqrt{x^2 + y^2 + z^2}}e^{-\lambda\sqrt{x^2+y^2+z^2}} + \frac{(\lambda x)x}{(\sqrt{x^2 + y^2 + z^2})^3}e^{-\lambda\sqrt{x^2+y^2+z^2}}$$

$$+ \left(\frac{-\lambda x}{\sqrt{x^2 + y^2 + z^2}}\right)^2 e^{-\lambda\sqrt{x^2+y^2+z^2}}$$

Using $r = \sqrt{x^2 + y^2 + z^2}$, we can rewrite this in the form

$$\frac{\partial^2\phi}{\partial x^2} = \left(-\frac{\lambda}{r} + \frac{\lambda x^2}{r^3} + \frac{\lambda^2 x^2}{r^2}\right)e^{-\lambda r} \qquad (14\text{--}53)$$

Since ϕ is completely symmetric in x, y, and z, being a function only of r, we can write down the other two second derivatives by replacing x everywhere in Eq. (14–53) by y or by z:

$$\frac{\partial^2\phi}{\partial y^2} = \left(-\frac{\lambda}{r} + \frac{\lambda y^2}{r^3} + \frac{\lambda^2 y^2}{r^2}\right)e^{-\lambda r}$$

$$\frac{\partial^2\phi}{\partial z^2} = \left(-\frac{\lambda}{r} + \frac{\lambda z^2}{r^3} + \frac{\lambda^2 z^2}{r^2}\right)e^{-\lambda r}$$

If we add the three derivatives, we obtain, because $x^2 + y^2 + z^2 = r^2$, the simple result

$$\frac{\partial^2\phi}{\partial x^2} + \frac{\partial^2\phi}{\partial y^2} + \frac{\partial^2\phi}{\partial z^2} = e^{-\lambda r}\left(-\frac{3\lambda}{r} + \frac{\lambda r^2}{r^3} + \frac{\lambda^2 r^2}{r^2}\right)$$

$$= \left(-\frac{2\lambda}{r} + \lambda^2\right)e^{-\lambda r}$$

Substituting into Eq. (14–52), we obtain

$$-\frac{\hbar^2}{2m}\left(-\frac{2\lambda}{r} + \lambda^2\right)e^{-\lambda r} - \frac{e^2}{4\pi\epsilon_0 r}e^{-\lambda r} = Ee^{-\lambda r} \qquad (14\text{--}53)$$

We see that the factor $e^{-\lambda r}$ cancels in each term; we are left with two terms which are constant and two terms which are proportional to $1/r$. Since Eq. (14–53) has to be correct for all r, we need

$$\frac{\hbar^2 \lambda}{mr} = \frac{e^2}{4\pi\epsilon_0 r}$$

and

$$-\frac{\hbar^2 \lambda^2}{2m} = E$$

Thus, we obtain

$$\lambda = \frac{e^2}{4\pi\epsilon_0} \frac{m}{\hbar^2}$$

and

$$E = -E_0 = -\frac{m}{2\hbar^2} \left(\frac{e^2}{4\pi\epsilon_0}\right)^2 \tag{14–54}$$

Comparing this with Eq. (14–50), we see that our ϕ is the wave function for the ground state $n = 1$. By comparison to Eq. (14–40), we see that

$$\frac{1}{\lambda} = a_0 = \text{classical Bohr radius of hydrogen atom}$$

Thus, the Bohr radius has a physical meaning: the ground state wave function is [Eq. (14–45)]

$$\psi = e^{-r/a_0}\, e^{+iE_0 t/\hbar}$$

where E_0 is given by Eq. (14–54). The probability of finding the electron at a point a distance r away from the proton is thus proportional to

$$|\psi|^2 = (e^{-r/a_0})^2 = e^{-2r/a_0}$$

We see that the probability of finding the electron at a point a distance a_0 away from the nucleus is only $e^{-2} = 13.5\%$ of the probability of finding it at the center of the atom. (Compare Fig. 14–18.) ■

Summary of Important Relations

Uncertainty relation	$\Delta p\, \Delta x \gtrsim \hbar$
De Broglie wave length	$\lambda = \dfrac{h}{mv} = \dfrac{h}{p}$
momentum	$p = \hbar k$
energy	$E = \hbar\omega = h\nu$
group velocity of wave packet	$v_g = \dfrac{d\omega}{dk}$

Bohr model of hydrogen   (n = positive integer)

radius $\qquad r_n = \frac{\hbar^2}{m}\frac{4\pi\epsilon_0}{e^2} n^2 = a_0 n^2$

energy $\qquad E_n = \frac{-m}{2\hbar^2}\left(\frac{e^2}{4\pi\epsilon_0}\right)^2\frac{1}{n^2}$

angular momentum $\qquad |\mathbf{L}| = n\hbar$

Schrödinger equation $\qquad i\hbar\frac{\partial\psi}{\partial t} = \frac{-\hbar^2}{2m}\left(\frac{\partial^2\psi}{\partial x^2} + \frac{\partial^2\psi}{\partial y^2} + \frac{\partial^2\psi}{\partial z^2}\right) + V\psi$

Solutions for hydrogen atom

$$E = -\frac{m}{2\hbar^2}\left(\frac{e^2}{4\pi\epsilon_0}\right)^2\frac{1}{n^2} \qquad n = 1, 2, 3, \ldots$$

$$|\mathbf{L}| = \sqrt{\ell(\ell+1)}\,\hbar \qquad \ell \text{ is integer } 0 \le \ell \le n-1$$

$$L_z = m\hbar \qquad m \text{ is integer } -\ell \le m \le +\ell$$

Questions

1. An electron beam has sharply defined momentum; but its position, and therefore its velocity, is completely uncertain. Considering that classically $\mathbf{p} = m\mathbf{v}$, is there a paradox? How would you *measure* momentum or velocity in such a beam?

2. What is the difference between stating "how things really are" and "how things appear," or "how things are described"? What do we mean by these statements? Note that there exists no general agreement on this problem. This question can be considered unresolved—and there may exist no answer.

3. Discuss the physical meaning of the wave function ψ.

4. What would be the uncertainty in the direction of angular momentum owing to the uncertainty principle for the smallest reasonable rotating macroscopic object? Make your own assumptions as to what is "reasonable."

5. Give several examples of physical variables that you have learned in classical physics to be continuous, but which are "quantized" in quantum mechanics. What exactly do we mean by "quantized"?

Problems

1. The distance between two hydrogen atoms in the hydrogen molecule is defined to within 0.1 Å. Compare the minimum kinetic energy of each hydrogen atom in the molecule to kT at room temperature.

2. At what kinetic energy (in eV) is the de Broglie wavelength of a proton the same as its diameter, $d = 1.6 \times 10^{15}$ m? Use relativistic kinematics.

3. A neutron has nearly the same mass as the proton. The deuteron is the nucleus of heavy water, consisting of one proton and one neutron. Estimate the kinetic energy of the two constituents of the deuteron, given the fact that the deuteron radius is $\sim 2 \times 10^{-15}$ m.

4. You are living through a nightmare, during which $\hbar = 1$ Joule-sec. A policeman stops you for going five miles an hour over the speed limit, 100 feet inside a village with speed limit of 30 mph. If your car weighs 2000 lbs, can you use the uncertainty principle as a defense?

5. In the same nightmare in which $\hbar = 1$ Joule-sec, you find yourself in the Wild West. A sharpshooter offers you $100 if you will hold up a quarter which he promises to shoot out of your hand with one shot from 20 meters away. You look over his gun; it has an 8 mm bore and the bullet weighs 15 g. You estimate the bullet velocity at 1000 m/sec. Should you accept the money?

6. A radio station emits trains of radio frequency pulses one second long at a frequency of 3×10^7 Hz. What is the maximum accuracy to which the frequency can be determined?

7. What is the de Broglie wavelength (in Å) of a neutron of kinetic energy 1/40 eV? What is the de Broglie wavelength of an electron of the same kinetic energy? Note that $1/40$ eV $\approx kT$ at room temperature.

8. The NAL (National Accelerator Laboratory) proton accelerator in Batavia, Illinois, accelerates protons to an energy of 500 GeV ($= 5 \times 10^{11}$ eV). What is the momentum of such a proton? What is the ratio of its de Broglie wavelength to the proton radius of 0.8×10^{-15} m?

9. What is the relative error of the momentum of a particle whose position uncertainty is equal to 10 times its de Broglie wavelength?

°10. The functions shown in Fig. 14–3 are of the type

$$f(x) = \sum_{n=0}^{N} \binom{N}{n} \cos (k_i x)$$

with $k_i = k_0 + \dfrac{n-N}{N} \alpha$ and $\alpha \approx 0.1 \, k_0$. Write a computer program which will calculate the above function and verify that

$$\alpha \, \Delta x \approx 1$$

if Δx is the half-width of the resulting packet. Use as Δx the value for which $g(x) = 1/2$ if we write

$$f(x) = g(x) \cos (k_0 \, x)$$

Vary α to convince yourself that the uncertainty principle $\alpha \Delta x \approx 1$ is really satisfied. The definition of $\binom{N}{n}$ is $\binom{N}{n} = \dfrac{N!}{n!(N-n)!}$.

11. Calculate the group velocity of a water wave packet if the average wavelength is $\lambda = 1$ m. Use Eq. (1–31); $\gamma = 7.3 \times 10^{-2}$ N/m and $h \gg \lambda$.

12. What is the energy required to remove an electron from the hydrogen atom from the level $n = 6$?

13. How many emission lines of the hydrogen spectrum have wavelengths between 4000 Å and 7000 Å? How many between 1500 Å and 4000 Å (ultraviolet)? How many between 7000 Å and 14000 Å (infrared)?

14. What is the size of the hydrogen atom if its principal quantum number is $n = 50$? What is the binding energy (in eV) of such an electron?

15. What is the wavelength of a photon whose energy is equal to the ionization energy of hydrogen, E_0?

°16. Using the Bohr model, calculate the period of circular orbits in the hydrogen atom. Show that if the principal quantum number n is larger ($n \gg 1$), the period of motion has a simple relation to the frequency of light emitted if the atom makes the transition $n \rightarrow n - 1$.

17. Using the Bohr model, calculate the energy levels in a three-dimensional harmonic oscillator in which a mass m is acted upon by the restoring force

$$\mathbf{F} = -k\,\mathbf{r}$$

where $\mathbf{r}$ is the radius vector from the origin. Consider only circular orbits.

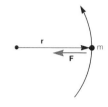

18. Using the Bohr model, calculate the energy levels in a spherically symmetric potential of the form

$$U = \alpha r^4 \qquad\qquad r = \sqrt{x^2 + y^2 + z^2}$$

Consider only circular orbits.

19. With the help of the Bohr model, show that in an attractive potential

$$U = -\frac{A}{r^3}$$

there exists no lowest energy bound state (ground state). This result is true also in the exact quantum mechanical theory.

°20. Using the Schrödinger equation, show that in the *one-dimensional* harmonic oscillator in which the potential energy is $U = \dfrac{k}{2}x^2$, the following two wave functions have sharply defined energies:

(a) $\psi = e^{-\lambda x^2}\, e^{-iE_1 t/\hbar}$

(b) $\psi = xe^{-\mu x^2}\, e^{-iE_2 t/\hbar}$

Determine the parameters λ and μ as well as the energies E_1 and E_2 of the states.

°21. Show that
$$\psi = xe^{-\lambda r}\, e^{-iEt/\hbar} \qquad (r = \sqrt{x^2 + y^2 + z^2})$$

is the wave function of one of the stationary states in hydrogen. Determine λ and the energy E. Can you write down two other wave functions which have the same energy? Which levels are these?

°22. Show that for any function $f(r) = f(\sqrt{x^2 + y^2 + z^2})$ the following identity holds:

$$\frac{\partial^2 f}{\partial x^2} + \frac{\partial^2 f}{\partial y^2} + \frac{\partial^2 f}{\partial z^2} = \frac{\partial^2 f}{\partial r^2} + \frac{2}{r}\frac{df}{dr}$$

°23. Using the result from Problem 22, show that

$$\psi = (1 - \alpha r)\, e^{-\lambda r}\, e^{-iEt/\hbar}$$

is the wave function of one of the hydrogen stationary states. Determine α, λ, and the energy E. What level does ψ represent?

CHAPTER 15

ELECTRONS IN ATOMS AND SOLIDS

1. ELECTRONS IN ATOMS

The hydrogen atom is the simplest possible atomic system, because it consists of a single negatively charged electron attracted by the positively charged heavy hydrogen nucleus (proton). The heavier atoms are more complicated entities. The central nucleus of charge $+Ze$ is surrounded by Z electrons, each of charge $-e$. The *atomic number* Z determines the chemical properties of the atom and is therefore used in classification charts of atoms—the periodic table of elements. Atomic numbers up to $Z=92$, the uranium atom, can be found in nature. Atoms with values of Z up to 106 can be made artificially; but their nuclei are unstable and decay, some of them within a few seconds or even milliseconds.

In an atom with $Z > 1$ electrons, each electron not only is subject to the attractive force of the central nucleus, but is also repelled by all the other electrons. Thus, the potential energy of one electron (let us call it electron number one) will be the sum of the potential energy due to the nucleus

$$U_1 = \frac{(Ze)(-e)}{4\pi\epsilon_0 r} = -\frac{Ze^2}{4\pi\epsilon_0}\frac{1}{r} \tag{15-1a}$$

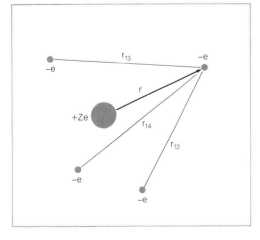

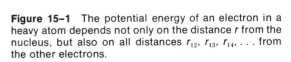

Figure 15-1 The potential energy of an electron in a heavy atom depends not only on the distance r from the nucleus, but also on all distances r_{12}, r_{13}, r_{14}, . . . from the other electrons.

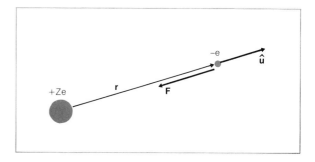

Figure 15-2 The effective force $\mathbf{F}(r)$ is always a central force; its direction is opposite to $\hat{\mathbf{u}} = \mathbf{r}/r$.

and of the potential energy due to the repulsion by all the other electrons

$$U_2 = \sum_{k=2}^{Z} \frac{(-e)(-e)}{4\pi\epsilon_0 r_{1k}} = \frac{e^2}{4\pi\epsilon_0} \sum_{k=2}^{Z} \frac{1}{r_{1k}} \qquad (15\text{-}1b)$$

where r_{1k} is the distance between the "number one" electron and the kth electron in the atom (Fig. 15-1). Nobody has yet succeeded in calculating such a complex problem exactly, either in classical or in quantum mechanics.

Fortunately, an exact calculation is not required. Since because of the uncertainty principle each electron is smeared over a region of space, each electron sees only the mean effect of all the other electrons. The overall force by which the electron is influenced is still nearly a central force (Fig. 15-2), because the other electrons are distributed in a nearly spherically symmetric way around the nucleus. Near the nucleus, the electron will feel the full attractive force of the nucleus

$$\mathbf{F} = -\frac{Ze^2}{4\pi\epsilon_0 r} \hat{\mathbf{u}}_r \qquad (15\text{-}2)$$

since all of the other electrons are further out; as we learned in Chapter 7, a spherically symmetric charge distribution has no effect on a charge inside it.

Far away from the nucleus, the positive charge of the nucleus will be screened by all the other electrons which are closer to the nucleus. The effective charge will be $Ze - (Z-1)e = e$, and thus the attractive force is

$$\mathbf{F} = -\frac{e^2}{4\pi\epsilon_0 r} \hat{\mathbf{u}}_r \qquad (15\text{-}3)$$

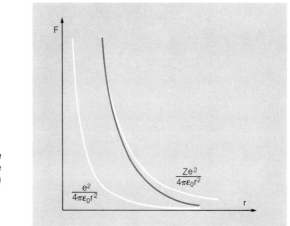

Figure 15-3 We approximate the effect of all the other electrons by an effective attractive force $\mathbf{F}(r)$ whose magnitude varies between $Ze^2/(4\pi\epsilon_0 r^2)$ and $e^2/(4\pi\epsilon_0 r^2)$.

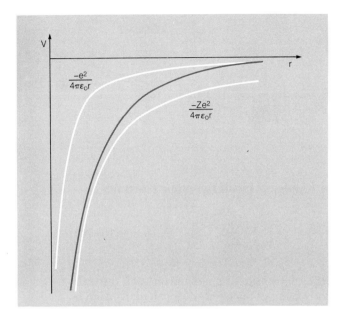

Figure 15-4 The effective potential is $V(r) = -\int_r^\infty F\, dr$; it is roughly equal to $-e^2/4\pi\epsilon_0 r$ for large r, and nearly equal to $-Ze^2/4\pi\epsilon_0 r$ for very small r.

At intermediate distances, the effective charge of the nucleus will vary between its maximum value Ze and the value e far away (Fig. 15–3). The effective potential can be obtained from the force by integration; we have sketched the shape of the potential in Fig. 15–4.

The stationary states in such a potential can be found using numerical methods on a computer. Fig. 15–5 shows how the levels in such a potential are arranged. They are similar to the hydrogen levels. However, states that have the same principal quantum number n, but different angular momentum quantum number ℓ, no longer have exactly the same energy, as was the case in the hydrogen atom. They still lie close to each other, so that the levels are arranged in shells separated by larger gaps. The lowest lying shell is frequently called the K-shell, the next one the L-shell, and so forth $(K, L, M, N, O, P, \ldots)$. Another way of labeling the levels is by their principal quantum number n and an

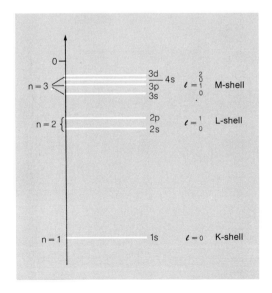

Figure 15-5 Energy levels in heavier atoms; note the splitting of levels, all of which had exactly the same energy in the hydrogen atom.

alphabetic symbol denoting the value of ℓ:[*]

$$\ell = 0 \quad s \text{ state}$$
$$\ell = 1 \quad p \text{ state}$$
$$\ell = 2 \quad d \text{ state}$$
$$\ell = 3 \quad f \text{ state}$$

Thus, the $2p$ state is the state in the L-shell ($n = 2$) with orbital angular momentum $|\mathbf{L}| = \sqrt{1\,(1+1)}\,\hbar = \sqrt{2}\,\hbar$, corresponding to $\ell = 1$.

2. ELECTRON SPIN AND THE PAULI PRINCIPLE

We have up to now ignored the fact that the electron possesses an intrinsic angular momentum, called the spin, which corresponds to an "ℓ value" of $1/2$. We say somewhat loosely that the electron has a spin of $(1/2)\hbar$; by this we mean that the absolute value of its angular momentum is

$$|\mathbf{S}| = \sqrt{\frac{1}{2}\left(\frac{1}{2} + 1\right)}\,\hbar = \frac{\sqrt{3}}{2}\hbar \tag{15-4a}$$

and that the spin component along any direction (e.g., the z-axis) can take on the values

$$S_z = +\left(\frac{1}{2}\right)\hbar$$

or

$$S_z = -\left(\frac{1}{2}\right)\hbar \tag{15-4b}$$

We say that two electrons have their spins *parallel* if both have a z-component of spin of $+(1/2)\hbar$ or $-(1/2)\hbar$; we say that their spins are antiparallel if one has $S_z = (1/2)\hbar$ and the other has $S_z = -(1/2)\hbar$.

The existence of the electron spin is very important, because it is closely connected to a fundamental principle which prevents many electrons from occupying the same state in the same atom. This so-called *Pauli principle*, which is valid for all particles with spin $(1/2)\hbar$, is that *at most two such particles can occupy the same stationary state* and *if two particles occupy the same state, their spins have to be antiparallel.*

If it were not for the Pauli principle, all electrons would occupy the lowest available level, the $1s$ level (Fig. 15–5), because this would yield the overall state of lowest energy and thus the most stable situation. However, because of the Pauli principle, only two electrons can be in the K-shell; the next two electrons have to occupy the $2s$ state. Note that there are three states if the orbital angular momentum has the ℓ-value $\ell = 1$ (i.e., the $2p$ state); these are the three states corresponding to azimuthal quantum numbers m lying between the limits given by Eq. (14–42a): $m = -1, 0, +1$. Thus, six electrons can occupy the $2p$ state.

The Pauli principle has a large effect on the properties of atoms. Let us discuss one typical consequence. From the Bohr model we can conclude that the mean radius of an atom is given by Eq. (14–35a). For a heavier atom we can replace e^2 by Ze^2, if we want only a rough estimate of the radius. If all electrons

---

[*] These letters, assigned by early spectroscopists, indicate characteristics of the associated spectral lines: s = sharp; p = principal; d = diffuse; f = fine.

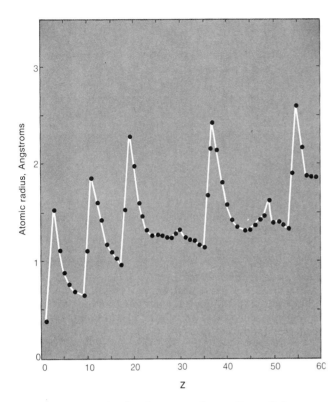

Figure 15-6 The atomic radii show rapid variations with Z because of the shell structure; the mean value increases slowly with Z.

were in the $1s$ state, the radius of the atom would rapidly decrease with atomic number:

$$r = \frac{\hbar^2}{m}\left(\frac{4\pi\epsilon_0}{Ze^2}\right) \approx \frac{1}{Z}$$

Thus, the uranium atom ($Z = 92$) would be smaller by a factor of 90 than the hydrogen atom. Experimentally, one finds (Fig. 15–6) that the atomic radii show rapid variations, but the average value changes only slowly with Z. This is so because as Z increases, so does the principal quantum number n of the outermost electrons. Thus,

$$r \approx \frac{\hbar^2}{m}\left(\frac{4\pi\epsilon_0}{Ze^2}\right)n^2$$

changes on the average only slowly with Z.

Example 1   Estimate the binding energy of a *K*-shell electron in copper.

Since the *K*-shell electron corresponds to $n = 1$ and thus is one of the innermost electrons, it will feel the full attractive force of the nucleus. There is a second electron in the same shell; however, for a rough estimate we can neglect its effect. We will thus approximate the potential energy by its form if there were only the nucleus of charge $+Ze$:

$$V = -\frac{Ze^2}{4\pi\epsilon_0 r} \tag{15-5}$$

Thus, it is as if we had a hydrogen atom in which the central nucleus had a charge Ze. From a periodic table of elements we find for copper $Z = 29$. Since for the

K-shell $n = 1$, we can use Eq. (14–36), setting $n = 1$ and replacing e^2 by Ze^2, to obtain the negative total energy of a K-shell electron. However, the binding energy is defined as the energy required to *remove the electron;* it will be positive:

$$E_B \approx \frac{m}{2\hbar^2}\left(\frac{Ze^2}{4\pi\epsilon_0}\right)^2 \tag{15-6}$$

Substituting the numerical values, we obtain

$$E_B = \frac{0.91 \times 10^{-30} \text{ kg}}{2 \times (1.055 \times 10^{-34} \text{ J-sec})^2}\left(\frac{29 \times (1.6 \times 10^{-19} \text{ coul})^2}{4\pi \times 8.85 \times 10^{-12} \text{ coul/V-m}}\right)^2$$

$$= 1.83 \times 10^{-15} \text{ J} = 11.4 \times 10^3 \text{ eV} = 11.4 \text{ keV}$$

Experimentally, one finds $E_B = 8.98$ keV; thus, our estimate is 25% too large. In Fig. 15–7 we have plotted the square root $\sqrt{E_B}$ of the binding energy against Z for the lighter elements. If Eq. (15–6) were exactly correct, all points would lie on a straight line through the origin. Instead, we see that for $Z \geq 10$ all points lie on a straight line, but the line does not go through the origin. The discrepancy can be explained if one takes into account that:

(a) there are two electrons in the K-shell; even close to the nucleus, the nuclear electric field is somewhat shielded by the other K-shell electron;

(b) while the outer electrons do not contribute to the force on the K-shell electrons, they do affect the binding energy – which is the energy required to remove the electron from its state and move it past the outer electrons to infinity. Far away, the nucleus is largely screened by the other electrons, thus reducing the electric field and therefore the binding energy.

One frequently approximates these effects by replacing Z in Eq. (15–5) by an experimentally determined *effective charge* Z_{eff}. For heavier atoms and for electrons in the K-shell, one has

$$Z - 3 \lesssim Z_{\text{eff}} \lesssim Z - 1$$

While the binding energies of the inner electrons are smooth functions of Z, the binding energy of the outermost electrons – called the *ionization energy* –

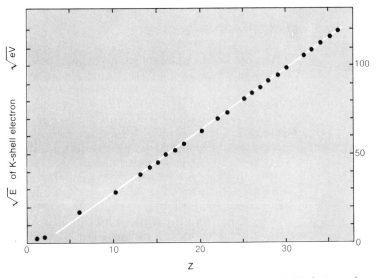

Figure 15-7 The binding energy of the innermost (K-shell) electrons is a smooth function of Z; indeed, to a very good approximation, $\sqrt{E} \propto Z$.

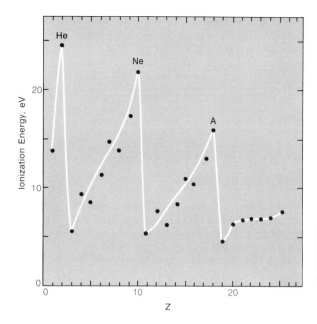

Figure 15-8 The ionization energy is the binding energy of the outermost electron; like the atomic radius, it shows large variations, peaking for those atoms for which a shell is completely filled.

varies rapidly with Z (Fig. 15–8). We see here—as in the atomic radii—the effect of the shell structure. The ionization energy has peaks for Z = 2, 10, and 18, corresponding to the elements He, Ne, and A. These fall into the category of "rare gases" (a misnomer—they are really not so rare in nature) which are chemically rather inert. If we go back to Fig. 15–5, we see that two electrons fill the K-shell. The filled L-shell will be occupied by $2 + 6 = 8$ electrons. Thus the neon atom, with Z = 10, will have both the K- and L-shells fully occupied. The next higher group of levels, the 3s and 3p states, again can accomodate 8 electrons. Thus, $2 + 8 + 8 = 18$ electrons again form a closed shell. Note that the 3d level, which in hydrogen has the same energy as the 3s and 3p levels, in heavier atoms has moved up above the energy of the 4s level. As for the inner electrons, one can define the effective charge Z_{eff} for the binding energy of the outer electrons; for the outermost electrons, Z_{eff} will range from 1 to 3.

3. ELECTRONS IN SOLIDS

Matter does not exist only in the form of isolated atoms or molecules, which we can best describe as small conglomerates of atoms having a well-defined size. Solids frequently exist in crystalline form; the individual atoms form a regular grid pattern, an example of which is shown in Fig. 15–9. The size of a crystal is arbitrary; although it may be difficult to make large crystals, there is no physical limitation to their size. Crystals of sodium iodide (NaI) 9 inches across have been made. Here we will not discuss the crystalline structure itself—we want to know instead why some of these materials are good conductors, while others are very poor conductors or even near-perfect insulators.

The inner electrons of each atom are quite closely tied to the nucleus. However, the outer electrons in *any crystalline structure* are common to all atoms; they provide the bonding force which holds the crystal together. Each of the outer electrons is under the influence of all the positive ions in the crystal.

The net potential energy of the electrons is sketched in Fig. 15–10a. There is a dip at the site of each ion because of its attractive force on the electrons. At the surface, the potential step is higher. This is a very complicated potential; but we can approximate it by the potential of a box with rigid walls (Fig. 15–10b).

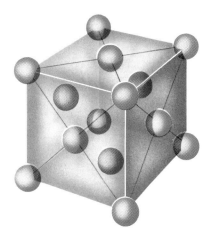

Figure 15-9 Unit cell of the face-centered cubic lattice.

The potential of our simplified model is zero inside the box of size $2L$ and infinitely large everywhere outside. Of course, the size of our box is arbitrary; any physical statement we make about real solids should be independent of the size of the box.

In a one-dimensional box of length $2L$ as shown in Fig. 15–10, we can calculate the energy levels easily. We know that a stationary state corresponds to a standing wave; the wave amplitude has to vanish at the two ends of the box. Thus, the length of the box has to be an integer multiple of $\lambda/2$, for standing waves on a string (see p. 49):

$$2L = n\frac{\lambda}{2} \qquad n = 1, 2, 3, \ldots \qquad (15\text{–}7)$$

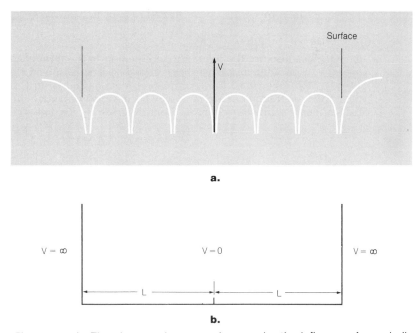

Figure 15-10 The electrons in a crystal are under the influence of a periodic potential due to the lattice ions. (b) We approximate this by a "box potential"; i.e., we assume the electron is free inside the crystal, but cannot leave it.

The particle momentum can now be calculated using the de Broglie relation [Eq. (14–3)]:

$$p = \frac{h}{\lambda} = \frac{h}{4L} n$$

From this we obtain the kinetic energy:

$$T = \frac{p^2}{2m} = \frac{h^2}{32mL^2} n^2 \qquad n = 1, 2, 3, \ldots$$

The kinetic energy is equal to the total energy, because inside the box the potential energy vanishes: $V = 0$. If we use in addition Dirac's constant $\hbar = h/2\pi$ instead of Planck's constant h, we obtain

$$E = \frac{\pi^2 \hbar^2}{8mL^2} n^2 \qquad n = 1, 2, 3, \ldots \tag{15–8}$$

The standing wave pattern in a cubic box like that shown in Fig. 15–11 can no longer be visualized so easily. The energy then is not given by one single integer quantum number n, but instead by three positive integers n_x, n_y, and n_z which have to be independently specified:

$$E = \frac{\pi^2 \hbar^2}{8mL^2} (n_x^2 + n_y^2 + n_z^2) \tag{15–9a}$$

where n_x, n_y, and n_z can independently take on any positive integer value:

$$\left. \begin{array}{c} n_x \\ n_y \\ n_z \end{array} \right\} = 1, 2, 3, 4, \ldots \tag{15–9b}$$

■ One can also use the Schrödinger equation in the form of Eq. (14–46) to obtain the energy. In one dimension we have already partially solved the problem in Example 14–8 (p. 458). We chose the coordinate system as shown in Fig. 15–12. Inside the box the particle is free; there are no forces because the potential energy is a constant. Thus, the stationary solutions will have the form given by Eq. (14–47)

$$\phi = A \cos kx + B \sin kx$$

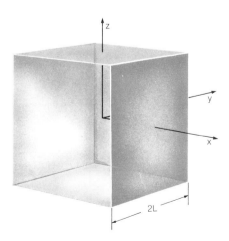

Figure 15–11 The three-dimensional "box." The electron potential energy is zero inside and infinite outside.

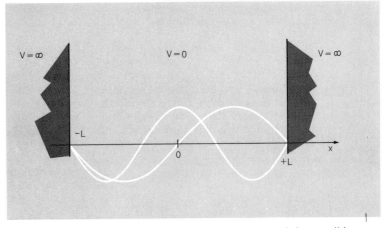

Figure 15–12 The one-dimensional "box" and two of the possible wave functions.

Because for $x > L$ and for $x < -L$ the potential energy is infinite, therefore there ϕ has to vanish. But since ϕ is continuous, it also has to vanish for $x = \pm L$.

$$\phi(x = \pm L) = 0 \tag{15–10}$$

This cannot happen if both A and B have nonzero values, because $\sin(-\alpha) = -\sin \alpha$ and $\cos(-\alpha) = \cos \alpha$ and therefore

$$\phi(-L) = A \cos(kL) - B \sin(kL) \neq \phi(+L)$$

Thus, either $A = 0$ or $B = 0$.[°] However, this is not sufficient; we have to require in addition that

$$\text{if } B = 0 \quad \text{then } \cos(kL) = 0 \quad \text{or} \quad kL = \frac{\pi}{2} + n\pi \quad n = 0, 1, 2, 3, \ldots$$

$$\text{if } A = 0 \quad \text{then } \sin(kL) = 0 \quad \text{or} \quad kL = n\pi \quad n = 1, 2, 3, \ldots$$

We can unite both conditions by stating that k has to take on one of the values

$$k = n \frac{\pi}{2L}, \qquad \text{where } n = 1, 2, 3, \ldots \tag{15–11}$$

which is equivalent to Eq. (15–7).

The reader could ask why only positive integers $n = 1, 2, 3, \ldots$ occur in Eq. (15–11). After all, Eq. (15–10) will also be satisfied if n is a negative integer. However, the complete solutions are either

$$\psi(x,t) = A \cos(kx) \, e^{-iEt/\hbar} \tag{15–12a}$$

or

$$\psi(x,t) = B \sin(kx) \, e^{-iEt/\hbar} \tag{15–12b}$$

Choosing a negative integer instead of a positive integer is equivalent to replacing k by $-k$. Eq. (15–12a) will then not change at all, because $\cos(-\alpha) = \cos \alpha$. Eq. (15–12b) will be replaced by its negative; but this itself does not give a different wave function, because the probability $|\psi(x,t)|^2$ is exactly the same as before at any time t. Thus, taking all positive integers in Eq. (15–11) yields all possible stationary states.

[°] But *not* both $A = 0$ and $B = 0$; this would yield the result $\phi = 0$ everywhere. The particle has to be *somewhere;* thus there has to be a region of space where $|\phi|^2 > 0$.

We can also calculate the energy of each stationary state. Substituting $A \cos kx$ or $B \sin kx$ into Eq. (14–47a) yields

$$E = \frac{\hbar^2}{2m} k^2 = \frac{\pi^2 \hbar^2}{8mL^2} n^2 \qquad n = 1, 2, 3, \ldots$$

which is identical to Eq. (15–8). The energy levels are shown in Figure 15–13.

We are now ready to tackle the full three-dimensional problem. In order to find all possible stationary states, we have to find all possible solutions to Eq. (14–46) for our special potential which is zero inside a cubic box of side $2L$ and infinitely large outside. Thus, inside the box we have

$$E\,\phi(x,y,z) = -\frac{\hbar^2}{2m}\left(\frac{\partial^2\phi}{\partial x^2} + \frac{\partial^2\phi}{\partial y^2} + \frac{\partial^2\phi}{\partial z^2}\right) \qquad (15\text{–}13)$$

Using the same argument which led us to Eq. (15–10), we conclude that ϕ has to vanish everywhere on the surface of the box. It is easy to write down a set of solutions which satisfy Eq. (15–13) and vanish on the surface of the box. It is harder to prove that the set of solutions presented below indeed includes all the possible solutions, i.e., that there are no other possible solutions. Thus, we merely assert that the most general solution to our problem is

$$\phi = \frac{\sin}{\cos}\,(k_x x) \times \frac{\sin}{\cos}\,(k_y y) \times \frac{\sin}{\cos}\,(k_z z) \qquad (15\text{–}14)$$

where each term in the product can be either the sine or the cosine. Since, for instance,

$$\frac{\partial^2\phi}{\partial x^2} = -k_x^2\,\phi$$

we obtain by substitution into Eq. (15–13)

$$E = \frac{\hbar^2}{2m}\,(k_x^2 + k_y^2 + k_z^2) \qquad (15\text{–}15)$$

The values of k_x, k_y and k_z are of course not arbitrary; the boundary condition is that $\phi = 0$ if $x = \pm L$ or $y = \pm L$ or $z = \pm L$. This will be true if

$$k_x = \frac{\pi}{2L}\,n_x$$

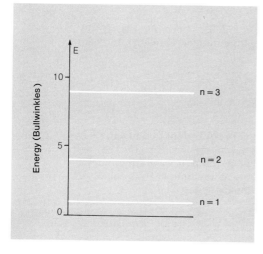

Figure 15–13 Energy levels in one-dimensional box.

and

$$k_y = \frac{\pi}{2L} n_y \qquad\qquad (15-16)$$

and

$$k_z = \frac{\pi}{2L} n_z$$

where n_x, n_y, n_z are arbitrary positive integers. Thus, all the energy levels in the box are given by

$$E = \frac{\hbar^2 \pi^2}{8mL^2}(n_x^2 + n_y^2 + n_z^2) \qquad\qquad (15-17)$$

as we had already stated in Eqs. (15–9). ■

Example 2   Calculate and sketch the lowest energy levels in a box of size $2L = 1$ cm.

We will calculate the energy in units of $\pi^2 \hbar^2 / 8mL^2$. That this is a unit of energy can be seen from Eq. (15–9a), since n_x, n_y, and n_z are integers, and thus pure numbers. For an electron of mass 0.91×10^{-30} kg in a box of size $2L = 10^{-2}$ meters, our unit is

$$1 \text{ Bullwinkle}^\circ = \frac{(1.055 \times 10^{-34} \text{ J-sec})^2 \times (3.14)^2}{8 \times (0.91 \times 10^{-30} \text{ kg}) \times \left(\frac{1}{2} \times 10^{-2} \text{ m}\right)^2}$$

$$= 6.04 \times 10^{-34} \text{ J} = 3.77 \times 10^{-15} \text{ eV}$$

This is an extremely small unit; but then our box is very large on the atomic scale.

The ground state will have $n_x = n_y = n_z = 1$; thus, its energy according to Eq. (15–9a) will be $E_0 = 3$ Bullwinkles. The next higher state will have $(n_x, n_y, n_z) = (1, 1, 2)$ or $(1, 2, 1)$ or $(2, 1, 1)$. The energy is $E_1 = (1^2 + 1^2 + 2^2) = 6$ Bullwinkles, and there are three states at this energy. The next higher energy is $E_2 = (1^2 + 2^2 + 2^2) = 9$ Bullwinkles, and there are again three states at this energy. Then come three states with energy $E_3 = (1^2 + 1^2 + 3^2) = 11$ Bullwinkles, and then only one state at $E_4 =$

° We stress that the unit is *not* standard; we chose the particular name in honor of the great physicist and friend of Rocky the Squirrel.

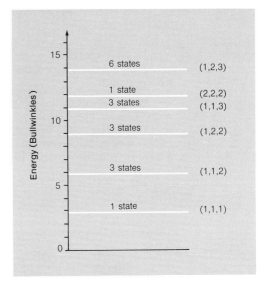

Figure 15–14 Example 2: The lowest energy states in the three-dimensional box.

$(2^2 + 2^2 + 2^2) = 12$ Bullwinkles, because $(n_x, n_y, n_z) = (2, 2, 2)$ allows no permutation. Then come six states at $E_5 = (1^2 + 2^2 + 3^2) = 14$ Bullwinkles, because there are six permutations of $(1, 2, 3)$. We see that the spectrum is quite complex (Fig. 15–14).

Also, the number of states obeys no simple law; it is equal to the number of ways we can express an integer as a sum of three squared integers. For example, at the energy of 86 Bullwinkles, there are 12 states, since the numbers (n_x, n_y, n_z) can be any of the six permutations of $(1, 6, 7)$, but also any of the six permutations of $(1, 2, 9)$.

In a crystal there are many electrons. The energy of each will be given by Eq. (15–9a) for some combination of integers (n_x, n_y, n_z). Because of the Pauli principle, only *two electrons* can be in a state characterized by the same combination, and the two electrons will have antiparallel spins. The electrons will be distributed among the energy levels in a statistical fashion; the exact distribution will depend on the temperature of our sample. However, if the absolute temperature T is zero, each electron will occupy the lowest available level. Thus, there will be two electrons in the lowest level, corresponding to $n_x = n_y = n_z = 1$; as we saw in Example 2, there are three states and therefore six electrons at the next available energy, and so forth. If the number of electrons in the box is large, some of the electrons will have an appreciable energy even at zero absolute temperature $T = 0$. One can calculate the energy of the most energetic electron in a box at zero temperature, called the *Fermi energy* E_F. One finds that the Fermi energy is dependent on both the size of the box and the number of electrons in it in a particular way, so that only the electron density

$$\rho_e = \frac{\text{number of electrons}}{\text{volume of box}}$$

enters the expression for the Fermi energy:

$$E_F = \frac{\hbar^2}{2m}(3\pi^2\rho_e)^{2/3} \tag{15–18}$$

For nearly all solid crystalline substances the Fermi energy is large compared to $kT \approx 1/40$ eV at room temperature; thus, even at room temperature nearly all energy levels below the Fermi energy are occupied by electrons, while nearly all states with energies $E > E_F$ are unoccupied.

Example 3 Calculate the Fermi energy of copper, assuming one free electron per atom.

We have calculated earlier (p. 274) the electron density in copper; but let us redo it here. The density of copper is 8.95×10^3 kg/m³ and its atomic weight is $A = 63.5$ kg/kg-mole. Since Avogadro's number, the number of atoms per kg-mole, is $N = 6.02 \times 10^{26}$ atoms/kg-mole, the number of free electrons will be the same. Thus, the electron density is the same as the atomic density:

$$\rho_e = \frac{\rho N}{A} = \frac{(8.95 \times 10^3 \text{ kg/m}^3) \times (6.02 \times 10^{26} \text{ atoms/kg-mole})}{(63.5 \text{ kg/kg-mole})} = 8.5 \times 10^{28} \text{ atoms/m}^3$$

Using Eq. (15–18), we then obtain for the Fermi energy

$$E_F = \frac{\hbar^2}{2m}(3\pi^2\rho_e)^{2/3} = \frac{(1.055 \times 10^{-34} \text{ J-sec})^2}{2 \times (0.91 \times 10^{-30} \text{ kg})}(3\pi^2 \times 8.5 \times 10^{28} \text{ m}^{-3})^{2/3} = 1.13 \times 10^{-18} \text{ J}$$

or, converted into electron volts,

$$E_F = 7.1 \text{ eV}$$

This is a large energy. We can calculate the temperature T_F at which a completely free electron would have the same mean kinetic energy:

$$\frac{3}{2} kT_F = E_F$$

where $k = 1.38 \times 10^{-23}$ J/°K is the Boltzmann constant. From this we obtain

$$T_F = \frac{2}{3} \frac{E_F}{k} \approx 55,000° \text{ K!}$$

The most energetic electron in copper at zero temperature has the same kinetic energy as a free electron in space would have at a temperature 10 times higher than that found at the surface of the sun. Quantum mechanics has a large effect even for the "free" electrons in a metal.

Because the Fermi temperature in metals is much higher than room temperature, the temperature has nearly no effect on the energy distribution of electrons in a metal. When we heat a metal, all of the additional energy supplied is transformed into thermal vibrations of the ions; the electrons are hardly affected at all. In Fig. 15–15 we show the function $f(E,T)$, the number of electrons per unit volume and unit energy, at zero temperature and at a finite temperature. At an absolute temperature T in a box of volume $V = 1$ m³, there are between the energy E and $E + \Delta E$

$$\Delta N = f(E,T) \, \Delta E$$

electrons. As we see from Fig. 15–15, there is little difference between the average kinetic energy of electrons at room temperature and at zero absolute temperature. Except for a narrow range of width kT around the Fermi energy, the energy distribution of the electrons in a metal is independent of temperature.

■ We can derive Eq. (15–18). To be precise, we want to find the energy of the most energetic electron at a temperature $T = 0$ in a cubic box of edge length $2L$, if there are N electrons inside the box. If we wanted to be exact, we would have to count off all the possibilities for (n_x, n_y, n_z) as we did in Example 2, including all permutations, until we found $N/2$ of them. Fortunately, however, N is a very large number—of the order of 10^{21}, if our box has a volume of 1 cm³. Thus, we can use an approximation which we will explain first in two dimensions: We can plot (Fig. 15–16) a combination (n_x, n_y), where

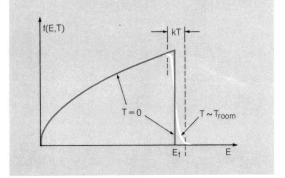

Figure 15–15 The number $\Delta n = f(E,T) \, \Delta E$ of electrons per unit volume having energy between E and $E + \Delta E$ at a temperature $T = 0$ and at room temperature.

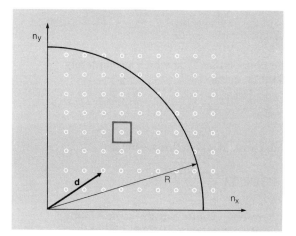

Figure 15–16 Counting off states labeled by the quantum numbers (n_x, n_y) in the two-dimensional box. Each state occupies a square of unit area in the n_x-n_y plane.

both n_x and n_y are positive integers, as a point in a quadrant of a plane with the coordinates n_x and n_y. The distance of each point from the origin is

$$d = \sqrt{n_x^2 + n_y^2}$$

If we now draw a circle of radius R, how many points will there be with a distance $d < R$? Obviously, these are all the points lying inside the circle. But since the points are spaced by one unit, there is one point per unit area inside the quarter circle. This is not true along the boundaries—the two axes and the circle itself—but the error we thus introduce will be small if R is very large. Thus, there are M points with $d < R$, where

$$M = \frac{\pi R^2}{4}$$

is the area of the quarter circle.

In three dimensions we can argue exactly the same way (Fig. 15–17). If we plot each point (n_x, n_y, n_z), with n_x, n_y, and n_z positive integers, as a point in space, they will all lie within an octant. There will be one such point per unit volume, and the distance of each point from the origin is equal to

$$d = \sqrt{n_x^2 + n_y^2 + n_z^2}$$

If we specify a maximum distance R, there will be M points with $d < R$, where

$$M = \frac{1}{8} \times \frac{4\pi}{3} R^3 \tag{15–19}$$

is the volume of the octant of radius R.

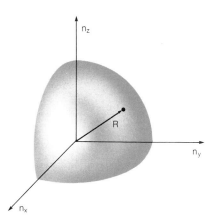

Figure 15–17 The quarter circle limiting all states of energy $E < E_0$ becomes an octant of a sphere in the three-dimensional case.

Let us apply this approximation to our model. From Eq. (15–9a) we obtain

$$d = \sqrt{n_x^2 + n_y^2 + n_y^2} = \frac{2L}{\pi\hbar}\sqrt{2mE}$$

The number of states M with $d < R$ is given by Eq. (15–19). Thus, the number M of states with energy E less than E_0 is given by

$$M(E_0) = \frac{\pi}{6}R^3 = \frac{\pi}{6}\left(\frac{2L}{\pi\hbar}\sqrt{2mE_0}\right)^3$$

However, according to the Pauli principle we have two electrons in each state. Thus, the number of electrons with energy $E < E_0$ is

$$N(E_0) = 2M(E_0) = \frac{8L^3}{3\pi^2\hbar^3}(2mE_0)^{3/2} \tag{15–20}$$

The factor $8L^3$ on the right-hand side in Eq. (15–20) is equal to $V = (2L)^3$, the volume of the box. We can therefore define

$$n(E) = \frac{N(E)}{V} = \frac{N(E)}{(2L)^3} \tag{15–21}$$

the number of electrons with energy less than E per unit volume of the box. Because the size of our box is quite arbitrary, $n(E)$ is the real physical quantity of interest. If we double the volume of the box, but keep the *electron density* the same, there should be no change in any physical result. From Eq. (15–20) we obtain

$$\boxed{n(E) = \frac{(2mE)^{3/2}}{3\pi^2\hbar^3}} \tag{15–22}$$

We are now able to answer our original question: What is the energy of the most energetic electron in the box? If the most energetic electron has an energy E_F, all other electrons have to have a lower energy. Thus, if the total electron density is ρ_e, then

$$\rho_e = \frac{(2mE_F)^{3/2}}{3\pi^2\hbar^3}$$

or

$$\boxed{E_F = \frac{\hbar^2}{2m}(3\pi^2\rho_e)^{2/3}} \tag{15–23}$$

which is exactly the desired expression [Eq. (15–18)] for the Fermi energy. The name was introduced in honor of Enrico Fermi (1901–1954), one of the pioneers of the theory of electrons in solids — as well as of nuclear physics and relativistic quantum mechanics.

Example 4   Calculate $f(E,T)$ at $T = 0$.

We are asked to calculate the number of electrons *per unit volume* and *per unit energy* at zero temperature. From Eq. (15–22) we know the number of electrons of energy less than E per unit volume. Between E and $E + \Delta E$ there will therefore be

$$n(E + \Delta E) - n(E) \approx \frac{dn}{dE}\Delta E$$

electrons per unit volume. Thus, per unit volume and per unit energy we have

$$f(E,T=0) = \frac{n(E + \Delta E) - n(E)}{\Delta E} \approx \frac{dn}{dE}$$

electrons. From Eq. (15–22) we obtain

$$f(E,T=0) = \frac{dn}{dE} = \frac{2m\,\sqrt{2mE}}{2\pi^2\hbar^3} \tag{15–24}$$

However, Eq. (15–24) will be correct only up to $E = E_F$; all electron states with energy $E < E_F$ are occupied by two electrons each, while all states with energy $E > E_F$ are empty. Thus, $f(E,T=0) = 0$ for $E > E_F$. This is the function labeled "$T=0$" in Fig. 15–15. ∎

4. CONDUCTORS AND SEMICONDUCTORS

Up to now we have completely neglected the fact that our "electrons in a box" model does not properly represent the potential energy in a crystal, with its deep minimum at the position of each ion as shown in Fig. 15–10. Our approximate model will be good as long as the de Broglie wavelength of the electron is large compared to the atomic separation a:

$$\lambda = \frac{h}{p} = \frac{2\pi\hbar}{p} >> a \tag{15–25}$$

where $p = \sqrt{2mE}$ is the momentum of the particle. Unfortunately, this approximation breaks down just in the neighborhood of the Fermi energy. If we have one free electron per atom, then the free electron density is

$$\rho_e \approx \left(\frac{1}{a}\right)^3$$

which leads to a Fermi energy of [Eq. (15–23)]

$$E_F = \frac{\hbar^2}{2m}\left(\frac{3\pi^2}{a^3}\right)^{2/3} \approx \frac{9\hbar^2}{2ma^2}$$

where we have used the approximation $\pi \approx 3$.

The wavelength of an electron with kinetic energy E_F is then

$$\lambda = \frac{2\pi\hbar}{\sqrt{2mE_F}} \approx \frac{2\pi\hbar}{\sqrt{9\hbar^2/a^2}}$$

or

$$\lambda \approx \frac{2\pi}{3}\,a \approx 2a$$

Thus, at the Fermi energy the condition (15–25) is only marginally satisfied for one-electron atoms; for atoms with two or more free electrons per atom, Eq. (15–25) will be wrong.

Calculating the energy levels for electrons in a real crystal is a major computational effort, even with today's large computers. Luckily, one finds that unless

$$\lambda \approx a, \quad \text{or} \quad a/2, a/3, \ldots \tag{15–26}$$

the free electron model gives the same energy levels as the exact theory. At energies corresponding to one of the wavelengths given in Eq. (15–26), or when the electron kinetic energy

$$E = \frac{p^2}{2m} = \frac{h^2}{2m\lambda^2}$$

is near to one of the values where Eq. (15–26) is satisfied

$$E \approx \frac{h^2}{2ma^2} \quad \text{or} \quad \frac{2h^2}{ma^2} \quad \text{or} \quad \frac{9h^2}{2ma^2} \cdots$$

then there can appear gaps in the energy levels. Some of the energy levels are higher and others lower than the energies given by the free electron model; thus, a region of width 0.1 to 2 eV can appear in which there are no energy levels at all, as shown in Fig. 15–18b. We stress that this does not have to happen; the question whether or not such *energy gaps* appear can be determined only by a detailed study of a particular crystalline lattice. Even if such gaps appear in the energy spectrum, the free electron model will be a good approximation as long as the Fermi energy is not near (within $\Delta E \sim kT$) an energy gap. This is the case in metals. The Fermi energy in metals (Fig. 15–19a) lies inside a *conduction band* far away from the nearest gap; each electron at the Fermi energy has many empty spaces nearby.

This is necessary if the material is to be a good conductor. The wave functions in a box are standing waves of well-defined wavelength (compare Fig. 15–12). Thus, each wave function is of the form

$$\psi = A \cos kx \, e^{-i\omega t} \quad \text{or} \quad B = \sin kx \, e^{-i\omega t}$$

However, moving electrons in a current-carrying conductor are represented by

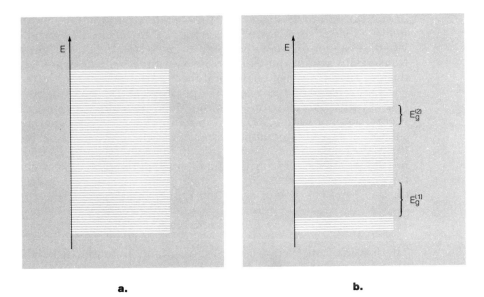

a.   **b.**

Figure 15–18 (a) In the "electrons-in-a-box" model there are states near any energy *E*. (b) In a real crystal there are gaps in the states whenever the ionic spacing is an integer multiple of the de Broglie wavelength λ.

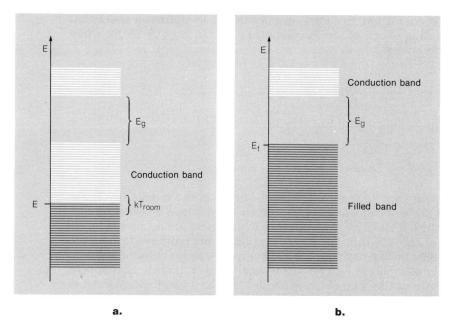

Figure 15-19 In a metal (a) the nearest gap is far away from the Fermi energy. Thus there are many empty states (in white) near the Fermi energy. In a semiconductor (b) all states below the gap are filled at $T = 0$, while all states in the conduction band above the gap are empty.

wave packets similar to those shown in Fig. 14–8. Taking the simplest form [Eq. (14–20)]

$$\psi = e^{i(kx - \omega t)} = \cos kx \, e^{-i\omega t} + i \sin kx \, e^{-i\omega t} \qquad (15\text{-}27)$$

one finds that because of the Pauli principle the electron can have this wave function only if neither the state $\cos kx$ nor the state $\sin kx$ is already occupied by two electrons. Thus, a current can occur only if there are free states so close in energy to occupied states that the smallest pertubation—an external electric field—will suffice to raise the energy of an electron to an unoccupied state, where it can form a wave function as given by Eq. (15–27). Only then can electrons easily acquire momentum in one direction; the material is a good conductor. The electrons behave as if they were free.

In a *semiconductor* at zero temperature, all the states below a gap are filled (Fig. 15–19b). An electron has to be given the large gap energy E_g before it can occupy another state. Thus, at zero temperature such a material is a perfect insulator. At room temperature, a small fraction of the electrons will be raised by thermal excitation into the conduction band. But as long as $kT \ll E_g$, this will indeed be only a very small fraction; most of the electrons cannot participate in the current. An exact theory says that the number of electrons in the upper conduction band is proportional to the Boltzmann factor

$$n(\text{conduction band}) \approx e^{-E_g/2kT} \qquad (15\text{-}28)$$

The material will conduct electricity, although poorly. It is what its name implies—a semiconductor.

5. ELECTRON MOBILITY

We now want to learn how an electron in the conduction band responds to an external electric field. We have learned earlier (Chapter 9) that the current density in most substances, defined as

$$\mathbf{j} = (-e)n_e <\mathbf{v}> \qquad (15\text{-}29a)$$

is proportional to the electric field

$$\mathbf{j} = \frac{1}{\rho}\mathbf{E}$$

From this we conclude that the electron velocity also has to be proportional to the electric field. We can define the mobility μ_e of the electrons by the relation:

$$-<\mathbf{v}> = \mu_e\mathbf{E} \qquad (15\text{-}29b)$$

We have added the factor (-1) in Eq. (15-29b), because the velocity $<\mathbf{v}>$ actually has a direction opposite to the direction of the electric field for negative charge carriers such as electrons. With our definition, the mobility is always positive.

We now want to relate the mobility μ_e to atomic properties. Let us first study the free electrons. We will use classical physics, because we are interested only in the average behavior of all the free electrons. Then quantum mechanics will affect our theory only through the Pauli principle—the electrons have an initial kinetic energy of the order of the Fermi energy E_F. Newton's equation tells us that

$$ma = m\frac{d\mathbf{v}}{dt} = (-e)\,\mathbf{E} \qquad (15\text{-}30)$$

and assuming the electron has the initial velocity $\mathbf{v}_0$, we obtain

$$\mathbf{v} = -\frac{e}{m}\mathbf{E}\,t + \mathbf{v}_0 \qquad (15\text{-}31)$$

The electrons will continue accelerating forever; we cannot define a mean velocity or a mobility. Given enough time, the electron velocity and thus also the current will become arbitrarily large. Even the smallest electric field will produce in the steady state an arbitrarily large current. We can say that the *free electron gas is a perfect conductor*: its resistivity ρ vanishes.

It can be shown that a perfectly periodic lattice potential does not affect our conclusion. A perfect crystal, if it has any free electrons at all, is a perfect conductor. However, no real crystal is perfect. At nonzero temperature $T > 0$ the individual lattice ions oscillate about their equilibrium positions (Fig. 15-20). The potential is no longer exactly periodic, and the random deviations of the potential will scatter the electrons. The electrons will not be accelerated indefinitely; after traveling a *mean free path* Λ, they will scatter (Fig. 15-21). One can assume that the direction of the velocity $\mathbf{v}_0$ immediately after scattering will again be random. We further assume—and this we will have to prove— that the magnitude $|\mathbf{v}_0|$ of the random velocity $\mathbf{v}_0$ is much larger than the additionally acquired velocity due to the electric field for all "reasonable" times:

$$\left|\frac{e}{m}\mathbf{E}\,t\right| << \left|\mathbf{v}_0\right| \qquad (15\text{-}32)$$

Then the *mean free time* τ can be deduced from Eq. (15-31). Because of Eq. (15-32), the magnitude of the velocity $\mathbf{v}$ is nearly equal to $\mathbf{v}_0$; thus, after the time

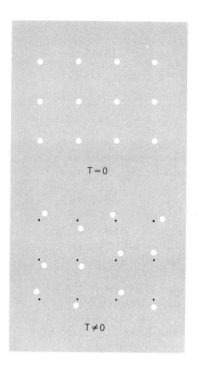

Figure 15-20 Thermal vibrations of the lattice ions are the main cause of electron scattering in metals at room temperature. Each ion performs oscillations about its equilibrium position; thus the lattice at any instant is not exactly periodic.

τ the electron will have traversed the distance

$$\Lambda = |\mathbf{v}_0|\tau$$

from which we immediately obtain

$$\tau = \frac{\Lambda}{|\mathbf{v}_0|} \tag{15-33}$$

Because after each collision the initial velocity is again randomized, the electron

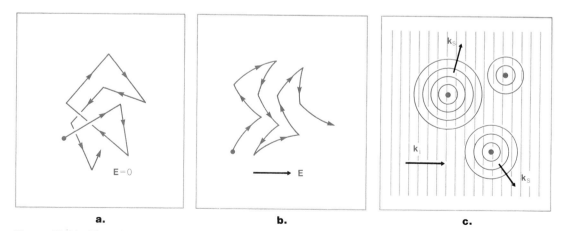

a.　　　　　　　　**b.**　　　　　　　　**c.**

Figure 15-21 Electrons in real crystal scatter off any deviation from a perfect lattice. (a) and (b) show the classical picture in the absence (a) and presence (b) of an external electric field. In (c) we show the quantum mechanical picture; the incident electron wave of wave vector $\mathbf{k}_i$ is scattered by the imperfections. Circular waves start at each imperfection, while the incident wave is attenuated.

velocity will be given by Eq. (15–31) only between $t = 0$ and $t = \tau$. After this, the electron scatters and the whole process starts all over.

We can also obtain the mean velocity of the electrons in the direction of **E**. First we note that the initial velocity $\mathbf{v}_0$ has completely random direction, and thus its average vanishes:

$$<\mathbf{v}_0> = 0$$

Next we note that the time-dependent part of the velocity, $-e/m\mathbf{E}\,t$, increases linearly with time; thus, its mean value will be one-half of its maximum value. Therefore, we have

$$<\mathbf{v}> = -\frac{e}{m}\mathbf{E}\frac{\tau}{2} \tag{15–34}$$

and using Eq. (15–29) we obtain for the mobility

$$\mu_e = \frac{e}{m}\frac{\tau}{2} = \frac{e}{m}\frac{\Lambda}{2|\mathbf{v}_0|} \tag{15–35}$$

Example 5   Show that the inequality (15–32) is satisfied for copper.

We have earlier (Example 3) calculated the Fermi energy of copper, $E_F = 7.1$ eV $= 1.13 \times 10^{-18}$ J. From this we obtain the Fermi velocity by noting that

$$\frac{mv_F{}^2}{2} = E_F$$

or

$$v_F = \sqrt{\frac{2\,E_F}{m}} = \sqrt{\frac{2 \times 1.13 \times 10^{-18}\ \mathrm{J}}{0.91 \times 10^{-30}\ \mathrm{kg}}} = 1.58 \times 10^6\ \mathrm{m/sec}$$

The mean magnitude of the electron velocity $|<\mathbf{v}_0>|$ will be of the same order of magnitude. Let us now assume a current density in copper of 1000 A/mm² $= 10^9$ A/m². This is a huge current density; if we tried to put this current through a copper wire, it would melt in a fraction of a second (1 to 2 A/mm² is a reasonably conservative current). We know from Example 3 the electron density, $n_e = 8.5 \times 10^{28}$ electrons/m³. Using Eq. (15–29a), we can then obtain the mean electron velocity

$$|<\mathbf{v}>| = \frac{|\mathbf{j}|}{e\,n_e} = \frac{10^9\ \mathrm{A/m^2}}{(1.6 \times 10^{-19}\ \mathrm{coul}) \times (8.5 \times 10^{28}/\mathrm{m^3})} = 0.074\ \mathrm{m/sec}$$

This is indeed very much smaller than the Fermi velocity v_F. On the other hand, $|<\mathbf{v}>|$ according to Eq. (15–34) is exactly the left-hand side of Eq. (15–32) for reasonable times — times of the order of the mean free time τ.

Example 6   Estimate the mean free path of free electrons in copper at room temperature.

We can find the resistivity of copper at room temperature in Table I of Appendix B: $\rho = 1.7 \times 10^{-8}$ Ω-m. We also have calculated in Example 3 the electron density, $n_e = 8.5 \times 10^{28}$ electrons/m³. From Eqs. (15–29a) and (15–29b) we can obtain a relation between mobility and resistivity of the electrons

$$\mu_e = \frac{1}{en_e\rho} = \frac{1}{(1.6 \times 10^{-19}\ \mathrm{coul})\,(1.7 \times 10^{-8}\ \Omega\text{-m})\,(8.5 \times 10^{28}\ \mathrm{m^{-3}})}$$

$$= 4.3 \times 10^{-3}\ \mathrm{m^2/V\text{-}sec}$$

On the other hand, using Eq. (15–35), we have

$$\Lambda = \frac{2m\mu_e v_0}{e}$$

and if we use the estimate of $v_0 = 1.6 \times 10^6$ m/sec from Example 5, we have

$$\Lambda = \frac{2 \times (0.91 \times 10^{-30} \text{ kg}) \, (4.3 \times 10^{-3} \text{ m}^2/\text{V-sec}) \, (1.6 \times 10^6 \text{ m/sec})}{(1.6 \times 10^{-19} \text{ coul})}$$
$$\approx 8 \times 10^{-8} \text{ m} = 800 \text{ Å}$$

As we see, the mean free path of electrons is only a few hundred atomic distances.

Let us now discuss briefly the physical processes which determine the mean free path Λ and thus the mobility of electrons in a solid. Quite generally, the electrons will scatter off any deviation from an exact periodic structure of the lattice. Such deviations occur because of thermal vibration. These deviations increase with temperature, and thus the mean free path decreases with temperature. An exact theory yields

$$\Lambda \approx \frac{1}{T} \tag{15–36}$$

From Eq. (15–36) we would expect that the mean free path becomes infinitely large at zero temperature. This does not happen for real materials, because there are imperfections in any real crystal (Fig. 15–22). There will always be impurities — small amounts of different atoms. In addition, occasionally a lattice site may be empty (forming a vacancy), or an atom can occupy an interstitial position in the lattice. There will be also dislocations — more or less sudden shifts of the crystalline lattice. All of these imperfections will scatter electrons; thus, for any real metal the mean free path at very low temperatures rarely exceeds $10 \, \mu = 10^{-5}$ m.

There are notable exceptions to this statement; these are the *super-con-ductors*. In many metals (e.g., mercury, lead, niobium), the electrons at low

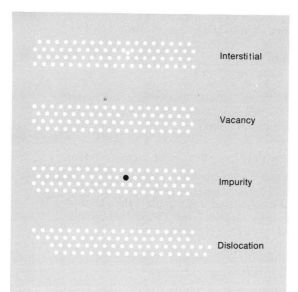

Interstitial

Vacancy

Impurity

Dislocation

Figure 15–22 Even at low temperatures, crystal imperfections cause electron scattering. We show schematically the various types of imperfections.

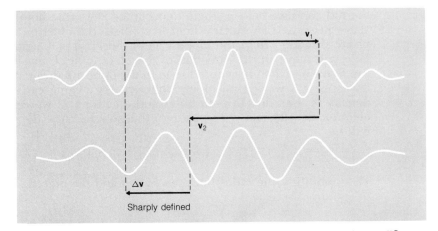

Figure 15-23 In some metals at low temperatures, two electrons form a "Cooper pair"; because of the wave nature, these pairs are spread over many lattice spacings. If one of the two electrons is near an imperfection, the other is usually far away.

temperatures (a few °K) form pairs which are called *Cooper pairs* (Fig. 15-23). In these pairs, the two electrons have an exact difference of velocities. Because of the uncertainty principle [Eq. (14-1)] this implies that the two electrons are very unlikely to be very close together; the Cooper pairs are extended over many atomic distances. Thus, when one electron is near a scattering center, the other one is far away; the nearby electron cannot scatter (i.e., change its velocity) without breaking the Cooper pair. This requires energy, because the Cooper pairs have a binding energy of the order of 10^{-3} eV. Since this energy at low temperatures is not available, no scattering takes place and the mean free path Λ is truly infinitely long. The resistivity of superconductors therefore rather suddenly vanishes at a transition temperature T_s, which ranges from 0.5° K (for titanium) to 20° K (some niobium alloys).

Let us now turn our attention to semiconductors; here the situation is more complex because the free electron density itself depends strongly on temperature, as we had said in Eq. (15-28). In addition, whenever an electron is lifted by thermal agitation from the full band into an empty conduction band, it leaves behind an empty state—a *hole* (Fig. 15-24). Thus, we have in the conduction band a state occupied by an electron, with only unoccupied states nearby. In

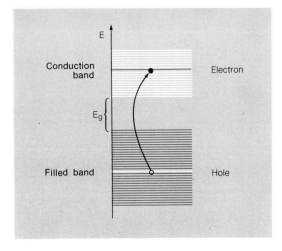

Figure 15-24 In a semiconductor at room temperature, sometimes an electron will be "excited" into the conduction band, leaving a hole behind. This happens very rarely; thus, semiconductors conduct poorly.

the full band we have a single unoccupied state (the hole), with only occupied states nearby. If an electric field is applied, the electron will move in the opposite direction from the electric field, because it is negatively charged. The hole will move in the direction of the electric field, because nearby electrons can move into the unoccupied state in the opposite direction. Thus, the hole behaves as if it were a positively charged free charge carrier. The total current density is then due to both holes and electrons. If we call n (= negative) the electron density and p (= positive) the hole density, the total current density will be

$$\mathbf{j} = (-e)n<\mathbf{v}_e> + ep<\mathbf{v}_p> \qquad (15\text{-}37)$$

The mobilities, the ratios between $|<\mathbf{v}>|$ and the electric field,

$$\mu_p = \frac{|\mathbf{E}|}{|<\mathbf{v}_p>|} \quad \text{and} \quad \mu_e = \frac{|\mathbf{E}|}{|<\mathbf{v}_e>|}$$

are in general of the same order of magnitude in most semiconductors, although they do not usually have exactly the same value.

The resistivity is given by the ratio of electric field and current density. From Eq. (15–37) we obtain

$$\rho = \frac{1}{e\,(n\mu_e + p\mu_p)} \qquad (15\text{-}38)$$

Example 7   The resistivity of pure silicon is $\rho_{20} = 3000$ Ω-m at room temperature (20° C) and $\rho_{100} = 30$ Ω-m at 100°C. Estimate from this the gap energy E_g in silicon in units of electron-volts.

We should note that this is a very rapid change in resistivity—a factor of 100 between room temperature and the temperature of boiling water. We cannot explain this by temperature variation of mobilities. Since $T_1 = 20°$ C = 293° K and $T_2 = 100°$ C = 373° K, we would expect the mobilities to change by a factor of the order of

$$T_2/T_1 = \frac{373}{293} = 1.27$$

Thus, practically all of the temperature variation of ρ is due to the change in carrier density, both of holes and of electrons. Since the density of holes and electrons is the same, we expect according to Eq. (15–28) that ρ is proportional to $1/n$:

$$\frac{\rho_{100}}{\rho_{20}} \approx \frac{1/n_{100}}{1/n_{20}} = \frac{e^{+E_g/2kT_2}}{e^{+E_g/2kT_1}}$$

Taking the natural logarithms on both sides, we have

$$\ln \frac{\rho_{100}}{\rho_{20}} \approx \frac{+E_g}{2k}\left(\frac{1}{T_2} - \frac{1}{T_1}\right)$$

or

$$E_g = \frac{2k \ln (\rho_{20}/\rho_{100})}{(1/T_1) - (1/T_2)}$$

We have $\ln(\rho_{20}/\rho_{100}) = \ln 100 = 4.61$. We can find k in Appendix B. Table III: $k = 1.38 \times 10^{-23}$ Joule/°K. We then obtain

$$E_g = \frac{2 \times (1.38 \times 10^{-23}\ \text{J/°K}) \times 4.61}{\left[\left(\dfrac{1}{293} - \dfrac{1}{373}\right)(\text{°K})^{-1}\right]} = 1.7 \times 10^{-19}\ \text{J}$$

or, expressed in electron volts,

$$E_g = 1.1 \text{ eV}$$

6. DONORS AND ACCEPTORS

The density of free electrons and holes in semiconductors is very strongly affected by impurities. At room temperature in silicon only a fraction

$$\frac{n_{\text{free}}}{n_0} \approx e^{-E_g/2kT} \approx e^{-21.6} \approx 10^{-9.4}$$

of the electrons find themselves in the conduction band. Thus, even a small amount of an impurity, if it can easily contribute conduction electrons, will have a drastic effect on the resistivity of the material.

We will discuss this phenomenon, using silicon as an example. The silicon atom has four electrons in its outermost shell. These electrons completely fill two bands, called the *valence bands* (Fig. 15–25). The gap between the upper valence band and the conduction band has the value $E_g = 1.1$ eV. If we add a small amount of arsenic, which has five electrons in the outer shell, only four of them will be in the valence band. The fifth electron is loosely tied to the $(\text{As})^+$ ion, with a binding energy of only $E_d = 0.05$ eV. We can plot these impurity levels (Fig. 15–26) as being below the conduction band by E_d because the electron freed by the arsenic atom is a free electron. If there are N_d such arsenic atoms per unit volume, then at the temperature T a fraction

$$n_e \approx N_d \, e^{-E_d/kT} \tag{15–39}$$

of the atoms will be ionized, each contributing one free electron to the material.*
Arsenic thus contributes donor sites of electrons. Because $E_d << E_g$, we have

$$e^{-E_d/kT} >> e^{-E_g/2kT} \tag{15–40}$$

---

* The estimate in Eq. (15–39) is an oversimplification of the real situation. The number of ionized donors depends in a nontrivial way on the temperature T, but also is quite different if there are acceptor atoms simultaneously present. Nevertheless, our conclusions do not depend on the specific form of Eq. (15–39), but simply on the fact that at room temperature a large fraction of the available donors is ionized.

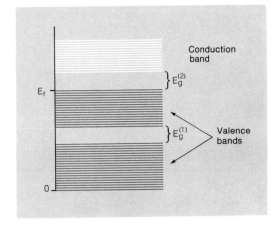

Figure 15–25 The band structure of silicon. The two valence bands are completely filled.

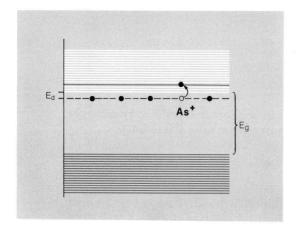

Figure 15-26 By adding arsenic to silicon, one creates "donor impurities," which require only an energy E_d to release an electron into the conduction band.

Even if the concentration of As is only a few parts per million, the room temperature electron density will be essentially determined by the As concentration alone.

If instead of arsenic we add a small amount of boron to the silicon, we have a different situation (Fig. 15–27). Boron has only three electrons in its outer shell. Since all surrounding silicon atoms have four valence electrons, only the small energy $E_a = 0.045$ eV is required to form the B$^-$ ion by capturing one electron from the valence band; this process leaves a hole (Fig. 15–27). *Boron is an acceptor of electrons.*

A semiconductor with donor impurities is called an *n*-type semiconductor, while acceptor impurities make it a *p*-type semiconductor. Of course, in reality there are always impurities of both types—we can never hope to start with perfectly pure materials. However, as long as one type of impurity strongly predominates, this does not matter. If there are 100 times more donor impurities than acceptor impurities, 1% of the donors will provide the electrons to form the negative acceptor ions, and the residual 99% will behave as if there were no acceptors around.

If one connects a *p*-type and an *n*-type conductor, one obtains a solid state diode. If we connect the *n*-type material to the positive terminal of an emf source and the *p*-type material to the negative terminal, the holes and electrons will move away from the junctions between the two materials (Fig. 15–28a). The As$^+$ and B$^-$ ions left behind in the *depletion region* constitute a net charge density. The electric field is concentrated in this region, but it is due to charges (the ions) which cannot move. It is as if we had a capacitor; no current will flow through the diode.

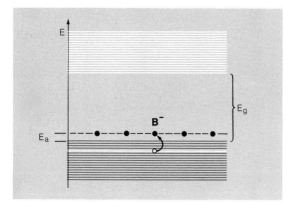

Figure 15-27 Boron in silicon acts as electron acceptor; electrons from the valence band are captured by the boron atom, leaving a mobile hole behind.

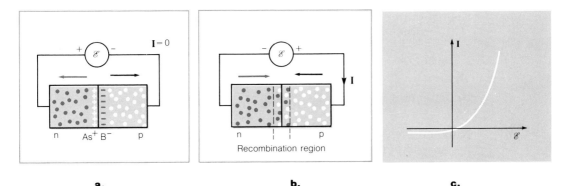

Figure 15-28   (a) The pn junction diode under reverse potential bias; the holes and electrons separate, leaving at the junction positive As$^+$ ions and negative B$^-$ ions. There is no current. (b) Forward biasing the diode; the holes and electrons move toward each other. In the recombination region in the center, holes and electrons recombine with each other. The result is a current whose magnitude depends strongly on the size of the applied voltage (c).

If the emf source terminals are reversed, then the holes and electrons will move toward each other (Fig. 15–28b). In a *recombination* region near the junction they will recombine; electrons in the conduction bands "fall into" the free states which the holes constitute. A large current can exist in the material. If we plot the current I against the voltage V for such a diode, we obtain a curve such as that shown in Figure 15–28c.

This discussion of electrons in atoms and solids had to be only qualitative; this is a penalty we have to pay whenever we discuss real phenomena instead of artificial examples. There are really two legitimate domains of physics. One investigates the fundamental laws of physics, which hopefully are simple. Experiments to test these fundamental laws have to be very carefully designed to minimize extraneous influences. Often a bold extrapolation from rather crude data is necessary. The physicist is "guessing" how nature would behave if a "pure experiment," testing only one law with perfect equipment, could be performed. After the fundamental laws, such as Coulomb's law and Ampère's law, have been found, the physicist still has to try to explain what happens in the real world. Imperfections of the experimental setup have to be understood, and frequently their investigation leads to new understanding of fundamental properties. The long, hard effort to make purer and purer materials in an attempt to understand the "ideal crystal" led to a more profound understanding of the theory of crystal impurities and imperfections and how they affect the conduction of an electric current. The development of the transistor, in which carefully controlled impurities are added to a semiconductor to create new electric properties, was possible only because a generation of physicists had fought to understand how a "perfect" semiconductor behaves.

Summary of Important Relations

Spin

    magnitude $\qquad\qquad |\mathbf{S}| = \dfrac{\sqrt{3}}{2}\hbar$

    sharp component $\qquad S_z = \pm\,{}^1\!/_2\,\hbar$

Energy of free electron
   in cubic box of edge 2L
$$E = \frac{\pi^2 \hbar^2}{8mL^2} (n_x^2 + n_y^2 + n_z^2)$$

where n_x, n_y, n_z are positive integers

Fermi energy
$$E_F = \frac{\hbar^2}{2m} (3\pi^2 \rho_e)^{2/3}$$

Number of electrons with energy less
   than E per unit volume
$$n(E) = \frac{(2mE)^{3/2}}{3\pi^2 \hbar^3}$$

Definition of electron mobility: $<\mathbf{v}> = -\mu_e \mathbf{E}$
   mean electron velocity

connection to mean free path Λ
$$\mu_e = \frac{e}{m} \frac{\Lambda}{2v_0}$$

Temperature dependence of resistivity:

  in metals $\rho \propto T$
  in pure semiconductors $\rho \propto e^{E_g/2kT}$
  in impure semiconductors $\rho \propto e^{E_0/kT}$

  where $E_0 = E_a$ if acceptors predominate
         $E_0 = E_d$ if donors predominate

QUESTIONS

1. A superconducting ring is inserted into a magnetic field. What will be the magnetic flux through the ring? (Do not forget Faraday's law!)

2. Would you expect the resistivity of thin metallic films at low temperatures to be larger or smaller than the resistivity of the bulk material at the same temperature? Explain.

3. How would you electrically test whether there are impurities in a silicon sample?

4. Would you expect the Hall coefficient (p. 327) to be temperature-dependent in metals? In semiconductors? Explain why.

Problems

1. Using the level scheme in Fig. 15–5, show that the electrons in the elements He, Ne, and A fill completely closed shells.

2. The ionization energy of lithium (Z = 3) is $I = 5.36$ eV. What is the effective charge Z_{eff} for the outermost electron?

3. Determine from Fig. 15–7 an equation for the effective charge Z_{eff} for K-shell electrons as a function of the real nuclear charge Z.

4. What is the energy of the $3d$ state in hydrogen?

5. N electrons ($N >> 1$) are in a one-dimensional box of length $2L$. Calculate the Fermi energy, i.e., the energy of the highest occupied level at $T = 0$.

6. Calculate the Fermi energy if N electrons are enclosed in a two-dimensional box, i.e., a square of sides a. Assume $N >> 1$.

7. Calculate the 20 lowest energy states in a two-dimensional box of side $a = 10$ Å.

8. Fifty electrons are in a cubic box of size $10 \times 10 \times 10$ Å. Calculate the Fermi energy (a) counting off states as in Example 2, (b) using the approximation formula Eq. (15–18).

9. Using a computer, determine how many electrons have to be in a cubic box if Eq. (15–18) is to be accurate to better than 3%. Be careful to write your program so as to minimize the CPU time required.

*10. Calculate the *average energy* of an electron gas of density ρ_e in a cubic box at $T = 0$. What is the ratio of the average energy to the Fermi energy? Use the approximation that the number of electrons is large.

11. An electron is confined to the inside of a box which is not a cube, but of dimensions 5 Å $\times 3$ Å $\times 7$ Å. Calculate the ground state energy.

12. Calculate the mobility of electrons in silver at room temperature, assuming one free electron per atom.

13. Assume a hypothetical material in which the Fermi energy vanishes, i.e., the mean initial velocity v_0 in Eq. (15–31) has negligible magnitude. Assume further that the electrons accelerated by the electric field stop and start out again from rest after traversing the distance Λ. Show that under these assumptions Ohm's law is not valid, but that instead the current density is proportional to the square root of the electric field.

14. The electrical conductivity of lead as a function of temperature is:

$t(°C)$	20	200	300	333
$\rho(10^{-8}$ Ω-m$)$	22	36	45	95

The melting point of lead is $327°$ C. (a) Calculate by extrapolation what would be the mean free path of electrons in lead at $333°$, if the lead was still solid at this temperature. (b) Compare this with the actual mean free path of electrons in liquid lead. Assume two free electrons per atom. By what factor does the mean free path change during melting?

15. The temperature dependence of the resistivity of copper agrees well with the theoretical prediction that it is proportional to the absolute temperature. You are to build a 10-km-long transmission line (2 wires) which carries the current of 1000 A; the voltage drop across the line should be less than 100 V. (a) Given that copper costs approximately \$2/kg, how much would the copper in the line cost? (b) Installing a cryogenic system to cool the whole line to liquid air temperature ($-196°$ C) would add \$300,000 to the cost of the line. Given that you can use less copper at low temperature, compare the capital costs of the two approaches.

16. An electromagnet with copper wiring rated at 1000 A maximum has a resistance of 0.5 Ω at room temperature. Although cooled, the magnet temperature rises to $80°$ C if a current of 1000 A flows through it. Calculate the maximum voltage which a power supply driving the magnet should be able to deliver.

17. At what temperature will the mean free path of very pure copper become larger than 1 $\mu = 10^{-6}$ m?

18. A wire-wound resistor has $1/10$ of the resistance of a sample of pure silicon at $20°$ C. At what temperature will the two resistors have equal resistance?

19. Germanium is a semiconductor with a gap between the full band and the conduction band of $E_g = 0.67$ eV. Estimate the ratio between the resistances of a pure germanium sample at $-40°$ C and at $+60°$ C.

20. A silicon rectifier (1N1124) has the following values of current at the given voltage:

V(volt)	0.4	0.5	0.6	0.7	0.8	0.9	1.0
I(A)	<0.02	0.08	0.15	0.3	0.8	2.5	7.5

(a) Plot $I(V)$. (b) Determine the maximum current through the diode if the maximum power dissipated can be 3 W. (c) If the diode is connected to a power supply of up to 10 volts through a resistor R, determine the value of R which will prevent damage to the diode.

CHAPTER 16

ELECTROMAGNETIC RADIATION

1. ELECTROMAGNETIC WAVES

We have shown in the last two chapters how the wave description of particles affects the properties of atoms and solids. In this chapter we want to study both the wave and the particle aspects of electromagnetic radiation.

We will first show that Maxwell's equations, which govern the time development of electromagnetic fields, allow solutions which have all the properties required of a wave: They obey the wave equation [Eq. (1–18)], and in one dimension (plane waves) they can be expressed in the general form of Eq. (1–14). We will use Maxwell's equations as they are valid in vacuum or in nonconducting materials of well-defined dielectric constant K and relative magnetic permeability μ_r. We have already written them down (p. 431), but we will repeat them here for easier reference. In vacuum, we have four equations satisfied by electromagnetic fields:

$$\int_{\substack{\text{closed} \\ \text{loop}}} \mathbf{E} \cdot d\mathbf{l} = - \int_{\text{inside}} \frac{\partial \mathbf{B}}{\partial t} \cdot d\mathbf{S} \tag{16–1}$$

$$\int_{\substack{\text{closed} \\ \text{loop}}} \mathbf{B} \cdot d\mathbf{l} = \epsilon_0 \mu_0 \int_{\text{inside}} \frac{\partial \mathbf{E}}{\partial t} \cdot d\mathbf{S} \tag{16–2}$$

$$\int_{\substack{\text{closed} \\ \text{surface}}} \mathbf{E} \cdot d\mathbf{S} = 0 \tag{16–3}$$

$$\int_{\substack{\text{closed} \\ \text{surface}}} \mathbf{B} \cdot d\mathbf{S} = 0 \tag{16–4}$$

In nonconducting materials Eqs. (16–1), (16–3), and (16–4) are unchanged; only the factor $\epsilon_0\mu_0$ in Eq. (16–2) is to be replaced by $(K\epsilon_0)(\mu_r\mu_0)$. We will search for solutions which represent waves traveling in the positive z-direction. We will thus assume that both the magnetic and the electric fields depend only on z

and t, but are independent of x and y. Specifically, we assume that

$$\mathbf{B} = B_x(z - ct)\,\hat{\mathbf{i}} \tag{16–5a}$$

and

$$\mathbf{E} = E_y(z - ct)\,\hat{\mathbf{j}} \tag{16–5b}$$

with an as yet unspecified wave velocity c. It is important to note that we have assumed both $\mathbf{B}$ and $\mathbf{E}$ to be perpendicular to the propagation direction of the wave, i.e., to the z-axis. Then Eqs. (16–3) and (16–4) are satisfied whatever exact functional form we choose for B_x and E_y. As we can see from Fig. 16–1, where we arbitrarily show $\mathbf{B}$, the magnetic field is the same at the two points P_1 and P_2; indeed, it has the same magnitude and direction anywhere in the narrow tube T. We have discussed in Chapter 7 the idea of flux through a surface and the analogy to the flow of a liquid; in our case the analogy is a liquid flowing with constant velocity along the tube T. Since we can consider the whole inside of the surface S as being filled with such tubes, we conclude that indeed $\int \mathbf{B} \cdot d\mathbf{S} = 0$ over the closed surface S. Exactly the same argument, of course, can also be used to show that Eq. (16–3) is also satisfied. Thus only Eqs. (16–1) and (16–2) give us any condition on the functional form of the unknown quantities E_y and B_x.

Let us therefore choose a rectangular loop, as shown in Fig. 16–2. The loop can be of arbitrary length a in the y-direction; but in the z-direction we assume it to be very narrow – of width dz. If we choose the direction of integration along the arrows, the loop integral

$$I = \int_C \mathbf{E} \cdot d\mathbf{l}$$

can be written as

$$I = [E_y(z - ct) - E_y(z + dz - ct)]a$$

because at the top and bottom $\mathbf{E}$ and $d\mathbf{l}$ are perpendicular to each other. Since E_y is a function of a single variable $(z - ct)$, and since dz is by definition a small quantity, we have

$$I = \int_C \mathbf{E} \cdot d\mathbf{l} = -a\,dz\,E_y' \tag{16–6}$$

where we denote by the prime the differentiation with respect to the single variable $(z - ct)$. According to Eq. (16–1), this has to be equal to

$$I = -\int_S \frac{\partial \mathbf{B}}{\partial t} \cdot d\mathbf{S} = -\frac{\partial B_x}{\partial t}(a\,dz) \tag{16–7a}$$

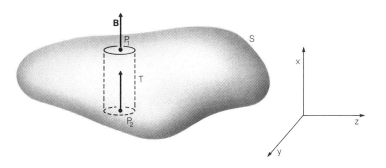

Figure 16–1 The integral $\int \mathbf{B} \cdot d\mathbf{S}$ over the closed surface S vanishes if $\mathbf{B}$ is constant along tube T; actually, it is sufficient if at equivalent points P_1 and P_2 the magnetic field is the same.

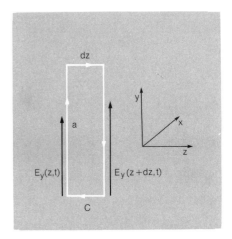

Figure 16-2 Calculating the loop integral ∫ **E**•*d***l** over a narrow rectangular loop.

where S is the inside of the loop. (The reader should convince himself that the sign convention between $d\mathbf{l}$ and $d\mathbf{S}$ is satisfied.) Over the small width dz, we can assume B_x and thus $\partial B_x/\partial t$ to be the same everywhere. We can write

$$\frac{\partial B_x}{\partial t} = B'_x\,(-c) \tag{16–7b}$$

where again the prime denotes differentiation with respect to the single variable $(z - ct)$. Combining the three equations (16–6) and (16–7a, b), we have, since the factor $a\,dz$ can be cancelled:

$$E'_y = -c\,B'_x \tag{16–8}$$

In exactly the same way, we can obtain from Eq. (16–2) the relation

$$B'_x = -\epsilon_0\mu_0 c\,E'_y \tag{16–9}$$

and we see that both equations will be satisfied if and only if $1/c = \epsilon_0\mu_0 c$ or

$$\boxed{c = \frac{1}{\sqrt{\epsilon_0\mu_0}}} \tag{16–10}$$

Note that we still can prescribe a completely arbitrary functional dependence to one of the two variables, for instance, to E_x; but once we have done this, B_y is completely determined.[°] Indeed, from either Eq. (16–8) or Eq. (16–9) we can conclude that B_x is at any point always proportional to E_y. In Fig. 16–3 we show a "snapshot" of the field distribution in such an electromagnetic wave.

We have thus shown that Maxwell's equations in vacuum allow the existence of waves traveling with the speed $c = 1/\sqrt{\epsilon_0\mu_0} = 3 \times 10^8$ m/sec. The solutions which we have seen have both the magnetic and the electric fields perpendicular to each other and to the direction of propagation. This is indeed a property obeyed by all electromagnetic waves.

[°] Actually, from Eq. (16–8) or Eq. (16–9), one can only conclude that E_y and $-c\,B_x$ differ from each other by at most an arbitrary constant. However, a constant (space- and time-independent) electric and magnetic field does not contribute to any transport of electromagnetic energy.

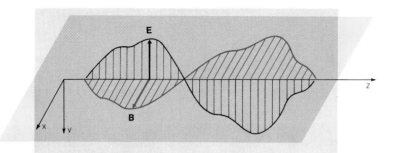

Figure 16–3 At any point, $|\mathbf{E}|$ and $|\mathbf{B}|$ in an electromagnetic plane wave are proportional to each other; $\mathbf{E}$, $\mathbf{B}$ and the propagation direction form a right-handed axis system.

Example 1   Show that there can be no component of the electric or magnetic field along the direction of propagation in an electromagnetic wave.

This can be shown using Eq. (16–3). If we have a z-component of the electric field

$$E_z(z - ct)\hat{\mathbf{k}}$$

then we can define as a closed surface a disk of height dz and area πr^2 (Fig. 16–4) and show that Eq. (16–3) cannot be satisfied. Indeed, we have

$$\int_S \mathbf{E} \cdot d\mathbf{S} = \pi r^2 [E_z(z + dz - ct) - E_z(z - ct)]$$

$$= \pi r^2\, dz \frac{dE_z}{d(z - ct)} = \pi r^2\, dz\, E'_z$$

which vanishes only if E_z is constant or independent of time and position. This, of course, does not imply that there cannot be a *time-independent* electromagnetic field along the direction of propagation. A light beam can pass through a capacitor, or through the gap of an electromagnet, and the direction of the static electric or magnetic field has nothing to do with the wave direction. But the time-dependent electric or magnetic field which makes up the wave can have no component along the direction of propagation.

We can also easily calculate the *intensity of the electromagnetic wave*. We have learned earlier [Eq. (8–6)] that the energy density* of the electric field

---

* Although the symbols are the same, the energy density ρ_e in this chapter has no connection with the electron density (also denoted ρ_e) in Chapter 15.

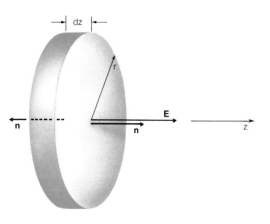

Figure 16–4 Example 1: A wave propagating in the z-direction can have no component along the propagation direction.

in vacuum is

$$\rho_e = \frac{\epsilon_0}{2} \mathbf{E}^2$$

and also [Eq. (12–43)] that the magnetic field $\mathbf{B}$ has the energy density

$$\rho_m = \frac{\mathbf{B}^2}{2\mu_0}$$

The total energy density will be the sum of the two. However, the magnitudes of $\mathbf{E}$ and $\mathbf{B}$ are not independent of each other. From Eq. (16–8) we can conclude that $|\mathbf{B}| = \frac{1}{c} |\mathbf{E}|$, and thus the total instantaneous energy density is

$$\rho = \frac{\epsilon_0}{2} \mathbf{E}^2 + \frac{\mathbf{B}^2}{2\mu_0} = \frac{\epsilon_0 \mathbf{E}^2}{2} + \frac{\mathbf{E}^2}{2c^2\mu_0}$$

and because according to Eq. (16–10) $1/c^2 = \epsilon_0 \mu_0$, we have

$$\rho = \epsilon_0 \mathbf{E}^2 \qquad\qquad (16\text{–}11)$$

If we have a harmonic wave, in which the electric field is

$$\mathbf{E} = \mathbf{E}_0 \cos (\mathbf{k} \bullet \mathbf{r} - \omega t)$$

then we can also define the average energy density

$$<\rho> = \frac{1}{2} \rho_{\max} = \frac{\epsilon_0 \mathbf{E}_0^2}{2} \qquad\qquad (16\text{–}12)$$

because the average of $\cos^2\alpha$ is equal to 1/2. The average intensity of the wave, according to Eq. (1–35), is then

$$\boxed{I = c <\rho> = \sqrt{\frac{\epsilon_0}{\mu_0}} \frac{\mathbf{E}_0^2}{2}} \qquad\qquad (16\text{–}13a)$$

We can also write, since the average of $\mathbf{E}^2$ is $<\mathbf{E}^2> = E_0^2/2$:

$$I = \sqrt{\frac{\epsilon_0}{\mu_0}} <\mathbf{E}^2> \qquad\qquad (16\text{–}13b)$$

2. THE SPEED OF LIGHT IN MATERIALS

In nonconducting materials, the basic Maxwell's equations (16–1) to (16–4) have to be modified by substituting

$$K\epsilon_0 \text{ for } \epsilon_0 \quad \text{and} \quad \mu_r\mu_0 \text{ for } \mu_0$$

in Eq. (16–2). Here K is the dielectric constant and μ_r is the relative magnetic permeability. We will limit ourselves here to transparent media, in which the absorption of the electromagnetic (E.M.) wave can be neglected. Since we are interested here mostly in E.M. waves having wavelengths of 0.1 to 10 microns (i.e., in the near ultraviolet, the visible region, and the infrared region), and since there are no ferromagnetic materials which are transparent at these wavelengths, we can set $\mu_r \approx 1$. For paramagnetic and diamagnetic substances, $|\mu_r - 1| \lesssim 10^{-4}$; thus, our approximation is certainly quite good.

The speed of light in transparent materials of dielectric constant K can be read from Eq. (16–10) by substituting $K\epsilon_0$ for ϵ_0:

$$v \text{ (in materials)} = \frac{1}{\sqrt{K\epsilon_0\mu_0}} = \frac{c}{\sqrt{K}} = \frac{c}{n} \tag{16–14a}$$

Here we have introduced the *refractive index*, which we already used in Chapter 4:

$$\boxed{n = \sqrt{K}} \tag{16–14b}$$

Since the dielectric constant is always larger than 1, we have $n > 1$ and $v < c$; the velocity of light in materials is always less than in vacuum.

However, the dielectric constant in Eq. (16–14b) is not exactly the same as the dielectric constant we measure by applying a static electric field. When we discuss visible (or infrared) light, we are discussing wavelengths of less than 10 microns or frequencies of

$$\nu = \frac{c}{\lambda} \gtrsim \frac{3 \times 10^8 \text{m/sec}}{10^{-5} \text{ m}} = 3 \times 10^{13} \text{ Hz}$$

At these frequencies it no longer is obvious that the proportionality factor between the polarization $\mathbf{P}$ and the electric field $\mathbf{E}$

$$\mathbf{P} = (K - 1)\epsilon_0\mathbf{E}$$

is the same as for a static electric field; the atomic constituents may not be able to rearrange themselves fast enough. We have derived earlier [Eq. (8–45)] the Clausius-Mosotti relation, which we can rewrite with the help of Eq. (16–14b) in the form

$$\frac{n^2 - 1}{n^2 + 2} = \frac{N}{3\epsilon_0}(\alpha_e + \alpha_i + \alpha_p) \tag{16–15}$$

We now want to investigate how the three types of polarizability (electronic, α_e; ionic, α_i; and orientational, α_p) depend on frequency. We will use a crude but rather simple model shown in Fig. 16–5. Consider a charge q of mass m tied to a spring with a spring constant k. If we apply a static electric field E, the mass

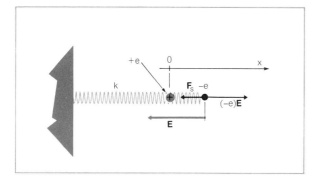

Figure 16–5 Crude model for polarizability of matter; an electron of charge $-e$ is fastened to a spring of spring constant k. A positive charge $+e$ is fixed at $x = 0$.

will move a distance

$$x = \frac{qE}{k}$$

because then the spring force $-kx$ cancels the electric force qE. The induced dipole moment (if we assume a charge $-q$ fixed at $x = 0$) is then

$$p = qx = \frac{q^2}{k}E \qquad (16\text{-}16a)$$

and we have a static polarizability

$$\alpha = \frac{q^2}{k} \qquad (16\text{-}16b)$$

Since experimentally both the electronic and ionic polarizabilities are of the order $\alpha \approx 10^{-40}$ coul-m²/V, we can even estimate the "effective spring constant"

$$k_{\text{eff}} = \frac{q^2}{\alpha} \approx \frac{(1.6 \times 10^{-19} \text{ coul})^2}{10^{-40} \text{ coul-m}^2/\text{V}} \approx 2.6 \times 10^2 \text{ N/m} \qquad (16\text{-}17)$$

where we have estimated the effective charge* to be of the order of one electron charge.

If the electric field is oscillating with time

$$E = E_0 \cos \omega t$$

we have to write out the full equation of motion

$$m \frac{d^2x}{dt^2} = -kx + qE_0 \cos \omega t$$

If we assume $x = A \cos \omega t$, we obtain [because of $d^2(\cos \omega t)/dt^2 = -\omega^2 \cos \omega t$]:

$$(-m\omega^2 + k)A \cos \omega t = qE_0 \cos \omega t$$

From this we obtain for the amplitude of oscillations:

$$A = \frac{qE_0}{(k - m\omega^2)}$$

Defining the resonant frequency

$$\omega_0 = \sqrt{\frac{k}{m}} \qquad (16\text{-}18a)$$

we can then write for $x = A \cos \omega t$:

$$x = \frac{qE_0 \cos \omega t}{m(\omega_0^2 - \omega^2)} = \frac{qE}{m(\omega_0^2 - \omega^2)} \qquad (16\text{-}18b)$$

---

* The effective charge q will be negative if α is the electronic polarizability. However, α in Eq. (16-16b) and in Eq. (16-19) depends only on q^2 and is thus the same for either q or $-q$.

In analogy to Eqs. (16–16a) and (16–16b), we then obtain the dynamic polarizability at the frequency ω

$$\alpha = \frac{q^2}{m(\omega_0{}^2 - \omega^2)}$$

(16–19)

which we have plotted in Fig. 16–6. Note that ω is plotted on a logarithmic scale; we are not interested in what happens if $\omega = 0.95\,\omega_0$, but rather in the difference between $\omega = 10^{-1}\,\omega_0$ and $\omega = 10\omega_0$. We see that for $\omega \lesssim 1/4\omega_0$ the polarizability has for all practical purposes its static value. For $\omega > \omega_0$ the polarizability reverses sign and very rapidly becomes very small. Thus, for $\omega \geq 10\,\omega_0$ the polarizability becomes negligibly small (less than 1% of its static value).

Of course, an atom is not an electron on a spring; also, we have completely neglected any quantum mechanical effects. But we want only a qualitative model which tells us how the polarizability changes at large frequencies of the electric field. Our model is qualitatively correct. The electrons in atoms are bound, although not by springs. Any ion in a crystalline ionic lattice (e.g., Na^+Cl^-) will also be subject to a restoring force if it moves from its equilibrium position (Fig. 16–7). The spring constant of our spring is just a measure of how large these forces are.

We can even estimate the resonant frequency ω_0 above which the polarizability vanishes. From Eqs. (16–17) and (16–18a) we obtain

$$\omega_0 = \sqrt{\frac{k}{m}} \approx \frac{q}{\sqrt{\alpha m}}$$

(16–20)

For electrons we have to use the electron mass $m_e = 0.9 \times 10^{-30}$ kg. Since $\alpha \approx 10^{-40}$ coul-m²/V, we have

$$\omega_0 \approx \frac{1.6 \times 10^{-19}\ \text{coul}}{\sqrt{10^{-40}\ \text{coul-m}^2/\text{V} \times (0.9 \times 10^{-30}\ \text{kg})}} \approx 1.7 \times 10^{16}\ \text{radians/sec}$$

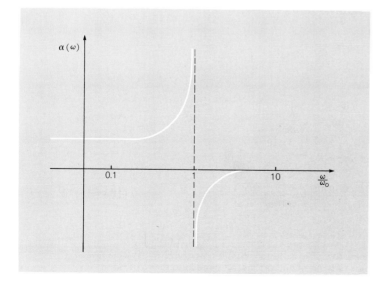

Figure 16–6 Polarizability as a function of electric field angular frequency for our model. Note the logarithmic scale for ω.

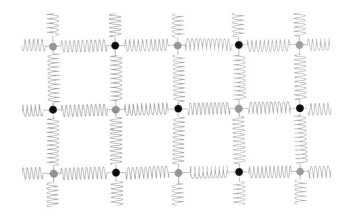

Figure 16-7 An ionic lattice resembles our crude model; one can think of the positive (color) and negative (black) ions as being connected by springs.

which corresponds to a frequency of

$$\nu_0 = \frac{\omega_0}{2\pi} = 2.7 \times 10^{15} \text{ Hz}$$

and a wavelength in vacuum of

$$\lambda = \frac{c}{\nu} \approx 10^{-7} \ m = 1000 \text{ Å}$$

This lies in the ultraviolet; we conclude that for visible light ($\lambda \approx 3500$ to 7000 Å) and in the infrared, the electronic polarizability of transparent materials will have effectively its static value.

On the other hand, the ionic polarizability is due to the much heavier ions. Their mass will be A times the proton mass $m_p = 1.67 \times 10^{-27}$ kg; taking[*] $A = 30$ we obtain a mass

$$m_A \approx 5 \times 10^{-26} \text{ kg}$$

and an angular frequency

$$\omega_0 \approx \frac{1.6 \times 10^{-19} \text{ coul}}{\sqrt{10^{-40} \text{ coul-m}^2/\text{V} \times (5 \times 10^{-26} \text{ kg})}} \approx 7 \times 10^{13} \text{ radians/sec}$$

which corresponds to a wavelength

$$\lambda = \frac{2\pi c}{\omega_0} = 2.6 \times 10^{-5} \ m = 26 \ \mu$$

and lies in the infrared. Thus, for visible light ($\lambda \approx 0.5 \ \mu$) the ionic polarizability will already have decreased to a practically negligible contribution.

The orientational polarizability, which is due to the permanent dipole moments of the molecules, drops off at even lower frequencies in the microwave range ($\omega_0 \approx 10^{10}$ to 10^{11} radians/sec). The large moment of inertia of a molecule will prevent it from adjusting its direction to a rapidly changing electric field.

---

[*] This value of A would correspond to an "average" value for rocksalt, NaCl, since sodium has $A \approx 23$ and chlorine has $A \approx 35$.

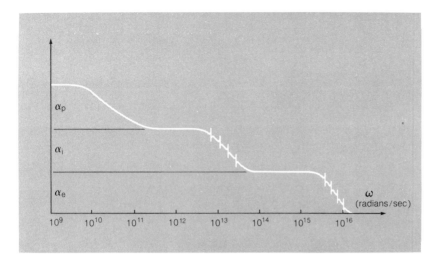

Figure 16–8   The polarizability of a typical substance against angular frequency. Not all substances will have all three types of polarizability present. α_e = electronic; α_i = ionic; α_p = orientational (permanent dipole) polarizability.

If we plot the refractive index of a medium against the logarithm of the frequency $\nu = \omega/2\pi$, we obtain in general a graph such as that in Fig. 16–8. There are many sharp spikes in the region of rapid falloff of n, because there are generally many resonant frequencies in a narrow region. Of course, one or another of the contributions to the refractive index can be absent for a substance; many molecules have no permanent electric dipole moments, and in even fewer are there separate positive and negative ions which can produce an ionic polarizability.

Even over the range of optical frequencies, the refractive index of most transparent substances is not constant. The refractive index generally increases with frequency (or decreases with wavelength; see Fig. 16–9) because the electronic polarizability has its resonances in the range of $\nu \approx 3 \times 10^{15}$ to 3×10^{16} Hz (or $\lambda = 100$ to 1000 Å), we are in the increasing part of the "resonance curve" shown in Fig. 16–6.

Another result of our discussion is that at wavelengths $\lambda \lesssim 100$ Å (or frequencies $\nu \gtrsim 3 \times 10^{16}$ Hz) the refractive index of all materials is equal to unity. Electromagnetic radiation of wavelengths down to 0.1 Å is produced by X-ray

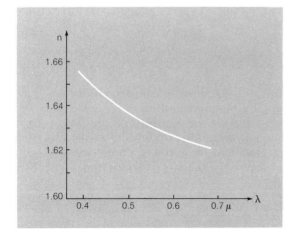

Figure 16–9   The refractive index of flint glass in the visible region.

machines; but since this radiation has the same speed in all materials, it is impossible to make lenses which would focus such radiation. Even mirrors of good reflectivity are impossible, because reflection also is caused by a sudden change (over less than a wavelength) of material properties. While such radiation does interact with individual atoms, and one can produce "mirrors" of very poor reflectivity using Bragg scattering (p. 94), conventional optical systems are useless for X-rays.

The refractive index of water for visible light is $n = 1.3$. What is the electronic polarizability of the water molecule?

From the Clausius-Mosotti equation we have

$$\frac{n^2 - 1}{n^2 + 2} = \frac{N}{3\epsilon_0}\alpha_e$$

Solving for α_e, we obtain

$$\alpha_e = \frac{3\epsilon_0}{N}\frac{n^2 - 1}{n^2 + 2}$$

The only unknown quantity to the right is the number N of molecules per cubic meter. However, we have

$$N = \frac{N_0\rho}{A}$$

where $N_0 = 6.02 \times 10^{26}$ atoms/kg-mole is Avogadro's number and $A = 18$ kg/kg-mole is the molecular weight of water. The density ρ of water is 10^3 kg/m³; thus we have

$$N = \frac{6.02 \times 10^{26} \text{ atoms/kg-mole} \times 10^3 \text{ kg/m}^3}{18 \text{ kg/kg-mole}} \approx \frac{1}{3} \times 10^{29} \text{ atoms/m}^3$$

and from this we obtain for the electronic polarizability

$$\alpha_e = \frac{3 \times 8.85 \times 10^{-12} \text{ A-sec/V-m}}{\frac{1}{3} \times 10^{29} \text{ m}^{-3}} \times \frac{0.69}{3.69} = 1.5 \times 10^{-40} \frac{\text{coul-m}^2}{\text{V}}$$

Note that the static dielectric constant $K = 81$ is much larger than $n^2 = 1.69$ at optical frequencies. The static dielectric constant is dominated by the large permanent dipole moment of the water molecule; but this has no effect on the optical refractive index.

3. POLARIZATION OF E.M. WAVES

Electromagnetic waves are transverse waves; by this we mean that both the electric field and the magnetic field are perpendicular to the direction of propagation, i.e., the wave vector **k**. Thus, even if the direction of propagation is uniquely defined, the electric field vector **E** can still point in an arbitrary direction perpendicular to the direction of propagation. We designate as the *polarization* **P** of the E.M. wave° a vector in the plane perpendicular to the wave vector

---

° The polarization of an E.M. wave has nothing to do with the polarization of a medium (i.e., the dipole moment/unit volume), in spite of the identical name and the same symbol **P** used. The name *polarization* in this context is used for purely historical reasons. We could just as well talk directly about the **E** vector, when discussing only electromagnetic waves. We use the symbol **P** to indicate that our discussion applies to any type of transverse waves.

k which is everywhere proportional to the electric field vector **E**. We call a *linearly polarized wave* a wave for which the polarization vector **P** always points along the same line as it oscillates. However, the phenomenon of polarization is not limited to electromagnetic waves. In Fig. 16–10a we show a linearly polarized electromagnetic wave, represented by its **E** vector. In Fig. 16–10b one can see a linearly polarized transverse wave on a stretched string. For any such linearly polarized wave one has

$$\mathbf{P} = \mathbf{P}_0 \cos (\omega t - kz)$$

where $\mathbf{P}_0$ is a constant vector perpendicular to the direction of propagation (in our case the z-direction). At a fixed point $z = z_0$, one has

$$\mathbf{P} = \mathbf{P}_0 \cos (\omega t - kz_0) = \mathbf{P}_0 \cos (\omega t - \phi_0) \qquad (16\text{–}21)$$

The superposition principle, which is valid for any wave, allows us to add plane polarized waves of different polarization directions. Specifically for electromagnetic waves, we can then discuss more complex types of waves than those discussed in Section 1. As a first example, consider two linearly polarized waves going in the z-direction (Fig. 16–11); at a certain point one has the polarization vector

$$\mathbf{P}_1 = A \cos \omega t \ \hat{\mathbf{i}} \qquad (16\text{–}22a)$$

and the second has

$$\mathbf{P}_2 = B \cos \omega t \ \hat{\mathbf{j}} \qquad (16\text{–}22b)$$

The superposition of the two waves will have the polarization vector

$$\mathbf{P} = \mathbf{P}_1 + \mathbf{P}_2 = (A\hat{\mathbf{i}} + B\hat{\mathbf{j}}) \cos \omega t \qquad (16\text{–}23a)$$

Thus, we have again a linearly polarized wave whose polarization angle with the

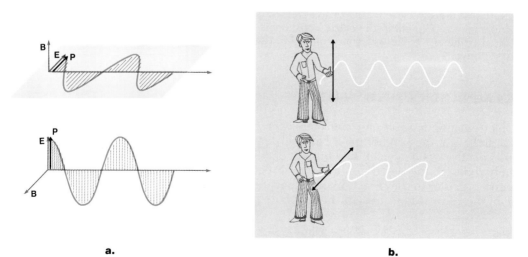

a.   **b.**

Figure 16–10 (a) Two possible linearly polarized waves; the polarization is per definition parallel to **E**. (b) Transverse waves on a string also show the phenomenon of polarization.

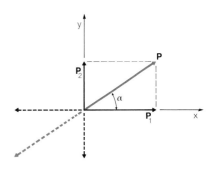

Figure 16-11 Adding linearly polarized waves; if both waves are in phase, the result is again a linearly polarized wave.

x-axis is

$$\alpha = \tan^{-1}(B/A) \qquad (16\text{-}23b)$$

However, the two polarization vectors do not have to be in phase, as they were in the previous example; we can have instead of Eq. (16-22b)

$$\mathbf{P}_3 = B \, \cos\left(\omega t - \frac{\pi}{2}\right)\hat{\jmath} = B \, \sin\omega t \, \hat{\jmath} \qquad (16\text{-}22c)$$

The superposition of the two waves represented by (16-22a) and (16-22c) has a polarization vector (Fig. 16-12)

$$\mathbf{P}' = \mathbf{P}_1 + \mathbf{P}_3 = A \cos\omega t \, \hat{\imath} + B \sin\omega t \, \hat{\jmath} \qquad (16\text{-}24)$$

The length of the vector now never vanishes:

$$|\mathbf{P}'| = \sqrt{A^2 \cos^2 \omega t + B^2 \sin^2 \omega t} > 0 \text{ for all } t$$

The polarization vector, instead of oscillating along a line, now moves around the circumference of an ellipse of axes A and B, as we can easily see if we note that the two components P'_x and P'_y of the polarization are

$$P'_x = A \cos\omega t$$
$$P'_y = B \sin\omega t$$

Dividing the first equation by A and the second by B, and then squaring and adding the two equations, we obtain

$$\frac{P'^2_x}{A^2} + \frac{P'^2_y}{B^2} = \cos^2 \omega t + \sin^2 \omega t = 1 \qquad (16\text{-}25)$$

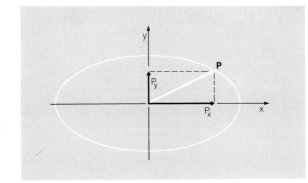

Figure 16-12 Adding two linearly polarized waves which are 90° out of phase results in an elliptically polarized wave.

which indeed is the equation of an ellipse with principal axes A and B. Such a wave is called *elliptically polarized*. If $A = B$, the ellipse becomes a circle and we say that the wave is *circularly polarized*.

Show that the superposition of two linearly polarized waves having polarization directions perpendicular to each other will always lead to an elliptically polarized wave, with circular and linear polarization being limiting cases.

We will assume that the propagation direction is the z-axis. The most general linearly polarized wave in the x-direction then has the form

$$\mathbf{P}_1 = A \cos (\omega t - \phi_1) \, \hat{\mathbf{i}} = P_x \, \hat{\mathbf{i}} \tag{16–26a}$$

with an arbitrary ϕ_1; and the most general form for a wave linearly polarized in the y-direction has the form

$$\mathbf{P}_2 = B \cos (\omega t - \phi_2) \, \hat{\mathbf{j}} = P_y \, \hat{\mathbf{j}} \tag{16–26b}$$

With the help of the addition theorems for trigonometric functions, we can calculate the components of polarization for the resulting wave:

$$P_x = A \cos (\omega t - \phi_1) = A \cos \omega t \cos \phi_1 + A \sin \omega t \sin \phi_1 \tag{16–27a}$$

$$P_y = B \cos (\omega t - \phi_2) = B \cos \omega t \cos \phi_2 + B \sin \omega t \sin \phi_2 \tag{16–27b}$$

Multiplying Eq. (16–27a) by $B \sin \phi_2$ and multiplying Eq. (16–27b) by $A \sin \phi_1$, and then subtracting the two, we obtain

$$P_x B \sin \phi_2 - P_y A \sin \phi_1 = AB \cos \omega t \, (\cos \phi_1 \sin \phi_2 - \sin \phi_1 \cos \phi_2)$$
$$= AB \sin (\phi_2 - \phi_1) \cos \omega t$$

from which we can obtain

$$\cos \omega t = \frac{P_x \sin \phi_2}{A \sin (\phi_2 - \phi_1)} - \frac{P_y \sin \phi_1}{B \sin (\phi_2 - \phi_1)} \tag{16–28}$$

In a similar way, we can also solve the two Eqs. (16–27) for $\sin \omega t$ and obtain

$$\sin \omega t = \frac{P_x \cos \phi_2}{A \sin (\phi_2 - \phi_2)} - \frac{P_y \cos \phi_1}{B \sin (\phi_1 - \phi_2)} \tag{16–29}$$

Squaring and adding Eq. (16–28) and (16–29), we obtain after a little algebra

$$\sin^2 (\phi_1 - \phi_2) = \frac{P_x^{\,2}}{A^2} + \frac{P_y^{\,2}}{B^2} - \frac{2 \, P_x P_y}{AB} \cos (\phi_1 - \phi_2) \tag{16–30}$$

which is indeed the equation of an ellipse. In the special case when $A = B$ and $\cos (\phi_1 - \phi_2) = 0$ (or $\phi_1 - \phi_2 = \pm 90°$), we then have the equation of a circle:

$$P_x^{\,2} + P_y^{\,2} = A^2$$

On the other hand if $\phi_1 = \phi_2$ then $\sin (\phi_1 - \phi_2) = 0$ and $\cos (\phi_1 - \phi_2) = 1$; we then obtain

$$\left(\frac{P_x}{A} - \frac{P_y}{B} \right)^2 = 0$$

which is the equation of a straight line. It is left to the reader to calculate the major and minor axes of the ellipse from Eq. (16–30) in the general case.

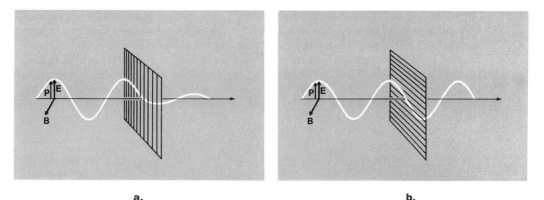

a.   b.

Figure 16–13 A polarizer consists of many parallel conducting "wires." If **E** is parallel to the wires, the wave is strongly attenuated; if **E** is perpendicular, no attenuation takes place.

We still have not discussed how to detect the polarization of E.M. waves— or how to produce linearly polarized waves. The methods used actually have to depend on the wavelength of the E.M. wave; no single piece of equipment can be used to span the entire range.

For the microwaves (wavelength $\lambda = 0.5$ to 10 cm) and for longer radio waves, a screen made of parallel conducting wires is an effective polarizer. If the electric field of the incoming wave is parallel to the wires (Fig. 16–13a), it will start currents flowing in them; this will use up the energy of the wave, and the wave thus will be strongly attenuated. On the other hand, if the electric field— or the polarization vector—is perpendicular to the wires (Fig. 16–13b), no currents can develop and the electromagnetic wave will pass through the screen with no attenuation. If the incoming wave is elliptically polarized—or linearly polarized in an arbitrary direction—only the component of polarization perpendicular to the wires will pass through unattenuated. A wire screen can thus be used either as a polarizer to produce linearly polarized waves, or as an analyzer to measure the direction of polarization of a linearly polarized wave. Indeed, if we can rotate the screen so that the wave is completely attenuated, we know that the incident wave is linearly polarized and that the direction of the polarization is parallel to the screen wires.

For such a screen to be fully effective, it is necessary that the wire spacings (and the wire thickness) be small compared to the wavelength of the wave; if the spacing is too large, the screen acts as a diffraction grating. The wire length, on the other hand, should be long enough that the current in the wires does not

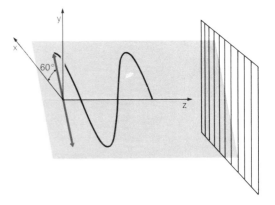

Figure 16–14 Example 4: The incident wave.

produce a significant charge at the ends of the wires over one period of oscillation of the E.M. wave. These conditions obviously prohibit the use of screens for light ($\lambda \approx 0.5\ \mu$); it is simply not possible to produce such fine wires. Luckily, however, there exist molecules which are very long compared to their thickness; they can act like a screen of wires if a plastic film is made in which all the molecules are parallel to each other. This is the principle of *polaroid filters*, which can be made very cheaply – they can be found in sunglasses of the drugstore variety.

Example 4   A polaroid screen (Fig. 16–14) is hit by linearly polarized light whose electric field vector **E** forms an angle of $\theta = 60°$ with the direction of maximum transmission. Calculate the fractional intensity loss in the filter. Assume that there is no attenuation if $\theta = 0°$. (This is not really true; a typical polaroid filter has some loss even if the polarization is perpendicular to its "wire molecules".)

We have to remember that according to Eq. (16–13b) the intensity is proportional to the mean square of the electric field **E**. We can decompose the incident wave into two linearly polarized waves, one parallel and one perpendicular to the direction of maximum transmission (Fig. 16–15):

$$P_x = |\mathbf{P}|\cos(60°) = \frac{1}{2}|\mathbf{P}|$$

$$P_y = |\mathbf{P}|\sin(60°) = \frac{\sqrt{3}}{2}|\mathbf{P}|$$

Only the parallel component P_x will be transmitted. Since **P** is proportional to **E**, we can write instead of Eq. (16–13b)

$$I = \text{constant} \times <\mathbf{P}^2>$$

Then we have for the transmitted intensity I_{out}

$$\frac{I_{\text{out}}}{I_{\text{in}}} = \frac{<P_x{}^2>}{<P_x{}^2> + <P_y{}^2>} = \cos^2(60°) = \frac{1}{4} = 25\%$$

and for the absorbed fraction of the incident intensity

$$\text{fractional loss} = \frac{I_{\text{in}} - I_{\text{out}}}{I_{\text{in}}} = \frac{<P_y{}^2>}{<P_x{}^2> + <P_y{}^2>} = \sin^2(60°) = \frac{3}{4} = 75\%$$

Only 25% of the incident intensity is transmitted; 75% is absorbed in the filter.

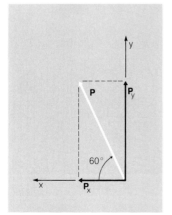

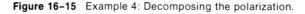

Figure 16–15 Example 4: Decomposing the polarization.

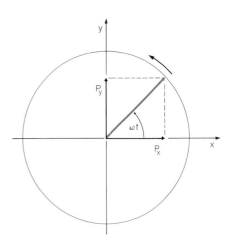

Figure 16–16 Example 5: Decomposing a circularly polarized wave into two linearly polarized waves.

Example 5   A circularly polarized light wave passes through a polaroid filter. What fraction of the incident intensity is transmitted?

Again we can decompose the circular polarized wave into two linearly polarized waves (Fig. 16–16):

$$P_x = |\mathbf{P}| \cos \omega t$$
$$P_y = |\mathbf{P}| \sin \omega t$$

Since we can always consider the polarization vector as another name for the electric field vector, we know that the incident light intensity is

$$I_{\text{in}} = \text{constant} \times <\mathbf{P}^2> = \text{constant} \times \mathbf{P}^2$$

while the outgoing intensity is equal to

$$I_{\text{out}} = \text{constant} \times <P_x{}^2> = \text{constant} \times <\mathbf{P}^2 \cos^2 \omega t>$$

But because for light one has $\omega \approx 10^{16}$ radians/sec, and no measuring equipment is able to measure such rapid intensity variations, we have to average over many periods. Since $<\cos^2 \omega t> = 1/2$, we have

$$I_{\text{out}} = \text{constant} \times \mathbf{P}^2 \times \frac{1}{2} = \frac{1}{2} I_{\text{in}}$$

so that 50% of the incident wave is transmitted.

If we were to study the light emitted by a light bulb, we would be unable to find any polarization, whether linear, circular, or elliptical. Nevertheless, visible light is an electromagnetic wave; thus, we have to explain why many sources emit *unpolarized waves*. An electromagnetic wave of infinitely sharp frequency always has a well-defined polarization. But there is no way to obtain such a perfectly defined frequency. The ultimate limit is given by the laws of quantum mechanics; light is emitted in wave packets, which are the individual photons. There is a range of wavelengths and thus also of frequencies involved in any wave packet. But any wave with a finite frequency spread will have a well-defined polarization only if *all frequency components* have the same polarization. To see this more clearly, consider the superposition of two waves with slightly different angular frequencies ω and $(\omega + \Delta\omega)$. For simplicity, we will assume them both to be linearly polarized, with the polarization planes

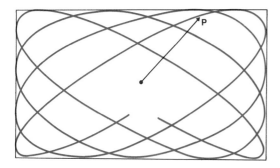

Figure 16-17 Superposition of two light waves of slightly different frequencies which are linearly polarized in perpendicular directions. The tip of the polarization vector moves along a curve which completely fills the rectangle.

perpendicular to each other. The total wave will have a polarization vector

$$\mathbf{P} = P_x \,\hat{\mathbf{i}} + P_y \,\hat{\mathbf{j}}$$

where

$$P_x = A \cos \omega t$$

$$P_y = B \cos \left[(\omega + \Delta\omega)t\right] = B \cos (\omega t + \Delta\omega t) = B \cos (\omega t + \phi(t))$$

If we neglect the relatively slow time dependence of $\phi(t)$, we have exactly the situation discussed in Example 3. Thus, if $\Delta\omega << \omega$, at any instant of time (for a time interval of a few periods $T_1 = 2\pi/\omega$) the resulting wave will have a well-defined polarization corresponding to the phase difference $\phi(t)$. But the polarization will change as $\phi(t) = \Delta\omega t$ changes; thus, over a time period of the *coherence time* $T_2 = 2\pi/\Delta\omega$, the polarization will go through all possible types, as sketched in Fig. 16–17. The net result is an *unpolarized wave*.

Example 6.   For visible light of wavelength $\lambda = 6000$ Å (yellow light), how long must the averaging time be in order to obtain unpolarized light, assuming $\Delta\omega/\omega = 10^{-5}$?

To the wavelength λ corresponds the frequency

$$\nu = c/\lambda = 0.5 \times 10^{15} \text{ Hz}$$

and the angular frequency

$$\omega = 2\pi\nu = \frac{2\pi c}{\lambda} = 3.14 \times 10^{15} \text{ radians/sec}$$

The frequency uncertainty then is

$$\Delta\omega = 10^{-5} \, \omega = 3.14 \times 10^{10} \text{ radians/sec}$$

and the coherence time is

$$T = 2\pi/\Delta\omega = 2 \times 10^{-10} \text{ sec}$$

To see things happening this fast is just barely possible with the fastest electronic methods available today. However, the human eye or any conventional detection equipment will always average over much longer periods of time.

4. THE PHOTON

As we had discussed in Chapter 14, electrons and other particles behave like waves. Their "wave equation," the Schrödinger equation, explains and describes their behavior. This particle-wave duality is not a feature only of massive particles such as electrons; the concept of the electromagnetic field is also only one side of the dual picture of electromagnetism. The other side is the concept of the photon. Experimentally, we know that electromagnetic energy of angular frequency ω (or frequency $\nu = \omega/2\pi$, or wavelength $\lambda = 2\pi c/\omega$) is always emitted in integer multiples of the energy quantum

$$E = \hbar\omega = h\nu = hc/\lambda \qquad (16\text{-}31)$$

We want to discuss here some of the properties of E.M. radiation which are due to the photon nature. They also contribute the most convincing evidence that electromagnetic radiation actually does consist of particles – the photons.

We have already briefly mentioned (p. 444) the photoelectric effect; now, with what we have learned about electrons in solids, we can give a more quantitative discussion.

If we shine visible or near ultraviolet light at the surface of a metal in vacuum (Fig. 16–18a), we can detect electrons leaving the surface. We can understand this if we remember that the "free" electrons in a metal are in a "box" of potential depth V_0 (Fig. 16–18b). In Chapter 14 we made the approximation $V_0 = \infty$; i.e., we assumed a completely impervious box. Now we have to take into account the fact that V_0 is finite. At zero absolute temperature, the most energetic electrons have a kinetic energy $T = E_F$, equal to the Fermi energy. If a photon of energy $E_\gamma = h\nu$ is absorbed by a single electron at the Fermi energy, its kinetic energy inside the metal will then be

$$T' = E_F + E_\gamma = E_F + h\nu \qquad (16\text{-}32)$$

However, if the electron leaves the metal, it has to overcome the potential energy step V_0. Thus, outside the metal the maximum possible kinetic energy of an electron will be

$$T_{\max} = E_F + E_\gamma - V_0 = h\nu - (V_0 - E_F) = h\nu - W \qquad (16\text{-}33)$$

where W is called the *work function* of the metal. In the metal, there are also electrons of kinetic energy less than the Fermi energy E_F. If one of these electrons of kinetic energy E_i absorbs a photon of kinetic energy $h\nu$ and leaves the

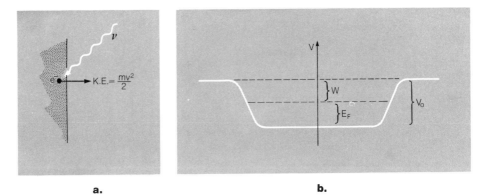

a.   **b.**

Figure 16–18 Photoeffect: (a) the physical process; (b) the energy differences which determine the kinetic energy of the ejected electron.

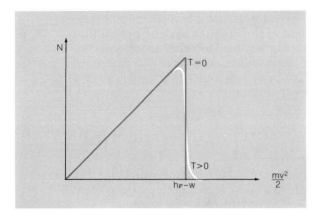

Figure 16-19 The fraction N of electrons leaving with a certain kinetic energy at zero temperature and room temperature.

metal, its kinetic energy outside the metal will be

$$E_k = E_i + h\nu - V_0 = h\nu - W - (E_F - E_i) \tag{16-34}$$

At room temperature, or at any nonzero temperature, some of the electrons in the metal will have a kinetic energy slightly larger than the Fermi energy

$$E \approx E_F + kT > E_F$$

and thus after absorption of a photon of energy $h\nu$ there will be some electrons outside the metal of slightly higher kinetic energy than that given in Eq. (16-33). However, since $kT \approx 1/40$ eV at room temperature is typically much smaller than E_F or than W, this is really quite a minor effect. The overall energy distribution of outgoing electrons, if light of a sharp frequency ν impinges on the surface, is sketched in Fig. 16-19.

A photon can be also absorbed by electrons in insulators or semiconductors. If the photon energy $h\nu$ is sufficient, the electron can possibly leave the material just as in a metal. However, if an electron in the full valence band absorbs a photon of energy $h\nu > E_g$ (= gap width), the electron will end up as a free electron in the conduction band (Fig. 16-20). Thus a semiconductor, when light of sufficiently high frequency ν —or sufficiently short wavelength λ — shines on it, will become a much better conductor; one calls this effect *photoconductivity*.

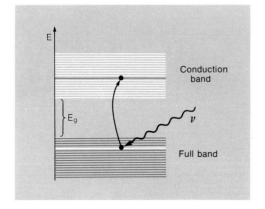

Figure 16-20 Photoconductivity in semiconductors: A photon can raise an electron from the full band into the conduction band.

Example 7   The work function of cesium is $W = 1.8$ eV. What will be the maximum kinetic energy of electrons ejected by the photoelectric effect by light of wavelength $\lambda = 4000$ Å? What is the longest wavelength which still enables photoelectrons to leave the metal?

We first have to calculate the photon energy in eV. We have

$$E_\gamma = h\nu = \frac{hc}{\lambda}$$

or

$$E_\gamma = \frac{6.63 \times 10^{-34} \text{ J-sec} \times 3 \times 10^8 \text{ m/sec}}{4 \times 10^{-7} \text{ m}} = 4.97 \times 10^{-19} \text{ J} = 3.1 \text{ eV}$$

Thus, the maximum kinetic energy of the electrons will be

$$E_{max} = 3.1 \text{ eV} - 1.8 \text{ eV} = 1.3 \text{ eV}$$

The longest wavelength which will still produce photoelectrons, according to Eq. (16–23), is given by

$$W = h\nu_0 = \frac{hc}{\lambda_0}$$

However, since we want λ_0 in meters and h is given in Joule-sec, we have to first convert W into Joules:

$$W = 1.8 \text{ eV} = 1.8 \times 1.60 \times 10^{-19} \text{ J} = 2.88 \times 10^{-19} \text{ J}$$

We then have

$$\lambda_0 = \frac{hc}{W} = \frac{(6.63 \times 10^{-34} \text{ J-sec}) \times (3 \times 10^8 \text{ m/sec})}{2.88 \times 10^{-19} \text{ Joules}} = 6.91 \times 10^{-7} \text{ m}$$

or about 6900 Å. Let us also note that at room temperature

$$kT_{room} = \frac{1}{40} \text{ eV}$$

is small compared to any of the energies considered. Our results are independent of temperature to within a few per cent.

The photoelectric effect occurs not only in crystalline lattices, but also in isolated atoms — or in solids for electrons which are tightly bound to the atom. Each of these electrons has a binding energy E_0. The atom can absorb a photon of energy $h\nu > E_0$; the electron will then be ejected with the kinetic energy

$$E_k = \frac{mv^2}{2} = h\nu - E_0 \tag{16–35}$$

Since the typical binding energies of the inner (K-shell) electrons in heavier atoms are 100 eV or larger, this effect will be most noticeable for photons in the far ultraviolet or for X-rays ($0.1 \text{ Å} \lesssim \lambda \lesssim 100 \text{ Å}$). If such photons traverse a layer of material of very small thickness dx, some of the photons will interact by the photoelectric effect. Let us call the number of interacting photons δ; we want to calculate δ in terms of the properties of the material layer. It will be proportional to the number of incident photons; if twice as many photons hit the layer, twice as many will interact by the photoelectric effect within it. But δ will also be proportional to the density ρ of the material — the more atoms are encountered by the incident photon, the more likely it is to interact. Finally, δ will also be proportional to the thickness dx of the material, because the thicker the material, the more atoms are in the path of each incident photon (Fig. 16–21). In consequence, one can write

$$\boxed{\delta = \mu N \rho \, dx} \qquad (16\text{--}36)$$

where the absorption coefficient μ depends on the atomic composition of the material and on the incident photon energy, but is independent of the macroscopic properties of the material, such as density or thickness. The units in which μ is measured can be read from from Eq. (16–36):

$$[\mu] = \frac{[\Delta]}{[N][\rho][dx]} = \frac{1}{1\frac{kg}{m^3}m} = m^2/kg$$

However, it is usually measured in units of

$$1 \text{ cm}^2/g = 0.1 \text{ m}^2/kg$$

It is an interesting fact that the absorption coefficient is greatest when the photon energy is very little above the binding energy E_0 of the ejected electron. Since there are many electrons in an atom, each with its own binding energy, the absorption coefficient of an element plotted against photon energy (Fig. 16–22) shows a sequence of sharp edges (steps), each corresponding to the binding energy of one of the atomic electrons. It is usual to label each of the edges by the shell from which the electron has been ejected; we thus talk about the K-edge, the L-edge, and so forth. Since the electrons in a single shell occupy levels of slightly different energies, many of the "edges" are actually groups of edges lying close together. An example of this can be seen in the "L-edge" for tin in Fig. 16–22.

Eq. (16–36) is valid only in the limit of a very (infinitely) thin layer of material. Obviously, even if the layer is arbitrarily thick, the number of interacting photons δ can never be greater than the number of incident photons N. However, we can take this into account if we note that $\delta = -dN$, the number of photons lost from the incident photon beam. We thus have

$$dN = -N\mu\rho \, dx$$

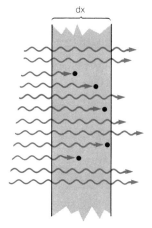

Figure 16–21 Attenuation of photons by atomic photoeffect. The number of interactions is proportional to the number of incident photons, the material thickness dx, and the density of the material.

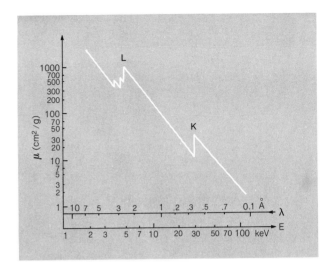

Figure 16-22 Atomic photoeffect absorption coefficient of tin for a range of photon energies. Note the sharp edges when electrons from an inner shell can be ejected.

The number N of incident photons has decreased by an amount $-\delta = dN$; thus, only $(N - dN)$ photons are impinging on the material beyond. We can rewrite the above equation:

$$\frac{dN}{N} = -\mu\rho\, dx$$

If the total thickness of material is x, we can obtain the number $N(x)$ of photons surviving the passage through the material by integration:

$$\int \frac{dN}{N} = \ln\left(N(x)\right) - \ln\left(N(0)\right) = -\mu\rho x$$

and if we call $N_0 = N(0)$ the number of incident photons, we have (Fig. 16-23)

$$\boxed{N(x) = N_0 e^{-\mu\rho x}} \qquad\qquad (16\text{-}37a)$$

Thus, of N_0 photons incident on material of thickness x, only $N(x)$ survive without interacting; the number of interactions by the photoelectric effect is

$$\Delta N = N_0 - N = N_0(1 - e^{-\mu\rho x}) \qquad (16\text{-}37b)$$

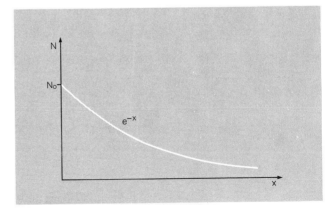

Figure 16-23 The number of photons passing through a finite thickness x of material obeys an exponential law.

Example 8 The binding energy of a K-shell electron in lead is $E_0 = 89$ keV. What is the wavelength of a photon which can just produce a photoelectron from a K-shell electron in lead?

The energy of the photon has to be at least equal to the electron binding energy. Thus, we have

$$h\nu = \frac{hc}{\lambda} = E_0$$

E_0 is given in electron volts; however, 1 eV $= 1.6 \times 10^{-19}$ J, so

$$E_0 = (8.9 \times 10^4) \times (1.6 \times 10^{-19} \text{ Joule}) = 1.42 \times 10^{-14} \text{ Joule}$$

The wavelength is then

$$\lambda = \frac{hc}{E_0} = \frac{(6.63 \times 10^{-34} \text{ Joule-sec}) \times (3 \times 10^8 \text{ m/sec})}{1.42 \times 10^{-14} \text{ Joule}}$$
$$= 1.40 \times 10^{-11} \text{ m} = 0.14 \text{ Å}$$

Note that this wavelength is about 30,000 times shorter than the wavelength of visible light!

Example 9 The absorption coefficient for X-rays in copper is $\mu_1 = 300$ cm²/g for wavelengths just shorter than its K-edge (1.377 Å), while it is only $\mu_2 = 38.5$ cm²/g for a slightly longer wavelength ($\lambda = 1.390$ Å). An incident X-ray beam has two components of equal intensity at the two mentioned wavelengths. If the beam traverses a foil of copper 1 mil $= 0.001$ inch thick, what will be the ratio of the two intensities afterwards?

From Eq. (16–37) we can calculate the ratio of intensities before and after the copper sheet for each wavelength component:

$$\frac{N}{N_0} = e^{-\mu\rho x}$$

We need to know the density of copper, which we can find in Appendix B:

$$\rho = 8.92 \times 10^3 \text{ kg/m}^3 = 8.92 \text{ g/cm}^3$$

The thickness of the foil is 1 mil $= 10^{-3}$ inch $= 2.54 \times 10^{-3}$ cm. Since the absorption coefficient is given in cm²/g (i.e., in CGS units), we might as well use CGS units throughout. We then have

$$\mu_1\rho x = \left(\frac{300 \text{ cm}^2}{g}\right) \times \left(8.92 \frac{g}{cm^3}\right) \times (2.54 \times 10^{-3} \text{ cm}) = 6.80$$

and

$$\mu_2\rho x = 38.5 \times 8.92 \times (2.54 \times 10^{-3}) = 0.872$$

The longer wavelength X-rays will be attenuated by a factor

$$\frac{I_2}{I_0} = e^{-\mu_2\rho x} = e^{-0.872} = 0.42$$

while the shorter wavelength will be attenuated by a factor

$$\frac{I_1}{I_0} = e^{-\mu_1\rho x} = e^{-6.80} \approx 1.1 \times 10^{-3}$$

Thus, after passage through the (very thin!) copper sheet, we have a beam of X-rays of wavelength 1.39 Å, with only a small contamination by X-rays of slightly different wavelength 1.38 Å. The contamination is

$$\frac{I_1}{I_2} = \frac{e^{-\mu_1 \rho x}}{e^{-\mu_2 \rho x}} = \frac{1.1 \times 10^{-3}}{0.42} = 0.27\%$$

5. THE COMPTON EFFECT

The photon has not only a well-defined energy $E = h\nu$, but also a momentum

$$p = \frac{h}{\lambda} = \frac{h\nu}{c} \tag{16-38}$$

We recognize in Eq. (16–38) the de Broglie relation [Eq. (14–3)]; the relation between wavelength and momentum is closely related to the uncertainty principle [Eq. (14–1)] and thus has to be true for photons as well as for massive particles such as electrons. Eq. (16–38) has to be correct also if the photon is to obey relativistic kinematics. According to the special theory of relativity, a particle moving with velocity **v** has[*]

$$\text{momentum} \quad \mathbf{p} = \frac{m_0 \mathbf{v}}{\sqrt{1 - (v^2/c^2)}} \tag{16-39a}$$

and

$$\text{energy} \quad E = \frac{m_0 c^2}{\sqrt{1 - (v^2/c^2)}} \tag{16-39b}$$

where m_0 is the rest mass and the energy in Eq. (16–39b) is not just the kinetic energy, but the sum of the kinetic energy and the rest mass energy

$$E_0 = m_0 c^2 \tag{16-39c}$$

which the particle can free only by annihilating (disappearing). We cannot directly apply these equations to the photon, because it is moving with the speed of light c. Thus, the square root in the denominator in Eqs. (16–39a) and (16–39b) vanishes. In order for E and $\mathbf{p}$ to be finite, it is necessary that $m_0 = 0$. Thus, the rest mass of the photon vanishes and the two equations have the form 0/0. However, we can divide Eq. (16–39a) by Eq. (16–39b) to obtain

$$\frac{\mathbf{p}}{E} = \frac{\mathbf{v}}{c^2} \tag{16-40}$$

and since the rest mass no longer appears explicitly, we can go to the limit $m_0 \to 0$ and $v \to c$ to obtain

$$\left(\frac{p}{E}\right)_{\text{photon}} = \frac{c}{c^2} = \frac{1}{c}$$

Since the energy of the photon is $E = h\nu$, we obtain for the photon

$$p = \frac{h\nu}{c} = \frac{h}{\lambda}$$

which is exactly Eq. (16–38).

---

[*] Compare Volume I, Chapter 9 and Supplement 6.

The vanishing rest mass of the photon is really quite "obvious." If we have a beam of electrons, we can slow them down by intercepting the beam with a block of some material. When the electron has stopped, we still have electrons at rest. The same thing cannot be done with photons. When we stop an X-ray beam (or a light beam) with some material, the material will heat up because of the absorbed energy; but we would search in vain for the "stopped photons" in the material.

The photon is always an extremely relativistic particle. When we discuss its properties, we have to take into account the complications due to both quantum mechanics and the special theory of relativity. One immediate consequence of relativistic mechanics is that there is no "conservation law" for the number of photons. If we have an isolated system of massive particles, all of which move with slow velocities, then the number of particles is not changed. But a photon is created whenever an atom emits it during a transition between two stationary states of the atom, and it is annihilated by some process such as the photoelectric effect. Neither before emission nor after absorption is there any photon around. Actually, even massive particles can be created or annihilated, if there is enough energy in a collision to supply the rest mass energy E_0 given by Eq. (16–39c).

Nevertheless, there exist processes in which a photon interacts with an electron without vanishing. Such a process is the *Compton effect*, in which a photon of energy $h\nu$ collides with a free electron at rest. The photon scatters off the electron without being absorbed; however, conservation of energy and momentum dictate that the photon must have a different energy and thus wavelength after collision. Since we do not want to assume that the photon energy is small compared to the electron rest mass energy, we have to use relativistic kinematics also for the electron. We can use a relation between energy and momentum which is easily derived from Eqs. (16–39a) and (16–39b). If we square the two equations and subtract them from each other, after multiplying the first one with c^2, we obtain

$$p^2 c^2 - E^2 = \frac{m_0^2 v^2 c^2}{1 - (v^2/c^2)} - \frac{m_0^2 c^4}{1 - (v^2/c^2)} = \frac{m_0^2 c^4}{1 - (v^2/c^2)}\left(\frac{v^2}{c^2} - 1\right)$$

or

$$\boxed{E = \sqrt{m_0^2 c^4 + p^2 c^2}} \tag{16–41}$$

We can now apply the laws of conservation of energy and of momentum to the situation shown in Fig. 16–24. Energy conservation implies that the sum of the incident photon energy $h\nu$ and the electron rest mass energy $m_0 c^2$ has to be equal to the sum of the energies of the photon and the electron after collision:

$$h\nu + m_0 c^2 = h\nu' + \sqrt{m_0^2 c^4 + p^2 c^2} \tag{16–42}$$

Momentum conservation gives us two relations; the one for the components of momentum along the incident photon direction is

$$\frac{h\nu}{c} = p \cos\theta + \frac{h\nu'}{c}\cos\phi \tag{16–43a}$$

while in the perpendicular direction the initial momentum vanishes and we have

$$0 = p \sin\theta - \frac{h\nu'}{c}\sin\phi \tag{16–43b}$$

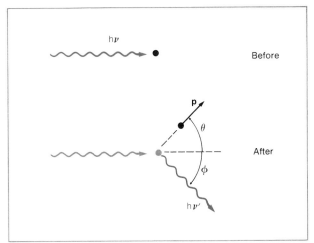

Figure 16–24 In the Compton effect the photon scatters off a free (or loosely bound) electron without being completely absorbed.

We can eliminate θ by calculating from Eq. (16–43b)

$$\cos \theta = \sqrt{1 - \sin^2 \theta} = \frac{\sqrt{p^2 - \left(\frac{h\nu'}{c}\right)^2 \sin^2 \phi}}{p}$$

If we substitute this into Eq. (16–43a), we obtain after a slight rearrangement

$$h\nu - h\nu' \cos \phi = \sqrt{p^2 c^2 - (h\nu')^2 \sin^2 \phi}$$

We can square both sides; noting that $\sin^2 \phi + \cos^2 \phi = 1$, we obtain

$$(h\nu)^2 - 2(h\nu)(h\nu') \cos \phi + (h\nu')^2 = p^2 c^2 \qquad (16\text{–}44)$$

But $p^2 c^2$ can also be obtained from Eq. (16–42) if we transfer $(h\nu')$ to the left and square:

$$(h\nu + m_0 c^2 - h\nu')^2 = m_0^2 c^4 + p^2 c^2$$

Multiplying out the square in the parentheses, we obtain

$$(h\nu)^2 - 2(h\nu)(h\nu') + (h\nu')^2 + 2(h\nu - h\nu')m_0 c^2 = p^2 c^2 \qquad (16\text{–}45)$$

Subtracting Eq. (16–45) from Eq. (16–44), we have

$$2(h\nu)(h\nu')(1 - \cos \phi) - 2(h\nu - h\nu')m_0 c^2 = 0$$

which we can rearrange to read

$$\nu - \nu' = \frac{h}{m_0 c^2} \nu\nu' \ (1 - \cos \phi) \qquad (16\text{–}45)$$

We can obtain a more transparent form if we divide Eq. (16–45) by $\nu\nu'$ and note that $1/\nu = \lambda/c$:

$$\boxed{\lambda' - \lambda = \frac{h}{m_0 c}(1 - \cos \phi)} \qquad (16\text{–}46)$$

Thus, the *difference in wavelength* before and after scattering is a simple function of the scattering angle ϕ of the photon. The quantity

$$\lambda_c = \frac{h}{m_0 c} = 0.02426 \text{ Å}$$

is called the *Compton wavelength* of the electron.

Example 10   A photon of energy $E_\gamma = h\nu = 10$ MeV (1 Mev $= 10^6$ eV) scatters off an electron backwards ($\phi = 180°$). Calculate the final energy of the electron and of the photon in electron volts.

We can best use Eq. (16–45); if we divide it by $h\nu\nu'$ we obtain

$$\frac{1}{E_\gamma'} - \frac{1}{E_\gamma} = \frac{1 - \cos\phi}{m_0 c^2} = \frac{2}{m_0 c^2}$$

because $h\nu' = E_\gamma'$ is the photon energy after collision and because $\cos\phi = -1$ for $\phi = 180°$. From this we have with a little algebra

$$E_\gamma' = \frac{E_\gamma m_0 c^2}{2E_\gamma + m_0 c^2}$$

Since E_γ is given in eV and the results are also wanted in electron volts, we calculate the electron rest mass energy also in eV. We have

$$m_0 c^2 = (0.91 \times 10^{-30} \text{ kg})(3 \times 10^8 \text{ m/sec})^2 = 8.2 \times 10^{-14} \text{ J}$$
$$= 5.11 \times 10^5 \text{ eV} = 0.511 \text{ MeV}$$

This is only 2.5% of the quantity $2\,E_\gamma = 20$ MeV $= 2 \times 10^7$ eV in the denominator; thus, to this accuracy we can neglect $m_0 c^2$ in the denominator and we have

$$E_\gamma' \approx \frac{E_\gamma m_0 c^2}{2E_\gamma} = \frac{m_0 c^2}{2} = 0.25 \text{ MeV}$$

The total electron energy E_e' after the collision can be obtained from the law of energy conservation:

$$E_\gamma + m_0 c^2 = E_\gamma' + E_e'$$

or

$$E_e' = E_\gamma - E_\gamma' + m_0 c^2 = 10 \text{ MeV} - 0.25 \text{ MeV} + 0.5 \text{ MeV} = 10.25 \text{ MeV}$$

Of course, the rest mass energy $m_0 c^2$ is included in E_e'. The kinetic energy of the electron is $m_0 c^2$ less, or 9.75 MeV.

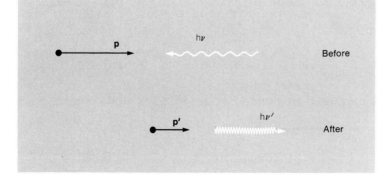

Figure 16–25 Inverse Compton effect: A fast electron collides with a low energy photon; after the collision the photon has a much larger energy.

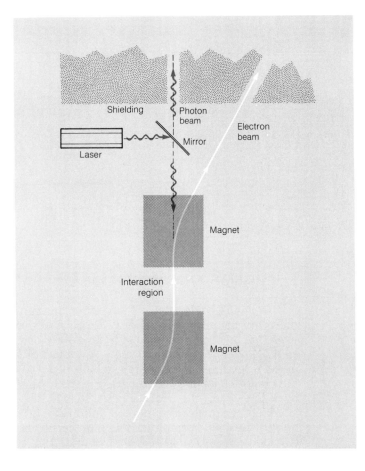

Figure 16–26 The SLAC inverse Compton effect arrangement: In the interaction region between two magnets, a light beam and an electron beam travel in opposite directions. Photons scattered by 180° pass through a hole in a thick shield which absorbs all photons scattered at different angles.

We can also invert the Compton reaction; a high energy ($E_e = 10^{10}$ eV) electron, colliding with a low energy (visible light, $h\nu \approx 3$ eV) photon, will transmit a substantial fraction ($\approx 20\%$) of its energy to the photon, if the scattering occurs under an angle $\theta \approx 180°$ (Fig. 16–25). This has a practical application (Fig. 16–26). An electron beam of very high intensity (10^{14} electrons in a burst 1 μsec long) is made to interact with a pulsed laser beam going in the opposite direction. The electron energy is 10^{10} eV = 10 GeV (giga electron volt). The backscattered photons are collimated by a hole in a thick lead wall. All the unscattered electrons are bent to the side by a powerful electromagnet. The net result is a "monochromatic" photon beam of a sharp energy of 2 GeV, something rather difficult to achieve otherwise. Since, in addition, the polarization of the photon is unchanged in the scattering, one can produce a *monochromatic linearly polarized* high energy proton beam simply by using polarized light from the laser. This method, beautiful by its simplicity, has been tried at the Stanford electron linear accelerator (SLAC) and can actually produce useful high energy photon beams (10^4 photons/sec) in spite of the fact that backward scattering between an electron and a photon is a rather rare occurrence.

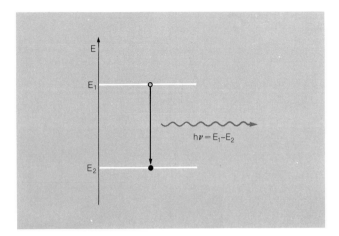

Figure 16-27 Spontaneous emission by an atom; an electron changes to a state of lower energy and the freed energy escapes in the form of a photon.

6. EMISSION AND ABSORPTION OF PHOTONS

In atoms, molecules, or any other quantum mechanical system of particles, there are usually many energy levels. According to the Pauli principle, each level can be occupied at most by two electrons, one with spin "up" and one with spin "down." However, the energy level of an electron can change. If there exists a level of lower energy, which is not already occupied by two electrons, the electron can make a *transition* to the lower level, emitting a photon of energy $h\nu = E_1 - E_2$ in the process (Fig. 16-27). The inverse process is also possible; an electron can absorb a photon and make the transition to a higher state—again only if the Pauli principle does not forbid this: there can never be more than two electrons in a single level. (Compare Fig. 16-32a.)

From Maxwell's equations, using only classical Newtonian mechanics, one can deduce that any accelerated charge, and thus any electron subject to the electric forces in an atom, has to radiate electromagnetic energy. This is not really surprising; a moving charge gives rise to a time-dependent electric field because of its Coulomb field and to a time-dependent magnetic field because it constitutes a time-dependent current. However, if the electron is moving with constant velocity **v**, the time-dependent electromagnetic field cannot produce a wave. This can be seen if one considers the electron from the point of view of observer $\mathscr{B}$ moving with the electron (Fig. 16-28). To observer $\mathscr{B}$ the electron is at rest; the *time-independent* electric field cannot give rise to an emanating wave. This argument fails if the electron is accelerated; both observers $\mathscr{A}$ and $\mathscr{B}$ then observe the same acceleration **a**, and thus there is no reason why the electron should not radiate energy away. We will only state without proof an exact classical theorem that if a point charge q is subject to the acceleration a,

Figure 16-28 As long as a charge q moves with constant velocity **v**, it is at rest to the inertial observer $\mathscr{B}$ also moving with **v**. The charge can radiate electromagnetic energy only if it is moving with respect to all possible inertial observers, i.e., if it is undergoing an acceleration **a**.

then the total *radiated power* (energy radiated away per unit time) is

$$S = \frac{q^2 \, |\mathbf{a}|^2}{6\pi\epsilon_0 c^3} \tag{16-47}$$

Example 11   In the classical picture of the hydrogen atom, the electron is in a circular orbit of radius $a_0 = 0.53$ Å. How long would it take—according to classical mechanics—before it would radiate away 1.0 eV?

The example is unrealistic because one cannot use classical mechanics on the atomic scale. Nevertheless, it illustrates the puzzle physicists felt at the turn of the century, before the discovery of quantum mechanics. Note that $r = 0.5$ Å corresponds to the main quantum number $n = 1$, and thus to the ground state of the hydrogen atom. Therefore, the atom in reality does not radiate at all.

We have to calculate the acceleration **a**. We know that

$$\mathbf{a} = \frac{\mathbf{F}}{m}$$

where **F** is the electrostatic force due to the hydrogen nucleus:

$$|\mathbf{F}| = \frac{1}{4\pi\epsilon_0} \frac{e^2}{a_0{}^2}$$

Substituting this into Eq. (16–47) yields, because the electron charge is e,

$$S = \frac{e^2}{6\pi\epsilon_0 c^3} \left(\frac{e^2}{4\pi\epsilon_0 m a_0{}^2} \right)^2$$

or, rewritten slightly,

$$S = \left(\frac{e^2}{4\pi\epsilon_0 c} \right)^3 \frac{2}{3m^2 a_0{}^4}$$

Numerically we obtain, because $1/4\pi\epsilon_0 = 9 \times 10^9$ V-m/coul

$$S = \left(\frac{(1.60 \times 10^{-19} \text{ coul})^2 \times (9 \times 10^9 \text{ V-m/coul})}{3 \times 10^8 \text{ m/sec}} \right)^3 \frac{2}{3(0.91 \times 10^{-30} \text{ kg} \times (0.53 \times 10^{-10}\text{m})^2)^2}$$

$$= 4.6 \times 10^{-8} \text{ W}$$

or expressed in eV/sec

$$S = 2.9 \times 10^{11} \text{ eV/sec}$$

Thus, the electron will require only the time

$$\tau \approx \frac{1}{2.9 \times 10^{11} \text{ sec}^{-1}} = 3.4 \times 10^{-12} \text{ sec}$$

to radiate 1 eV away! The binding energy of the electron in the hydrogen atom is 13.6 eV; thus, according to classical mechanics, the whole atom would collapse in a few times 10^{-11} sec.

In quantum mechanics, a single atom cannot continuously radiate energy; it can only emit the energy in the form of a single photon, if a stationary state of lower energy is available. If there are many atoms, all in the same initial state, many photons will be emitted. While we cannot predict the instant when any single atom will emit a photon, we can determine the probability that it will do so in some future time interval.

Consider the simple situation shown in Fig. 16–27. At $t = 0$ the electron is at the level of energy E_1; then sometime in the future it will emit a photon. Let us now assume that at the instant t we have a large number N of identical atoms, all at the energy E_1. We cannot predict when each individual atom will decay; but each has the same probability that it will decay within a time interval Δt in the future. Thus, a number δ of them will *decay* (i.e., emit a photon) in the small time interval Δt. The number δ of decays will be proportional to the time interval Δt and also to the number N presently existing at the energy E_1:

$$\delta = AN\,\Delta t$$

The quantity A is called the *spontaneous transition rate* or the *decay rate* between the levels E_1 and E_2. Its value depends on the electron wave functions in both the initial state E_1 and the final state E_2. It also depends on the energy difference $\Delta E = E_1 - E_2$; in atomic transitions, A is proportional to $(\Delta E)^3$. Since $(-\delta)$ is the loss dN of atoms at the upper level in the short time period Δt, we can write, if we go to the limit $\Delta t \to dt \to 0$:

$$-\frac{dN}{dt} = AN$$

This can be easily integrated [see the derivation of Eq. (16–37)] to yield

$$\boxed{N = N_0 e^{-At} = N_0 e^{-t/\tau}} \tag{16–48}$$

where

$$\tau = \frac{1}{A}$$

is the *lifetime* of the atomic transition and N_0 is the number of atoms at the upper energy E_1 at the time $t = 0$. Of course, the electron may not stay in the final level. The final level will also be *unstable* if there is a third available level even lower in energy.

■   Fig. 16–29a shows the possible transitions between the levels in the hydrogen atom; we see that, for example, the $3p$ level decays into the $2s$ level or the $1s$ level, but not into the $2p$ level. This is a rather general situation in atoms; only those transitions for which the orbital angular momentum quantum number ℓ changes by ± 1 occur in atoms. Thus, the $2s$ level cannot decay into the $1s$ level, because both have $\ell = 0$. However, as we show in a blowup in Fig. 16–29b, the levels of the main quantum number $n = 2$ in hydrogen do not lie at exactly the same energy because of the "spin-orbit" coupling. An observer at rest to the electron will see a magnetic field $\mathbf{B}$ (compare p. 414) besides the electric field of the nucleus. The magnetic dipole moment $\mathbf{m}_e$ of the electron interacts with this magnetic field; the additional energy (compare p. 330)

$$\Delta E = -(\mathbf{m}_e \bullet \mathbf{B}) = +\mu_B(\mathbf{S} \bullet \mathbf{B}) \tag{16–49}$$

will be different if the electron spin points along the magnetic field than if it points opposite to the magnetic field. Thus, there are really two $2p$ levels of slightly different energy: the $2p_{3/2}$ state with total angular momentum

$$|\mathbf{J}| = |\mathbf{L} + \mathbf{S}| = \sqrt{\frac{3}{2}\left(\frac{3}{2} + 1\right)}\,\hbar \qquad\qquad \begin{matrix} \mathbf{S} \uparrow \\ \mathbf{L} \uparrow \end{matrix} \Big\} \mathbf{J} = \mathbf{L} + \mathbf{S}$$

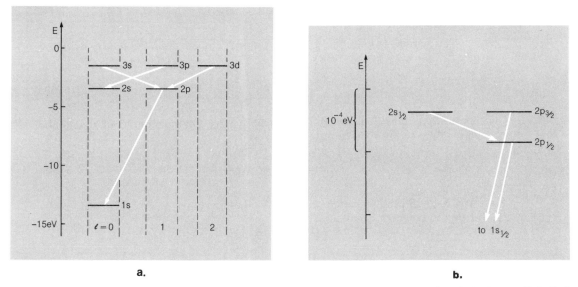

Figure 16–29 (a) The lowest energy levels and transitions between them in hydrogen. Note that the 2s state has seemingly "nowhere to go." (b) Detail of hydrogen level structure; in reality there are two 2p levels, and the 2s level decays into one of them. Because the energy difference is very small, the 2s level is metastable.

for which the two angular momenta are parallel; and the $2p_{1/2}$ state° with

$$|\mathbf{J}| = |\mathbf{L} - \mathbf{S}| = \sqrt{\frac{1}{2}\left(\frac{1}{2} + 1\right)}\,\hbar$$

$$\left.\begin{array}{c} \mathbf{S} \\ \mathbf{L} \end{array}\right\} \mathbf{J} = \mathbf{L} - \mathbf{S}$$

for which the orbital and spin angular momenta are antiparallel to each other. The 2s state can decay into the lower-lying $2p_{1/2}$ state; however, since the energy difference is only 5×10^{-5} eV, the 2s level is "metastable" with a lifetime $\tau \cong 10^{-4}$ sec, about a factor 10^4 larger than the typical atomic lifetimes of 10^{-8} sec.

Example 12   A state of energy E_1 decays into the state E_2 with lifetime τ_1; E_2 in turn decays into the state E_3 with lifetime $\tau_2 = 2\tau_1$ (Fig. 16–30). Initially, all

---

° If the total angular momentum $\mathbf{J}$ matters, states are labeled by the suffix j, where j is $\ell + 1/2$ if the spin is parallel to $\mathbf{L}$ and $\ell - 1/2$ if it is antiparallel; the magnitude of $\mathbf{J}$ is then $\sqrt{j(j+1)}\,\hbar$.

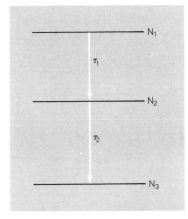

Figure 16–30 Example 11: Successive decay between levels.

atoms (N_0 in total) were in the state E_1. Calculate for any instant t the number N_2 of atoms which are in the state of energy E_2.

The state E_2 is being fed by decays from E_1, and is depleted in turn by decays into E_3. Thus, we have

$$dN_2 = +N_1\frac{dt}{\tau_1} - N_2\frac{dt}{\tau_2} \tag{16-50a}$$

because the transition rates $A_{1,2}$ are equal to $1/\tau_{1,2}$. On the other hand, the number N_1 of atoms in the state E_1 only decreases:

$$dN_1 = -N_1\frac{dt}{\tau_1} \tag{16-50b}$$

From Eq. (16–50b) we obtain, if we remember that $N_1(t=0) = N_0$, the total number of atoms:

$$N_1 = N_0\, e^{-t/\tau_1}$$

Inserting this into Eq. (16–50a), we obtain after a slight rearrangement of terms

$$\frac{dN_2}{dt} = \frac{N_0}{\tau_1}e^{-t/\tau_1} - \frac{N_2}{\tau_2} \tag{16-51a}$$

Let us assume for N_2 the form

$$N_2 = Ae^{-t/\tau_1} + Be^{-t/\tau_2} \tag{16-51b}$$

with A and B as yet unknown constants. Inserting this into Eq. (16–51a), we have

$$-\frac{Ae^{-t/\tau_1}}{\tau_1} - \frac{Be^{-t/\tau_2}}{\tau_2} = \frac{N_0}{\tau_1}e^{-t/\tau_1} - \frac{Ae^{-t/\tau_1}}{\tau_2} - \frac{Be^{-t/\tau_2}}{\tau_2} \tag{16-52}$$

Both those terms having the factor e^{-t/τ_1} and those having the factor e^{-t/τ_2} have to cancel separately, because Eq. (16–52) has to be valid for all times. Thus, we have

and

$$-\frac{A}{\tau_1} = \frac{N_0}{\tau_1} - \frac{A}{\tau_2} \tag{16-53a}$$

$$-\frac{B}{\tau_2} = -\frac{B}{\tau_2} \tag{16-53b}$$

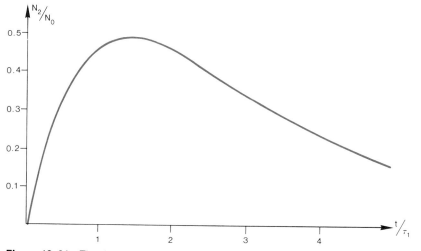

Figure 16–31 The time dependence of the population of the middle level.

From Eq. (16–53a) we obtain, because $\tau_2 = 2\tau_1$,

$$A = \frac{\tau_2 N_0}{\tau_1 - \tau_2} = -2N_0$$

while Eq. (16–53b) is an identity and imposes no limit on B. However, we know that at $t = 0$ we had $N_2(t = 0) = 0$; from Eq. (16–51b) we thus have

$$B = -A = +2N_0$$

In consequence, using the fact that $\tau_2 = 2\tau_1$,

$$N_2 = 2N_0(e^{-t/2\tau_1} - e^{-t/\tau_1})$$

We have plotted $N_2(t)/N_0$ in Fig. 16–31 versus t/τ_1.   ■

Transitions between atomic levels do not occur only spontaneously. If an electromagnetic wave — many photons — of frequencies close to ν passes through atoms which have two levels E_1 and E_2 such that

$$E_1 - E_2 = h\nu$$

then transitions will be produced. Let us assume that there are N_1 and N_2 atoms at energies E_1 and E_2 respectively. Then there will be two types of transitions. In the first type, atoms in the state E_2 can absorb a photon and make the transition to the state E_1 (Fig. 16–32a). The number of atoms absorbing photons in the very short time dt will be

$$dN_{2 \to 1} = u(\nu) N_2 B \, dt \qquad (16\text{–}54a)$$

where $u(\nu)$ is the incident electromagnetic wave intensity per unit frequency range, i.e., the energy arriving per unit time at a unit area and within a unit of frequency:

$$u(\nu) = \frac{dE}{dt \, dA \, d\nu} \qquad (16\text{–}54b)$$

The total intensity, the energy arriving per unit time and unit area, is then

$$I = \int_{\nu=0}^{\infty} u(\nu) \, d\nu$$

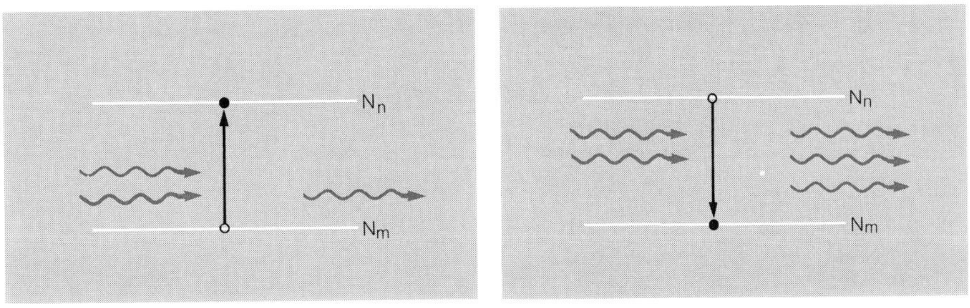

a.   b.

Figure 16–32 Induced absorption (a) and emission (b) of photons. The incident light beam has to have a photon energy corresponding exactly to the energy difference between the two states.

The *induced transition rate B* is proportional to the spontaneous transition rate A which, as we discussed earlier, governs the spontaneous emission of a photon from the level E_1:

$$B = \frac{c^2}{8\pi h \nu^3} A$$

(16−54c)

However, induced transitions will exist not only from the level E_2 to the level E_1 but, under the influence of the external radiation, the atom in the state E_1 can also make a transition to the state E_2. Thus, the number of photons emitted in a time interval dt will be equal to the sum of the numbers of spontaneous and induced transitions from E_1 to E_2

$$dN_{1 \to 2} = N_1 A \, dt + N_1 \, u(\nu) \, B \, dt$$

(16−55)

The photons emitted owing to the two terms in Eq. (16−55) will behave differently; the spontaneously emitted photons are emitted in a random direction, while the photons emitted by the induced transition leave the atom in exactly the same direction as the incident electromagnetic wave; they increase the intensity of the incident wave. Thus, the intensity of a light beam of correct frequency $\nu = (E_1 - E_2)/h$ will be affected by the transitions between the two levels E_1 and E_2. The absorption, whose rate is given in Eq. (16−54), will attenuate the beam; the induced emission [second term in Eq. (16−55)] will increase its intensity. Thus, the net change in intensity will be proportional to

$$\left(\frac{dN_{1 \to 2}}{dt} - \frac{dN_{2 \to 1}}{dt} \right)\Bigg|_{\text{induced}} = (N_1 - N_2) \, u(\nu) B$$

(16−66)

The incident wave will be attenuated if $N_1 < N_2$ and will increase in intensity if $N_1 > N_2$. Unfortunately, however, we normally find $N_1 < N_2$ because in statistical thermal equilibrium

$$\frac{N_1}{N_2} = \frac{e^{-E_1/kT}}{e^{-E_2/kT}} = e^{-(E_1 - E_2)/kT} < 1$$

(16−67)

because $E_1 > E_2$. One has to produce a "population inversion" (i.e., $N_1 > N_2$) artificially before one can increase the incident light intensity (Fig. 16−33). This

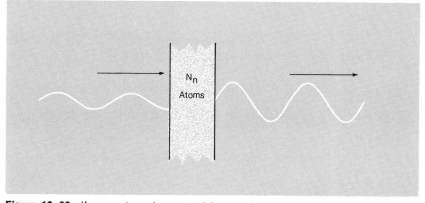

Figure 16−33 If more atoms in a material are at the upper energy state, an incident wave of correct frequency will be amplified by the induced emission of photons.

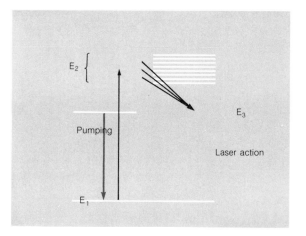

Figure 16-34 The three level laser; the state at E_3 has to be metastable.

is produced in the *laser*, an instrument whose name comes from its function: Light Amplification by Stimulated Emission of Radiation. Lasers have been built using ruby, neon, carbon dioxide, various dyes, and other materials. Any material can be used in a laser if it has the basic level structure shown in Fig. 16-34.

A group of levels (the more the better) at energy E_2 decays spontaneously either into the ground state E_1 or into a *metastable* level E_3. Light of a range of frequencies close to

$$\nu_{12} = \frac{E_2 - E_1}{h}$$

is absorbed when it is incident on the material. Some fraction of the atoms at E_2 will decay spontaneously with emission of a photon of energy $h\nu = E_2 - E_3$, thus increasing the population N_3 of the state E_3. Since the level E_3 is metastable — its spontaneous decay rate is very small — the population inversion ($N_3 > N_1$) can be achieved if this *optical pumping* from E_1 to E_3 is sufficiently strong.

The material is set between two exactly parallel mirrors (Fig. 16-35). A rare (E_3 is metastable!), spontaneously emitted photon of frequency $\nu_3 = (E_3 - E_1)/h$ will be reflected by the mirrors back into the material; during its passage back and forth, it will by stimulated emission generate more photons, all of them again bouncing back and forth between the two mirrors. In practice, one of the two mirrors is kept partially transparent, so that some of the light can escape, forming a light beam of a very well-defined frequency ν_3 and very well-defined direction.

The level structure of the "three-level laser" shown in Fig. 16-34 is suit-

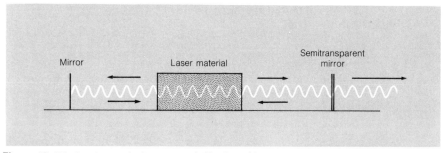

Figure 16-35 In a laser, the laser material is placed between two accurately aligned mirrors so that a standing light wave forms between them.

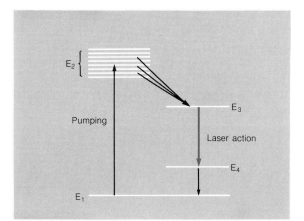

Figure 16-36   The four-level laser.

able only for *pulsed lasers.* Once laser action starts, the population inversion very quickly disappears; laser action stops once the populations N_3 of the level E_3 and N_1 of the ground state are equal. This happens after a very short time (typically 10^{-10} to 10^{-8} sec). Continuous lasers usually have a level arrangement like that shown in Fig. 16–36. Initially all atoms are in the ground state E_1. Optical pumping—through the group of levels at E_2—populates the metastable level E_3. Laser action takes place for the transition between the state E_3 and the short-lived state E_4. Since none of the atoms were initially at E_4, population inversion is achieved at once. Also, the number N_4 always stays very low, because any atom at E_4 immediately decays by spontaneous transition to the ground state E_1. Thus, laser action will take place as long as the pumping goes on. It is not even necessary to excite the group of levels at E_2 by absorption; in an electrical discharge, the levels at E_2 will be populated by collision of the fast free electrons in the discharge with the gas atoms.

   Example 13   Proton magnetic resonance. The proton (hydrogen nucleus), like the electron, has a spin of

$$|\mathbf{S}| = \sqrt{\frac{1}{2}\left(\frac{1}{2}+1\right)}\,\hbar = \sqrt{\frac{3}{2}}\hbar$$

and a magnetic moment $\mathbf{m}_p$ (which is, however, much smaller than that of the electron). Its value is

$$\mathbf{m}_p = 2.790\,\frac{e\hbar}{2M_p}\left(\frac{2S}{\hbar}\right) \tag{16–68}$$

in units of the nuclear magneton

$$\mu_N = \frac{e\hbar}{2M_p} = 5.051 \times 10^{-27}\ \text{A-m}^2 \tag{16–69}$$

The unit of the nuclear magneton—because the proton mass M_p instead of the electron mass m_e appears in Eq. (16–69)—is 1820 times smaller than the Bohr magneton

$$\mu_B = \frac{e\hbar}{2m_e} = 0.923 \times 10^{-23}\ \text{A-m}^2$$

In a magnet of magnetic field $\mathbf{B}$, the proton spin will align itself either along or

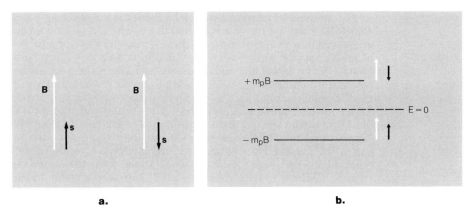

Figure 16–37 (a) In a magnetic field, the proton spin will be either parallel or antiparallel to the field because angular momentum is quantized. (b) Because of the interaction of the proton magnetic dipole moment with the magnetic field, there are two levels depending on whether the spin is parallel or antiparallel to **B**.

opposite to the magnetic field (Fig. 16–37). Thus, we have

$$S_z = \pm \frac{1}{2}\hbar$$

and the potential energy of the proton magnetic moment in the magnetic field is

$$W = -(\mathbf{m}_p \bullet \mathbf{B}) = \pm 2.790 \, \mu_N B$$

We thus have two energy levels separated by

$$\Delta E = 5.58 \, \mu_N B$$

Photons of energy $h\nu = \Delta E$ will thus be absorbed by the hydrogen-containing material (e.g., water, H_2O). In a magnetic field of 1 weber/m², the frequency is

$$\nu = \frac{5.58 \, \mu_N B}{h} = \frac{5.58 \times (5.05 \times 10^{-27} \, \text{A-m}^2) \times (1 \, \text{Wb/m}^2)}{(6.63 \times 10^{-34} \, \text{J-sec})}$$
$$= 4.25 \times 10^7 \, \text{Hz}$$

Figure 16–38 Schematic arrangement of a nuclear magnetic resonance (NMR) setup; the driving coil produces photons which, if the magnetic field is exactly correct, are absorbed by the sample material. The pickup coil detects the spontaneously re-emitted electromagnetic energy.

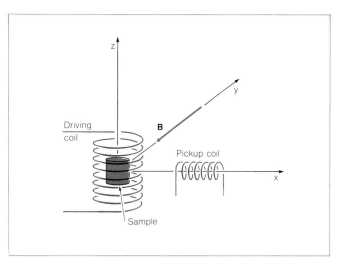

or 42.5 MHz, a frequency easily achievable by an ordinary LC circuit. Note that 42.5 MHz corresponds to a wavelength $\lambda = c/\nu \approx 7$ meters! Quantum mechanical effects can thus be important even for macroscopic wavelengths. A practical method of measuring the absorption (Fig. 16–38), known as nuclear magnetic resonance (NMR), is to detect the radiation emitted when the proton decays back spontaneously from its higher energy state to the lower energy state. The detector and the source of the photons are ordinary coils of a few turns. One coil is driven by an LC circuit and radiates the photons. The second coil, the pickup coil, is perpendicular to the first so that there is no mutual inductance between the two coils. This is necessary; the pickup coil receives radiation at the same frequency as the driving coil. Thus, even a very small mutual inductance would result in a signal many times larger than the extremely weak signal from the water sample. When the magnetic field B is now slightly varied, one sees a maximum pickup signal "at resonance," i.e., when

$$\nu = \frac{5.58\mu_N B}{h} = (42.5 \text{ MHz/ Wb /m}^2) \times B$$

Summary of Important Relations

Velocity of light

in vacuum

$$c = \frac{1}{\sqrt{\epsilon_0 \mu_0}}$$

in material of dielectric constant K

$$v = \frac{c}{\sqrt{K}} = \frac{c}{n}$$

Refractive index

$$n = \sqrt{K}$$

Polarization of light = direction of **E**
    linearly polarized light   **P** oscillates along line $\perp$ propagation direction
    circularly polarized light   **P** moves along a circle

Photon
    energy $E = h\nu$

    momentum $p = h/\lambda = E/c$

Compton effect

$$\lambda' - \lambda = \frac{h}{m_0 c}(1\text{-}\cos\phi)$$

Energy radiated by point charge

$$S = \frac{q^2 a^2}{6\pi\epsilon_0 c^3}$$

Spontaneous transition rate

$$A = -\frac{1}{N}\frac{dN}{dt}$$

Induced transition rate $$B = \frac{c^2}{8\pi h \nu^3} A$$

Questions

1. Assume a plane electromagnetic wave in which both **E** and **B** are perpendicular to the direction of propagation, but parallel to each other. Show that Maxwell's equations preclude such an assumption.

2. There are no transparent metals. Explain why.

3. Blue light is bent more by a glass prism than is red light. Explain why.

4. In optics, one frequently talks about "partially polarized light." How would you describe such light?

5. Why can a photon not be absorbed by a free electron? Why then can it be absorbed by a free atom?

6. Is it possible to produce a laser in which all (or at least most) of the pumping energy is converted into laser light? If not, why not?

7. Discuss why the two mirrors in Figure 16–35 have to be separated by an exact multiple of the wavelength λ of the emitted light.

Problems

1. A fluorescent light bulb delivers at a point in space an intensity of 1 watt/cm². Calculate the average electric and magnetic fields at the same point.

2. Some power lasers (neodymium glass lasers) can deliver an energy of 1000 Joules in one pulse lasting 10^{-9} seconds. Calculate the peak electric field and magnetic field, if the laser beam has a diameter of 2 mm.

3. At what light intensity in glass of refractive index $n = 1.5$ (or dielectric constant $K = n^2$) will the peak electric field be 10^7 V/m?

4. Show that the group velocity of light in material of refractive index n is

$$v_g = \frac{c}{n}\left(1 + \frac{\lambda}{n}\frac{dn}{d\lambda}\right)$$

Does this impose a limit on how fast n can change with λ?

5. The static dielectric constant of NaCl is $K = 6.12$; its refractive index for visible light is $n = 1.54$. Calculate the ionic and electronic polarizability of NaCl. The density is $\rho = 2.2 \times 10^3$ kg/m³.

6. A lens is made of flint glass; its nominal focal length is $f = 10$ cm at $\lambda = 5500$ Å. What will be its focal length at $\lambda = 4500$ Å? This effect is called chromatic aberration. Consult Fig. 16–9 for numerical values of the refractive index.

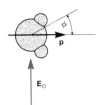

*7. Let us estimate at what frequency the orientational polarization of water no longer can follow the oscillating electric field. For this purpose, consider a free water molecule of dipole moment $p = 0.62 \times 10^{-29}$ coul-m and moment of inertia

$$I \approx 2 \times (1 \text{ proton mass}) \times (1 \text{ Å})^2$$

under the influence of an electric field

$$\mathbf{E} = \mathbf{E}_0 \cos \omega t$$

with $\mathbf{E}_0 = 10^6$ V/m. Assume that initially $\mathbf{p}$ was perpendicular to $\mathbf{E}$.

(a) At what frequency will $\mathbf{p}$ change its direction by less than 0.5° over one period of the electric field?

(b) What will be the maximum directional change if ω is ten times larger than the value calculated in (a)?

(c) What is the average angle ϕ for a static electric field of the same magnitude at room temperature? (Compare p. 261.)

8. Linearly polarized light is incident on a polaroid filter. After the filter, the light intensity is 1/3 of the original intensity. By what angle was the polaroid filter rotated from the direction of complete absorption?

9. What fraction of the intensity of an unpolarized light wave is transmitted by a polaroid filter?

10. Unpolarized light of intensity I_0 is impinging on a sequence of three polaroid filters; the first and the last have their directions of optimum transmission rotated by $\pm 45°$ relative to the center filter. Calculate the light intensity after passage through all three filters. What happens if the center polaroid filter is removed?

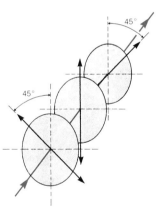

11. Good reception of a radio signal implies that the power received by the antenna is 10^{-6} W = 1 μW. Assume a frequency of 10^7 Hz—the typical frequency for AM radio.

(a) How many radio frequency (RF) photons are absorbed by the antenna per second?

(b) How many photons are absorbed in one period of the wave?

(c) Comment on why the photon nature—the quantization of energy in quanta of $h\nu$—does not matter in this case.

12. If light of $\lambda = 5500$ Å is used, the maximum kinetic energy of electrons emitted by the photoelectric effect by a certain material is 0.5 eV. What is the work function of the material? What will be the maximum kinetic energy of the emitted electrons, if light of wavelength 4000 Å is used?

13. The longest wavelength of light with which one can eject electrons from tungsten by the photoelectric effect is $\lambda = 2700 \text{ Å}$. What will be the maximum ejected electron velocity at a wavelength $\lambda = 2000 \text{ Å}$?

14. What is the longest wavelength photon which can produce an electron by the photoelectric effect from hydrogen (in its ground state)?

15. Using Fig. 15–7 and a table of elements, calculate the photon energy and wavelength at the K-edge of aluminum.

16. A small radioactive source emits 10^{10} photons/sec isotropically. How many photons arrive per second at a detector of 1 cm² area which is 2 meters away? The detector faces the source.

17. The attenuation coefficient in most materials for photons of 1 to 2 MeV energy is ~ 0.1 cm²/gram. How thick a layer of concrete (density $\rho = 2.31 \times 10^3$ kg/m³) is required to obtain a reduction by a factor 10^{10} in photon flux?

°18. A radioactive source (Co⁶⁰) used for cancer treatment emits $N_0 = 10^{15}$ photons of energy ~ 1.2 MeV per second. The source size is very small (it can be made as small as a few millimeters across). Safety rules require that the source be stored in a lead shield; just outside the shield, the maximum tolerable flux is 2 photons/cm² sec. The absorption coefficient in lead for photons of energy 1 to 1.5 MeV is $\mu = 0.08$ cm²/g. How thick does the spherical shield have to be and what is its mass? Note that both distance and lead thickness reduce the photon flux.
[Hint: You will end up with an equation which cannot be solved algebraically. Use a first guess and then improve on it. The density of lead is $\rho = 11.4$ g/cm³.]

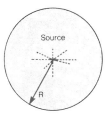

19. A laser beam of intensity $I = 10$ W hits a mirror at an angle of 45° and is reflected. What is the force on the mirror?

20. A photon of energy $E = 1$ MeV collides with an electron at rest. After the collision, the angle between the photon and electron directions is 90° in the laboratory. Calculate the energy of the scattered photon.

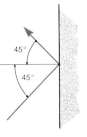

21. An electron of energy $E_0 = 2 \times 10^{10}$ eV collides with a laser photon of energy 2 eV. After the collision, the photon has exactly reversed its direction. (Compare Figs. 16–25 and 16–26.) Calculate the energy of the photon after the collision.

22. (a) In what units is the intensity per unit frequency range $u(\nu)$ measured?
    (b) Given an energy difference between the ground state and excited state of an atom of $\Delta E = 3.0$ eV, at what value of $u(\nu)$ will 1/3 of the atoms be in the excited state in stable equilibrium between spontaneous decay, induced absorption, and induced emission?

23. A point light source is emitting isotropically 10^{20} photons per second; each photon has a random energy between zero and 6 eV. (This means that it is equally probable that the next photon has an energy between 0.5 and 0.6 eV or an energy between, say, 5.0 and 5.1 eV.) Calculate:

    (a) the total power of the source.

    (b) the intensity one meter away.

    (c) the intensity per unit frequency range $u(\nu)$ at $\nu = 5 \times 10^{14}$ Hz one meter away.

UNITS

I. UNITS OF PHYSICAL OBSERVABLES IN MKSC SYSTEM

OBSERVABLE	SYMBOL	DIMENSION	MKSC UNIT
Length, position	$\mathbf{r}, x, \ell, \ldots$	L	*meter* (m)
Time	t, τ	T	*second* (sec)
Mass	m, M	M	kilogram (kg)
Area	A, S	L^2	m^2
Volume	V, τ	L^3	m^3
Density	ρ	ML^{-3}	kg/m^3
Frequency	ν	T^{-1}	*Hertz* (Hz) = sec^{-1}
Angle	θ, ϕ	—	*radian* (rad)
Solid angle	Ω	—	*steradian* (sr)
Velocity	$\mathbf{v}, v, w$	LT^{-1}	m/sec
Acceleration	$\mathbf{a}$	LT^{-2}	m/sec^2
Angular frequency	ω	T^{-1}	rad/sec
Angular acceleration	α	T^{-2}	rad/sec^2
Force	$\mathbf{F}$	MLT^{-2}	*newton* (N) = $kg\text{-}m/sec^2$
Energy, work	E, W, U	ML^2T^{-2}	*Joule* (J) = N-m
Power	P	ML^2T^{-3}	*watt* (W) = J/sec
Momentum	$\mathbf{p}$	MLT^{-1}	kg-m/sec
Angular momentum	$\mathbf{L}, \mathbf{J}$	ML^2T^{-2}	$kg\text{-}m^2/sec^2$
Moment of inertia	I	ML^2	$kg\text{-}m^2$
Torque	$\mathbf{T}, \tau$	ML^2T^{-2}	N-m
Pressure	P	$ML^{-1}T^{-2}$	$N/m^2 = kg/m\text{-}sec^2$
Temperature	T	—	*degree Kelvin* (°K)
Heat	Q	ML^2T^{-2}	J
Entropy	S	$ML^2T^{-2}/°K$	J/°K
Electric charge	Q, q	C	*coulomb* (coul)
Charge density	ρ, ρ_e	CL^{-3}	$coul/m^3$
Surface charge density	σ	CL^{-2}	$coul/m^2$
Current	I	CT^{-1}	*ampere* (A) = coul/sec
Current density	$\mathbf{j}$	$CL^{-2}T^{-1}$	A/m^2
Electrostatic potential	V	$ML^2T^{-2}C^{-1}$	*volt* (V) = J/coul
Electric field	$\mathbf{E}$	$MLT^{-2}C^{-1}$	V/m = N/coul
Polarization	$\mathbf{P}$	CL^{-2}	$coul/m^2$
Resistance	R	$ML^2T^{-3}C^{-2}$	*ohm* (Ω) = V/A
Resistivity	ρ	$ML^3T^{-3}C^{-2}$	Ω-m
Dielectric constant	K	—	—
Electric dipole moment	$\mathbf{p}$	CL	coul-m
Magnetic field	$\mathbf{B}$	$MT^{-1}C^{-1}$	*tesla* (T) = $V\text{-}sec/m^2$
Magnetic flux	ϕ	$ML^2T^{-1}C^{-1}$	*weber* (Wb) = V-sec
Magnetic dipole moment	$\mathbf{m}$	CL^2T^{-1}	$A\text{-}m^2$
Magnetization	$\mathbf{M}$	$CL^{-1}T^{-1}$	A/m
Magnetizing field	$\mathbf{H}$	$CL^{-1}T^{-1}$	A/m
Capacitance	C	$C^2M^{-1}L^{-2}T$	*farad* (F) = coul/V
Inductance	L	$ML^2T^{-2}C^{-2}$	*henry* (H) = V-sec/A = Ω-sec

II. CONVERSION FACTORS

The conversion factors between frequently used units are given.

Length   (Main unit = meter = m)

1 centimeter (cm) = 10^{-2} m	1 Angstrom (Å) = 10^{-10} m = 10^{-8} cm
1 kilometer (km) = 10^3 m	1 Fermi (f) = 10^{-15}m = 10^{-13} cm
1 inch (in) = 0.0254 m	1 light-year = 9.4600×10^{15} m
1 foot (ft) = 0.3048 m	1 parsec = 3.084×10^{18} m
1 mile (mi) = 1609 m	1 mil = 10^{-3} in = 2.54×10^{-5} m

Area   (Main unit = square meter = m²)

The conversion factors for area are the square of those for length, e.g., 1 (inch)² = $(0.0254)^2$ m² = 6.452×10^{-4} m². Other units used:

1 acre	= 43,600 ft² = 4050 m²
1 barn	= 10^{-24} cm² = 10^{-28} m² (nuclear physics)
1 millibarn (mb)	= 10^{-27} cm² = 10^{-31} m²

Volume   (Main unit = cubic meter = m³)

The conversion factors for volume are the cube of those for length, e.g., 1 cubic foot = $(0.3048)^3$ m³ = 2.832×10^{-2} m³. Other units used:

1 liter (ℓ) = 1.000028×10^{-3} m³ = volume of 1 kg of water at 4° C

1 U.S. fluid gallon = 4 U.S. fluid quarts = 8 U.S. pints = 128 U.S. fluid ounces = 231 in³ = 3.785 ℓ = 3.785×10^{-3} m³

1 British Imperial gallon = 277.4 in³ = 4.546 ℓ = 4.546×10^{-3} m³ = 1.201 U.S. gallon

Mass   (Main unit = kilogram = kg)

1 gram (g)	= 10^{-3} kg
1 slug	= 14.59 kg
1 atomic mass unit (amu)	= 1.660×10^{-27} kg

1 pound	= 0.4536 kg	These are actually units of weight; however, 1 pound = 0.4536 kg means that 0.4536 kg weighs 1 pound under standard conditions of gravity (g = 9.8067 m/sec²).
1 short ton	= 907.2 kg	
1 ounce	= 0.02835 kg	

1 long ton	= 1016 kg	
1 eV/c^2	= 1.782×10^{-36} kg	$(m = E/c^2)$
1 MeV/c^2	= 1.782×10^{-30} kg	

Time   (Main unit = second = sec)

1 minute (m)	= 60 sec
1 hour (h)	= 3600 sec
1 day (d)	= 86,400 sec = 8.6400×10^4 sec
1 year (y)	= 3.156×10^7 sec

Speed   (Main unit = m/sec)

1 cm/sec	= 10^{-2} m/sec
1 ft/sec	= 0.3048 m/sec
1 km/h	= 0.2778 m/sec
1 mi/h (mph)	= 0.4470 m/sec

II. CONVERSION FACTORS (continued)

The conversion factors between frequently used units are given.

Force   (Main unit = newton = N = kg-m/sec²)

    1 dyne   = 1 g-cm/sec² = 10^{-5} N
    1 pound   = 4.448 N
    1 kg-force = 9.807 N (= weight of 1 kg at standard gravity, g = 9.807 m/sec²)

Pressure   (Main unit = N/m² = kg/m-sec²)

    1 atmosphere   = 1.013×10^5 N/m²
    1 dyne/cm²   = 10^{-1} N/m²
    1 inch of water (4°C)   = 249.1 N/m²
    1 centimeter of mercury (0°C) = 1.333×10^3 N/m²
    1 pound/in²   = 6.895×10^3 N/m²
    1 bar   = 10^5 N/m²

Energy   (Main unit = Joule = J = kg-m²/sec²)

    1 erg   = 1 g-cm²/sec² = 10^{-7} Joule
    1 British thermal unit (Btu) = 1055 Joules
    1 watt-second   = 1 Joule
    1 kilowatt-hour (kWh)   = 3.6×10^6 Joules
    1 calorie   = 4.186 Joules
    1 electron volt (eV)   = 1.602×10^{-19} Joule
    1 MeV   = 10^6 eV = 1.602×10^{-13} Joule
    1 GeV (giga electron volt)   = 1 BeV (billion electron volt)
                               = 10^9 eV = 1.602×10^{-10} Joule
    1 kg-c^2 = 8.987×10^{16} Joules ($E = mc^2$)
    1 Megaton $\approx 4 \times 10^{15}$ Joules (nuclear explosions)

Power   (Main unit = watt = W = Joule/sec = kg-m²/sec³)

    1 calorie/sec   = 4.186 W
    1 kilowatt (kW)   = 10^3 W
    1 megawatt (MW) = 10^6 W
    1 horsepower   = 745.7 W
    1 Btu/hour   = 0.2930 W

Temperature   (Main unit = °C or °K)

    0 °C   = 273.16 °K
    X °C   = (X + 273.16) °K
    0 °C   = 32 °F
    100 °C = 212 °F
    X °C   = (1.8 X + 32) °F
    Y °K   = (Y − 273.16) °C
    Y °F   = $\left(\dfrac{Y - 32}{1.8}\right)$ °C

III. OTHER ELECTROMAGNETIC UNITS

There are two other systems that are frequently used, besides the MKSC system that is used in this book; both use the CGS system (centimeter, gram, second) as the purely mechanical units.

a) The Gaussian (electrostatic) CGS system has as its basic definition $\epsilon_0 = 1/4\pi$ statampere-sec/statvolt-m.
b) The absolute (electromagnetic) CGS system, in which $\epsilon_0 = (2.998 \times 10^{10})^2/4\pi$ abamp-sec/abvolt-m.
   Note that the numerical value of ϵ_0 is c^2, the speed of light squared, in CGS units (cm/sec)^2.

However, care should be taken that the basic equations of electromagnetism (Maxwell's equations, force laws, etc.) are written differently in the various systems. Thus, in the Gaussian system the force on a charge in a magnetic field is

$$\mathbf{F} = \frac{q}{c}(\mathbf{v} \times \mathbf{B})$$

while in both the MKS system as well as the absolute system one has

$$\mathbf{F} = q\,(\mathbf{v} \times \mathbf{B})$$

Frequently, particularly in advanced books, the "scientific Gaussian" system is used with $\epsilon_0 = 1/4\pi$, a pure number. Then only three basic units are defined; the unit of charge is then a derived quantity, 1 unit $= 1$ cm$^{1/2}$-g$^{1/2}$/sec. Numerically, the value of any quantity is identical to its value in the ordinary Gaussian system.

CONVERSION BETWEEN CGS UNITS AND MKS UNITS

The electrostatic units generally have the prefix "stat"; the electromagnetic units have the prefix "ab" (absolute).

Charge

1 statcoulomb	$= 3.336 \times 10^{-10}$ coulomb
1 abcoulomb	$= 10$ coulomb

Current

1 statampere	$= 3.336 \times 10^{-10}$ ampere
1 abampere	$= 10$ ampere

Potential

1 statvolt	$= 299.8$ volt
1 abvolt	$= 10^{-8}$ volt

Resistance

1 statohm	$= 8.987 \times 10^{11}$ ohm
1 abohm	$= 10^{-9}$ ohm

Capacitance

1 statfarad	$= 1$ "cm" $= 1.113 \times 10^{-12}$ farad
1 abfarad	$= 10^{9}$ farad

CONVERSION BETWEEN CGS UNITS AND MKS UNITS **(continued)**

The electrostatic units generally have the prefix "stat"; the electromagnetic units have the prefix "ab" (absolute).

Magnetic field

1 statunit	$= 2.998 \times 10^6$ weber/m²
1 gauss (abs)	$= 10^{-4}$ weber/m²

Magnetizing field

1 statunit	$= 2.654 \times 10^{-9}$ A/m
1 oersted (abs)	$= 79.58$ A/m

Magnetic flux

1 statunit	$= 299.8$ weber
1 maxwell (abs)	$= 10^{-8}$ weber

Magnetic susceptibility

$$\chi \text{ (stat)} = \frac{1}{4\pi} \chi \text{ (MKS)}$$

$$\chi \text{ (abs)} = \frac{1}{4\pi} \chi \text{ (MKS)}$$

Inductance

1 statunit	$= 8.988 \times 10^{11}$ henry
1 abunit	$= 1$ "cm" $= 4\pi \times 10^{-7}$ henry

PROPERTIES OF MATERIALS AND FUNDAMENTAL CONSTANTS

These tables are more extensive than required for the solution of the problems in this textbook. It is hoped that they will be useful to the student outside the physics course.

TABLE I. PROPERTIES OF METALS

Density, melting and boiling point, electrical resistivity ρ_e and its temperature coefficient α for metals. The temperature coefficient is always at 20°C.

METAL	SYMBOL OR COMPOSITION	DENSITY 10^3 kg/m³	MELTING POINT °C	BOILING POINT °C	ELECTRICAL RE-SISTIVITY ρ_e		$\alpha = \dfrac{1}{\rho_e} \dfrac{d\rho_e}{dt}$ %/°C
					10^{-8} Ω-m	AT °C	
Aluminum	Al	2.702	660	2470	2.82	20	0.39
Antimony	Sb	6.684	631	1380	41.7	20	0.36
Arsenic (black)	As	5.724	—	615 subl.	33.3	20	0.42
Barium	Ba	3.5	725	1140	36	0	0.65
Bismuth	Bi	9.80	271	1560	120	20	0.004
Brass	Cu 61% Zn 36% Pb 3%	8.6	~900	—	~7	20	0.2
Cadmium	Cd	8.642	321	767	7.6	20	0.38
Calcium	Ca	1.55	842	1487	4.0	0	0.42
Cesium	Cs	1.87	28.5	690	19.0	0	0.50
Chromium	Cr	7.20	1890	2480	1.5	0	—
Cobalt	Co	8.9	1495	2900	9.8	20	0.33
Constantan	Cu 54% Ni 45% Mn 1%	8.9	1200	—	49	20	0.001
Copper:	Cu						
Annealed		8.92	1083	2595	1.72	20	0.39
Hard drawn		8.92	—	—	1.77	20	0.38
Gold	Au	19.3	1063	2966	2.44	20	0.34
Indium	In	7.3	156	2000	8.2	0	0.51
Iridium	Ir	22.4	2410	4527	4.93	0	0.41
Iron (99.98%)	Fe	7.86	1535	3000	10	20	0.56
Lead	Pb	11.3	327	1744	22	20	0.39
Magnesium	Mg	1.74	651	1107	4.6	20	0.4
Manganese	Mn	7.20	1244	2097	185	20	—
Manganese steel	Fe 86% Mn 13% C 1%	7.81	1510	—	70	20	0.1
Mercury	Hg	13.5	−39	357	95.8	20	0.09
Molybdenum	Mo	10.2	2610	5560	5.7	20	0.4
Nickel	Ni	8.90	1455	2732	6.14 6.84	0 20	—

TABLE 1. PROPERTIES OF METALS (continued)

METAL	SYMBOL OR COMPOSITION	DENSITY 10^3 kg/m^3	MELTING POINT °C	BOILING POINT °C	ELECTRICAL RE-SISTIVITY ρ_e		$\alpha = \dfrac{1}{\rho_e}\dfrac{d\rho_e}{dt}$ %/°C
					10^{-8} Ω-m	AT °C	
Osmium	Os	22.5	3000	5000	60.2	20	—
Palladium	Pa	12.0	1550	2930	11	20	0.33
					7.17	−78	
Platinum	Pt	21.45	1769	3800	10	20	0.3
Potassium	K	0.86	64	774	6.1	0	—
Rubidium	Rb	1.53	38.9	688	11.6	0	—
Silver	Ag	10.5	961	2212	1.47	0	0.38
Sodium	Na	0.97	98	892	4.3	0	—
Strontium	Sr	2.6	769	1384	23	20	—
Tantalum	Ta	16.6	2996	5425	15.5	20	0.31
Thallium	Tl	11.85	302	1457	17.6	0	0.4
Tin (white)	Sn	7.28	232	2270	11.5	20	0.42
Titanium	Ti	4.5	1800	>3000	42	~20	—
Tungsten	W	19.3	3410	5900	5.6	20	0.45
Zinc	Zn	7.14	419	907	5.75	20	0.37

TABLE II-a PHYSICAL PROPERTIES OF SOME NONMETALLIC SOLID SUBSTANCES

MATERIAL	DENSITY 10^3 kg/m$_3$	RESISTIVITY ρ_e AT 20°C Ω-m	DIELECTRIC CONSTANT K AT 20°C
Amber	1.05	$\geq 10^{18}$	2.8
Barium titanate	6.1	—	~1500
Bakelite (polystyrene)	1.07	$\sim 10^7 - 10^{14}$	2 − 6
Diamond	3.52	—	5.7
Germanium	5.36	10	15.8
Glass	2.2 − 2.8	$10^{10} - 10^{13}$	4 − 10
Ivory	1.8 − 1.9	2×10^6	6
Lucite	1.2	$\sim 10^{13}$	3 − 3.6
Marble	2.5 − 2.8	10^8	8 − 12
Mica	~3	$10^{12} - 10^{15}$	6 − 8
Nylon	1.08 − 1.14	4×10^{14}	3.5
Paper	0.7 − 1.2	$10^{12} - 10^{16}$	1.2 − 3
Paraffin	0.8 − 0.9	$\sim 10^{16}$	1.9 − 2.2
Phosphorus	1.8 − 2.7	—	3.6 − 4
Polyethylene	~1.0	10^{14}	2.3
Porcelain	2.5	3×10^{12}	6 − 8
Quartz	2.6	3×10^{14}	4.1
Rubber, hard	1.1 − 1.2	10^{16}	2.5 − 3.5
Ruby	4.0	—	11.3 − 13.3
Selenium	4.8	—	6.6
Silicon	2.4	3×10^3	11.7
Sulfur	2.0	10^{14}	3.6 − 4.3
Teflon	2.2	$> 10^{14}$	2
Wood (dry)	~0.8	$10^9 - 10^{13}$	2.5 − 7

TABLE II-b PHYSICAL PROPERTIES OF SOME LIQUIDS

LIQUID	COMPOSITION	DENSITY AT 20°C 10^3 kg/m$_3$	MELTING POINT °C	BOILING POINT °C	DIELECTRIC CONSTANT K AT 20°C
Aniline	C_6H_7N	1.022	−6.2	184.4	7.0
Acetone	$(CH_3)_2CO$	0.791	−95	56.1	21
Benzene	C_6H_6	0.879	5.53	80.1	2.3
Bromine	Br_2	3.12	−7.3	58.8	3.1
Chloroform	$CHCl_3$	1.49	−63.6	61.3	4.8
Ethyl alcohol	C_2H_5OH	0.789	−117.3	78.3	21
Formic acid	HCOOH	1.22	8.4	100.7	58
Glycerin	$C_3H_8O_3$	1.26	18	290	41.1
Methyl alcohol	CH_3OH	0.791	−97.7	65	34
Toluene	C_7H_8	0.867	−95.0	110.6	2.38
Water	H_2O	0.998	0.0	100.0	80.3
Heavy water	D_2O	1.105	3.8	101.4	79.7
Xylene *(m-)*	$C_6H_4(CH_3)_2$	0.868	−47.9	139.1	2.37

TABLE II-c PHYSICAL PROPERTIES OF GASES

GAS	COMPO-SITION	DENSITY AT 0°C AND 760 mm Hg (kg/m³)	BOILING POINT °C	DIELECTRIC CONSTANT $(K\text{-}1) \times 10^6$ AT GIVEN °C AND 760 mm Hg	
Ammonia	NH_3	0.771	−33.4	6590	25°C
Argon	A	1.784	−185.8	506	25
Acetylene	C_2H_2	1.175	−84	1217	25
Hydrogen chloride	HCl	1.64	−85	3790	21
Ethane	C_2H_6	1.357	−88.6	1380	25
Ethylene	C_2H_4	1.26	−103.7	1328	25
Helium	He	0.178	−268.9	66	25
Hydrogen	H_2	0.090	−252.8	254	20
Carbon dioxide	CO_2	1.98	−78.5 subl.	985	0
Methane	CH_4	0.72	−161.4	804	25
Neon	Ne	0.90	−246.0	123	25
Nitrogen	N_2	1.25	−196	548	20
Air (dry)	N_2 80% O_2 20%	1.293	−193	536	20
Oxygen	O_2	1.429	−183.0	495	20
Xenon	Xe	5.9	−107.1	1238	25

TABLE III. PHYSICAL CONSTANTS

	SYMBOL	VALUE
A) Fundamental Constants		
speed of light	c	2.998×10^8 m/sec
elementary charge	e	1.602×10^{-19} coul
Planck's constant	h	6.626×10^{-34} J-sec
Dirac's constant	$\hbar = \dfrac{h}{2\pi}$	1.055×10^{-34} J-sec
Avogadro's number	N_0	6.022×10^{26} molecules/kg-mole
		6.022×10^{23} molecules/g-mole
universal gas constant	R	8.314 J/g-mole-°K
Boltzmann's constant	$k = R/N_0$	1.381×10^{-23} J/°K
gravitational constant	G	6.673×10^{-11} N-m²/kg²
dielectric permittivity	ϵ_0	8.854×10^{-12} coul/V-m
of vacuum	$1/4\pi\epsilon_0$	8.988×10^9 V-m/coul
magnetic permeability of	μ_0	$4\pi \times 10^{-7}$ V-sec/A-m
vacuum		$= 1.257 \times 10^{-6}$ V-sec/A-m
B) Atomic and Nuclear Constants		
electron mass	m_e	0.911×10^{-30} kg
proton mass	M_p	1.673×10^{-27} kg
electron charge/mass ratio	e/m_e	1.759×10^{11} coul/kg
Bohr radius	a_0	0.529 Å $= 0.529 \times 10^{-10}$ m
Bohr magneton	$\mu_B = \dfrac{e\hbar}{2m_e}$	9.274×10^{-24} A-m²
ionization energy of hydrogen	E_0	2.179×10^{-18} J
		13.60 eV
nuclear magneton	$\mu_N = \dfrac{e\hbar}{2M_p}$	5.051×10^{-27} A-m²
electron magnetic moment	μ_e	1.00115 Bohr magnetons
proton magnetic moment	μ_p	2.793 nuclear magnetons
neutron magnetic moment	μ_n	-1.913 nuclear magnetons
C) Terrestrial and Solar Data		
Mean acceleration of gravity	g	9.81 m/sec²
Mean radius of earth		6.37×10^6 m
Mass of earth		5.98×10^{24} kg
Mean density of earth		5.52×10^3 kg/m³
Mean distance moon-earth		3.84×10^8 m
Mass of moon		7.35×10^{22} kg
Mean distance sun-earth		1.49×10^{11} m
Mass of sun		1.99×10^{30} kg
Mean distance sun-Mercury		$5.8 \ \times 10^{10}$ m
-Venus		1.08×10^{11} m
-Mars		2.28×10^{11} m
-Jupiter		7.78×10^{11} m
-Saturn		1.43×10^{12} m
D) Miscellaneous Data		
Standard volume of ideal gas		2.24×10^{-2} m³/g-mole
Density of air at 0°C and 1 Atm		1.293 kg/m³
1 radian		57.296 degrees
e (basis of natural logarithms)		2.7183
π		3.1416
$\log_{10} e$		0.4343
$\ln 2$		0.6931

TABLE IV. PERIODIC TABLE OF ELEMENTS

1	2	3	4	5	6	7	8	9	10	11	12	13	14	15	16	17	18
1 H 1.00797 ±0.00001																	2 He 4.0026 ±0.00005
3 Li 6.939 ±0.0005	4 Be 9.0122 ±0.00005											5 B 10.811 ±0.003	6 C 12.01115 ±0.00005	7 N 14.0067 ±0.00005	8 O 15.9994 ±0.0001	9 F 18.9984 ±0.00005	10 Ne 20.183 ±0.0005
11 Na 22.9898 ±0.0005	12 Mg 24.312 ±0.0005											13 Al 26.9815 ±0.00005	14 Si 28.086 ±0.001	15 P 30.9738 ±0.00005	16 S 32.064 ±0.003	17 Cl 35.453 ±0.001	18 Ar 39.948 ±0.0005
19 K 39.102 ±0.0005	20 Ca 40.08 ±0.005	21 Sc 44.956 ±0.0005	22 Ti 47.90 ±0.005	23 V 50.942 ±0.0005	24 Cr 51.996 ±0.001	25 Mn 54.9380 ±0.00005	26 Fe 55.847 ±0.003	27 Co 58.9332 ±0.00005	28 Ni 58.71 ±0.005	29 Cu 63.54 ±0.005	30 Zn 65.37 ±0.005	31 Ga 69.72 ±0.005	32 Ge 72.59 ±0.005	33 As 74.9216 ±0.00005	34 Se 78.96 ±0.005	35 Br 79.909 ±0.002	36 Kr 83.80 ±0.005
37 Rb 85.47 ±0.005	38 Sr 87.62 ±0.005	39 Y 88.905 ±0.0005	40 Zr 91.22 ±0.005	41 Nb 92.906 ±0.0005	42 Mo 95.94 ±0.005	43 Tc (99)	44 Ru 101.07 ±0.005	45 Rh 102.905 ±0.0005	46 Pd 106.4 ±0.05	47 Ag 107.870 ±0.003	48 Cd 112.40 ±0.005	49 In 114.82 ±0.005	50 Sn 118.69 ±0.005	51 Sb 121.75 ±0.005	52 Te 127.60 ±0.005	53 I 126.9044 ±0.00005	54 Xe 131.30 ±0.005
55 Cs 132.905 ±0.0005	56 Ba 137.34 ±0.005	57 La 138.91 ±0.005	72 Hf 178.49 ±0.005	73 Ta 180.948 ±0.0005	74 W 183.85 ±0.005	75 Re 186.2 ±0.05	76 Os 190.2 ±0.05	77 Ir 192.2 ±0.05	78 Pt 195.09 ±0.005	79 Au 196.967 ±0.0005	80 Hg 200.59 ±0.005	81 Tl 204.37 ±0.005	82 Pb 207.19 ±0.005	83 Bi 208.980 ±0.0005	84 Po (210)	85 At (210)	86 Rn (222)
87 Fr (223)	88 Ra (226)	89 Ac (227)	104 Ku (257)	105 Ha (260)													

Lanthanum Series

58 Ce 140.12 ±0.005	59 Pr 140.907 ±0.0005	60 Nd 144.24 ±0.005	61 Pm (147)	62 Sm 150.35 ±0.005	63 Eu 151.96 ±0.005	64 Gd 157.25 ±0.005	65 Tb 158.924 ±0.0005	66 Dy 162.50 ±0.005	67 Ho 164.930 ±0.0005	68 Er 167.26 ±0.005	69 Tm 168.934 ±0.005	70 Yb 173.04 ±0.005	71 Lu 174.97 ±0.005

Actinium Series

90 Th 232.038 ±0.0005	91 Pa (231)	92 U 238.03 ±0.005	93 Np (237)	94 Pu (242)	95 Am (243)	96 Cm (247)	97 Bk (247)	98 Cf (249)	99 Es (254)	100 Fm (253)	101 Md (256)	102 No (253)	103 Lw (257)

Atomic Weights are based on $C^{12} = 12.0000$ and Conform to the 1961 Values

The number above each element is Z, the charge of the nucleus, in units of e. The number below the element is the atomic mass, the mass (in kg) of $N_0 = 6.02 \times 10^{26}$ atoms.

MATHEMATICAL FORMULAE

1. SYMBOLS USED

= equals

≠ is not equal to

≈ equals approximately

∝ is proportional to

> is greater than (>> much greater)

< is less than (<< much less)

∑ sum of

2. TRIGONOMETRY

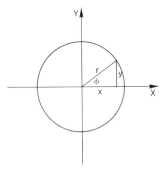

$$\sin \phi = \frac{y}{r} \qquad \csc \phi = \frac{r}{y}$$

$$\cos \phi = \frac{x}{r} \qquad \sec \phi = \frac{r}{x}$$

$$\tan \phi = \frac{y}{x} \qquad \cot \phi = \frac{x}{y}$$

$$\sin^2 \phi + \cos^2 \phi = 1$$

Addition Theorems

$$\sin (\alpha \pm \beta) = \sin\alpha \, \cos\beta \pm \cos\alpha \, \sin\beta$$

$$\cos (\alpha \pm \beta) = \cos\alpha \, \cos\beta \mp \sin\alpha \, \sin\beta$$

$$\tan (\alpha \pm \beta) = \frac{\tan\alpha \pm \tan\beta}{1 \mp \tan\alpha \, \tan\beta}$$

Inverse Addition Theorems

$$\sin\alpha \pm \sin\beta = 2 \, \sin\left(\frac{\alpha \pm \beta}{2}\right) \cos\left(\frac{\alpha \mp \beta}{2}\right)$$

$$\cos\alpha + \cos\beta = 2 \, \cos\left(\frac{\alpha + \beta}{2}\right) \cos\left(\frac{\alpha - \beta}{2}\right)$$

$$\cos\alpha - \cos\beta = 2 \, \sin\left(\frac{\alpha + \beta}{2}\right) \sin\left(\frac{\alpha - \beta}{2}\right)$$

Law of Sines:

$$\frac{\sin\alpha}{a} = \frac{\sin\beta}{b} = \frac{\sin\gamma}{c}$$

Law of Cosines:

$$c^2 = a^2 + b^2 - 2ab \cos \gamma$$

Connection to Complex Numbers

$$\left.\begin{array}{l} e^{ix} = \cos x + i \sin x \\ e^{(x+iy)} = e^x \left(\cos y + i \sin y\right) \end{array}\right\} \quad x, y \text{ in radians } (180° = \pi)$$

$$e^{2\pi i} = 1, \qquad e^{\pi i} = -1, \qquad e^{\pm \pi i/2} = \pm i$$

3. SERIES EXPANSIONS

$$(1 \pm x)^n = 1 \pm nx + \frac{n(n-1)}{2!} x^2 \pm \frac{n(n-1)(n-2)}{3!} x^3 + \dots \qquad \begin{array}{l} n \text{ integer} > 0 \\ \text{all } x \end{array}$$

$$(1 \pm x)^{-n} = 1 \mp nx + \frac{n(n+1)}{2! +} x^2 \mp \frac{n(n+1)(n+2)}{+ 3! +} x^3 + \dots \qquad \begin{array}{l} n \text{ integer} \\ |x| < 1 \end{array}$$

$$\frac{1}{1 \pm x} = 1 \mp x + x^2 \mp x^3 + x^4 \mp x^5 + \dots \qquad |x| < 1$$

$$(1 \pm x)^\alpha = 1 \pm \alpha x + \frac{\alpha (\alpha - 1)}{2!} x^2 \pm \frac{\alpha(\alpha - 1)(\alpha - 2)}{3!} x^3 + \dots \qquad \begin{array}{l} \alpha \text{ real} \\ |x| < 1 \end{array}$$

$$\sqrt{1 \pm x} = 1 \pm \frac{x}{2} - \frac{x^2}{8} \pm \frac{x^3}{16} - \frac{5x^4}{128} \pm \dots \qquad |x| < 1$$

$$e^{\pm x} = \lim_{n \to \infty} \left(1 \pm \frac{x}{n}\right)^n = 1 \pm x + \frac{x^2}{2!} \pm \frac{x^3}{3!} + \frac{x^4}{4!} \pm \frac{x^5}{5!} + \dots \qquad \text{all } x$$

$$e^1 = e = \lim_{n \to \infty} \left(1 + \frac{1}{n}\right)^n = 1 + \frac{1}{1} + \frac{1}{2!} + \frac{1}{3!} + \dots = 2.71828 \dots$$

$$\sin x = x - \frac{x^3}{3!} + \frac{x^5}{5!} - \frac{x^7}{7!} + \dots \qquad \text{all } x$$

$$\cos x = 1 - \frac{x^2}{2!} + \frac{x^4}{4!} - \frac{x^6}{6!} + \dots \qquad \text{all } x$$

$$\tan x = x + \frac{x^3}{3} + \frac{2x^5}{15} + \frac{17x^7}{315} + \frac{62x^9}{2835} + \dots \qquad |x| < \frac{\pi}{2}$$

$$\log_e (1 + x) = x - \frac{x^2}{2} + \frac{x^3}{3} - \frac{x^4}{4} + \dots \qquad |x| < 1$$

For any function $f(x) = f(a + \Delta x)$,

$$f(a + \Delta x) = f(a) + f'(a) \Delta x + \frac{f''(a)}{2!} (\Delta x)^2 + \frac{f'''(a)}{3!} (\Delta x)^3 + \dots$$

where $f' = \dfrac{df}{dx}, f'' = \dfrac{d^2 f}{dx^2}, \dots$ are all assumed to exist and to be continuous in the interval $a < x < a + \Delta x$.

4. SUMS OF INTEGERS

$$\sum_{k=1}^{n} k = 1 + 2 + 3 + \ldots + n = \frac{n(n + 1)}{2}$$

$$\sum_{k=1}^{n} k^2 = 1^2 + 2^2 + 3^2 + \ldots + n^2 = \frac{n(n + 1)(2n + 1)}{6}$$

$$\sum_{k=1}^{n} k^3 = 1^3 + 2^3 + 3^3 + \ldots + n^3 = \frac{n^2(n + 1)^2}{4}$$

5. VECTOR ALGEBRA

$\mathbf{a}, \mathbf{b}, \mathbf{c}, \ldots$ vectors

$\alpha, \beta, \gamma, \ldots$ angles

$p, r, s, \ldots$ numbers

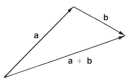

a) Addition of Vectors

$$\mathbf{a} + \mathbf{b} = \mathbf{b} + \mathbf{a}$$

$$p(\mathbf{a} + \mathbf{b}) = p\,\mathbf{a} + p\,\mathbf{b}$$

$$(p + q)\,\mathbf{a} = p\,\mathbf{a} + q\,\mathbf{a}$$

$$(\mathbf{a} + \mathbf{b}) + \mathbf{c} = \mathbf{a} + (\mathbf{b} + \mathbf{c}) = \mathbf{a} + \mathbf{b} + \mathbf{c}$$

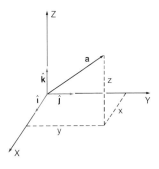

$\hat{\mathbf{i}}, \hat{\mathbf{j}}, \hat{\mathbf{k}}$ unit length vectors along x, y, z axes

$\mathbf{a} = x\hat{\mathbf{i}} + y\hat{\mathbf{j}} + z\hat{\mathbf{k}}$

$|\mathbf{a}| = \sqrt{x^2 + y^2 + z^2}$ (length formula)

$\mathbf{a} + \mathbf{b} = (x_1\hat{\mathbf{i}} + y_1\hat{\mathbf{j}} + z_1\hat{\mathbf{k}}) + (x_2\hat{\mathbf{i}} + y_2\hat{\mathbf{j}} + z_2\hat{\mathbf{k}})$

$\qquad = (x_1 + x_2)\hat{\mathbf{i}} + (y_1 + y_2)\hat{\mathbf{j}} + (z_1 + z_2)\hat{\mathbf{k}}$

b) Scalar Product

$$\boxed{\mathbf{a} \cdot \mathbf{b} = |\mathbf{a}|\,|\mathbf{b}|\,\cos\gamma}$$

$$\mathbf{a} \cdot (\mathbf{b} + \mathbf{c}) = \mathbf{a} \cdot \mathbf{b} + \mathbf{a} \cdot \mathbf{c}$$

$$(p\mathbf{a}) \cdot \mathbf{b} = p\,(\mathbf{a} \cdot \mathbf{b})$$

$$\mathbf{a} \cdot \mathbf{b} = \mathbf{b} \cdot \mathbf{a}$$

In coordinates: if $\mathbf{a} = x_1\hat{\mathbf{i}} + y_1\hat{\mathbf{j}} + z_1\hat{\mathbf{k}}$

$\qquad\qquad\qquad \mathbf{b} = x_2\hat{\mathbf{i}} + y_2\hat{\mathbf{j}} + z_2\hat{\mathbf{k}}$

then $\boxed{\mathbf{a} \cdot \mathbf{b} = x_1 x_2 + y_1 y_2 + z_1 z_2}$

c) *Vector Product*

The vector product **a** × **b** is defined as a vector of length

$$|\mathbf{a} \times \mathbf{b}| = |\mathbf{a}| \cdot |\mathbf{b}| \cdot \sin \gamma$$

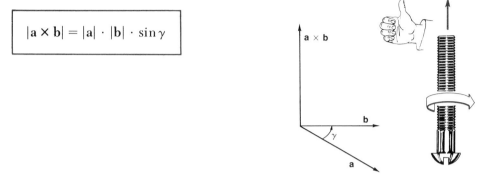

which is perpendicular to the plane formed by the vectors **a** and **b**. The direction of **a** × **b** is given by the right-hand rule.

$$\mathbf{a} \times \mathbf{b} = -\mathbf{b} \times \mathbf{a}$$

$$\mathbf{a} \times (\mathbf{b} + \mathbf{c}) = \mathbf{a} \times \mathbf{b} + \mathbf{a} \times \mathbf{c}$$

$$(p\mathbf{a}) \times \mathbf{b} = p\,(\mathbf{a} \times \mathbf{b})$$

$$(a_x\hat{\mathbf{i}} + a_y\hat{\mathbf{j}} + a_z\hat{\mathbf{k}}) \times (b_x\hat{\mathbf{i}} + b_y\hat{\mathbf{j}} + b_z\hat{\mathbf{k}})$$
$$= (a_xb_y - a_yb_x)\hat{\mathbf{i}} + (a_yb_z - a_zb_y)\hat{\mathbf{j}} + (a_zb_x - a_xb_z)\hat{\mathbf{k}} = \begin{vmatrix} \hat{\mathbf{i}} & \hat{\mathbf{j}} & \hat{\mathbf{k}} \\ a_x & a_y & a_z \\ b_x & b_y & b_z \end{vmatrix}$$

d) *Multiple Products*

$$\mathbf{a} \times (\mathbf{b} \times \mathbf{c}) = (\mathbf{a} \cdot \mathbf{c})\mathbf{b} - (\mathbf{a} \cdot \mathbf{b})\mathbf{c}$$

$$\mathbf{a} \times (\mathbf{b} \times \mathbf{c}) + \mathbf{c} \times (\mathbf{a} \times \mathbf{b}) + \mathbf{b} \times (\mathbf{a} \times \mathbf{c}) = 0$$

$$(\mathbf{a} \times \mathbf{b}) \cdot \mathbf{c} = \mathbf{a} \cdot (\mathbf{b} \times \mathbf{c}) = \begin{vmatrix} a_x & a_y & a_z \\ b_x & b_y & b_z \\ c_x & c_y & c_z \end{vmatrix} = -\mathbf{c} \cdot (\mathbf{b} \times \mathbf{a}) = -\mathbf{a} \cdot (\mathbf{c} \times \mathbf{b})$$

6. DIFFERENTIATION

$$y = f(x)$$

$$\frac{dy}{dx} = \lim_{\Delta x \to 0} \frac{\Delta y}{\Delta x} = \lim_{\Delta x' \to 0} \frac{f(x + \Delta x) - f(x)}{\Delta x}$$

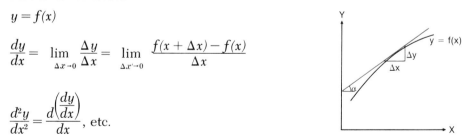

$$\frac{d^2y}{dx^2} = \frac{d\left(\frac{dy}{dx}\right)}{dx}, \text{ etc.}$$

If we plot $y(x)$ in a Cartesian x-y coordinate system, then the angle α between the tangent to the curve and the x-axis is given by $\tan \alpha = \dfrac{dy}{dx}$. Other ways of writing derivatives: $\dfrac{dy}{dx} = y'$, $\dfrac{d^2y}{dx^2} = y''$, ... Frequently, for functions of time $y(t)$, one also uses $\dfrac{dy}{dt} = \dot{y}$, $\dfrac{d^2y}{dt^2} = \ddot{y}$, ...

$y = f(x)$	$\dfrac{df}{dx} = \dfrac{dy}{dx}$	
a = constant	0	
x^n	nx^{n-1}	n integer
x^α	$\alpha x^{\alpha-1}$	α real
e^x	e^x	
$e^{\lambda x}$	$\lambda e^{\lambda x}$	λ real or complex
a^x	$(\ln a)a^x$	
$\sin x$	$\cos x$	
$\cos x$	$-\sin x$	
$\ln x$	$\dfrac{1}{x}$	
$u(x) \cdot v(x)$	$u\dfrac{dv}{dx} + v\dfrac{du}{dx}$	
$\dfrac{u(x)}{v(x)}$	$\dfrac{v\dfrac{du}{dx} - u\dfrac{dv}{dx}}{v^2}$	
$a\,\phi(x)$	$a\dfrac{d\phi}{dx}$	
$\tan x = \dfrac{\sin x}{\cos x}$	$1 + \tan^2 x = \dfrac{1}{\cos^2 x}$	
$\cot x = \dfrac{\cos x}{\sin x}$	$-(1 + \cot^2 x) = -\dfrac{1}{\sin^2 x}$	
$f\left(\phi(x)\right)$	$\dfrac{df}{d\phi} \cdot \dfrac{d\phi}{dx}$	
$e^{\phi(x)}$	$e^{\phi(x)}\dfrac{d\phi}{dx}$	
$\ln \phi(x)$	$\dfrac{1}{\phi(x)}\dfrac{d\phi}{dx}$	
$(a + bx)^\alpha$	$\alpha b(a + bx)^{\alpha-1}$	α real
$x = \phi(y)$	$\dfrac{dy}{dx} = 1/(d\phi/dy)$	
arc $\sin x$	$\dfrac{1}{\sqrt{1 - x^2}}$	
arc $\cos x$	$\dfrac{-1}{\sqrt{1 - x^2}}$	
arc $\tan x$	$\dfrac{1}{1 + x^2}$	

7. INTEGRATION

The indefinite integral is defined as the inverse of the derivative:

$$\int f(x)\,dx = F(x) \quad \text{is by definition the same as} \quad \frac{dF(x)}{dx} = f(x).$$

The indefinite integral is defined only up to an arbitrary additive constant. One frequently writes

$$\int f(x)\,dx = F(x) + C$$

because for any constant C one has

$$\frac{d[F(x) + C]}{dx} = \frac{dF(x)}{dx}$$

The definite integral of a function $f(x)$ between $x = a$ and $x = b$ is defined as the area under the curve between a and b (compare figure):

$$\int_a^b f(x)\,dx = \lim_{\Delta x_i \to 0} \sum_i f(x_i)\,\Delta x_i$$

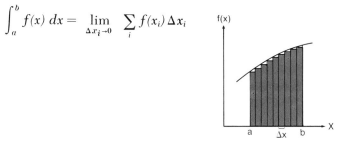

where $\sum_i \Delta x_i = b - a$.

Connection between the definite and indefinite integral: If $dF(x)/dx = f(x)$, then

$$\int_a^b f(x)\,dx = F(b) - F(a)$$

TABLE C2: SOME INDEFINITE INTEGRALS

$f(x)$	$\int f(x)\,dx$ (arbitrary additive constant omitted)
1	x
x^n	$\dfrac{x^{n+1}}{n+1} \qquad n \neq -1$
$\dfrac{1}{x}$	$\ln x$
x^α	$\dfrac{x^{\alpha+1}}{\alpha + 1} \qquad \alpha \neq -1$
$e^{\alpha x}$	$\dfrac{1}{\alpha} e^{\alpha x} \qquad \alpha \neq 0$

TABLE C2: SOME INDEFINITE INTEGRALS **(continued)**

$f(x)$	$\int f(x)\,dx$ (arbitrary additive constant omitted)
a^x	$\dfrac{a^x}{\ln a}$
$\dfrac{1}{x^2 + a^2}$	$\dfrac{1}{a}\text{ arc tan }\dfrac{x}{a}$
$\dfrac{1}{x^2 - a^2}$	$\dfrac{1}{2a}\ln\left(\dfrac{x-a}{x+a}\right) \qquad x > a$
$\dfrac{1}{\sqrt{x^2 \pm a^2}}$	$\ln(x + \sqrt{x^2 \pm a^2}) \qquad x^2 \pm a^2 > 0$
$\dfrac{1}{\sqrt{a^2 - x^2}}$	$\text{arc sin}\left(\dfrac{x}{a}\right) = \dfrac{\pi}{2} - \text{arc cos}\left(\dfrac{x}{a}\right)$
$u(x)\dfrac{dv(x)}{dx}$	$uv - \displaystyle\int v\dfrac{du}{dx}dx$
$\dfrac{1}{a + bx}$	$\dfrac{1}{b}\ln(a + bx)$
$\dfrac{x}{a + bx}$	$\dfrac{1}{b^2}(a + bx - a\ln(a + bx))$
$\dfrac{x}{(a + bx)^2}$	$\dfrac{1}{b^2}\left(\dfrac{a}{a + bx} + \ln(a + bx)\right)$
$\dfrac{1}{\sqrt{(x^2 \pm a^2)^3}}$	$\dfrac{\pm x}{a^2\sqrt{x^2 \pm a^2}}$
$\dfrac{x}{\sqrt{(x^2 \pm a^2)^n}}$	$\dfrac{-1}{(n-2)\sqrt{(x^2 \pm a^2)^{n-2}}}$
$\cos x$	$-\sin x$
$\sin x$	$\cos x$
$\tan x$	$-\ln\cos x$
$\cot x$	$\ln\sin x$
$\sin^2 x$	$\dfrac{1}{2}x - \dfrac{1}{4}\sin 2x$
$\cos^2 x$	$\dfrac{1}{2}x + \dfrac{1}{4}\sin 2x$
$\dfrac{1}{\sin x}$	$\ln\tan\dfrac{x}{2}$
$\dfrac{1}{\cos x}$	$\ln\tan\left(\dfrac{\pi}{4} + \dfrac{x}{2}\right)$
$\dfrac{1}{\sin^2 x}$	$-\cot x$
$\ln x$	$x\ln x - x$
$x^n e^{ax}$	$\dfrac{x^n e^{ax}}{a} - \dfrac{n}{a}\displaystyle\int x^{n-1}e^{ax}dx$
$\dfrac{\ln x}{x}$	$\dfrac{1}{2}(\ln x)^2$

TABLE C3: SOME DEFINITE INTEGRALS

$$\int_0^{2\pi} \sin^2 x \, dx = \int_0^{2\pi} \cos^2 x \, dx = \pi$$

$$\int_0^\infty e^{-\lambda x} dx = \frac{1}{\lambda}$$

$$\int_0^\infty e^{-\lambda x^2} dx = \sqrt{\frac{\pi}{4\lambda}}$$

$$\int_0^\infty \frac{dx}{x^2 + a^2} = \frac{\pi}{2a}$$

Given a function $V(x,y,z) = V(\mathbf{r})$, one can study its change if only x changes by a small amount. One defines the partial derivative of V with respect to x:

$$\frac{\partial V}{\partial x} = \lim_{\Delta x \to 0} \frac{V(x + \Delta x, y, z) - V(x,y,z)}{\Delta x}$$

and $\dfrac{\partial V}{\partial y}$ and $\dfrac{\partial V}{\partial z}$ are defined analogously.

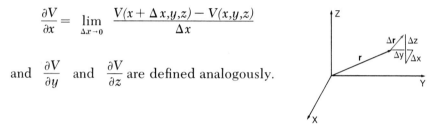

The symbol ∂ (instead of d) indicates that V depends also on other variables. Multiple derivatives are defined the same way:

$$\frac{\partial^2 V}{\partial x^2} = \frac{\partial}{\partial x}\left(\frac{\partial V}{\partial x}\right); \qquad \frac{\partial^2 V}{\partial x \partial y} = \frac{\partial}{\partial x}\left(\frac{\partial V}{\partial y}\right); \text{ etc.}$$

The order of differentiation is immaterial:

$$\frac{\partial^2 V}{\partial x \partial y} = \frac{\partial^2 V}{\partial y \partial x}; \qquad \frac{\partial^2 V}{\partial x \partial z} = \frac{\partial^2 V}{\partial z \partial x}; \qquad \frac{\partial^2 V}{\partial y \partial z} = \frac{\partial^2 V}{\partial z \partial y}$$

If $\Delta \mathbf{r} = \Delta x \hat{\mathbf{i}} + \Delta y \hat{\mathbf{j}} + \Delta z \hat{\mathbf{k}}$ is a small (infinitesimal) vector, then one can write approximately

$$V(x + \Delta x, \, y + \Delta y, \, z + \Delta z) \approx V(x,y,z) + \frac{\partial V}{\partial x}\Delta x + \frac{\partial V}{\partial y}\Delta y + \frac{\partial V}{\partial z}\Delta z$$

The approximation becomes exact in the limit $|\Delta r| \to 0$. Using the definition of the gradient of V:

$$\boxed{\text{grad } V = \frac{\partial V}{\partial x}\hat{\mathbf{i}} + \frac{\partial V}{\partial y}\hat{\mathbf{j}} + \frac{\partial V}{\partial z}\hat{\mathbf{k}}}$$

one can write

$$V(x + \Delta x,\ y + \Delta y,\ z + \Delta y) - V(x,y,z) \approx \text{grad } V \cdot \Delta \mathbf{r} \qquad \text{(scalar product)}$$

or

$$V(\mathbf{r} + \Delta \mathbf{r}) - V(\mathbf{r}) \approx \text{grad } V \cdot \Delta \mathbf{r}$$

Special Values for Gradient

$$(r = \sqrt{x^2 + y^2 + z^2},\ \mathbf{r} = x\hat{\mathbf{i}} + y\hat{\mathbf{j}} + z\hat{\mathbf{k}})$$

$$\text{grad } r = \frac{\mathbf{r}}{r}$$

$$\text{grad }(f(r)) = \frac{df}{dr} \cdot \frac{\mathbf{r}}{r}$$

$$\text{grad }(V_1(\mathbf{r}) \cdot V_2(\mathbf{r})) = V_1 \text{ grad } V_2 + V_2 \text{ grad } V_1$$

$$\text{grad }(aV) = a \text{ grad } V \qquad a \text{ an arbitrary constant}$$

$$\text{grad }(V_1 + V_2) = \text{grad } V_1 + \text{grad } V_2$$

9. LINE INTEGRALS

Given a curve C going from a point A to a point B in space, and given a vector field $\mathbf{v}(x,y,z)$ defined at least for any point on the curve C, one defines the line integral along C from A to B as:

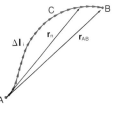

$$\int_A^B \mathbf{v} \bullet d\mathbf{l} = \lim_{|\Delta l_i| \to 0} \sum_{i=1}^{N} (\mathbf{v}_i \bullet \Delta \mathbf{l}_i)$$

$$= \lim_{\Delta l \to 0} \sum |\mathbf{v}_i|\, |\Delta \mathbf{l}_i|\, \cos \alpha_i$$

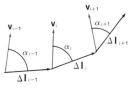

where the $\Delta \mathbf{l}_i$ cover the curve C:

$$\sum_{i=1}^{N} \Delta \mathbf{l}_i = \mathbf{r}_{AB}$$

and $\sum\limits_{i=1}^{n<N} \Delta \mathbf{l}_i = \mathbf{r}_n$ lies on a point on C for any $n < N$.

The second condition is necessary to ensure that the $\Delta \mathbf{l}_i$ follow the curve C as closely as possible.

If the direction of integration is reversed, the integral changes sign:

$$\int_{\substack{A \\ C}}^{B} \mathbf{v} \cdot d\mathbf{l} = -\int_{\substack{A \\ C}}^{B} \mathbf{v} \cdot d\mathbf{l}$$

If $\mathbf{v} = \operatorname{grad} V$, then for any curve C connecting points A and B,

$$\int_{\substack{A \\ C}}^{B} \mathbf{v} \cdot d\mathbf{l} = \int_{\substack{A \\ C'}}^{B} \mathbf{v} \cdot d\mathbf{l} = V(B) - V(A)$$

The points A and B can be the same point, if C is a closed curve (loop). The integral

$$\int_{\substack{\text{closed} \\ \text{loop } C}} \mathbf{v} \cdot d\mathbf{l}$$

will in general not vanish; its value depends on the particular curve C. The starting (and ending) point A is immaterial; any point on the curve C can be taken as start point without affecting the value of the integral. The following theorem holds:

The integral $\int \mathbf{v} \cdot d\mathbf{l}$ vanishes for all closed loops in a region of space exactly if (= if and only if) there exists a scalar function $V(x,y,z)$ such that $\mathbf{v} = \operatorname{grad} V$.

10. VOLUME INTEGRAL

Given a scalar function $f(x,y,z)$ and a volume of space enclosed by a surface of arbitrary shape, one defines the volume integral of $f(x,y,z)$ over the volume to be

$$\int_{\text{volume}} f \, d\tau = \lim_{\Delta\tau_i \to 0} \sum f_i \, \Delta\tau_i$$

where $\Delta\tau_i$ is the volume of a small cube (or any other small volume) and f_i is the value of the function at a point inside $\Delta\tau_i$. It is, of course, required that $\Delta\tau_i$ fill the whole volume:

$$\lim_{\Delta\tau_i \to 0} \sum \Delta\tau_i = \text{volume}$$

11. SURFACE INTEGRAL

A surface element $d\mathbf{S}$ is defined as a differential (infinitesimally small) vector whose value (length) is the area of the element dA and whose direction

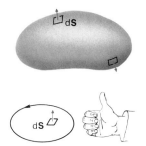

is the normal to the surface at the point in question. Since for an arbitrary surface the direction of the normal is only defined to within a $\pm$ sign, the following convention is adopted:

(a)   For a closed surface, $d\mathbf{S}$ always points outwards:

(b)   For a surface bordered by a single loop C, the direction of $d\mathbf{S}$ follows the right-hand rule: if the fingers point in the direction of the loop, then the thumb points in the direction of $d\mathbf{S}$. Note that $d\mathbf{S}$ is defined only if a direction on the bordering loop is also defined.

The integral over a surface is then defined again as the limit of a sum. For a vector field,

$$\int_{\text{surface } S} \mathbf{v} \cdot d\mathbf{S} = \lim_{|\Delta S_i| \to 0} \sum \mathbf{v}_i \cdot \Delta\mathbf{S}_i = \lim_{|\Delta S_i| \to 0} \sum |\mathbf{v}_i|\,|\Delta\mathbf{S}_i| \cos \alpha_i$$

This integral is a scalar, being a sum of scalar products. On the other hand, for a scalar function $f(x,y,z)$, the integral

$$\int_{\text{surface}} f\, d\mathbf{S} = \lim_{|\Delta S_i| \to 0} \sum f_i \Delta\mathbf{S}_i$$

is a vector. In particular, for any closed surface,

$$\int_{\substack{\text{closed}\\\text{surface}}} 1\, d\mathbf{S} = 0$$

12. CURVILINEAR COORDINATE SYSTEMS

a) Cylindrical coordinates r, ϕ, z

$$r = \sqrt{x^2 + y^2} \qquad x = r \cos \phi$$

$\phi = \text{arc tan } y/x \quad y = r \sin\phi$

Volume element: $r \, d\phi \, dr \, dz$

Surface element
on cylinder: $r \, d\phi \, dz$

Surface element
perpendicular
to cylinder axis: $r \, d\phi \, dr$

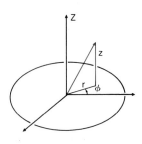

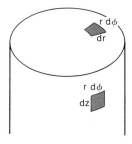

b) Spherical coordinates r, θ, ϕ

$r = \sqrt{x^2 + y^2 + z^2}$ $\qquad x = r \sin\theta \, \cos\phi$

$\theta = \text{arc cos } z/\sqrt{x^2 + y^2}$ $\qquad y = r \sin\theta \, \sin\phi$

$\phi = \text{arc tan } y/x$ $\qquad z = r \cos\theta$

Volume element: $r^2 \sin\theta \, dr \, d\theta \, d\phi$

Surface element
on sphere: $r^2 \sin\theta \, d\theta \, d\phi$

Surface element
in (x,y) plane: $r \, d\phi \, dr$

Surface element
in (x,z) plane: $r \, d\theta \, dr$

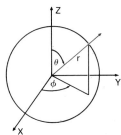

ANSWERS TO SELECTED PROBLEMS

CHAPTER 1

4. 1.5×10^{18} Hz
7. 82°
8. 0.42 mm
9. 8 min 20 sec
17. < 620 m/sec
19. 1.71 cm
23. 0.88 W/m²
25. 0.6 mm

CHAPTER 2

1. 150 m
6. 0.25% at 400 Hz
   0.067% at 1500 Hz
7. (a) 34.0
   (b) 9.0
11. 0.95 kW
13. 0.93 m
16. 968 N
19. ≈ 1 m
23. (a) 0.29 Hz
    (b) 1.0 Hz

CHAPTER 3

2. 0.24 W
3. 15.4 W
5. 7 $\lambda/2$
7. $\theta = 0°$ or 90°
11. ≈ 580 km
14. 55°
15. 6.7 mm
19. 21°40'
24. 1.9 Å

CHAPTER 4

1. 0.29 cm
2. 0.69 cm
4. 28.2°
6. 22.4 cm
9. 11°
12. 0.77 cm
13. 51.2 cm
22. ≈ 370 m

CHAPTER 5

2. 3 km
4. 22.5 N
6. 1700 N
9. 1.76×10^{13} m/sec²
11. 7.7×10^{-7} coul
14. 5 points
15. 1.52×10^{10} N/coul
21. $E_z = Q/(2\pi^2 \, \epsilon_0 \, R^2)$

CHAPTER 6

2. ≈ 7×10^{-3} Å
4. $11.0 \, Q_0/(4\pi\epsilon_0 \, a)$
5. $-13.0 \, Q^2/(4\pi\epsilon_0 \, a)$
8. $V = \dfrac{Q}{4\pi\epsilon_0 \, a} \ln \left(1 + \dfrac{a}{x}\right)$
10. 3.54 eV
11. (i) 5.56 eV
    (ii) 7.58 eV
21. $p = 4 \, Q_0 \, R/\pi$
    p points up
27. $\left| \mathbf{E} \right| = \dfrac{3 \, Q \, a^2}{\pi\epsilon_0 \, r^4};$ **E** points down

CHAPTER 7

3. 332 N
5. 4.5×10^5 coul
7. (a) $+Q, -Q, +Q, -Q, +Q$
   (b) 70.5 V
9. (a) 5.56×10^{-4} coul
   (b) 278 J
10. 1.8×10^{-5} coul
11. 33 pF
15. 43 pF
18. 400 V

CHAPTER 8

3. 1.7×10^{-13} J $\approx 0.1\%\ mc^2$
6. $\approx 0.06\ \mu$F
7. 6.3×10^{-6} coul/m^2
9. -3.3%
11. -420 V
17. $K = 2.09$
18. $p < 5.7 \times 10^{-30}$ coul-m
20. $\alpha_e = 7.6 \times 10^{-40}$ coul-m^2/V
    $p = 3.8 \times 10^{-30}$ coul-m

CHAPTER 9

2. 8.3×10^{-11} A/m^2
3. 4.3×10^{-4} coul/m^3
5. 78 m
8. 33.2 W
11. 7.96 Ω
14. 37 sec
16. $(\sqrt{3} - 1)\ R_0$
18. $I_3 = 7$ A

CHAPTER 10

1. 2×10^{-3} Wb/m^2
5. $|\mathbf{B}| = \mu_0 I \ln 2/(2\pi a)$
7. 1.13 Gauss
9. $|\mathbf{B}| = \mu_0 N Q/R$
13. $T = 7.1 \times 10^{-8}$ sec
    $r = 21$ cm
19. 3.13×10^{22} m^{-3}
22. 0.5 A-m^2
25. 12.4 Wb/m^2

CHAPTER 11

1. 8.4×10^{-4} A-m^2
4. 2.06×10^{-2} N
8. 3.8×10^{-6} N-m
12. 4.0 m
15. 0.50 A
16. 0.63 Wb/m^2

CHAPTER 12

2. 5000 Gauss
4. 0.24 mV
7. one turn
8. 5.66 V
15. 0.98 H
16. ≈ 2.2 H
21. 2.53×10^{-4} H
23. $I_0 = 0.40$ A
    $\phi = 88.5°$

CHAPTER 13

1. 6.2×10^{-26} N-m
2. 10^3 m/sec^2
7. 2.7×10^{-9} coul/m^2
11. $B = 12.4$ Wb/m^2
    $\Delta U = 1.43 \times 10^{-3}$ eV
14. (a) no
    (b) $v = 10^6$ m/sec
15. 1.1 mm
18. 6.0×10^{-7} Wb/m^2

CHAPTER 14

2. 280 MeV
4. no
9. 1.6%
11. 1.02 m/sec
14. $r = 1320$ Å
    $E = -5.5 \times 10^{-3}$ eV
15. 910 Å
17. $E = n\ \hbar\ \sqrt{k/m}$
20. (a) $\lambda = \sqrt{km}\ /2\hbar,\ E = \hbar\ \sqrt{k/m}\ /2$
    (b) $\mu = \sqrt{km}\ /2\hbar,\ E = 3\hbar\ \sqrt{k/m}\ /2$

CHAPTER 15

2. $Z_{eff} = 1.26$
4. 1.51 eV
10. $<E> = 3\,E_F/5$
11. 6.5 eV
16. 600 V
17. $\approx 23°$ K
19. 150 : 1

CHAPTER 16

2. $E_0 = 1.55 \times 10^{10}$ V/m
   $B_0 = 52$ Wb/m²
3. 2.0×10^{11} W/m²
6. 9.8 cm
8. 35.2°
10. $I = I_0/8$
14. 910 Å
16. 2000 photons/sec
19. 4.7×10^{-8} N

INDEX

PHYSICAL CONSTANTS

	SYMBOL	VALUE
Fundamental Constants		
speed of light	c	2.998×10^8 m/sec
elementary charge	e	1.602×10^{-19} coul
Planck's constant	h	6.626×10^{-34} J-sec
Dirac's constant	$\hbar = \dfrac{h}{2\pi}$	1.055×10^{-34} J-sec
Avogadro's number	N_O	6.022×10^{26} molecules/kg-mole
		6.022×10^{23} molecules/g-mole
universal gas constant	R	8.314 J/g-mole-°K
Boltzmann's constant	$k = R/N_O$	1.381×10^{-23} J/°K
gravitational constant	G	6.673×10^{-11} N-m²/kg²
dielectric permittivity	ϵ_O	8.854×10^{-12} coul/V-m
of vacuum	$1/4\pi\epsilon_O$	8.988×10^9 V-m/coul
magnetic permeability of	μ_O	$4\pi \times 10^{-7}$ V-sec/A-m
vacuum		$= 1.257 \times 10^{-6}$ V-sec/A-m
Atomic and Nuclear Constants		
electron mass	m_e	0.911×10^{-30} kg
proton mass	M_p	1.673×10^{-27} kg
electron charge/mass ratio	e/m_e	1.759×10^{11} coul/kg
Bohr radius	a_O	0.529 Å $= 0.529 \times 10^{-10}$ m
Bohr magneton	$\mu_B = \dfrac{e\hbar}{2m_e}$	9.274×10^{-24} A-m²
nuclear magneton	$\mu_N = \dfrac{e\hbar}{2M_p}$	5.051×10^{-27} A-m²
electron magnetic moment	μ_e	1.00115 Bohr magnetons
proton magnetic moment	μ_p	2.793 nuclear magnetons
neutron magnetic moment	μ_n	-1.913 nuclear magnetons
Miscellaneous Data		
Standard volume of ideal gas		2.24×10^{-2} m³/g-mole
Density of air at 0°C and 1 Atm		1.293 kg/m³
1 radian		57.296 degrees
e (basis of natural logarithms)		2.7183
π		3.1416
$\log_{10} e$		0.4343
$\ln 2$		0.6931